Apple
The First 50 Years

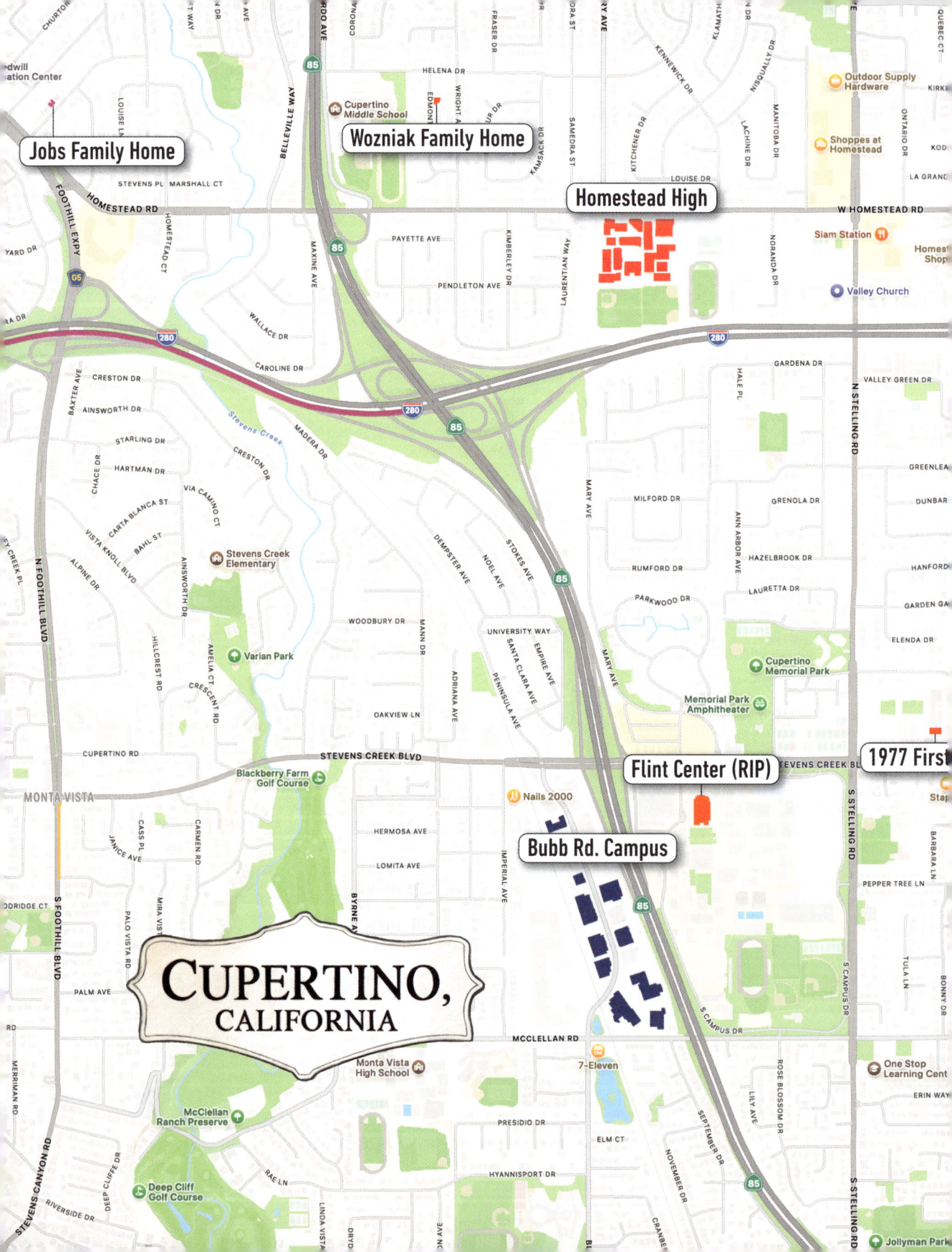

ALSO BY DAVID POGUE

How to Prepare for Climate Change

Mac Unlocked

iPhone Unlocked

Mac OS X: The Missing Manual

Windows 10: The Missing Manual

iPhone: The Missing Manual

iMovie: The Missing Manual

iPhoto: The Missing Manual

GarageBand: The Missing Manual

Pogue's Basics: Tech

Pogue's Basics: Money

Pogue's Basics: Life

Abby Carnelia's One and Only Magical Power

The World According to Twitter

Macworld Mac Secrets (with Joseph Schorr)

Macs for Dummies

Opera for Dummies (with Scott Speck)

Classical Music for Dummies (with Scott Speck)

Magic for Dummies

Apple
The First 50 Years

David Pogue

SIMON & SCHUSTER

London · New York · Amsterdam/Antwerp · Sydney/Melbourne · Toronto · New Delhi

First published in the United States by Simon & Schuster, an imprint of Simon & Schuster, LLC, 2026

First published in Great Britain by Simon & Schuster UK Ltd, 2026

Copyright © David Pogue, 2026

The right of David Pogue to be identified as the author of this work has been asserted
in accordance with the Copyright, Designs and Patents Act, 1988.

1 3 5 7 9 10 8 6 4 2

Simon & Schuster UK Ltd
1st Floor
222 Gray's Inn Road
London WC1X 8HB

For more than 100 years, Simon & Schuster has championed authors and the stories they create.
By respecting the copyright of an author's intellectual property, you enable Simon & Schuster and the author
to continue publishing exceptional books for years to come. We thank you for supporting the author's
copyright by purchasing an authorised edition of this book.

No amount of this book may be reproduced or stored in any format, nor may it be uploaded
to any website, database, language-learning model, or other repository, retrieval, or artificial intelligence
system without express permission. All rights reserved. Enquiries may be directed to Simon & Schuster,
222 Gray's Inn Road, London WC1X 8HB or RightsMailbox@simonandschuster.co.uk

www.simonandschuster.co.uk
www.simonandschuster.com.au
www.simonandschuster.co.in

Simon & Schuster Australia, Sydney
Simon & Schuster India, New Delhi

The authorised representative in the EEA is Simon & Schuster Netherlands BV,
Herculesplein 96, 3584 AA Utrecht, Netherlands. info@simonandschuster.nl

The author and publishers have made all reasonable efforts to contact copyright-holders for permission, and
apologise for any omissions or errors in the form of credits given. Corrections may be made to future printings.

The lyrics on p. 139 are from THE TIMES THEY ARE A-CHANGIN' by Bob Dylan Copyright © 1963, 1964
by Warner Bros. Inc.; renewed 1991, 1992 by Special Rider Music

Simon & Schuster strongly believes in freedom of expression and stands against censorship in all its forms.
For more information, visit BooksBelong.com.

A CIP catalogue record for this book
is available from the British Library

Hardback ISBN: 978-1-3985-6196-0
eBook ISBN: 978-1-3985-6197-7

Book text design by Paul Dippolito

Printed and Bound in Italy

Contents

	Introduction	1

PART 1: STARTUP

1.	Two Steves	5
2.	Apple I	12
3.	Apple Computer Company	18
4.	Apple II	27
5.	In Business	35
6.	Apple III	49
7.	Lisa	60
8.	Crashes	80
9.	Macintosh	85
10.	Sculley	104
11.	Software	113
12.	Marketing the Mac	125
13.	Insanely Great	137
14.	Macintosh Office	154
15.	Rift	163

PART 2: INTERREGNUM

16.	New Ideas	177
17.	Heresies	184

18.	High/Right, Low/Left	197
19.	System 7	207
20.	Newton	218
21.	Moonshots	232
22.	PowerBooks	238
23.	Spindler	249
24.	Clones	259
25.	Amelio	266
26.	NeXT	280
27.	Transition	287

PART 3: STEVE 2.0

28.	Turnaround	299
29.	Ive	316
30.	iMac	325
31.	Keynotes	335
32.	Millennium	343
33.	Retail	351
34.	iPod	357
35.	Back to the Macs	380
36.	iPhone	395
37.	Apps	416
38.	Sequels	421
39.	iPad	435
40.	China	446
41.	Apple Park	452
42.	Loss	457

PART 4: TIM

43. Services — 463
44. Watch — 475
45. Vision Pro — 488
46. Silicon — 501
47. Apple Intelligence — 506
48. Headwinds — 514
49. Apple: The Next 50 Years — 520
50. Throughlines — 530

Acknowledgments — 549
Sources — 551
Notes — 555
Photo Credits — 565
Index — 569

Introduction

"I've worked at five companies," says Myra Haggerty. "They were just all called Apple."

Haggerty, who's been at Apple since 1993 and is now vice president of Sensor Software & Prototyping, has a point. In its first half century, Apple has had three logos, four headquarters, and five CEOs. In 1976, Apple was two grungy guys named Steve hawking circuit boards. In 2026, its annual revenue approaches $400 billion a year—more than Meta, Netflix, and Intel combined.

Is Apple even the same company it once was?

Steve Jobs is a big part of the answer. He was at Apple for less than half of its first 50 years, but most of Apple's corporate throughlines can trace their origins to him: Simplicity, beauty, and elegant design. Making the whole widget. Focus on very few products. Closed systems. Secrecy.

The formula seems to work. Today, 27 percent of the entire global population uses Apple products—2.2 billion people. If they were a country, it would be the most populous in the world, bigger than India or China.

In 1976, the two Steves were thrilled to sell 150 of those Apple I boards. In 2026, Apple sells 220 million iPhones a year, bringing in $1 million every 90 seconds.

Those are impressive stats, but they become truly inconceivable next to the miracle that Apple exists at all. During the dark times without Steve Jobs, the company fell apart. At one point, less than six weeks from bankruptcy, its executives desperately tried to *sell* Apple, offering this once-revered brand at flea market prices. Nobody would touch it.

Everything was wrong. The products, the corporate structure, the distribution channel, the marketing, the board of directors, the bozo explosion within the ranks—all rotten.

And then Jobs returned to Apple and fixed all of it in a single year.

Whatever you think of Steve Jobs—his genius, his volatility, his passion, his pettiness—you have to admit that he pulled off the greatest turnaround in American history.

You've probably heard what came next: iMac, iPod, iTunes, iPhone, iPad. App Store, Apple Stores. The remaking of the music industry. And then, under Tim Cook, AirPods. Apple Watch. Growing empires in payments, health tech, streaming, chip design, digital services. Under Cook's watch, Apple became the first company ever to attain a $1 trillion market cap (the total value of all its stock). Then $2 trillion. Then $3 trillion. In 2025, it hit *$4 trillion.*

The company nobody wanted has become one of the most valuable in the world.

Origins

To tell Apple's story, I wound up interviewing 150 key players. (Apple did not see this book before publication, but did offer access to many of its current executives and designers.)

As you can imagine, there was a certain amount of credit-claiming and score-settling, which didn't interest me. There were also some fantastic origin stories, which did. I learned, along the way, a few eyebrow-raisers:

- Apple didn't start in a garage.
- Apple was Steve Wozniak and Steve Jobs's *fourth* business venture.
- There was a third founder.
- Steve Jobs was not Apple's first CEO—nor its second, third, or fourth.
- Steve Jobs did not originate or name the Macintosh; in fact, he forced out the man who did.
- John Sculley did not fire Steve Jobs.
- The tablet came before the phone.
- Jobs never fired anyone he'd just met in the elevator.
- And he didn't write the "Think Different" ad.
- The Newton saved Apple.

Apple's story is an epic tale of frenetic all-nighters and creative rebellion. Of titanic successes (iPods, iPhones, iPads) and instructive failures (Lisa, Apple III, MobileMe). Of funny, idealistic, scary-smart workaholics—coming up on three generations of them—who want to make things better by making *things* better.

It's about management, marketing, and strategy—and also about creativity, drive, and obsession. It's about the steady shrinking of the computer, from machines that filled buildings to machines that fill your ear canals. It's about how Steve Jobs founded Apple twice and left it twice, ultimately leaving behind one of the most successful, influential, controversial, resented, and loved companies in human history. It's about how Apple changed not only the devices we use, but *us*: how we communicate, consume, and create, in complicated ways we're still trying to understand and control.

It's the story of an infinite number of companies, all called Apple.

PART 1

Startup

1. Two Steves

In August 1950, the future Apple employee #1 was born—with a typo on his birth certificate.

Margaret and Francis Wozniak had intended to name their first child Stephen. But on that day in San Jose, California, a hospital worker spelled it "Stephan." 70 years later, he would legally change his name to Steve, but the world would always know him as Woz.

In 1957, the family moved to Sunnyvale, in the swath of former Northern California fruit orchards that would become known as Silicon Valley. Woz's father, who went by Jerry, was a Lockheed engineer, so the house was always full of electrical parts, and the whole region was a sprawling meadow of electronics companies and shops, thanks to the missile facilities nearby (Lockheed, NASA, Westinghouse) and all the people who worked there. So it was no surprise that at age ten, Woz announced that he, too, wanted to be an engineer.

In fifth grade, he built a computerized tic-tac-toe game that couldn't lose; in eighth grade, his digital adding machine won first place in the Cupertino science fair.

In high school, Woz spent late nights looking over the manuals for popular minicomputers,

Woz's Adder/Subtractor at the science fair.

designing and redesigning the same machines on paper to require fewer and fewer chips. Fewer chips could make a computer faster, less expensive, and simpler to repair.

In 1971, he and his across-the-street buddy, Bill Fernandez, developed an actual computer. It was little more than a tiny circuit board in a small black chassis with eight mechanical switches, eight red lights, and an Enter button. And the programs it ran were not exactly Microsoft Office. They were more like "Beep once every three seconds" or "Multiply *17* by *3* and flash the lights to indicate the result."

Fernandez and the Cream Soda circuit board.

(Today, that first computer is known as the Cream Soda computer, ostensibly because the two boys chugged Cragmont cream soda as they worked. Fernandez notes, however, that the boys themselves just called it "our computer." The term "Cream Soda computer" was made up, years later, by the author of a book about early Apple.)

That first computer didn't last long. During an interview with the *San Jose Mercury News*, arranged by Woz's mom, the reporter accidentally stepped on the power supply and fried the computer in a puff of smoke.

That Woz would ever meet Steve Jobs was by no means a sure thing. They lived 15 minutes apart and went to the same high school—but because Woz was four years older than Jobs, they didn't overlap.

Jobs had begun building his reputation for both brilliance and disruption before he was even out of elementary school. "You should have seen us in third grade—we basically destroyed our teacher," Jobs would say. "We would let snakes loose in the classroom and explode bombs." (His name rhymes with "Iobs," not "lobes.")

He skipped fifth grade, landing at Crittenden Middle School at ten years old. It was a tough place to be for an underage loner. Some kids brought knives to school; school administrators routinely called the police to break up fights. By seventh grade, he was bullied and miserable. He gave his parents an ultimatum: If they made him stay at Crittenden, he would quit school altogether.

Clara and Paul Jobs had adopted Steve a few days after his birth, and would do anything for him. Paul was a carpenter, car mechanic, and machinist—not exactly in the top tax bracket—but somehow, he pulled together $21,000 for a house in Los Altos, three miles away, in the excellent Cupertino school district. (Today, the house, at 2066 Crist Drive, is a tourist selfie hot spot. Jobs's sister Patti owns it.)

The town of Los Altos has designated Jobs's teenage home a "historic resource," which means that nobody can renovate it without the town's approval.

Jobs entered Homestead High in 1968. He was thin and good-looking, still a lone wolf—not laid-back enough for the hippies, not athletic enough for the jocks, not nerdy enough for the techies. He was curious, driven, and relentless.

At various points in his high school career, he was on the swim team, joined the water polo team, played trumpet in the marching band, read Shakespeare, wrote poetry, learned guitar, devoured art films, experimented with drugs, operated the lights for a school jazz band's concerts, and started a little club for executing pranks. One prank involved cementing a gold-painted toilet seat onto a planter. For another, they hauled a Volkswagen Beetle to the cafeteria roof.

But Jobs was also entranced by the new world of electronics. He hung out at the Explorers Club, a Tuesday-night gathering of high school kids held at the Hewlett-Packard cafeteria, where HP engineers gave talks about their latest projects.

When Jobs needed parts for an Explorers project he was working on—a homemade frequency counter—he cold-called Bill Hewlett, the CEO of HP. Hewlett chatted with the teenager for 20 minutes, agreed to give him some parts, and even got him a

Jobs in high school.

summer job on the HP factory line, assembling frequency counters. "Well, 'assembling' may be too strong; I was putting in screws," Jobs said. "It didn't matter; I was in heaven."

Perhaps inevitably, Jobs had become friends with Bill Fernandez. "We were deep thinkers. We were very smart. We were not interested in the social inanities," says Fernandez. "We were pretty much two lonely islands in the sea of teenage whatever-it-was that went on there in teenage cliques."

They took long walks after school, talking about girls, music, philosophy, and the meaning of life. And then, one day in 1971, Fernandez told Jobs that he should really meet his other friend Woz.

The Woz Prank Hall of Fame

If you want to make Woz's face light up, ask him about his pranks. His greatest hits could fill a book of their own—but here are some of his classics.

The TV jammer. As a college freshman, Woz built a tiny remote control that could make the student-lounge TV go fuzzy on command. Sitting in the back of the room, he'd time the static to the other kids' efforts to fix the picture.

At first, he made them believe that banging the TV, or fiddling with the tuning knob, cleared the interference. Eventually, the picture seemed to clear up only when someone stood on a chair with his hand *pressed against the middle of the screen*. They finally just gave up watching TV in that room.

Pan American international desk. Woz has always loved phone numbers with repeated digits, like 777-7777 and 888-8888. At one point, his number was (408) 221-1111—only an area code away from Pan Am's reservations 800 number. Customers mistakenly called him at all hours.

Instead of explaining that they'd misdialed, Woz pretended to be a Pan Am agent, often proposing ridiculously expensive and inconvenient itineraries. "I told some callers they could fly 'freight,'" he says, "but they had to wear warm clothing."

The ticking locker. In twelfth grade, Woz built a phony bomb out of some batteries and an electronic metronome, and stuffed the whole thing into his buddy's locker.

The school principal heard the ticking, ripped open the locker, and, clutching the apparatus to his chest to contain the blast, ran out to the playing fields to dismantle it. When the police arrived and realized that they'd all been hoaxed, Woz was in big trouble; he was sentenced to a night in a juvenile detention center.

The book from hell. Years later, Woz's mother-in-law began reading *The Exorcist*. She said it was the most evil book she'd ever read—so horrible that she threw it off the end of a pier into the ocean. So Woz bought another copy, ran it under the faucet, and left it by her bed. ("I'm going to hell," he says.)

"So one day, when Jobs came over to visit, I took him across the street," Fernandez says. "I saw Woz washing his car in front of the Wozniak family house, and I introduced them to each other on the sidewalk."

Woz showed Jobs his little computer. Jobs, in his usual outfit of torn jeans, T-shirt, and Birkenstock shoes, was fired up. He was beholding an actual computer, built from scratch by a couple of buddies with no money.

The two Steves wound up sitting on the sidewalk, becoming instant friends. Together, they chased down bootlegged Bob Dylan tapes, contemplated the lyrics—and, of course, teamed up on pranks.

The Blue Box

After his sophomore year of college, Woz, 21, took a gap year, hoping to earn enough money to attend Berkeley in the fall of 1971. But the night before his departure for Berkeley, he found an October 1971 *Esquire* article called "Secrets of the Little Blue Box."

It profiled a nationwide community of hackery nerds, known as phone phreaks, who developed what they called a blue box: a handheld machine that could generate the tones that controlled Bell phone-switching equipment. It gave them free, unlimited phone calls anywhere in the world.

If any two things can get Woz excited, it's pranks and electronics. Free, anonymous phone calls? The pranking possibilities were endless!

By midnight that very day, Woz and Jobs had built a prototype blue box of their own. It didn't actually work—the analog tone generator they'd used wasn't stable enough. But once at Berkeley, Woz built a compact, elegant *digital* box, incorporating a quartz-crystal oscillator like the ones in high-end watches. "It had some of the cleverest, most off-the-wall design techniques I've ever put into anything I've ever built, even to this day," he says.

Woz testing his blue box.

Woz used his box to call dial-a-joke lines all over the world, but Jobs had bigger plans: He wanted to turn the blue box into a product. Each box cost $75 in parts and took Woz four hours to build. They went door-to-door, selling the blue box to fellow college students for $150.

The business wasn't what you'd call a gold mine. Jobs remembers selling about a hundred of the boxes before he went off to college himself. Eventually, the phone networks became more sophisticated soon thereafter, blue boxes stopped working, and Woz let the whole thing peter out.

But the four-month experiment had established the magic that resulted from the pairing of their two very different personalities: Woz, the shy genius who could design circuitry like a prodigy, and Jobs, the charismatic marketer.

"I don't think there would have ever been an Apple computer had there not been blue boxing," Jobs said.

The Breakout Collaboration

In late 1975, Atari founder Nolan Bushnell developed what he considered a promising offshoot of Pong called Breakout and asked Jobs and Woz to work on a Breakout arcade machine. "Nolan had been complaining that the Atari games were going higher and higher in chip count," Woz says. Woz, Bushnell was sure, would solve *that* problem.

A project of this size should have taken months. But Jobs told Woz that the game had to be finished in four days. They didn't sleep, and they both caught mono, but they did it, with Jobs wire-wrapping the pegs of a prototype circuit board as Woz designed each circuit. Woz's design required only 45 chips. Jobs gave Woz half of the $700 fee that Bushnell had promised. (Or maybe it was $1,400. Accounts differ.)

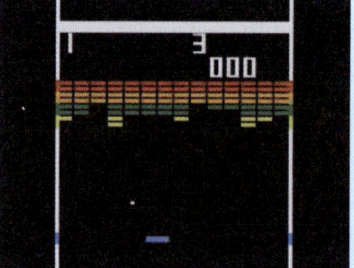

The story has two confounding postscripts. First: In the end, Atari didn't use Woz's design; it was so complex and tricky that the company couldn't manufacture it. Second: Years later, someone showed Woz a 1984 book called *Zap!: The Rise and Fall of Atari*. It referenced a $1,000 bonus for every chip they saved below 50. Jobs hadn't mentioned any bonus to Woz. "I was hurt," Woz says.

To his dying day, Jobs denied that he ever shortchanged Woz, and Woz has occasionally conceded the possibility that the account was wrong. "It's so ancient that maybe it didn't happen," he says.

In any case, within five years, both men would be vastly wealthy.

Educations

Reed College, near Portland, Oregon, is small, beautiful, academically rigorous, and expensive. When Steve Jobs arrived in the fall of 1972, the aftershocks of the civil rights movement, the war in Vietnam, and the countercultural revolution were impossible to miss.

He wore torn clothing, had long hair, and used no deodorant. He had few belongings, he hitchhiked to get around, and for months, he subsisted primarily on Dr. Jackson's Roman Health Meal, a hot breakfast cereal made of whole grain wheat, rye, bran, and flaxseed. (He was very thin.)

Then, after only a few months, Jobs dropped out. "I had no idea what I wanted to do with my life and no idea how college was going to help me figure it out," he said. "And here I was spending all of the money my parents had saved their entire life."

But he didn't leave; he lived in friends' dorm rooms and, with the dean's permission, attended the classes that interested him. The calligraphy class, in particular, left a deep impact on him. "If I had never dropped in on that single course in college, the Mac would have never had multiple typefaces or proportionally spaced fonts."

In the winter of 1974, Jobs returned home to Los Altos and found a job at Atari. Two years earlier, it had created Pong, the world's first commercially successful video game. Jobs made $4 an hour as a master of miscellany: testing, debugging, adding sounds, and so on.

At Reed, Jobs and fellow first-year Dan Kottke had become best friends. They read books about Buddhism, visited love festivals, and experimented with vegetarian diets. Now, only a few months into his job, Jobs left Atari to join Kottke on a spiritual quest to India.

In the end, Jobs didn't find peace there; the guru he hoped to meet had, in fact, already died, and he and Kottke spent much of the time sick with dysentery. But by the time he returned in early 1975, with a shaved head and a deep brown tan, Jobs was even more committed to the pursuit of Zen, to seeking inner calm, and to living life by intuition.

Woz, meanwhile, took another gap year to earn money for his *senior* year. His old high school pal Allen Baum, now an intern at Hewlett-Packard, helped Woz get a job there.

Woz was now making $24,000 a year designing the scientific calculators he adored. The HP culture was loose and creative, his colleagues were brilliant and friendly, and the coffee and donuts were free. He was 22, with the greatest job imaginable at the greatest company on earth. He fully believed that he would never leave.

2. Apple I

"I can tell you almost to the day when the computer revolution as I see it started," Woz says. "It happened at the very first meeting of a strange, geeky group of people called the Homebrew Computer Club in March 1975."

Computers in the early sixties were anything but personal. The machines Woz had read about as a kid in engineering journals were room-sized behemoths like the ENIAC, used to calculate missile trajectories and nuclear-weapon blast zones. Minicomputers, the size of microwave ovens, came along in the late sixties, but they still had no keyboards or screens—you entered programs on paper tape or by pressing sequences of switches—and still cost as much as $50,000. Only corporations, universities, and government agencies could afford them.

But on that rainy night in 1975, the 30 scruffy Homebrew hobbyists who met in an engineer's home garage had a larger mission: to democratize computers, to make them cheap enough and simple enough for anyone to use—a philosophy that aligned precisely with the dream Woz had nurtured for years.

The club members were buzzing about the MITS Altair 8800, a $400 build-it-yourself microcomputer kit. The Altair had appeared on the January 1975 cover of *Popular Electronics*—and there, tonight, they got a demo of the real thing.

Considering all the buzz it got in 1975, the Altair wasn't much of a computer. It was a set of parts that you were supposed to solder onto a circuit board yourself, housed in a plain rectangular enclosure. No keyboard, no disks, no screen.

Programming it was tedious and error-prone, as it had been on Woz's Cream Soda computer five years earlier. But there was a world-changing difference: Whereas his machine's CPU (its "brain") was constructed from a bunch of chips elaborately wired onto a circuit board, the Altair's came on a *single* chip, the Intel 8080 microprocessor.

Well, wait a second. If *they* could build a computer around a microprocessor, why couldn't he? Why should Woz pay $500 for an Altair kit when he could order a microprocessor and build his own computer around it?

"Oh my God," he thought. "I could build my own computer, a computer I could own and design to do any neat things I wanted to do with it for the rest of my life." It could have a keyboard for entering text. It could display results on a TV!

That very night, Woz grabbed a sheet of paper and drew the schematics for this dream machine. He didn't know it at the time, but the design he sketched would soon become Apple's first product.

WESCON

The Intel 8080 was too expensive for Woz's taste. Just one of these chips would cost $300, the equivalent of a month's rent—*if* Intel's reps would sell a single chip to a hobbyist. (They wouldn't.) The alternative was Motorola's 6800 microprocessor: $40 with the HP discount.

But in the fall of 1975, Woz spotted an engineering magazine ad he could scarcely believe: "Come to the WESCON, and we'll sell you a microprocessor for $25!"

The two new microprocessors, from a company called MOS Technology, were pin-for-pin the same as Motorola's chip. But they were faster, simpler, and less expensive.

On September 16, 1975, Woz and Allen Baum drove to WESCON, the Western Electronics Show and Convention, at Brooks Hall, the underground exhibit space of the San Francisco Civic Auditorium convention center.

The 6501 and 6502 were the buzz of the show—at $20 and $25, how could they not be?—and the MOS booth was mobbed. Chuck Peddle, the chip's chief designer, manned the booth. Woz bought one of each chip and a manual, and headed home.

*MOS Technology raised a lot of eyebrows (and pulses)
by advertising a $25 microprocessor.*

Woz's life was already packed full. He still had a full-time job at HP. He was running a dial-a-joke service, which required him to record a new joke every day (mostly from a Polish joke book). He had his first girlfriend, a dial-a-joke customer named Alice Robertson. But he became consumed with finishing his own personal computer, based on the MOS 6502 chip he'd bought.

Fortunately, he had a huge head start. A few months earlier, as a side project, he'd built a simple "TV terminal": a circuit board with keyboard—no processor or memory—connected to a TV. He used it to connect to ARPANET, the Department of Defense's ancestor of the internet. At the time, a few dozen computers were on it, mostly at universities like MIT and UCLA.

Woz realized that to build his self-contained new computer, he could use the same TV-terminal circuitry—but instead of sending commands over the phone lines, it would talk to a processor and some memory right there on the same machine. "Very simple step," he says.

For the next few months, he split his time between HP and HP. That is, after finishing his work on advanced calculators, he'd dash home, scarf down a frozen dinner or plate of spaghetti, and then

Conversor 4000

Alex Kamradt was a Lockheed engineer and entrepreneur. His business, Call Computer, rented out time on his minicomputer to local hobbyists, programmers, and businesses who couldn't afford to own such a machine themselves.

But reaching Call Computer required a teletype machine, which printed the computer's responses on paper, or a "glass terminal," which displayed them on a screen. Both were cumbersome $1,000 machines. Kamradt figured he'd hire some genius Homebrew engineer to create a *smart* terminal, good-looking and compact, purpose-made for accessing Call Computer; you'd just connect a keyboard and a TV.

When Kamradt asked around, Woz's name kept coming up.

Jobs pitched Kamradt on using Woz's existing TV terminal as the heart of a new $200 terminal, to be called the Computer Conversor 4000. He delivered the prototype in the early fall of 1975. Jeff Moffatt, who worked at Call Computer, isn't sure where the plastic housing came from, but says it looked awesome. "It looked like an IBM Selectric typewriter at about three-quarters scale."

But when Kamradt needed tweaks, fixes, and improvements, the Steves seemed to have moved on. Woz had been barely even aware of the venture, and Jobs had bigger fish to fry.

The Conversor 4000, in other words, was the shortest-lived business venture in the history of the two Steves.

return to his cubicle at HP, which had the testing equipment and spare parts he needed, to work on his home-brewed computer.

Once his prototype board was laid out, Woz scrounged up a power supply, 32 1 KB static RAM (memory) chips, and a 9-inch black-and-white screen from the HP labs.

Finally, late one night in October 1975 at his HP cubicle, he turned the thing on, typed a couple of letters, and saw them appear on the monitor.

It was nothing more than a circuit board, 9 x 15.5 inches, lying on a desk. Its socketed chips were exposed to the air. But it was a great-*looking* circuit board. It had four tidy rows of chips—two rows that did the computing and two that created the screen display. And because he'd cut every wire to exactly the right length, it looked polished and commercial. It was a big improvement over the usual hobbyist ritual of "wire-wrapping" a board—wrapping wires around posts on the back of the board, which looked like a snarled nest of fibers.

The Apple I did not set the world on fire.

"I would hold the wire down by the pin, and take the soldering iron with my right hand, and the wire in my left hand, and I'd have the solder *in my mouth*," Woz would say later, grinning. "And all the lead in that solder never af-af-af-affected me!"

Still, you might not have even recognized this early Woz machine as a computer. One wire trailed off to the screen, another to the keyboard, and a third to the power supply. It had only 256 *kilobytes* of storage, barely enough to store this sentence. (A typical laptop today has about 4.2 million times as much.)

And Woz's computer had no name. There was no entity called Apple yet, let alone a product name like Apple I.

But Woz had built a computer with a screen—even a TV would work—which was already a leap above other computer kits. And because he included a ROM chip (read-only-memory—a chip whose contents are preserved even when the power is turned off), his computer started up in under a minute. Compared to the 30-minute wait for the Altair, that was instantaneous.

Apple, in other words, was not born in a garage. It was born on Woz's desk at HP.

Entrepreneurs

By the end of 1975, the Homebrew Computer Club's membership had grown into the hundreds; hobbyist computing had caught fire all over the Valley. For comfort and capacity reasons, it moved to the auditorium of the Stanford Linear Accelerator Center in Menlo Park.

Homebrew was all about the exchange of plans, ideas, and software. At the beginning of each meeting, club members set up card tables in the lobby to show off whatever computer kits or software they wanted to explain or sell.

Woz began bringing his circuit board computer to Homebrew meetings in his beat-up Ford Pinto. Most weeks, he brought two young buddies with him: Randy Wigginton, whom he'd met at Call Computer, and Chris Espinosa, a fellow Homebrewer. Both were 14, still in high school, a good ten years younger than Woz, but they loved writing software and were dazzled by Woz's creations.

Steve Jobs sometimes came along, too.

Woz handed out photocopies of his computer's schematics to anyone who was interested. He was intensely proud of how much the machine could do with only 30 chips, and he hoped to spread his ideas far and wide.

Steve Jobs's mind, though, was turning. He was no engineer; in fact, much of the nerdy-obsessive conversation at Homebrew bored him. But he'd noticed that while many Homebrewers picked up Woz's schematics, few *did* anything with them. To Jobs, that was a marketing opportunity.

And so, one day in November 1975, Jobs asked Woz: "Why don't we build and sell the printed circuit boards?"

The two Steves set up the Apple I prototype at the Homebrew Computer Club.

By that, he meant the green, plastic/resin boards *before* chips and other parts were attached. "Printed" meant that the copper traces—electrically conductive pathways—would already be etched into place. Holes would be predrilled to accommodate the computer's components.

The painstaking work of wiring the board would already be done. You could have yourself a working computer in a matter of days instead of weeks. All a hobbyist would have to do was buy the components (processor, memory, power supply, and so on), solder sockets for them onto the board, and then snap the components in.

Woz was dubious. Having the board professionally designed would cost $600, and hiring a company to do the "printing" would be another $1,000. If they sold the boards for $50 each, as Jobs was proposing, Woz didn't think there were enough potential buyers at Homebrew to break even.

Furthermore, under no circumstances did Woz want to jeopardize his future at HP. He *loved* that job. His parents weren't enthusiastic, either. Jerry, his dad, wondered why his son should be a half partner with Steve Jobs, a non-engineer "who hadn't done anything."

But even at 20, Jobs was persuasive and relentless. First of all, he explained, he'd take care of manufacturing and selling the boards. Woz wouldn't have to leave HP. It wasn't going to be some big operation; maybe they'd build 50 of these boards, sell them off, and clear a nice $1,000 profit. Above all, Jobs stressed, "We'll have a company. For once in our lives, we'll have a *company*!"

At this point, Woz had already joined Jobs in three ventures: the phone blue boxes, the Conversor 4000, and the Breakout game for Atari.

But those were one-off projects. They weren't official. This idea was different. "To be two best friends starting a company. Wow," he says. "I knew right then that I'd do it. How could I not?"

3. Apple Computer Company

One day in March 1976, Jobs was returning from one of his visits to the All One Farm, a semi-commune on a farm/orchard near Portland. Woz picked him up at the San Jose airport.

"I have a great name: Apple Computer!" Jobs announced.

The two Steves had been batting around techie-sounding names for their little company for months: Power Computing. Executex. Matrix Electronics. Personal Computers Inc. None of them seemed great.

Apple, though, was a *great* name. For Jobs, it fit nicely with his fruitarian diet, and with the orchards at the All One Farm. "It sounded fun, spirited, and not intimidating," he remembered. "'Apple' took the edge off the word 'computer.' Plus, it would get us ahead of Atari in the phone book."

Woz had only one concern: that the name was already taken. "What about Apple Corps, the Beatles' record company?" he said.

"They're a record company, and we're a computer company," Jobs said dismissively.

"That's all it takes?" Woz said.

"Yeah!"

(It wasn't. Two years later, Apple Corps sued over the name. Apple settled for $80,000 and a promise not to enter the music business. But in 1986, when Apple added recording features to the Mac, the Beatles sued again; this time, the settlement cost Apple $26 million. A third suit followed when Apple opened the iTunes Store online; the Beatles lost that one. Finally, in 2007, Apple bought the forever rights to the trademark "Apple"—for $500 million.)

But before Woz could agree to market his computer, his religion of honesty compelled him to offer it first to his employer.

When Woz showed his HP bosses his designs, however, they had no interest. They were dubious about the notion of using a TV as a monitor, and didn't have the bandwidth to undertake a new project. Apple had the green light to proceed.

At this point, the Steves' business plan was still vague: Sell 50 circuit boards. But they still needed $1,600 to get them made. They needed some startup capital.

Woz sold his beloved HP-65 calculator for $500, consoled by the prospect of a new model that would be coming out soon. Jobs sold his beat-up Volkswagen bus for $1,500. Neither transaction went smoothly—Woz's calculator buyer coughed up only half the agreed-upon price, and Jobs's

The Steves show off one of the first finished Apple I boards.

Volkswagen buyer demanded reimbursement when its engine died two weeks later. Still, when the two Steves pooled their savings, they had scraped together $1,300—enough to begin moving forward.

There was, however, one final hurdle.

According to Jobs, Woz would no longer own his engineering designs once Apple was launched; the *company* would. But Woz foresaw a disaster: What if HP someday wanted him to incorporate one of the design tricks he'd used on the Apple board?

Jobs couldn't create Apple without Woz and his genius, but he was exasperated by Woz's resistance to the idea of corporate ownership. He asked Ron Wayne for advice.

Wayne, a 41-year-old senior designer Jobs had met at Atari, was an inventor, model maker, and tinkerer, with several patents to his name. He'd created, sold, and been cheated out of several companies in his long and varied career.

Ron Wayne
Born: May 17, 1934, Cleveland, OH
Schooling: San Fernando Valley Junior College
Before Apple: Bally, Atari
Apple: 1976
After Apple: Thor Electronics, Ocean Design

In 2000, years after their last encounter, Jobs invited Wayne to attend the San Francisco Macworld Expo as his guest—first-class airfare and chauffeur included, just for fun.

When Jobs described his impasse with Woz, Wayne suggested that both young men join him at his apartment; maybe he could mediate the disagreement.

"Jobs sat on one end of the sofa and Wozniak on the other end," Wayne remembers. "For over an hour, I verbally tiptoed over, through, and around the fact that the new company needed to entirely own the rights to these circuits."

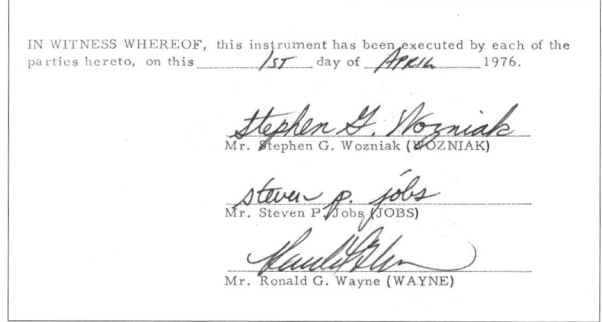

The three cofounders' signatures.

Eventually, Woz gave in. Okay, fine. Apple would own his designs.

Jobs was so thrilled by the outcome of Wayne's diplomacy that right there, on the spot, he proposed making Wayne a cofounder. Jobs and Woz would each own 45 percent of Apple; Ron Wayne would own 10 percent. If there were any future disputes, Wayne's vote could be the tiebreaker.

Wayne sat at his typewriter and typed up a two-page partnership agreement. "NOW THEREFORE," it said, "WOZNIAK shall assume both general and major responsibility for the conduct of Electrical Engineering; JOBS shall assume general responsibility for Electrical Engineering and Marketing, and WAYNE shall assume major responsibility for Mechanical Engineering and Documentation."

The next day, Wayne took the documents to the Santa Clara County registrar, who filed it as certificate number 20443.

It was April 1, 1976. Apple was born on April Fools' Day.

Jobs, delighted, immediately gave Wayne some assignments. Among the most urgent: a logo. Wayne, an experienced draftsman, came up with an old-timey pen-and-ink drawing of Isaac Newton sitting beneath the apple tree whose falling fruit inspired his theory of gravitation. A flowing "APPLE COMPUTER CO." ribbon floated through the scene. In smaller type, wrapping around the frame, Wayne incorporated the last line of a William Wordsworth sonnet, which he thought was appropriate for the new venture: "NEWTON ... A MIND FOREVER VOYAGING THROUGH STRANGE SEAS OF THOUGHT, ALONE."

Ron Wayne's original Apple logo.

Wayne's logo was too ornate to work as a modern corporate insignia; it lasted only a year. Still, it served the startup well at a couple of computer shows, where it adorned Apple's booth as a banner.

Byte Shop

In April 1976, the Steves proudly held up the first Apple product for the 500 members of the Homebrew club to see: a printed circuit board for a desktop computer. If you looked closely at the left margin of the board, you could read, in white sans serif lettering: "Apple Computer I."

They told the crowd that you could enter programs with a keyboard, rather than using, as Woz put it, "a stupid, cryptic front panel with a bunch of lights and switches." Once you'd bought your chips and soldered them in, you could have 8 KB of RAM—32 times more than the Altair!—and a MOS 6502 microprocessor. (RAM, random access memory, is where a computer holds the data it's actually working on; it's fast and expensive.)

Once you connected a security monitor or a TV with a video input, you would see 24 rows of text, with 40 typed characters across each one. The machine could display the letters *A* through *Z* (capital letters only), the digits *0* through *9*, plus 26 punctuation marks and symbols. They'd throw in a free copy of Woz's BASIC interpreter—all for $50.

Their presentation didn't set the room on fire. The Apple I used an off-brand processor. It couldn't display color. (By this point, you could get a color add-on board for the Altair.) Above all, nothing the two scruffy dudes were showing off was anything new. They'd been bringing the Apple I to Homebrew for six months.

One Homebrewer, however, approached the Steves after the meeting. He introduced himself as Paul Terrell and told them that he had recently opened a computer store—in fact, the world's first computer store—called the Byte Shop. Much of his business involved buying computer kits like the Altair, having his staff assemble them, and voilà: ready-to-use computers for customers who didn't have the time or the skill to build their own. His ambition was to build the Byte Shop into a national franchise. (And sure enough: By 1977, when Terrell sold the chain, there were 58 shops.)

The Apple I struck him as a perfect addition to the store's offerings. But he didn't want to sell it as a naked printed board that required customers to hand-solder their own chips—easily an eight-hour process. Instead, he suggested, what if the two Apple guys supplied their boards already assembled, with all the chips in place? That would make the Apple I attractive to a much bigger audience, and could command a much higher price.

It wasn't what the two Steves had had in mind. But at the Byte Shop the next morning, Terrell made a stunning offer to Jobs: He would buy 50 Apple I boards, "fully assembled and tested," for $500 each. The total order: $25,000. (That'd be $138,000 today.)

Jobs and Woz were elated. "Nothing in subsequent years was so great and so unexpected," Woz remembers.

The two Steves with a working Apple I in April 1976, just after founding Apple.

But where would they get the $15,000 they'd need to buy the processors, memory chips, and other components?

Allen Baum and his dad loaned them their first $5,000, but the next $10,000 seemed impossible. Banks turned them down. Jobs's old bosses at Atari turned him down.

Finally, Jobs began approaching parts suppliers directly. "He would call up the distributors of electronic parts in the area: Hamilton/Avenet, Elmar Electronics," remembers Bill Fernandez. "He would talk to their salespeople and say, 'Look, we're going to be great. And if you want our business in the future, you need to do this for me now.'" It was the business world's first exposure to his freakish abilities of persuasion.

Jobs carried around Terrell's purchase order for the 50 boards so he could prove to suppliers that the Byte Shop deal was real, that payment would come once the boards were built. He showed it to the financial guy at Cramer Electronics, at the time the nation's second-largest components supplier.

After a final due-diligence phone call to Terrell, Cramer gave Apple $20,000 credit, net 30 days. "I've always claimed that that purchase order was the seed capital that started Apple," says Terrell. "I'm just sorry I didn't ask him for stock!"

Jobs, now 21 years old, had no idea what "net 30 days" even meant. But it was exactly the arrangement he needed: getting parts for free, with 30 days to pay for them.

Building Apple I

If you've seen any of the three movies about the origins of Apple, you know the scene that comes next: a homemade assembly line in the Jobs family garage, working under the 30-day Cramer deadline.

In real life, the operation also took over the Jobs kitchen, a bedroom, and the living room. Jobs paid his pregnant sister, Patti, $1 per board to plug components into the boards. His college pal Daniel Kottke soon joined the brigade.

Jobs's dad, Paul, brought a workbench into the garage and set up tidy rows of drawers for the parts. He let them use the family's TV as a monitor, asking only to have it back every now and then so he could watch football. Jobs rigged phone cord extensions all the way from his bedroom to the garage so that he could continue trying to sell Apple I's to local computer stores over the phone.

Woz tested each board before gently placing it into the box bound for the Byte Shop. Unfortunately, the assembly frenzy was wreaking havoc on his home life; Alice, his new bride, never saw him. She was the first romantic partner to feel neglected during the push to meet a tight Steve Jobs deadline, but she would hardly be the last.

Jobs, fueled by a diet of carrot juice, raw carrot salad, cherries, dates, and almonds, was in constant motion. "He had that hustle from the beginning," says Fernandez. " 'I need to talk to this guy.' 'I need to find out how to do this.' 'I need to find a place to buy parts.' "

Soon, Jobs was ready to deliver the first batch of finished boards to the Byte Shop. Terrell, however, was taken aback. He'd expected to receive complete computers, but Jobs had brought only assembled circuit boards: no keyboards, monitors, cases, or power supplies. "What the hell is this?!" he said.

Paul Terrell discovers that Jobs and company have delivered only assembled boards, not complete computers. (A still from the 2013 movie Jobs.*)*

"But this is much better for you than fully assembled," Jobs argued. "Now you can sell those other parts to your customers, too. You're a computer store, aren't you?"

In the end, Terrell accepted the boards. He paid the promised $500 for each one, generating $280 profit for the two Steves. "Suddenly, our little business was making more than I was making at HP," Woz says.

Jobs wanted to expand. With the first profits, he opened a post office box, hired an answering service, took out ads in a couple of hobbyist magazines, and, in February 1977, hired Apple's very first permanent, full-time employee: Bill Fernandez.

Jobs ordered a second batch of 50 boards, which he planned to sell directly to Homebrew members and friends, and a third set of 100 for selling to other electronics stores.

He wasn't sure, however, what to charge. He proposed marking up the computers one-third over the wholesale cost ($500), which would bring the retail price to $666.66. Woz loved it; it scratched his lifelong itch for repeating numbers.

Neither of the Steves had seen *The Omen*, the hit horror movie the previous year, or read the book of Revelation in the Bible. It wasn't until they started getting angry phone calls—and endured a boycott in Los Angeles—that they learned that *666* is supposedly a symbol of the devil.

Terrell continued pressing Jobs and Wozniak to make the Apple I more complete. At least build a case for it, he complained. And for God's sake, give us cassettes!

The Apple I's biggest pain point was that it had no storage. Every time you wanted to run a

The Apple I as it would finally appear,
complete with audiocassettes for data storage.

program, you had to type it in from scratch—hundreds of lines of code. Woz spent 40 minutes keying in the code for his version of BASIC from his notebook before every demo at the Homebrew club. Every. Damn. Time.

Floppy disks and hard drives were still years away. What most hobbyist computers used for

> ### Ron Wayne Bows Out
>
> When Ron Wayne helped the two Steves found Apple, they offered him 10 percent of the new company. His first tasks included writing a manual for the Apple I, creating a diagram of its schematics, and designing Apple's first warranty card.
>
> But after a week or so, Wayne learned of the deals that Jobs had engineered with the Byte Shop and Cramer Electronics. Cramer had advanced Jobs $15,000 worth of parts, which Jobs would have to pay for once *he'd* been paid by the Byte Shop.
>
> A friend had told Wayne, however, that the Byte Shop did not have a sterling reputation for paying its bills. Jobs, Wayne says, "had committed the company to a substantial obligation. Since our enterprise was a company and not a corporation, it was an obligation that put me personally on the hook for $1,500" if the Byte Shop didn't come through.
>
> Only a few years earlier, Wayne's own startup—making digital slot machines for bars and casinos—had gone disastrously wrong. He'd spent years buying back all of its stock and paying off his creditors. At this point, he didn't have the stomach for another roller-coaster ride.
>
> What he couldn't know, of course, is that Apple would one day become the world's most valuable company, making his 10 percent worth more than $350 billion.
>
> What he *did* know was that he was now a named partner in a tiny startup with a flimsy first creditor. And so, on April 12, 1976—only 11 days after Apple's founding—Ron Wayne returned to the Santa Clara registry office to file an amendment to the partnership agreement he'd written. It said: "By virtue of a re-assessment of understandings by and between all parties, Wayne shall hereinafter cease to function in the status of 'Partner.' "
>
> "I have never regretted my action," Wayne says. "I was 42 years old at the time—the old man of the group—while the two Steves were in their twenties. Jobs, in particular, was an absolute whirlwind. It was like having a tiger by the tail. It took little consideration for me to conclude that if I stayed with the enterprise, I'd probably wind up the richest man in the cemetery."
>
> Jobs and Wozniak were gracious about his exit; they offered $800 to buy him out. A few months later, Apple's law firm sent him another check, for $1,500, along with a letter making clear that his acceptance of the money would more formally terminate his stake in Apple.
>
> "So to answer for the last time, 'Did I make the right decision?'" Wayne says. "I have no idea, but there is this: While it is true that in all of my life I've never been rich, I've never been hungry, either."

storage in 1976 was ordinary audiocassettes. Yes, the magnetic tape in a cassette could record music, but it was also perfectly capable of storing data.

You'd run two cables to your computer. One plugged into your cassette player's headphone jack, another into its microphone jack. On the Apple I, you'd type a command telling the computer to start reading from the tape; then you'd insert a tape and press Play. At that point, you had a painfully slow, *mostly* reliable method of storing and retrieving data without having to retype your code every day.

At last, Woz came up with an add-on card with cassette recorder connections—a $75 option.

As for cases: Jobs approached Charles Pfister, a local cabinetmaker, about building wooden enclosures for the Apple I. It would have to be sleek, Jobs said. No visible screws.

"It was me, my dad, and Steve at a little restaurant called Frankie, Johnnie, & Luigi Too, right next to the Byte Shop," says Pfister's son, Chas. "My dad bribed me to come along for pizza and french fries. I drafted this mockup. It was one of those napkin deals."

According to popular lore—and even auction-house listings to this day—the Pfisters' cases were made of rare Hawaiian koa wood. "That's hilarious," Chas says. "None of them are koa wood. They're walnut! Walnut is a beautiful wood. We just used a simple Watco oil finish on it, two, three coats, and that was it."

Paul Terrell would refer customers to the Pfisters' shop, where they could order the case customized. Jobs told the Pfisters that they'd be making a couple hundred of these cases. Ultimately, they sold only 25—partly because of the $140 price, and partly because the hobbyists of the day needed frequent access to the board for tinkering.

Today, only seven of the original cases survive, and they're extremely valuable. In a 2021 auction, an Apple I in an original Pfister case—advertised as, yes, Hawaiian koa wood—went for $500,000.

In the end, the Apple I didn't change the world. Only about 175 people bought them (about 15 working units still exist). Chances are excellent that you've never even seen an Apple I; most people have never even heard of it.

But building the Apple I gave Jobs and Wozniak an essential education in business finance, introduced them to vendor relationships, and established Apple as a company.

The Pfisters' sleek walnut Apple I Case

But even before the Apple I was done, Woz was already designing its sequel.

4. Apple II

Woz knew that the Apple I's successor would need a color display. Computer companies had begun selling accessory boards that added color to their machines' video output. But Woz wanted color built right in. You'd run a cable from the Apple to your color TV's antenna connector screws, tune the TV to channel 3, and presto: The computer's screen image would appear.

Second, he wanted to consolidate his own chip design. The Apple I, having been adapted from his TV terminal, maintained separate rows of chips for its two different functions: computing and displaying. The Apple II would fit all of that into the same memory. To Woz's immense satisfaction, that meant using fewer chips, which would offer better speed, a lower price, and better reliability. (In the end, the Apple II used *half* as many chips as the Apple I.)

The new machine would have sound circuitry, a speaker, faster dynamic RAM, and BASIC built right in. All told, "It wasn't just twice as good," Woz says. "It was, like, ten times better."

"Every few days, Woz would say, 'God, I've made an incredible breakthrough! I've saved a few chips here and there,'" Jobs said. "And it ended up, of course, being legendary."

Jobs, meanwhile, firmly believed that the Apple II should be a self-contained, fully packaged computer, with a case and a keyboard. "It was clear to me that for every hardware hobbyist who wanted to assemble his own computer, there were a thousand people who couldn't do that but wanted to mess around with programming, just like I did when I was ten," Jobs said. "My dream for the Apple II was to sell the first real packaged computer."

The Power Supply

The computer would also need a power supply—the circuit that converts the alternating current (AC) from the wall outlet to the direct current (DC) of the machine. Its descendant today is the modern power brick on a laptop's power cord. (Internal power supplies are too big and hot for laptops.)

The power supply was an analog project, outside of Woz's domain. Fortunately, Jobs knew a guy at Atari, a brilliant analog engineer named Rod Holt—"a chain-smoking Marxist who had been through many marriages and was an expert on everything," as Jobs described him—who'd designed power supplies.

Woz's Color TV Trick

Computer geeks and electrical engineers speak in hushed tones about the engineering genius of the Apple II. One trick in particular loomed over Woz's other design triumphs: fooling an ordinary TV into displaying color images from black-and-white Apple II circuitry.

At the time, computers that displayed color cost thousands of dollars. They relied on a whole array of video chips to produce a color signal. But the Apple II *had* no color circuitry. It was a black-and-white machine. It used only a single, $1 chip for its video output. So how on earth did Woz make it display color?

The inspiration came to him in late 1975 when he was attempting to design Breakout for Atari in only four days. He and Jobs did very little actual sleeping during that sprint. But now and then, Woz lay on the floor and closed his eyes for a few minutes. In the half-asleep twilight of uninhibited mental activity, an idea came to him.

The inside of a TV's glass screen was coated with phosphors—chemical compounds that light up when struck with a beam of electrons. They were arrayed in repeated dots of red, green, and blue.

The electron gun could strike only one dot at a time. To create a full-screen image, it had to paint the screen in *rows*, left to right, top to bottom, just the way your eyes scan text. It happened so fast (about 30 times a second) that you perceived it as a steady picture.

What Woz discovered was that he didn't have to send signals to the TV that said, "Paint this dot red, and this one blue." Since the electron beam scanned across those repeated sets of pixels, he just had to adjust the *timing* of the gun's firing; the color would take care of itself.

He determined that if the signal arrived a fraction of a second after the gun began its sweep across the screen, it would illuminate a green pixel; a hair after that, it would light up a red one; and so on.

Woz figured he could use four bits of data to code for 16 different delays. Through trial and error, he learned that if he programmed the four bits as 0101 (in binary), for example, he'd get green; if it were 0011, he'd produce purple (the gun would strike blue and red simultaneously). This trick generated an odd set of six colors—purple, green, blue, orange, black, and white—and worked only on American (NTSC) televisions.

Still, programmers were delighted to learn that they didn't have to worry about the details. In Woz's BASIC language, they could simply choose, for example, "blue"; the Apple II did the rest of the math for them.

In short, Woz controlled an analog TV with a digital signal. He created a color image from a chip capable of sending only monochrome (black-and-white) signals. And he left the Apple II's competitors, like the TRS-80 and the Commodore PET, in the black-and-white dust.

Jobs gave him a call. "This guy convinced me that the Apple II, if we could make it work, would change the world," says Holt. "Now *how* he did that—I will not, to this day, testify how he did it. But he did."

In 1977, most computers had *linear* power supplies, which throttle down the excess voltage by turning it into heat. But the Apple II couldn't afford to heat up inside, and Jobs was adamant that it would not have a fan. Fans are noisy, and noise is annoying.

Working after hours at Atari, Holt designed a *switching* power supply, which instead regulated the voltage by turning it off and on thousands of times per second. At that point, switching power supplies existed mostly in bulky minicomputers. Holt's was smaller, lighter, and cooler than linear units—"and we manufactured that thing for less than $30," he says.

Slots

Slots are thin, rectangular connectors on the main board that can accommodate optional add-on boards. Each such card adds a new feature to your computer. One might let you attach a printer; another could contain a memory upgrade; another might accommodate a dial-up modem.

Woz wanted lots of slots, convinced that they would unleash unlimited potential for the Apple II. Jobs, on the other hand, couldn't see why anyone would need more than two: one for a printer, one for a modem. More slots, he said, would add size, cost, and complexity.

What ensued was the Steves' very first argument.

Ordinarily, Steve Wozniak was, as many have described him, a teddy bear: easygoing, gentle. But from the core of his soul, he believed that slots were doorways to cool features of the future. He threw down an ultimatum: "If that's what you want, go get yourself another computer."

Reluctantly, Jobs gave in. The Apple II would come with eight slots.

Making the Case

The Apple II was the first inexpensive computer with color, the first to come complete and ready to run, the first that was usable by people who weren't engineers.

But it was also the first computer whose *design* wasn't an afterthought, whose case wasn't just a box to keep you from touching the innards. Steve Jobs felt strongly, intuitively, that its looks should affect you before you'd even turned it on. It should convey friendliness and competence.

He wandered the aisles at Macy's, studying the design of existing consumer products—in particular, Cuisinart's stylish food processors—and decided that the Apple II should be housed in plastic.

At the time, *nobody* was using plastic. In 1976, all commercial computers were made of metal. Metal was cheap. Metal was easy to shape. Metal provided shielding from electromagnetic interference. Custom tooling and molding of plastic was expensive, fussy, and slow.

But to Jobs, metal screamed "industrial." Consumer goods, products that looked at home at *home*, were made of plastic. "I got a bug up my rear that I wanted the computer in a plastic case," he said. He needed a professional industrial designer.

Two Valley design companies turned him down, considering Apple too iffy a client. Undaunted, Jobs turned to a 33-year-old designer named Jerry Manock. Manock had graduated from Stanford's product-design program, and had worked as a designer at HP; only a month earlier, he had decided to go freelance.

Once he heard Jobs's pitch—at, of course, the Homebrew Computer Club—Manock agreed to take on the project for $1,800. "But I did ask to be paid in advance," he says.

Manock's design called for a low-slung plastic case: tall enough to accommodate boards in the slots, yet compact enough to be approachable. All edges were rounded, smooth, and inviting—a radical departure from the sharp-cornered metal boxes of the time.

The original design had recessed handles on the sides. Jobs told Manock to get rid of them: Why would anyone want to move a computer once it was set up?

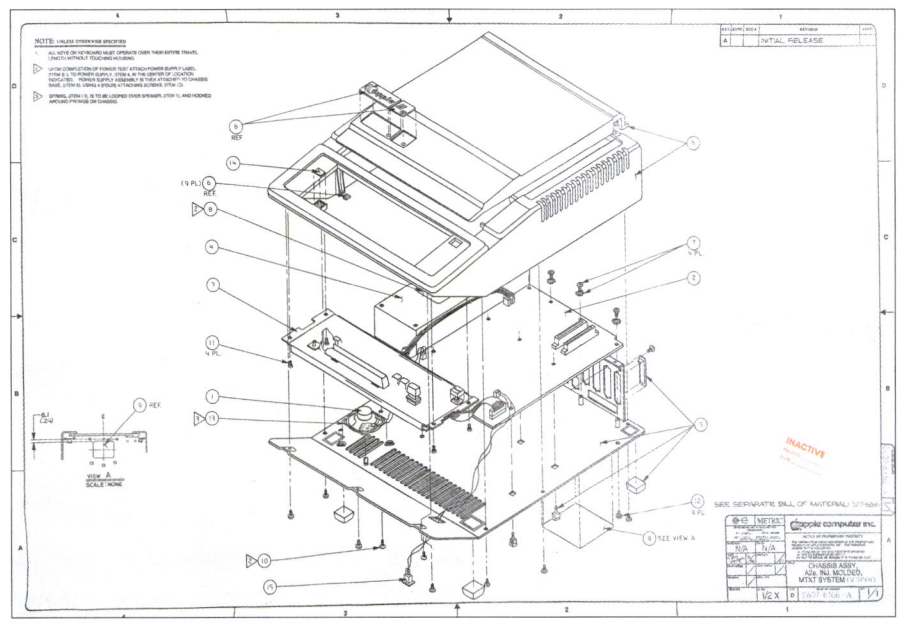

On Manock's Apple II design, the keyboard sloped, suggesting the familiar shape of a typewriter.

Jerry Manock
Born: 1944, Los Angeles, CA
Schooling: Stanford (BS and MS)
Before Apple: HP, Telesensory
Apple: 1977–1984
After Apple: Manock Comprehensive Design

Manock, an accomplished photographer, exhibited his nature shots from the Sierra Nevadas in various galleries.

At one point, Jobs asked Manock if it would be possible to chrome-plate the interior of the computer, "to look pretty when users opened the case," Manock says. Eventually, he talked Jobs out of that one. But it would not be the last time that Steve Jobs exhibited his passion for making the *insides* of his machines look beautiful, even if few people would ever see it.

Regis McKenna

There weren't many avenues for marketing computers to potential customers in 1977. The internet was still 16 years away, and no company could justify buying TV ads for such a rarefied, geeky product. That left only one channel: computer magazines.

As Jobs pored through *Interface Age* magazine, the ads for Intel jumped out at him. Instead of tables of "speeds and feeds," they popped with quirky photos and witty text. Jobs loved them.

Jobs cold-called Intel to ask who had created the ads. An Intel rep told him: Regis McKenna, Inc. Jobs dialed the firm, hoping to reach either Mr. Regis or Mr. McKenna.

In fact, Regis McKenna was just one person. He'd been an in-house publicist at National Semiconductor; in 1971, he left to found his own ad and PR agency. His first clients included Intel—and Paul Terrell's Byte Shop.

In December 1976, McKenna agreed to meet with the Steves. "They both had cutoffs on, and Birkenstocks, and long hair. Steve had what I called a Ho Chi Minh beard," remembers McKenna, who himself was groomed and well-dressed.

Woz had written an article about the Apple II's advances, which he hoped McKenna could get published. But McKenna considered it much too technical for a general audience, and offered to rewrite it. Woz was offended. He had been reluctant to attend this meeting in the first place—why did Apple need to spend money on some slick ad agency?—and he definitely didn't want his words mangled by a PR flack.

When Woz bristled, McKenna ended the meeting.

Regis McKenna, with Jobs in 1984, worked with Apple until well into the nineties.

But Jobs wouldn't give up. "Steve called back probably 40 times that night," McKenna remembers. Jobs wanted another meeting, this time without Woz.

The second meeting took place. And a third, and a fourth. The two wound up spending hours in the basement of McKenna's home in Sunnyvale, talking about PR, advertising, strategy, design, and business. Eventually, without any starting date either could pinpoint—without even a contract—McKenna began working with Apple.

In the short term, McKenna agreed to design a new logo, develop some advertising, and prepare a "backgrounder": a story about the origins and goals of a new company that he would send out to editors and reporters. He proposed placing Apple's first ads in a magazine with broader appeal than the computer rags: *Playboy*.

McKenna's team wound up creating the public voice of Apple: cheeky, playful, confident. His first brochure was uncannily true to the heart of Apple's philosophy for decades to come: "Simplicity is the ultimate sophistication." Jobs was elated.

As payment, Jobs offered McKenna 20 percent of the company; McKenna handed Jobs a memo declining the deal, and has never stopped regretting it. "I am always reminded of what I turned down because my letter is on display at Apple's headquarters," he says.

McKenna remained Apple's marketing firm until 1995; eventually, Jobs made McKenna himself part of Apple's elite executive team—extraordinary status for a nonemployee. As recently as 2010, Jobs called on McKenna when he needed advice.

Birth of a Logo

McKenna asked art director Rob Janoff to replace Ron Wayne's charming but ornate Newton logo. Janoff came up with an apple with a bite taken out of the right side, striped in six colors to suggest the Apple II's six-color palette.

"The bite in the apple was initially meant to indicate that it was an apple, and not something else," Janoff says, noting that many fruits could have a stem and a leaf. "Also, metaphorically, the bite indicated biting into all the knowledge users would get out of this computer."

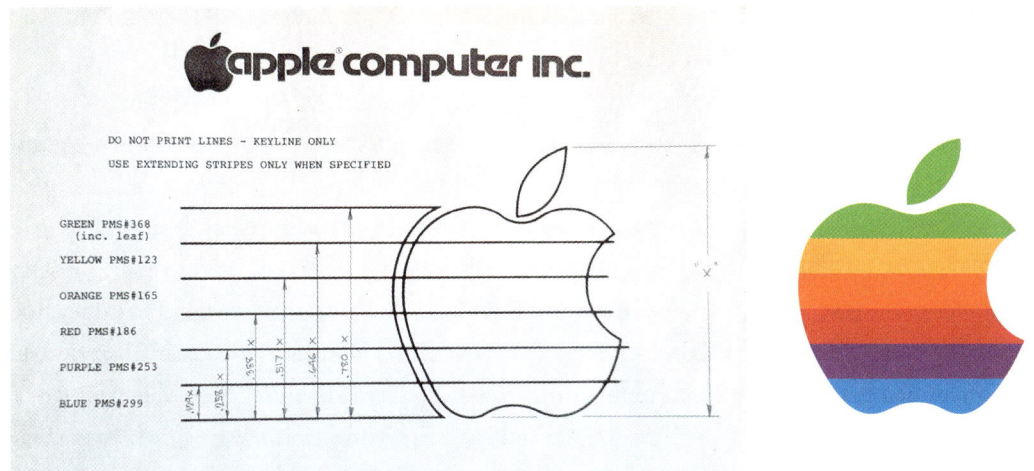

An early spec document for the new logo, and the result.

Janoff hadn't even intended the pun: A *byte* was a unit of computer data. "I wasn't computer-literate enough to see that initially. And I was like, 'There's a bit of wit that will last!'"

But *printing* the six stripes—green, yellow, orange, red, purple, and blue—would be tricky. Avoiding ink overlap would require a costly alignment of the ink plates.

McKenna proposed separating the color plates with thin metal bars. "But Steve said the colors had to meet precisely with no line and no overlap," he says. That kind of precision would, of course, cost more. Jobs, of course, didn't care.

The logo appeared on every Apple product for 22 years and became one of the most recognized logos in human history.

Finding Financing

Bringing a world-changing machine to life would require more than a spectacular design. Apple would need engineers, designers, marketing experts, and finance people. It would need an office, a warehouse, a factory. And to get all of that, Apple would need money.

One quick way to raise cash would be offering an investment, or even selling Apple outright. Jobs offered Atari president Joe Keenan a third of Apple for $50,000. "Not only are we not going to buy this thing, but get your feet off my desk," he said.

They also gave a demo to Chuck Peddle, whom Woz had met two years earlier at WESCON. Peddle, now working for Commodore, invited the Steves to present the Apple II to his bosses. They, too, declined; the Apple II seemed too similar to the company's own upcoming machine, the Commodore PET, a computer without color, sound, or graphics. (Had they bought Apple, Commodore would have realized a 2.5 billion percent return on its investment.)

Discouraged, Jobs pitched one venture capital firm after another: Kleiner Perkins, Sutter Hill, Institutional Venture Partners. They all passed. "How can you use a computer at home—put recipes on it?" said Asset's founder.

Jobs even returned to Atari, this time trying to persuade its founder, Nolan Bushnell, to invest. "He asked me if I would put $50,000 in and he would give me a third of the company," Bushnell remembers. "I was so smart, I said no. It's kind of fun to think about that—when I'm not crying."

But Bushnell did suggest that Jobs talk to Don Valentine, founder of Sequoia Capital, whose investment had helped turn Atari's home Pong game into a blockbuster. When Valentine saw the Apple II demo, he said he'd be willing to help out—on one condition. Apple's tiny team needed someone with experience in management, marketing, and distribution.

"Fine," said Jobs. "Send me three people."

Valentine introduced Jobs to three candidates. One of them, Jobs didn't like; one of them didn't like Jobs.

But the third man would not only invest in Apple; he'd become its first chairman. He would write Apple's first business plan, incorporate the company, hire its first CEO, and assemble its first board. He would serve as Apple's adult supervision for 20 years.

And to this day, nobody outside the tech world knows his name.

5. In Business

Armas Clifford Markkula Jr. is the youngest in a line of Finnish-descended inventors and engineers. His grandfather had five patents to his name. "You know the three-clawed gizmo you use to pull weeds in your garden?" Markkula says. "He invented that." His father was an inventor, too, and a builder of leg braces.

Actually, nobody calls Armas "Armas"; he goes by Mike—Mike Markkula—or just ACM. (Markkula rhymes with "SPARKLE-uh." It's definitely not "mar-COOL-ah," as the characters weirdly say it in the 1999 movie *Pirates of Silicon Valley*.)

Markkula got his bachelor's and master's degrees at the University of Southern California. After graduation, he worked as an engineering marketer for four years at Fairchild, a semiconductor pioneer set among the fruit orchards of Palo Alto, and then at Intel. Intel, he was delighted to learn, offered stock options to all employees—and soon after his arrival, the company went public. Overnight, Markkula was a millionaire.

By 1975, he was ready to retire. He was 32.

He'd had enough of the corporate life. He had plenty of money. He had a bucket list—and now, at last, he'd have the time to tackle it.

The list was 52 items long. Some of them were about pleasure and self-improvement, like learning to read music and getting into woodworking. Some were about sharing his expertise with his community, like teaching elementary school math.

One item on the list, though, was dedicated to helping startups. "I missed bright, fiery-eyed, fire-in-the-belly people wanting to accomplish things," he says. So he put out the word: Every Monday, he'd make himself available to young entrepreneurs as an adviser—no charge.

That's how, in early 1976, Markkula found

Before his tech career, Markkula had worked at a gas station, an auto-body shop, and a catering company for movie shoots.

himself on the phone with Don Valentine. "There's two guys over in this garage in Los Altos who could really use the kind of help you'd provide," Valentine said.

Markkula met the Steves at the Jobs family home. "The two of them did not make a good impression," Markkula says. "They were bearded, they didn't smell good. They dressed funny. Young, naive."

But the Apple II demo blew him away. "It was the world's first single-board computer—the world's first. It was the world's first single-board computer that had slots. It was the world's first computer that had BASIC in ROM," he says. "This was one elegant, beautifully crafted design that Woz had done. And I'm a circuit designer—I know!"

Jobs and Woz were dazzled in turn by Markkula. "I thought he was the nicest person ever," says Woz. "He had a beautiful house in the hills overlooking the lights of Cupertino, this gorgeous view, amazing wife, the whole package."

Markkula didn't view the Apple II as another nerd computer. He told the Steves that it could become an everyday home appliance. It could sell millions. It could change the world.

Markkula gave the Steves some homework: Write up a business plan. After a few days, however, it became clear that neither had any interest in that kind of assignment. "So, I finally said, '*I'll* write a business plan,'" Markkula says, "because I really wanted to see this thing go."

According to his calculations, with $142,000 in startup cash, Apple could become cash-flow positive in nine months, and could become a Fortune 500 company in less than five years. (In the end, it took seven.)

To Jobs's amazement, Markkula put up the money himself: a $250,000 line of credit. "I thought it was unlikely that Mike would ever see that $250,000 again," he'd say.

Don Valentine offered an additional $150,000. For the first time, Apple Computer was no longer scrambling for capital.

Leaving HP

"Okay, Steve. You know you have to leave Hewlett-Packard," Markkula said to Woz one day.

Woz couldn't understand why. After all, he'd been able to design the Apple I and II without leaving HP. They'd been side projects. "Well, I can work part-time," Woz said.

"No. That doesn't work, either."

When Markkula gave him a deadline for making a final decision, Woz was tormented. HP was his dream job. He'd pictured himself working there for the rest of his life. Apple, on the other hand, was a wobbly new enterprise that would sell *maybe* a thousand computers. Why on earth would he sacrifice such a sweet setup for a shaky startup?

Nor did he have any interest in management. In fact, he actively loathed the idea of having to order people around.

A few days later, Woz and Jobs met Markkula in his poolside cabana. Woz announced that he'd made his decision: He just couldn't leave HP.

Jobs was upset. Without Woz, there was no Apple; without Apple, there was no revolution. Over the next few days, Jobs worked the phones, setting up something like an intervention. He persuaded Woz's friends, parents, and brother to call Woz, to argue that he was making a terrible mistake.

The call that did the trick came from Woz's high school buddy Allen Baum, who argued that if Apple didn't work out, HP would likely love to rehire him. Besides: Becoming a manager didn't need to be part of Woz's future. "Think about it," Baum told him. "You can be an engineer and become a manager and get rich, or you can be an engineer and *stay* an engineer and get rich."

It was exactly what Woz needed to hear. He gave notice at HP the next day.

Apple Inc.

Markkula rented Apple's first office: Suite 3C at 20863 Stevens Creek Boulevard, a business district thoroughfare that traces Cupertino's horizontal center line. It would become known as the Good Earth building, after Jobs's favorite natural foods restaurant nearby.

Jobs, Woz, Randy Wigginton, and Chris Espinosa set up some desks; from Jobs's garage, they brought over the workbenches they'd used for Apple I board assembly. On their first day in the office, they ran around, buzzing each other on the newly installed phone system and cracking up.

Nobody would guess at the future of this rental building's occupants.

"It was basically one big space, with carpet on one half and linoleum on the other half. On the carpet were half a dozen desks, for sales, marketing, and administration," remembers Espinosa. "On the linoleum half, there were six lab benches, and that was engineering and manufacturing."

Although Apple had been *founded* in April 1976, it hadn't yet been incorporated. It wasn't yet a legal entity that, among other things, would protect the founders from exactly the financial liability that had terrified Ron Wayne a year earlier.

So on January 3, 1977, the two Steves met with Markkula at his cabana and signed the paperwork, dissolving the partnership they'd signed with Wayne. They now worked at Apple Computer, *Inc.*

"Each of us is going to have 26 percent, so any two of us can throw the other one out," Markkula explained. "So if somebody goes off the deep end, we can take care of that." The remaining 22 percent, he suggested, should be set aside for future investors and attracting talent.

Markkula and Jobs turned out to be an astonishingly good match. Jobs was the driven, exacting, passionate visionary; Markkula was older and wiser, but equally fired up about the future. "Mike really took me under his wing," Jobs would say. "His values were much aligned with mine."

Those values included focusing on the product, not the money; having the strength to say no to distracting opportunities; and making beautiful machines. "People DO judge a book by its cover," Markkula wrote in an early memo called "The Apple Marketing Philosophy." "We may have the best product, the highest quality, the most useful software etc.; if we present them in a slipshod manner, they will be perceived as slipshod; if we present them in a creative, professional manner, we will impute the desired qualities."

"When you open the box of an iPhone or iPad," Jobs would say later, "we want that tactile experience to set the tone for how you perceive the product. Mike taught me that."

The vibe in the Stevens Creek office was pure startup: "Incredible activity all the time," Espinosa says. "Customers, vendors, venture capitalists, engineers, Mike Markkula, Steve Jobs, Regis McKenna, running around all the time. Very few set schedules, no meeting rooms."

Markkula's first job was to hire a CEO.

It couldn't be Markkula; he had no intention of sticking around. "I was going to help them write a business plan, and that would be that. I wasn't going to be running off to spend the next 20 years of my life building a company!" he says, laughing. (He was quite mistaken on that point.)

Nor could it be Jobs, who was slammed with projects, had no managerial experience, and was becoming "increasingly tyrannical and sharp in his criticism," as Markkula remembers it. "He would tell people, 'That design looks like shit.' " In fact, one of the new CEO's most important duties would be managing Jobs's volatility and his hygiene.

Markkula proposed hiring Mike Scott, who was then running a manufacturing facility for National Semiconductor.

The two Mikes had a lot in common. They had joined Fairchild on the same day; they shared a cubicle there; and they were both February 11 babies, born a year apart. Every year, they celebrated their birthdays together.

"He's a Caltech physics grad, just a really, really brilliant guy, and fun to work with," Markkula says. "I wouldn't have to worry about the manufacturing end of it, because he was just really good at that stuff."

Jobs didn't love the idea—"Apple was my baby, and I didn't want to give it up"—but gave in. Mike Scott, swallowing a 50 percent pay cut from his Nat Semi job, became Apple's first CEO.

Apple's leadership now consisted of two Steves and two Mikes. To prevent a hell of miscommunication, they decided to call their new boss Scotty.

Mike Scott
1945–2025
Schooling: California Institute of Technology
Before Apple: Fairchild, National Semiconductor
Apple: 1977–1981
After Apple: Starstruck, Seattle Opera

Scott was an expert (and author) on colored gemstones. The dark-blue gem known as Scottyite (barium copper silicate) is named for him.

The new CEO was a no-nonsense, intensely practical manager. He dove into securing office space and office machines, putting together a sales operation, and hiring administrative staff.

Almost immediately, though, he began clashing with Jobs. Often, the disagreements pitted Jobs's perfectionist and artistic instincts against Scott's pragmatism and frugality. They fought over how rounded the Apple II's corners should be, about which color benches to order for the engineering tables, and about the length of the Apple II's warranty. (Jobs wanted a year; Scotty wanted 90 days.)

The West Coast Computer Faire

In January 1977, the Steves and Mikes decided that the Apple II needed a splashy unveiling party. A perfect opportunity was coming up: the first-ever West Coast Computer Faire, to be held in San Francisco's Brooks Hall. (Its organizers, Jim Warren and Bob Reiling, were members of—what else?—the Homebrew Computer Club.)

Woz had been on the Brooks Hall show floor two years earlier to buy his processor chips for the Apple I, at the MOS Technology booth. This time, *he* would be the man in the booth.

The Faire was in April, which gave them only four months to build working prototypes of the Apple II. There was not an instant to lose—but Jerry Manock encountered a showstopping problem almost immediately.

The housings for most plastic consumer products are made using injection molding, in which melted ABS plastic is injected into a metal mold and hardens under pressure. But creating the metal molds can take months and cost up to $100,000—and the Computer Faire was only weeks away.

Instead, Manock would have to make the first cases using reaction injection molding (RIM). In this method, two polyurethane-based chemicals are mixed and injected into an epoxy mold, where they harden on contact. This technique is faster and cheaper than injection molding, but prone to imperfections and bumps. Manock hired a Mountain View plastics company and crossed his fingers.

Only hours before the Computer Faire was to open, Manock delivered 20 Apple II cases to the Apple booth. The cases looked great in their fresh coat of light sandy-beige paint—Pantone 453. (Pantone is a color-matching system for print work, not for specifying paint; the beige of early Apple machines wasn't consistent. Later, Manock moved to something called the Munsell system, which offered numerical control over the color specification.)

All the team had to do was install the Apple II's circuitry and keyboard into the empty cases.

The West Coast Computer Faire was mobbed by computer fans.

The Faire Prank

One company that *wasn't* at the 1977 West Coast Computer Faire was MITS—the maker of the Altair computer that had ignited the hobbyist movement in the first place. Steve Wozniak thought MITS's absence was odd—and the perfect setup for a prank.

Following the success of the Altair, a chipmaker called Zilog had been marketing an Intel-compatible processor called the Z80. "At the time, there were lots of hobby computers coming out that were built around it. They were called the Z-this and the Z-that," Woz says. "So I came up with the Zaltair, a made-up computer that was also built around the Z80."

He spent $400 to print up 8,000 brochures on colored paper for this nonexistent new computer. He wrote up some deliberately terrible ad copy. "Imagine the computer surprise of the century, here today," it said. "Imagine an exquisitely designed cabinet that will add to the décor of any living room." The new computer, the brochure said, came with *18 slots*.

On the back of the flyer, a bogus comparison table of popular computers appeared, full of vague specs. The Zaltair got a perfect 1 in every category.

The Apple Prank Team (Woz, Espinosa, and Wigginton) stuffed the flyers all over the Faire—tables, shelves, booths, the pay phone area. Soon, hundreds of people were gawking at the leaflet.

The next day, Steve Jobs himself found one of the flyers. "Hey, we rank better than some of the others!" he said, reading the feature chart on the back. It was all Woz could do not to burst out laughing.

Woz had every intention of revealing his hoax to Jobs—but it would take eight years for him to find the perfect opportunity.

But to their horror, nothing fit properly. The local vendor's wooden molds resulted in so many bubbles and defects that the lid of one case wouldn't fit the opening on another.

With the clock ticking, they frantically sanded down some parts, built up others with Bondo resin, and then repainted to hide the damage. By the time the Faire actually opened, they had only six computers to show off.

The show was an explosion: the coming-out party for the personal computer. This one conference introduced what *Byte* magazine would call the "Trinity of Home Computing": the new Commodore PET, Tandy/Radio Shack's TRS-80, and the Apple II. 12,000 attendees—twice the number the organizers had expected—swarmed the 180 booths. "We didn't know what we were doing, and the exhibitors didn't know what they were doing, and the attendees didn't know what was going on," said organizer Jim Warren, "but everybody was excited and congenial and undemanding, and it was a tremendous turn-on."

Jobs had rented the best booth available. "When you walked in that front door to the Computer Faire, all you could see was Apple," Markkula says.

Apple was the darling of the show. Long tables filled the booth, draped in black and backed by a clear, illuminated sheet of Plexiglas bearing the new Apple logo. Jobs had ordered 20,000 slick brochures. People shoved to get through the crowd to watch, on Sony Trinitron screens, the Apple II demos offered by Chris Espinosa, Jobs, Scotty, and Randy Wigginton: the BASIC language, a couple of games (Breakout and Star Trek), and flashy color graphics.

One visitor, in particular, would become a lasting force at Apple: Arthur Rock, one of the heaviest hitters in Valley venture capital. He'd helped launch both Fairchild and Intel. Markkula encouraged him to see the Apple II at the Faire—and its effect on the public. Rock, 49 years old and wearing a suit, had no trouble finding the Apple booth. "I couldn't even get next to it," Rock remembers.

That fired-up crowd was all he needed to see. He kicked in $57,600, and became Apple's third investor.

The Faire was a sky-high success for Apple. Fans marveled at Woz's clever design: everything on one board, eight slots easily accessible. They gawked at the gorgeous case. "Compared to the primitive stuff on view elsewhere at the Faire, our finished plastics blew everyone away," Manock says.

By the end of the two-day conference, orders for the Apple II were pouring in; when an ad for it appeared in *Scientific American*, the orders blew up.

On Sale

Jobs, Markkula, and Scotty had fervently hoped to price the Apple II under $1,000. But dealers insisted on a 50 percent markup over their wholesale costs. Under that arrangement, Apple would net about $660 per computer, which wouldn't cover its costs.

And so, on June 10, 1977, the Apple II went on sale for $1,298 ($6,500 in today's dollars). That was much more than its two Computer Faire rivals, the Commodore PET ($800) and the TRS-80 ($600)—but those machines didn't have color, sound, slots, or graphics.

The standard Apple II had 4 KB of RAM, expandable to 48; a MOS 6502 processor running at 1 MHz; a built-in keyboard; eight expansion slots; a headphone jack for connecting to an audio-cassette player (for storage); and a built-in BASIC interpreter.

It could display color graphics on a monitor or TV in either of two modes: low resolution (40 x 48 pixels, up to 16 colors) or high resolution (280 x 192 pixels, but, on the first model, only four colors). It had a speaker, but no audio circuitry; all the sounds had to be generated in software. Once games came along, video had to pause while sound effects played.

The reviews were glowing. "I was able to turn on power and begin using the computer within

The Apple II slots were easy to reach (left). The box included simple paddles for controlling games: one button, one knob.

five minutes of receipt," raved *Byte*'s critic. Demand was practically melting Apple's phone lines. Within weeks, Apple was selling $84,000 worth of Apple II's a month, on track to make $1 million a year. Sales hit every milestone in Markkula's business plan. By year's end, Apple had sold 77,000 computers, generating $2.7 million. It had been in business for a year and a half.

Staffing Up

Apple was doubling the number of Apple II's it manufactured every three months. Every computer shop in America wanted to become a dealer. For Mike Scott, the challenge was hiring enough people to keep the Apple machine running without burning people to a crisp; every employee was already working 20 hours a day. "We couldn't add people fast enough. We'd bring in temporary workers, then we'd pay the headhunter fees to convert them over to Apple employees," he said. "For four or five years, Apple doubled in size every three months."

Apple's tiny engineering team now included Woz, Rod Holt, and Fairchild veteran Wendell Sander. They threw themselves into designing optional add-ons for the Apple II's slots: printer cards, communications cards, ROM cards, and so on.

Bandley 1

By the end of 1977, the company could no longer squeeze into its original two-room office on Stevens Creek Boulevard. Scotty and Markkula found and rented a much bigger office space on

The Bandley buildings were Apple's headquarters for 15 years.

Bandley Drive, about a mile away: a single-story, low-slung building with Spanish mission–tile roofing. On January 2, 1978, the Apple workforce of 30 people moved in.

Within months, Apple also took over the building next door, and then the one across the street. By the time Apple was finished renting and, eventually, constructing its own new buildings, it would occupy what it called Bandley 1 through Bandley 6—numbered in order of acquisition, not location.

Apple II Evolves

By 1978, the Apple II had become a cultural phenomenon: hailed in the press, available even in department stores, and adored by the public. It was the bestselling computer in history—by far. User groups—volunteer-run clubs where owners could exchange information and tips—sprang up in big and small cities. Job applications arrived by the mailbag full. Boards of education began equipping their schools with rows of Apple II's, to Steve Jobs's delight.

Freshly hatched software companies began selling Apple II programs. Most of them were, as Woz says, "a single guy in his house who figured out how to write a neat game," copied onto cassette and sold through specialty computer stores.

Mike Markkula was concerned, however, about the audiotape-as-storage thing. Transferring programs to the computer that way was still feeble, slow, and unreliable. As the Apple II manual put it: "To find the right volume setting, you will use a trial-and-error method. You will play a tape

> ### Badge Numbers
>
> When CEO Mike Scott opened a business account with Bank of America, he learned that everyone on the payroll would need an employee *number*. And in assigning these numbers to Apple's earliest staff, he created a now-infamous tussle between the Steves.
>
> "Each Steve wanted number *1*. I didn't give it to Jobs, because I thought that would be too much," Scott says. "That would stoke his ego even more."
>
> Jobs was furious. He argued that Apple would not even have come into existence without him. *He* should be #1. Finally, he proposed a workaround: Woz could remain Employee #1—if *he* could be Employee *Zero*!
>
> Scotty, rolling his eyes, gave in. Jobs's badge would say "Employee 0."
>
> Scotty himself was technically the fifth employee, but he thought it would be cool to see "007" on his badge, so he designated himself Employee #7.
>
> In the end, then, the first ten Apple employee badges were:
>
> 0. Steve Jobs
> 1. Steve Wozniak
> 3. Mike Markkula
> 4. Bill Fernandez
> 5. Rod Holt
> 6. Randy Wigginton
> 7. Mike Scott
> 8. Chris Espinosa
> 9. Jim Martindale (manufacturing manager)
> 10. Sherry Livingston (secretary)
>
> The bank's system, however, wouldn't accept zero as an employee number. To Bank of America's computer, Jobs would always be Employee #2.

softly to the computer and see if the information got in OK. If it doesn't work, you will try the tape again, a little louder this time. If that doesn't work, you will make it a little louder still. Eventually the volume will be just right for the APPLE and it will say so."

Dealers regularly complained. Apple II owners complained. And most important, Markkula complained. He was, after all, a coder himself. Every time he wanted to use the checkbook program he'd written, it took two minutes to load from the cassette player and another two minutes to load the data. He was losing his mind.

And so, in December 1977, Markkula called an executive board meeting. On a big sheet of paper, he listed Apple's future goals—and number one was "floppy disk."

The first floppy drive, introduced by IBM in 1971, was a hulking component whose disks were 8 inches across and intended for use in corporate minicomputers. But Jobs had just learned of a much more compact version: the Shugart Associates 5¼-inch floppy drive. "Steve was always looking for new technologies that had an advantage and were likely to be the trend," Woz says.

Inside each square 5¼-inch sleeve was a disk of magnetically coated Mylar plastic, spinning at 300 rotations per minute. Like the needle on a record player, a magnetic head moved across the concentric circles, recording or retrieving information. Each disk could store a whopping 90 KB of data, just enough to hold about 60 pages of text.

Shugart's drive struck Markkula as a perfect solution to the Apple II's data-storage problem—and, because it was a hundred times faster than audiotape, to his checkbook problem.

The Consumer Electronics Show (CES), an annual gathering famous for new-product launches, would be the ideal place to unveil a new drive. Over the years, pocket radios, the VCR, and the CB radio had all made their debuts at CES.

Unfortunately, CES was only four weeks away.

Disk II

Neither Steve Wozniak nor Rod Holt had any meaningful experience with disk storage. But Woz was no ordinary engineer. This challenge was his favorite kind: a ridiculously big engineering project on a ridiculously tight schedule. He agreed to adapt Shugart's floppy drive for the Apple II.

The function of a drive's controller card is to divide the disk's surface into concentric tracks, to subdivide each track into individual *blocks* of data, and then to control the magnetic head as it hunts, reads, and records data. Shugart's controller card required 22 chips to do the job.

Woz considered that design bulky and overengineered. In short order, he decided that 20 of the 22 chips weren't necessary; he could do much of their work in software, right on the Apple II.

Jobs, meanwhile, visited Shugart CEO Don Massaro, urging him to give Apple a better price than $390 per drive. After all, given the quantity of drives Apple would sell, surely Shugart would eventually make a lot of money even at a lower wholesale price. Jobs politely suggested a price of $100. Massaro politely declined.

But after speaking to Woz, Jobs made a different offer. He wanted to buy only the skeletons of the drives: a motor for spinning the disk, and a stepper motor to move the "needle." Woz and Rod Holt would take care of the controller card.

Massaro agreed. Jobs got his mechanism-only drives—for $100 apiece.

Woz and Randy Wigginton worked straight through Christmas, straight through New Year's, sleeping very little, intent on making the disk drive work and writing a disk operating system to control it. It was, Woz recalls, "as tricky as code gets"—but they made the CES deadline, and got to go to Vegas on Apple's dime.

"Randy was 17 years old, but I taught him to play craps and he won $35," says Woz. "You remember the important things."

In the end, the technical solutions Woz devised became electrical-engineering legend. The final product fit almost 45 percent more data on each disk than Shugart's and used eight chips instead of 40.

Two Disk II's made disk copying easier.

Apple's Disk II went on sale in July 1978 for $495—the least expensive floppy drive ever sold. And yet, because it cost Apple only $140 in parts, it became a cash cow the size of Jupiter.

Woz's floppy drive was also an Apple II sales accelerator. Because it made data storage so much simpler, faster, and more reliable than audiotapes, it made the Apple II even more attractive.

Two years passed before an Apple competitor introduced a floppy drive of its own.

Apple wound up buying $25 million worth of floppy-drive mechanisms every year, becoming Shugart's biggest customer. Throughout the years, Apple continued to make small improvements and adjustments in the drive, but all of them used the heart of Woz's original circuit design. By most assessments, Woz's floppy drive was as important a breakthrough as the Apple II itself.

VisiCalc

Woz's Disk II design was only one of the two mythical strokes of good fortune that befell the first Apple II. The other was VisiCalc.

In 1978, there was no such thing as spreadsheet software. There was only commercial financial-forecasting software for time-share computers. You couldn't select a cell, type a new number, and watch the totals change throughout the rows and columns in real time; you had to redo each calculation manually. And if you *didn't* subscribe to a time-share service, you did your bookkeeping in books.

Then one day, 26-year-old grad student Dan Bricklin and programmer Bob Frankston, a buddy from his college years at MIT, wrote a revolutionary new program called VisiCalc: a *digital* ledger book.

It should look familiar to anyone who's ever used Microsoft Excel: a screen full of cells, in numbered rows and lettered columns. If you changed the number in a cell, all of the formulas instantly recalculated. You could try out different prices to see their effect on your profits, or experiment with sales commissions to see how they influenced your bottom line.

It was intended for use by noncorporate mortals, which meant writing for one of three personal computers: the Commodore PET, Tandy TRS-80, or Apple II. But only one of those candidates had an available floppy drive, and only one could be expanded to have enough memory (32 KB or more). "That's why VisiCalc was written for the Apple II. It was the only computer that could hold it," Steve Jobs remembered.

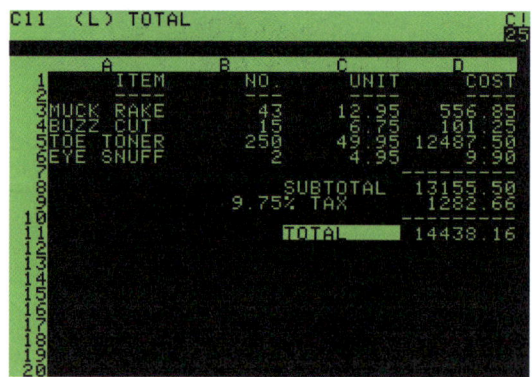

VisiCalc was the world's first "killer app."

VisiCalc went on sale in October 1979 for $100; it was an instant sensation. For business owners, the leap from ledger books to the spreadsheet felt like the jump from handwriting to the printing press. "The most exciting and influential piece of software that has been written for any microcomputer application," said *Byte*. It was the world's first "killer app": a piece of software so important that you buy the computer just to run it.

For an entire year, VisiCalc was available only for the Apple II. Thanks to VisiCalc, Apple went from selling 1,000 computers a month to *10,000*. "If VisiCalc had been written for some other computer," Jobs told a journalist in 1990, "you'd be interviewing somebody else right now."

6. Apple III

In two years—September 1977 to September 1979—Apple's annual sales grew from $774,000 to $17 million. Its sole product was the Apple II and its accessories. An entire card industry sprouted up to fill Woz's expansion slots, accommodating floppy drives, hard drives, modems, printers, memory upgrades, mice, game controllers, sound cards, accelerators, and so on.

By its third birthday, Apple had a revolutionary hit product, torrential cash flow, and a glowing reputation. What it did not have, however, was a supply chain or logistics department. Nobody was forecasting sales so that the company could have the parts they'd need. There was no purchasing department, no inventory tracking, and no factory. A local woman named Hildy Licht assembled the power supplies—"she got a bunch of housewives to do this on their kitchen tables when the husbands were away at work," recalls Bill Fernandez—and as late as 1979, workers were still putting Apple II's together *right in Bandley 2*, a few steps away from executives' desks, by hand.

"Oh, it was a mess," laughs Del Yocam, who arrived in November as director of materials. "There were product pieces, all kinds of stock, everywhere. There were all kinds of Apple II's sitting there, not moving because of shortage of parts."

Yocam, then 35, was a details guy. After working at Ford and two minicomputer companies, he was well-versed in parts-intensive companies. As he left his job interview with Mike Scott in Bandley 1, Yocam encountered Steve Jobs in a hallway. "What are you doing here?" Jobs asked.

"Interviewing for a materials manager job," Yocam said.

"No, no, no. What is your *purpose*? What are you *doing* here?"

To this day, Yocam doesn't know why the next words came out of his mouth: "I want the chance to change the world."

He got the job.

Yocam's first priority was taking swift and drastic action to address the hundreds of boxes of Apple II parts that were strewn all over the Bandley buildings. Over a weekend, he hired a crew of temp workers to build a room-sized enclosure of chain-link fencing, lined with shelves. The only way in or out was through a lockable door.

Then it was Operation Roundup. "We went through both buildings, scavenging all parts that were there, even if they were in half-mast production, and we put them in that cage," he says. His

Del Yocam
Born: 1943, Long Beach, CA
Schooling: California State University, Fullerton (BA), California State University, Long Beach (MBA)
Before Apple: Ford, Control Data, Fairchild, Computer Automation
Apple: 1979–1989
After Apple: Tektronix, Borland

Yocam was an avid investor in Broadway musicals. He did well on *Sunset Boulevard* and *Phantom of the Opera*—less well on *Annie 2*.

goal was to add some accountability and order to the flow of components. From now on, if you wanted parts, you filled out a request form, which had to be signed by a manager.

The engineers were furious. But within two weeks, the flow of finished computers resumed, in part because Yocam bought sleeping bags for the engineers to accommodate their night-owl habits.

Soon, the team was assembling a hundred Apple II's a day. The dad joke making the rounds was "That's a pretty hot Apple turnover!"

The Successor

Nobody knew how long it might take for rivals to leapfrog the Apple II. But Steve Jobs, now vice president of new product development, couldn't risk a competitor stealing the II's thunder. Even as its sales climbed, he was already thinking about its successor.

The Apple II had won the hearts of hobbyists all over the country. But what about businesses? Every time business owners walked into a computer store to buy an Apple II, they also bought its floppy drive, more memory, a monitor, VisiCalc, and one of the add-on cards that expanded the Apple II's text display from 40 characters across to 80.

Right there, Jobs had the concept for Apple II's business-oriented successor: Build all of that stuff right in.

Its code name was Sara, after the daughter of its designer, Wendell Sander. Its final name was Apple III.

The Apple II had been the product of Woz's sole genius and passion. The Apple III would have input from an archipelago of departments that didn't exist when the Apple II was born, such as marketing, engineering, and sales.

By the end of 1978, they'd identified the goals of the new machine. It would be twice as fast as the Apple II. It would play more audio than simple beeps. It would handle both uppercase and lowercase letters. The new keyboard would have a number keypad and cursor keys (arrow keys). In fact, the cursor keys would be two-level touch-sensitive: The cursor would move faster the harder you pushed down.

The Apple III's keyboard would introduce the now-standard raised bumps (on the D and K keys, and the 5 key on the number pad), to help your fingertips find their position by touch. To make a key repeat ("XXXXX" or "!!!!!!!"), you wouldn't have to invoke some special mode; you'd just hold it down.

The floppy drive would be built in now, and Apple introduced an option no personal computer had ever had before: a 5 MB *hard drive*. The drive, called the Apple ProFile, cost $3,500, almost as much as the computer itself. But it was an enhancement—nearly endless storage!—that no other personal computer had.

The Apple III would be the first computer to include a continuously operating clock, too, so that it could correctly time-stamp your files as you saved or changed them.

For the first time, an Apple machine would have a proper operating system. It was called Apple SOS—short for Sophisticated Operating System and pronounced "applesauce." It offered a hierarchical filing system (you could have folders within folders), advanced memory management (fewer crashes), and device independence (apps could communicate with peripherals like drives and printers with a single, uniform set of commands).

Best of all, the Apple III would be backward-compatible with the Apple II. Four slots would handle Apple II cards, and an emulation mode (simulator) would run any of thousands of Apple II programs.

From the beginning, though, project leader Wendell Sander had to confront two colossal engineering and design challenges.

The Fan and the FCC

One challenge came from within: Steve Jobs's unshakable belief that the new machine could not contain a fan.

To the day he died, Jobs hated fans in computers. He felt that a fan created tacked-on, junky, industrial-looking ugliness both inside the computer and outside, because moving air requires holes in the computer's case. Worst of all, fan noise is distracting and irritating.

The "no fan" edict meant extra work for the engineers. Each prototype made its way to a windowless room at the back of the lab. To study airflow, the engineers amassed a collection of incense sticks. "We'd light the incense stick, and stick it in by where the product circuit board was, and watch how the heat coming off the board would flow," remembers Jerry Manock. "Well, you can imagine two or three hours of doing this in a closed room. When visitors opened the door, it was—'What are they *doing* in there?!'"

The second engineering challenge came from much farther away: Washington, D.C.

Throughout the seventies and early eighties, the trickle of electronic products became a tsunami. There were CB radios, garage door openers, video games, walkie-talkies, TVs, and, of course, computers. Every electrical circuit in every one of these gizmos could produce a maelstrom of radio waves. Those signals could turn radio and TV broadcasts to static, interfere with aircraft and ship communications, and disrupt emergency signals.

All of this potential radio frequency interference (RFI) alarmed the FCC, the government agency in charge of the radio spectrum. It handed a mandate to the computer industry: Eliminate the interference by January 1981, or we'll shut you down.

By most analyses, the resulting regulation—Title 47, Part 15—is a bureaucratic success story. It lets today's personal unlicensed gadgets (Wi-Fi, Bluetooth, microwave ovens) coexist happily with critical licensed services (radio, TV, cell networks). But for Apple, it was a colossal headache. "The standard they wrote, it was very tight, very difficult," Sander says. "So the aim of the Apple III was to develop a computer that could meet the Part 15. That was one of its primary objectives."

Jerry Manock and industrial designer Dean Hovey came up with a feature that they hoped would solve both problems—heat and radio interference—in one swoop: a heavy, cast-aluminum chassis. In principle, this massive metal block would serve both as a powerful heat sink *and* a cage for the radio waves.

To confirm that this design would work, the engineers wanted to conduct its measurements in a "quiet" radio environment, far from civilization—

The Apple III weighed 25 pounds.

and so it was that Apple bought a farm in Pescadero, about an hour away. To simulate a house or a business, Apple built a shack in the middle of the field. Apple engineers then walked around the property with radio detectors, testing each prototype's emissions at specified distances.

The news was good: The Apple III would not violate Part 15.

Omens

The decision to let the Apple III run Apple II software came late in the game. The plan seemed simple: You'd start up your Apple III with the included Apple II Emulator disk, and you'd have yourself an Apple II. You'd lose the Apple III's advanced features, but you'd be able to run most of your Apple II software.

The problem was Woz's original Apple II design. It was a marvel of clever, out-of-the-box thinking, but it was also idiosyncratic. For example, in high-resolution mode, dots in even-numbered columns of pixels had to be purple, dots in odd-numbered columns had to be green, and white dots always had to be double width. The Apple III had no such limitations—any dot could be any of 16 colors. For Wendell Sander's team, then, the question was: How far should they go to simulate the Apple II's quirks?

Their answer: Far. They wound up adding chips to the Apple III's circuit board expressly to add the Apple II's quirks back in.

Unfortunately, the main board had no room for these extra components. Sander's clever solution: expand vertically. He moved the Apple III's memory chips off the main board onto a *daughterboard*—a smaller, secondary board—suspended above the main board, like the second floor of a house, on a set of four aluminum pegs. They were both posts for the memory board *and* its electrical connection to the main board. This design meant that a customer could easily remove or upgrade the memory card, and Apple could sell the machine in different memory configurations.

The ten-month development process for the Apple III that Jobs had envisioned blossomed into an 18-month marathon—and it wasn't over. Still, by May 1980, the machine was far enough along to enjoy a grand unveiling at the National Computer Conference in Anaheim, California.

Apple published an ad in the NCC show newspaper, letting showgoers know that if they found their way to the Apple booth, they'd get free tickets to nearby Disneyland, which Apple had rented for the evening. 7,000 of them wound up riding Apple's chartered British double-decker buses to the amusement park for free rides and band performances.

The Apple III, the company announced, would go on sale in November 1980, starting at $4,340 (about $16,625 in today's money)—pricey, but far less than the minicomputers at the time. It would come with a word processor, a spreadsheet program, enhanced BASIC—and, of course, Apple SOS.

To the world at large, it certainly looked as though Apple had another hit on its hands.

IPO

When Mike Markkula incorporated Apple in 1977, his original plan was to keep the company private. The employees, not stockholders, would own Apple.

What he didn't envision was Apple's immediate, white-hot success. Its earnings had screamed up 1,500 percent in three years.

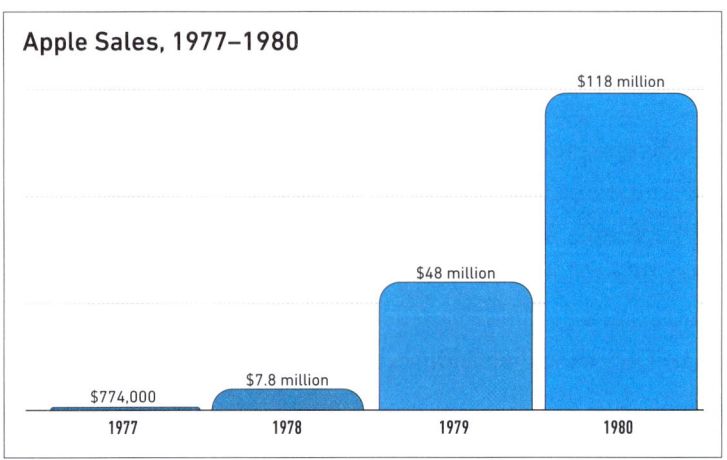

On the other hand, Apple was also incurring tremendous expenses. The company now employed about 1,000 people; its 1980 advertising alone cost $4.5 million.

Apple's board of directors included Markkula, Jobs, and Scotty, along with Arthur Rock, Teledyne founder (and early Apple investor) Henry Singleton, Macy's executive Philip Schlein, and, representing the East Coast, Venrock venture capitalist Peter Crisp. They could see that Apple's growth could go stratospheric—if it had the funds.

When a company issues stock to the public for the first time, it hires an investment bank to determine the starting price, market the shares to early buyers, and underwrite the offering. (Underwriting means buying all of the shares in advance at a discount—in Apple's case, 7 percent—removing all risk for the company going public. The bank then sells the shares to the public, usually making a decent profit.)

Plenty of investment banks wanted Apple's business; a successful IPO would bring in millions to such a bank. One of them was a boutique San Francisco firm called Hambrecht & Quist. As a gesture of thanks for an early investment in Apple, Jobs offered H&Q the gig—but also lined up the far more experienced Morgan Stanley to comanage the IPO.

When Apple filed with the SEC in October, the investment world went crazy, expecting it to be one of the biggest IPOs ever. "Not since Eve has an Apple posed such temptation," noted the *Wall Street Journal*.

The only hitch was the company's presentation in Chicago, one stop on its road show of mutual funds, pension funds, and hedge funds to sell them on the company and its future. (Morgan Stanley's new business manager, Jack Wadsworth, had to buy Jobs his first suit and nice shoes.) CEO Mike Scott, having arrived at the Chicago Club in a limousine, opened the street-side door—and nearly got creamed by a passing taxi. The impact ripped off the limo door so hard that it skittered, clattering, all the way down the block. "Another second, and we'd have been minus a president," Wadsworth remembers.

Wadsworth's original idea had been to underprice Apple shares at $18 apiece, so that the price would rocket upward at the opening bell. The institutional investors who'd been lucky enough to get access to Apple shares would enjoy a huge, instantaneous profit.

Jobs may have been new to the IPO business, but he smelled a rat. "You know, some people have told me that it's liable to trade at 26, 27, 28 dollars a share," he said.

"Well, yes, that's quite possible," Wadsworth replied.

"Then why $18? Why do I pay you seven percent to give cheap stock to your best customers?" Jobs had realized that the underpricing would benefit the banks and their rich clients—at Apple's expense.

Cornered, the bankers agreed to revisit the opening price. They settled on $22 a share.

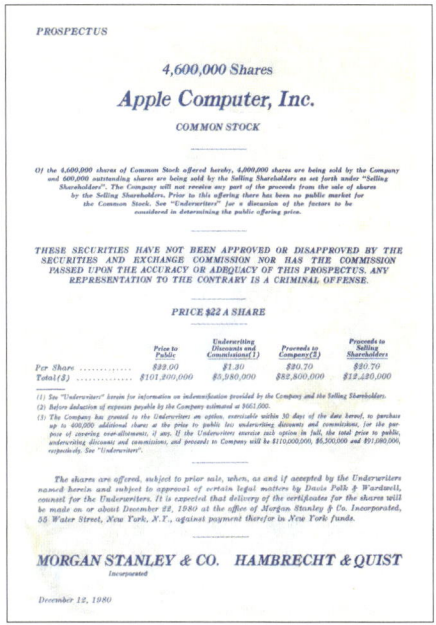

Copies of the original IPO prospectus fetch $5,000 or more today.

On December 12, 1980, within minutes of the bell, Apple had sold all 4.6 million available shares of the new AAPL stock. (If you're trying to compare the $22 share price with today's AAPL price, the opening share price would be ten cents, because the stock has split five times over the years.) By the end of day one, the share price had surged to $29 a share. The fledgling company was worth about $1.8 billion, making it the most successful IPO since Ford went public in 1956.

Steve Jobs poses with his IPO check from Mike Markkula.

The Woz Plan

Some original Apple team members remember Apple's IPO with more bitterness than affection.

Apple's policy was to offer stock options to full-time employees, but not to hourly contractors. Unfortunately, that latter category included many of the guys who had helped the two Steves create their first machines. Randy Wigginton, Chris Espinosa, and Bill Fernandez received no stock at all.

Most shocking was the case of Daniel Kottke: Jobs's best friend at Reed, his travel companion in India, Apple's very first paid employee, working in the Jobs garage to assemble Apple I boards. And after all of this, "Steve just cut me. He cut me off. He iced me," Kottke says. "Everyone I worked with became a millionaire. Everyone."

Wozniak, horrified, responded by devising the Woz Stock Plan. It had two parts. First, he offered up to 2,000 of his own shares at his own option price—$5, far below the $22 opening-bell price—to about 40 marketing executives, engineers, and others who'd been issued only small amounts of stock. "Almost everybody who participated in the Woz Plan ended up being able to buy a house and become relatively comfortable," Woz says. "I'm glad of that."

The other part of the Woz Plan covered those earliest Apple employees, that small circle of his best and oldest friends. "These employees weren't just around," Woz says. "I thought of them as part of the family, the family that had helped me design the Apple I and Apple II computers."

Over the discomfort of Apple's lawyers and accountants, Woz gifted them thousands of his shares outright, worth about $1 million per person.

To this day, Chris Espinosa hangs on to the pink "While You Were Out" slip from the Bandley 4 receptionist. It said that a broker at Morgan Stanley needed to talk to him. "An account had been created in my name with a notable amount of shares of Apple stock, and he needed my Social Security number," he says. "I ended up using that as a down payment on a house."

Woz is the first to tell you that the Woz Plan barely impacted his own wealth; he and his wife, Alice, bought a lovely house in Scotts Valley, in the middle of the Santa Cruz Mountains. But to this day, tech historians hold up the legend of the Woz Plan as a rare example of a tech founder spreading the wealth instead of hoarding it.

That night, the bankers, accountants, and Apple executives convened in a shuttered restaurant they'd rented. They'd hired a caterer, a decorator, and a piano player. There were toasts, speeches, and a song parody written by some of the H&Q bankers. Some younger Apple execs supplemented the decor with a detached car door they'd picked up at a junkyard, a loving testament to Scotty's near-taxi experience in Chicago.

It had been a week of emotional whiplash for Steve Jobs. On December 8, his favorite Beatle, John Lennon, was murdered in New York. And now, four days later, Jobs was worth $256 million.

He was 25 years old.

He never hired a chauffeur, owned a yacht, or bought an ostentatious mansion. In fact, when he did eventually buy a house, he left it mostly unfurnished. "I went from fairly poor—which was wonderful, because I didn't have to worry about money—to being incredibly rich, when I also didn't have to worry about money," Jobs said later. "I never did it for the money."

About a hundred of his coworkers had become overnight millionaires on paper—including, of course, Mike Scott ($95.5 million), Woz ($116 million), and Markkula ($203 million).

Scotty wheeled crates of champagne into Bandley 1. Exuberant pandemonium reigned for days. The stock market became an obsession for engineers and designers who'd never paid it any attention before.

Heat

The Apple III's design had looked spectacular on paper. But from the moment it landed on dealer shelves in March 1981, it was trouble. The forces of marketing, the FCC, shipping deadlines, and thermodynamics had produced a deeply flawed machine.

Critics dinged it for a shortage of available peripherals and programs, but that was only the beginning. The clock chip—the one that would time-stamp your files—stopped working after a few minutes. (Apple had bought the chips from National Semiconductor, which failed to conduct standard burn-in tests.)

The Apple II emulator was an amazing achievement, capable of running thousands of Apple II programs. But in Apple II mode, you couldn't use many popular Apple II cards, you couldn't run any games that relied on quirky disk formats and hardware tricks, and couldn't access cassette storage. And no matter how much memory your Apple III had, the simulated Apple II could "see" only 48 KB of it.

The hardware problems also included chips that came loose in their sockets and traces (copper electrical pathways etched onto the board) that were so crowded, they sometimes shorted out.

But the biggest problem of all was the spontaneous crashing. Customers complained to the dealers, and dealers complained to Apple. Apple set up 50 Apple III's in a crisis burn-in room, running them around the clock to study the crashing. They soon uncovered the primary problem: heat.

In the absence of a fan or ventilation slots, the Apple III's powerful electronics cooked the air inside as high as 220 degrees Fahrenheit. Some customers reported pulling floppy disks out of the drive, only to find that their plastic jackets had *melted*.

The crashes, it turned out, were a side effect of Sander's design hack: putting the memory chips on a daughterboard above the motherboard. As the memory board heated up and expanded, it

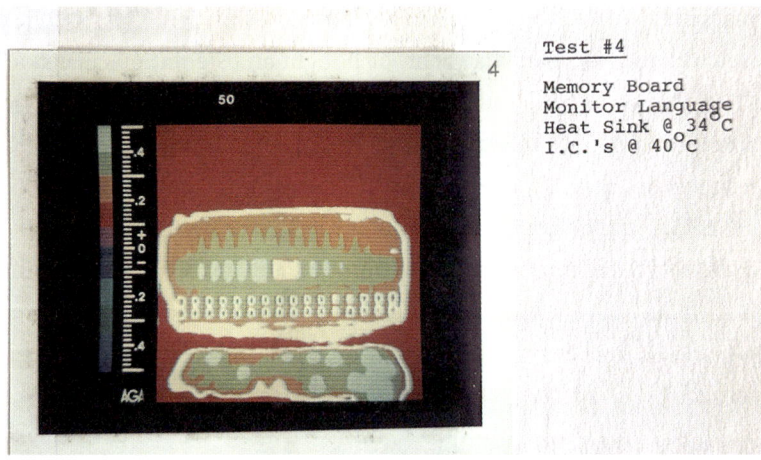

Apple hired a consulting firm to help solve the Apple III's crashing problems. It found that the chips (integrated circuits, or ICs) routinely baked at 104 degrees Fahrenheit or hotter.

scooted upward on its four aluminum connector pins, exposing them to the air and allowing them to oxidize. If the oxidation prevented even a bit or two of data from traveling down the pin to the motherboard, the Apple III crashed.

Apple issued a bulletin to its dealerships: If customers' Apple III's were spontaneously crashing, they should lift the front edge of the machine 2 or 3 inches and *drop it onto the desk.*

"Totally true," Sander says. "Because it was a connector problem! All you had to do is just jar the connectors, and you were going to fix it."

Markkula launched an all-hands campaign to identify and fix every design problem. "It would be dishonest for me to sit here and say it's perfect," he told reporters. "We'll know in December or January if Apple III is truly a reliable product."

In February 1981, Apple stopped including the faulty clock chip—no more date-stamping or calendar features. As an apology, it offered a $50 rebate to every past buyer, and cut the machine's price by $150. (Even once the clock chip problem was resolved, it was no masterpiece. It never did recognize February 29 in leap years.)

By the Apple III's first birthday, December 1981, Apple was confident that it had fixed the initial problems. The revised Apple III consumed less power, thereby lowering the amount of heat; incorporated wider traces on the motherboard; and replaced the aluminum memory board posts with a metal fork design. "The problems went away. I mean, that just fixed it," Sander says.

The biggest step in the cleanup was the Apple III Reintroduction Plan: Apple recalled every single Apple III sold in the first year, replacing each one with the revised model at no charge. As a thank-you, each replacement machine came with doubled memory: 256 KB. This first-anniversary Apple III was stable, fast, and capable. The reports of random crashes evaporated. Within Apple itself, it became a workhorse.

In December 1983, Apple tried one more rescue, by introducing the Apple III Plus. The clock chip finally made its permanent return, along with a completely redesigned motherboard, an improved keyboard, more connectors on the back, and a lower price: $3,000.

It was no good. The Apple III never outran its reputation as a stinker. By the time Apple officially discontinued the machine in April 1984, Apple had sold only 120,000 of them. (The Apple II had sold nearly two million in that time.)

The Apple III with monitor and drives.

It was a humbling blow: Apple's first failure. If it had had a fan, or hadn't attempted Apple II mode, or had had gold connectors, the Apple III would likely have been a home run. But, in the oft-quoted words of Randy Wigginton, "The Apple III was kind of like a baby conceived during a group orgy, and everybody had this bad headache and there's this bastard child, and everyone says, 'It's not mine.'"

For Woz, the heartbreak was the Apple II engineering projects that had been shut down, with the expectation that the III would soon replace it.

"We probably put $100 million in advertising, promotion, and research and development into a product that was 3 percent of our revenues," Woz says. "In that same time frame, think what we could have done to improve the Apple II." Indeed, it was the Apple II that was keeping the company alive. In 1980, Apple became the first company in history to sell a million computers—Apple II's. The company was now 1,500 people strong, with factories in Cupertino, L.A., San Jose, Dallas, and—for the European market—Cork, Ireland.

Steve Jobs acknowledged that the company had spent "infinite, incalculable amounts" on the Apple III. "But that's life. I think we emerged from that experience much stronger."

And, indeed, key technologies from Apple SOS found their way into the Apple II's 1983 software, ProDOS, which served well for years. The ProFile hard drive went straight to the Lisa, and the built-in floppy drive gave Apple a head start in designing the Lisa and the Macintosh.

Above all, Apple never forgot the lessons it learned about thermal design. The "no fans" edict didn't last long.

7. Lisa

Project Sara (the Apple III) was only one of the bets that Steve Jobs had placed on a computer for the business market. Project Lisa was another.

If anyone asked, Apple would say that Lisa stood for Local Integrated Software Architecture; industry commentators joked that it stood for "Let's Invent Some Acronym." (Years later, Jobs would admit that he named it after his baby daughter.)

Today, non-nerds may not even remember an Apple computer called the Lisa. But it was a historic, foundational machine that introduced the public to the computing elements we take for granted today: the mouse, windows, icons, folders, menus, fonts. A small team of designers hammered out thousands of design and interaction conventions over three years, detail by detail, establishing the modern computer interface of every Mac and Windows machine that followed.

68000

People adored the Apple II, but few would say it was intuitive to use; you still had to type out commands. Jobs intended the Lisa to be revolutionary in its simplicity. He asked John Couch, whom he'd hired from HP to oversee new projects, to lead the Lisa effort.

John Couch
Born: 1947, Chicago, IL
School: University of California, Riverside, University of California, Berkeley
Before Apple: HP
Apple: 1978–1983, 2002–2019
After Apple: Lightyear, Santa Fe Christian School, Pangea, DoubleTwist

Couch's first encounter with a computer was in a college horticulture class.

The Lisa would be a $2,000 workstation for office workers, managers, and secretaries. The Apple II and III had built-in keyboards and separate monitors, but the Lisa would flip that logic. It would have a *built-in* screen and *detached* keyboard. Jobs wanted to be able to use the keyboard in his lap.

Very Special Sushi

The Lisa's processor, Jobs hoped, would be Motorola's dazzling new 68000: a 16-bit chip, meaning that it could drive a computer with much more memory than 8-bit machines like the Apple II—and that meant speed.

Motorola's CEO, Robert Galvin, knew that his processor was Apple's only realistic option. "It was single-source," says Couch. "They had us by the balls, man." So he told Jobs that the price would be $125 per chip.

But Jobs had an ace up his sleeve. Hitachi was gearing up to produce its own 68000 clone—and Jobs had gotten his hands on one.

One night, Jobs met with Galvin for dinner at a Japanese restaurant to continue their price negotiations. When Galvin wouldn't budge, Jobs gently placed Hitachi's chip onto a piece of sushi and slid it across the table.

"It basically said, 'Look, you're no longer single-sourced,'" says Couch. "Here comes the Hitachi. What kind of price can you give us?"

Suddenly, Galvin was willing to haggle. He and Jobs settled on a price of $55 apiece for the first batch of chips, dropping to $35 thereafter.

"And that's how we went to the 68000," says Couch. This processor and its descendants would drive every Macintosh for the next 15 years.

Interface

The original plan was that the Lisa would have a standard character-based screen: green lettering on a dark background. But Jobs was emphatic that customers should not have to memorize and type commands. So the Lisa team planned to display a row of *soft keys*—buttons with commands like MOVE, COPY, and DELETE—across the bottom of the screen.

You could not, of course, tap these buttons with your finger. Instead, your keyboard would have a row of blank *physical* keys, aligned beneath the on-screen soft keys. To "press" the DELETE soft key, for example, you'd press the corresponding physical key below it.

Soft keys had worked fine on the computers that the HP veterans, like Couch and engineer Rich Page, had worked on at HP. But Bill Atkinson, who'd joined Apple in 1978 and was working on the Lisa's user interface, had doubts.

"In our testing, we found that those were very error-prone," he said. The soft keys' labels could change while you worked; one of them could suddenly say DELETE. If you weren't paying attention, you could accidentally nuke some of your work.

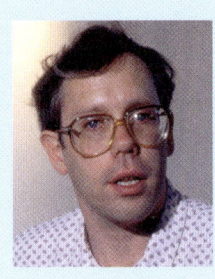

Bill Atkinson

1951–2025
Schooling: University of California San Diego
Apple: April 1978–June 1990
After Apple: General Magic, Numenta, PhotoCard

Atkinson first developed his passion for photography when he was ten, admiring the pictures in *Arizona Highways* magazine.

Jobs backed him up. He'd never wanted the Lisa to be yet another text-based, command-line machine anyway. He wanted something much easier to use, more advanced, and more beautiful.

The Atkinson Polaroids

Creating the Lisa was an exhausting process, especially for Bill Atkinson. "Bill spent every night programming, and the next day we would do testing and arguing," said project engineer Larry Tesler. "I don't know when he slept, actually."

Atkinson found it hard to get work done at the office. "People bustling around, talking all the time—you know, you need to write code," he said. "You need to put blinders on and focus, right?"

So he decided to work at his home in Santa Clara. On his desk, he built a proto-Lisa: an Apple II modded with a special bitmap graphics card to drive his monitor.

But now he had a problem: How could he show his work to the rest of the Lisa team? He couldn't haul the prototype back and forth to Apple; it was much too fragile for transport.

He wound up taking Polaroid self-developing photos of each new design idea, then riding his motorcycle to the office to show the team.

Atkinson preserved those Polaroids in a leatherette-bound photo album. Today, Atkinson's 135 Polaroids serve as a time machine, a glimpse of the earliest days of the user interfaces we now take for granted.

Bitmapped Graphics

The computers of the late 1970s and early '80s—like the Apple II and Commodore 64—displayed text in one-character chunks. When you pressed the A key, or 7 key, or # key, the software issued only a single, 1-byte code to beam the corresponding character onto your screen.

But in American research labs, something better was under development, described in hushed tones by engineers who'd seen them: bitmapped displays. On these systems, the screen picture is made up of pixels—tiny, individual dots—that the software could control *individually*. In these early systems, each pixel was mapped to a single bit of memory: black or white.

On a computer with a bitmapped screen, you weren't limited to fixed characters. You could have a choice of gorgeous fonts. You could have graphics, photos, and charts.

Of course, bitmapped displays lived only in labs for a reason: They made the computer slow and expensive. It takes a lot of pixels to fill a screen, a lot of memory to store the status of each pixel, and a lot of dollars to buy that memory.

But Atkinson was thrilled by the possibilities. On the pseudo-Lisa he'd set up at home, he spent hours writing routines that could show off the graphic possibilities of bitmapping. Bitmapping would also bring one of Steve Jobs's obsessions to a computer screen: *proportional* type.

On character-based computers, every letter is the same width. An *I* and a *W* occupy the same horizontal space. These monospaced fonts were easy and fast for a computer to pump out, but ugly.

Books, newspapers, and magazines, on the other hand, use *proportional* spacing, where each letter takes up only as much horizontal space as it needs. It's much easier to read.

Atkinson created one of the world's first proportionally spaced computer fonts.

The more he experimented with bitmaps, though, the more he longed for another radical change to the screen. "If you're going to do graphics, you have to use a white background! Because you're going to print on white paper; you're never going to print white text on black paper," he said.

The Lisa hardware team had a thousand objections. A white screen would require too much power, would flicker, would be bad for the eyes, would cost more. But Jobs didn't care. The Lisa would display black type on a white background.

At the end of 1979, the Lisa sat in a half-finished limbo. It had a white screen, bitmapped graphics, and soft keys at the bottom. It didn't seem especially revolutionary to Jobs—nor to Jef Raskin.

Raskin, a writer, artist, opera conductor, and computer-interaction expert, had written the Apple II user manual in 1976. Now, as Apple's director of publications, he remembered having seen something inspiring a few years earlier during a visit to Xerox Corporation's Palo Alto Research Center.

Until the Lisa, personal computers displayed only monospaced type (left), where every character gets the same amount of space.

PARC's headquarters was a magnet for pioneering computer engineers.

PARC, as tech fans now know, was a thrilling think tank of computing experimentation. Xerox chief scientist Jack Goldman had founded PARC in 1969, hoping to lead scientists to develop the next generation of technology, something like AT&T's Bell Laboratories.

The computers in its labs teemed with computing elements that are still with us today: mice, overlapping windows, menus, and fonts. They had ethernet networks and shared laser printers. In *1979*.

Bill Atkinson, Raskin's friend and former student, was a PARC admirer, too. He devoured every article and paper that PARC produced. When Jobs kept pushing his team to do something spectacular with the Lisa, Raskin and Atkinson encouraged him to visit PARC for inspiration.

Eventually, as the whole world knows now, Jobs did.

Xerox PARC

Apple's visit to Xerox PARC is one of the most told, most tantalizing tales in all of tech. It's been repeated in a thousand books, articles, speeches, and movies, and it usually goes like this:

> By the mid-seventies, PARC's geniuses, working in total secrecy, had essentially invented the modern computer: the mouse, menus, overlapping windows, scroll bars, fonts, icons and folders, copy and paste, bold/italic/underline, and so on. But the PARC engineers failed to understand how revolutionary their inventions were, and never brought them to market.
>
> In December 1979, Steve Jobs sweet-talked his way into a visit and a demo. In a flash, he saw the future. He stole PARC's ideas, reproduced them on the Mac, and changed history.

Well, at least they got the *year* right.

Almost everything else in that telling, though, is either misleading or wrong:

- **". . . invented the modern computer."** It's true that PARC's scientists had brought the modern *graphic* user interface to life, in a prototype called the Alto. (About 1,500 Altos wound up in universities and other Xerox labs, but they were never sold to the public.)

But a decade earlier, Doug Engelbart, a computing pioneer working at the Stanford Research Institute, had developed a radical thought: Maybe, instead of having workers at terminals that accessed room-sized, shared, mainframe computers, there should be one small computer on *each worker's* desk.

On December 9, 1968, Engelbart gave a now-famous 90-minute demonstration of this computer of the future. It included the first mouse, windows, on-screen graphics, video chat on a split screen, word processing, hyperlinks, networking, and remote collaboration. The shell-shocked audience gave him a standing ovation for what became known as "the mother of all demos."

Doug Engelbart demonstrating his vision for future computers.

The seeds of some of Xerox PARC's early ideas, in other words, had been sown ten years earlier.

The Alto's page-shaped screen.

- **". . . in total secrecy."** Actually, PARC published papers about its work, and over the years had given thousands of demos to colleagues, researchers, and famous people.

- **"The mouse, menus, overlapping windows with scroll bars, fonts, icons and folders, copy and paste, bold/italic/underline, and so on."** PARC's Alto machine did have a blocky plastic mouse, a bitmapped screen, and black type on a white background, with a choice of typefaces. Its Gypsy word processor had Cut, Paste, and Undo. (It had no Copy command; if you wanted to copy, you cut something and then pasted it back.) And unlike command-line computers, it was WYSIWYG: What You See [on the screen] Is What You Get [when printed].

  But by today's standards, the Alto was unfinished. There were no icons (for programs or documents, for example). And without icons, the Alto didn't have drag-and-drop. You couldn't, for example, drop a file into the Trash; there *was* no Trash.

  There was no menu bar. No dialog boxes. No checkboxes or

Lisa • 65

radio buttons. You could cut and paste within a single app, but not between apps. You couldn't drag to move or resize a window. You were still greeted, upon startup, by a command-line prompt.

- **". . . the PARC engineers failed to understand how revolutionary their inventions were."** They understood, all right. They spent years begging their overlords in Rochester, New York, to *do* something with their breakthroughs—to commercialize them.

    But Xerox was in the copier business. Its gigantic photocopiers were too expensive for businesses to own outright, so Xerox charged its clients a couple of cents per copy made. The company's top executives, mostly former copy-machine salesmen, had no experience, conception, or inclination to become a computer company.

- **". . . never brought them to market."** Two years after Apple's visit, Xerox did indeed commercialize its work. It introduced the Xerox Star, a descendant of the Alto. The price was $75,000 for the base unit, and $16,000 for each additional workstation. In today's dollars, that starter set would set you back about $320,000. Shocker: It flopped.

- **"Steve Jobs sweet-talked his way into a visit."** Actually, Xerox's top brass were thrilled to offer the visit to Jobs, because he was offering something valuable in return.

    At the time, Apple was white-hot. It was the kind of startup that made investors' palms sweat. In the summer of 1979, Apple intended to raise money by offering private shares to 16 lucky investors. Everyone knew that Apple would be going public the following year, making its shares worth a fortune. Everyone wanted in on this deal.

    "I will let you invest a million dollars in Apple," Jobs told Abraham Zarem, the head of Xerox's investing arm in West Hollywood, "if you will open the kimono at PARC."

    Zarem took the deal. Xerox paid Apple $10.5 million for 100,000 private shares of Apple stock, and Jobs got his PARC visit. (Zarem also agreed to start selling Apple II's in Latin America, where Xerox had a sales force and Apple had nobody.)

    The deal was hardly a rip-off for Xerox. Within a year, stock splits turned Xerox's Apple stock into 800,000 shares, worth $17.6 million. In today's dollars, that's about $73 million.

- **"He stole PARC's ideas."** Doug Engelbart's "mother of all demos" audience in 1968 included a contingent of executives from, of all places, Xerox, who brought some of what they saw that night to Xerox PARC. In fact, several Engelbart team members wound up working at PARC themselves.

    The truth is, the modern graphic interface evolved through a succession of inspirations. PARC was inspired by some of Engelbart's ideas, but took it in new directions; Apple was inspired by PARC and moved them even closer to the modern graphic interface.

———

Everyone involved in Apple's PARC visit remembers it slightly differently. Was it December 1979? Or November?

Most agree, though, that Apple visited PARC twice. The first time, the Apple group—Jobs, Couch, Scotty, Atkinson, marketing head Trip Hawkins, engineering head Tom Whitney, and engineers Ken Rothmuller and Rich Page—saw the same canned demo that PARC had given hundreds of visitors before them: the Alto and its mouse, a word processor with fonts, and a few programs written in Smalltalk, the advanced programming language that made the Alto's magic possible.

Jobs thought that even these technologies were astounding. "You're sitting on a goldmine," he told the PARC reps. "Why aren't you doing something with this technology? You could change the world!"

But after hearing the report of the visit, Jef Raskin realized that PARC's staff hadn't actually opened the kimono; they'd barely lifted it to the ankles. They'd offered a minimal demo not worth the lucrative Xerox stock deal.

Raskin suggested that Apple request a second demo. This time, he'd brief the team on what to look for. So Jobs and the Lisa team returned—and this time, they were armed. "They had read every paper we'd published," said PARC engineer Larry Tesler, who was one of the presenters. "It was clear to me that they understood what we had a lot better than Xerox did."

Tesler describes the meetings as uncomfortable all around. Some of the Apple crew considered the whole PARC thing a distraction, and some of the PARC staff felt that the demos were revealing too much. Either way, Jobs felt struck by lightning.

"It was one of those sort of apocalyptic moments," Jobs said. "I remember within ten minutes of seeing the graphical user interface stuff, just knowing that every computer would work this way some day; it was so obvious once you saw it."

"We'd still shown them only, like, one percent of what PARC was doing," Tesler remembered, "but it was enough that they got really excited. They fell in love with the mouse, and that changed everything. They decided they were going to retarget the Lisa."

Larry Tesler
1945–2020
Schooling: Stanford
Before Apple: Stanford AI Lab, Xerox PARC
Apple: 1980–1997
After Apple: Amazon, Yahoo, 23andMe

Tesler's view on artificial intelligence has come to be known as Tesler's Theorem: "AI is whatever hasn't been done yet."

Lisa Reset

The Lisa team returned to Cupertino and ripped up their plans. The Lisa would not be a green-text derivative of all the computers that had come before it. They'd wanted something revolutionary, and they'd found it: the graphic user interface.

But they weren't the only ones who'd been affected. PARC's Larry Tesler couldn't get over how fully the Apple team had grasped the potential of what he'd helped to create—and he wanted to help them bring it to the world. Within six months, he had left PARC to join the Lisa team. Eventually, 15 more PARC scientists would leave to join Apple.

It soon became clear, though, that the PARC visit had opened up more questions than it answered. How many buttons should the mouse have? What should the cursor look like? How should menus work, and where should they be? And what about scroll bars, windows, documents, programs, resize boxes, tool panels, preference settings, dialog boxes, error messages, text selections, and folders?

When Tesler joined Apple in July 1980, Steve Jobs planned to ship the Lisa in six months. Answering those questions would take three more years.

Taco Towers

The earliest work on Lisa took place in a historic 12-person office building on Stevens Creek Boulevard—historic in the Apple story, anyway: It was the Good Earth building, Apple's very first office.

As the project grew, the Lisa team moved into two office suites nearby, and began a Lisa-team tradition of giving their offices affectionate nicknames. The suite that housed the hardware team, all of whom were smokers, became known as "Scorched Earth"; the software engineers filled the adjacent "Salt of the Earth" suite.

In Apple's 1980 reorganization, the Lisa project became a separate division, housed in a new building a short walk from the main Apple building, at the corner of Lazaneo and De Anza Boulevard. The team wryly referred to it as Taco Towers, because it was the first *two-story* building Apple had ever occupied, and because, one theory goes, it suggested the thrilling architecture of a Taco Bell.

The Button Wars

When engineer Rod Holt saw the mouse at PARC, he thought it was an abomination. "Jesus, oh, God, you can't manufacture this thing!" he said in Jobs's silver Mercedes, on their way back from the PARC visit. "Any secretary or anybody else who's working on a word processor or spreadsheet or anything else—are they going to take their hands off the keyboard and reach over and grab a mouse and come around and watch the pointer? No—they're working!"

But Jobs had made up his mind. The Lisa would have a mouse.

Over the months, Jerry Manock produced 150 different designs, from perfect rectangles to, in one case, a little sphere. He orchestrated "wine-tasting sessions," where he'd lay out 25 models at a time for novices to try out. But the tougher question wasn't just about the mouse's shape; it was how many buttons it should have.

The Xerox Alto's mouse had three—red, yellow, and blue—and their functions changed depending on the program you were using, or even what *part* of the program. It was all hard for beginners to remember, but power users enjoyed the efficiency of the three buttons once they'd memorized their functions.

Different factions on the Lisa team made their cases for one, two, or three mouse buttons. One ex-PARC engineer actually argued for studding the mouse with *six*. "Bartenders have six buttons on those drink dispensers, and they can handle it," he'd say.

In the end, they decided to settle the issue with guinea pigs.

User Testing

In the summer of 1980, Tesler set up a testing room with a one-way mirror, an overhead microphone, and two video cameras: one trained on the computer screen, and the other on the keyboard and mouse. Synchronized VCRs recorded both cameras. A test subject would try out the latest element of the Lisa's design, guided by a team member and sometimes observed by psychologists.

The volunteers were anyone the team could round up: spouses or parents of Apple employees, custodial staff, visitors, or new Apple employees who weren't already computer savvy. Then the researchers gave the guinea pigs an instruction, such as "Edit this document, and then save it into this folder."

Those sessions finally settled the mouse-button wars. "As long as it was more than one button, they spent more time looking at the mouse than at the screen," Atkinson says. But when they tried a one-button mouse—presto! "They just stopped looking."

In the months to come, the user-testing room was pure gold for the designers. "They would immediately say, 'Why don't you just do it *this* way?'" said Tesler. "We all thought, 'Why didn't we think of that?'—and then we did it that way." In the end, he says, "We blundered into the Lisa user interface one design decision at a time."

> ### How "OK" Was Born
>
> The user-testing room became the beating heart of the Lisa development process. As ordinary non-geeks tried out each new feature, Atkinson says, "We asked them to mutter under their breath, a stream of consciousness. What are they thinking about as they do something?"
>
> One day in 1982, a dialog box appeared on the screen. Reading it, the test subject seemed visibly offended. "DoIt? I'm not a doIt!" he growled. Had the software implied that he was somehow stupid?
>
> When Tesler and Atkinson studied the video later, they saw the problem. The dialog box offered two buttons for moving ahead: "Do It" and "Cancel."
>
> But in the Lisa's system font, there wasn't quite enough space between the words in "Do It."
>
> The team hastily changed its dialog boxes to say "OK" and "Cancel"—labels that are with us to this day.

Keyboard Shortcuts

The invention of menus meant that you no longer had to memorize commands to use your Lisa. Every available action was neatly listed before you. But for the sake of power users, the Lisa team included keyboard equivalents for those menu commands. In fact, they gave the Lisa a dedicated key just for triggering those shortcuts: the key, a successor to the open-logo key on the Apple III.

The team gave a delightful logic to the key assignments: -Z for Undo, -X for Cut, -C for Copy, and -V for Paste. Not only are Z, X, C, and V all in a tidy row at the bottom of the keyboard, but they came with satisfying mnemonics. The Z for Undo made sense, because you were rewinding the *last* thing you'd done. Z was also the closest key to the key, "because we figured you'd undo a lot," said Tesler.

X, for Cut, looked like a pair of scissors. "We did V [for Paste] because it pointed down, and you were inserting; it was like an upside-down caret. And C for Copy—that was easy."

Atkinson actually argued that the keyboard should have special *keys* labeled Cut, Copy, Paste, Undo, and Again. "It never happened," he said. "I didn't win that one."

No Modes

As advanced as the Xerox Alto was, most of its programs still relied on operational *modes*: distinct software states where your mouse and keyboard activities take on different meanings.

In PARC's Bravo word processor, for example, you couldn't just start typing. You first had to type the letter *A* to enter Append mode . . . then click where you wanted to add the text . . . then do your typing . . . and finally press the Esc key to *exit* Append mode.

People routinely changed modes without realizing it—with disastrous consequences. Imagine, for example, what might happen if you accidentally entered Command mode. Now suppose that you began to type the word "edit." Typing the first letter, *e*, would drop you into *Entire* mode, meaning you'd select everything you'd typed. When you next typed *d*, you'd *delete* everything currently selected—the whole document! The next letter in the word "edit" is *i*, which triggered *Insert text* mode—and when you typed *t*, you would indeed insert that letter. You'd now have lost all your work except for the letter *t*.

Tesler couldn't *stand* modes. He undertook what he called "a near-fanatical campaign to eliminate modes from the face of the earth."

The Lisa, he vowed, would be modeless. You'd be able to type, scroll, and preview a document simultaneously. And to ensure that the world knew about his convictions, he ordered special license plates for his car.

Tesler died in 2020, but his family maintains his website—at *www.nomodes.com*, of course.

Menus

In PARC's Smalltalk programs, menus appeared only when you clicked a certain spot with a certain mouse button. "It was all mysterious and hidden," says Atkinson. So his first thought was to make the commands visible all the time—in a menu *bar* attached to the window.

User testing soon revealed the problem with that design, however: If the window was narrow, menu titles stuck out goofily beyond the window's frame. And if the window was short and near the bottom of the screen, the list of commands was chopped off.

Tesler proposed a radical new approach: Put the menus at the top of the *screen*, rather than inside the windows. The only significant downside was that if your cursor wasn't already near the top of the screen, moving it up there could feel like a long trek.

For a hot second, Atkinson experimented with letting the cursor wrap: When it passed the bottom of the screen, it would reemerge at the top. ("Stupid," he'd say later.)

Instead, he came up with cursor acceleration. Instead of moving the cursor one inch when you moved the mouse one inch, the cursor moved disproportionately faster when you sped up your mouse movement—faster on the screen than on the table. "A quick flick would get you to the top," Atkinson said.

> **The Apple Mouse**
>
> The mouse at Xerox PARC was a marvel in concept, but not in execution. When you rolled it on the desk, two wheels underneath tracked its position. One wheel spun forward and back; the other, side to side. It was basically a deconstructed Etch A Sketch. If you attempted a diagonal line, it produced a jagged line—a disaster on a computer designed, in part, for artists.
>
> In April 1981, Jobs asked design consultant Dean Hovey to come up with a better mouse. It should cost no more than $15 to manufacture, he said, "and I want to be able to use it on Formica and my blue jeans."
>
> Once he found out what a mouse *was*, Hovey drove to Walgreens to scavenge for parts. He started by snapping up one of every roll-on deodorant bottle on the shelves. His plan was to pry out the applicator ball from each bottle, on the theory that a ball would move the cursor more smoothly than the two right-angle wheels. In housewares, he bought some butter dishes, whose rounded, palm-sized covers he could use as potential mouse bodies.
>
> And that's how, in a single weekend, he cobbled together a simple prototype of the first Apple mouse. "In 90 days, we had a mouse that could be built for 15 bucks that was phenomenally reliable," Jobs said later.
>
> For the historical record: The winning ball came from Ban Regular Scent.

A Thousand Details

The Lisa interface of June 1982 would look distinctly familiar today. After hundreds of user-testing sessions, many person-years of debate, and thousands of memos, the Lisa team had developed a blizzard of now-standard interface details. The I-beam cursor when you edit text. The Clipboard, a window that holds whatever you've most recently cut or copied.

And scroll bars. Unless all you write are tweets and haikus, the computer screen is too small to show your entire document at once. But there was no clear answer to the question: *Which way should the scroll bar move your document?*

One philosophy held that dragging the scroll handle *down* should move the page *up*. You're saying, "Move my porthole on the document down, *relative to the page.*"

> ### Atkinson in the Hot Tub
>
> Bill Atkinson was fired up by the possibilities of the graphic interface he'd seen at Xerox PARC—especially the details. Take, for example, the illusion of overlapping windows, like real-world documents. When you moved a window, you'd reveal whatever was in the window "underneath" it. No computer had ever performed such a graphic feat.
>
> Unfortunately, re-creating that effect on the Lisa was a devilishly tough challenge. The machine didn't have enough memory to store the images of all your windows' contents at once. And even if it did, when you moved a front window, there'd be a lag and a white flicker as the window behind it redrew itself. Atkinson drove himself to distraction: *How had the PARC engineers created the effect?*
>
> It took weeks to crack the code. First, he devised a system of describing the overlapped window areas using only their corner coordinates. The Lisa wouldn't have to memorize the entirety of every window's contents; it could just calculate which regions of background windows were visible.
>
> Second, in his suite of graphics algorithms (QuickDraw), Atkinson came up with a way to capture a snapshot of what was in a window (a QuickDraw *picture*)—a method a hundred times faster than forcing each program to repeatedly recalculate everything in its window. With his regions and his QuickDraw pictures, Atkinson had finally found a way to move or resize overlapping windows with no delay and no flicker.
>
> But it's the epilogue to this story that has made it a classic bit of Atkinson lore.
>
> Weeks later, he was sitting in a hot tub at a naturist (nudist) club. "I'm sitting there naked next to this other engineer, swapping notes," Atkinson said. "This guy says: 'That was so brilliant, what you did with region clipping on the Lisa!'"
>
> Atkinson was dumbfounded. His soaking companion worked at Xerox, where Atkinson remembered seeing region clipping demonstrated in the first place. "Wait—I thought you guys did that!"
>
> "No, no, we didn't! We never solved it!" was the reply.
>
> Atkinson had misremembered. He only *thought* he'd seen non-flickering overlapping windows at Xerox PARC; in fact, its scientists had never found a solution. Atkinson solved the problem on his own, assuming that he'd already seen it solved.
>
> "And so because I 'knew' it could be done, I was enabled to do it," he said.

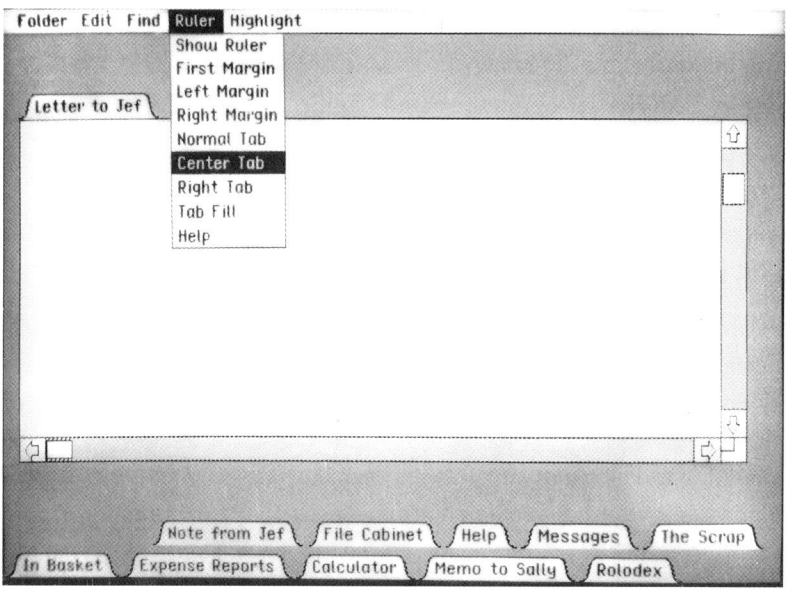

This Atkinson Polaroid shows how the original Lisa interface treated documents like folders. The option to turn an open window into a tab resurfaced 17 years later—as a feature in Mac OS 8.

The other view was that dragging the scroll handle *down* should move the page *down*. You're saying, "Slide the page down, as though I'm dragging it downward on my desk with my finger." In user testing, the guinea pigs guessed wrong half the time.

In the end, the team went with option 1: Scrolling *down* moves the document *up*. But decades later, the invention of the iPhone changed Atkinson's mind. "Now we realize: seeing more of what's later in the document has to be an up arrow, because you're going to shove the page up," he said. "My mistake—I made that choice."

(These days, on the Mac, you can choose which direction logic you prefer in Settings.)

For a long stretch of Lisa development, the team called windows "folders." If you dragged a window to the bottom of the screen, it collapsed down into a folder tab peeking up. "A document and a folder were kind of the same thing," says Atkinson. "We got confused about the difference."

What the interface really needed, of course, was *icons* that would easily differentiate files from folders. They would come.

The Desktop Metaphor

By October 1981, the Lisa team believed that they were creating a machine that was infinitely easier and more joyous to use than any computer ever released. In trials they ran at a testing center in Daly City, people who'd never even seen a computer up close could go from opening the box to word processing in *ten minutes*.

By late 1982, the software design was frozen—no further changes were allowed except for bug fixes. All was prepared for the Lisa's unveiling in early 1983.

But something was bothering programmer Dan Smith: the Filer, the screen you'd use to create, delete, rename, and organize files. He and fellow Filer author Frank Ludolph had built a dialog box that walked you through a series of queries: What document did you want to operate on? What did you want to *do* with it (Refile, Cross-File, Discard)? Where did you want to refile it? Among the engineers, its interrogation format earned it the nickname "20 Questions."

"There has to be a simpler way to do this," Smith announced to Atkinson. They pored through published papers for inspiration. In an IBM research proposal, they read about an idea, never implemented, for a graphical office system called Pictureworld. It was a fairly literal on-screen office, complete with icons for file cabinets, in and out trays, and a wastebasket.

Intrigued, Smith and Atkinson coded up a prototype of what Mac fans today call the Finder (or the File Explorer on Windows). It displayed icons for your files and folders that you could drag around. Documents zoomed open and closed from their icons. And when you dropped an icon

Before the desktop, there was the "20 Questions" filing dialog box.

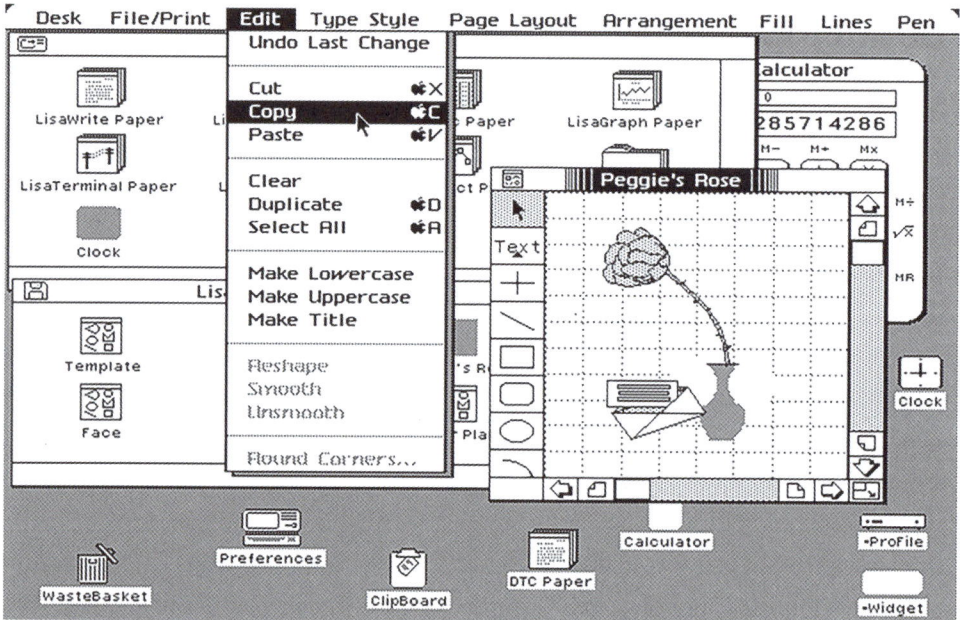

The Lisa was the first computer with a menu bar, one-button mouse, Clipboard, and Trash.

onto the trash can, its lid popped off, and tiny animated *flies* appeared, buzzing around. ("I maybe went a little too far with the Trash," Atkinson said.)

It was exhilarating. You were *directly* interacting with your files and folders, almost touching them.

Engineering manager Wayne Rosing loved it, too, but he couldn't risk a delay. So he made them a deal. They could work on their prototype after hours. But if it wasn't stable and ready to go in two weeks, they'd abandon the whole thing. And they must not, under any circumstances, show their idea to Steve Jobs. If he hated it before it was finished, it was as good as dead.

The next day, Rosing called an all-hands meeting. Smith, Ludolph, Atkinson, and Rosing showed up in newly printed purple T-shirts that said "Rosing's Rascals." The idea was to show the rest of the team that there had been no mutiny; Rosing had sanctioned the new features.

In the end, testers loved the new desktop, and so did Jobs. The Lisa had a new home base.

Hello World

When it was ready for release in 1983, the Lisa weighed 48 pounds. It contained a Motorola 68000 microprocessor running at 5 MHz; 1 MB of RAM (250 times as much as the first Apple II), expandable to 2 MB; two 5¼-inch floppy disk drives, with the world's first Eject button; the one-button mouse; and a built-in 12-inch black-and-white bit-mapped screen (720 x 364 pixels). It came with seven preinstalled apps: LisaWrite, LisaDraw, LisaCalc, LisaGraph, LisaProject, LisaList, and LisaTerminal. The ProFile hard disk was an option.

Its design, by Jerry Manock's successor Bill Dresselhaus, carried over the rounded corners and beige plastic case of Manock's Apple II and III designs, but seemed more businesslike.

It had the world's first multitasking windowing system for a personal computer. You could run multiple apps simultaneously and even

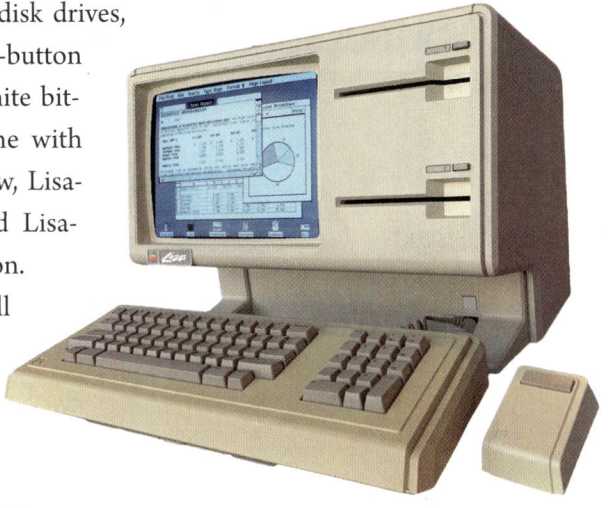

The Lisa in its original configuration: two floppy-drive slots, no hard drive.

copy and paste among them—a feat even Xerox PARC hadn't achieved. It was also *object oriented*, meaning that its software was modular and extensible, and had *protected memory*: Each app was walled off from the others, which prevented one crashing program from affecting others. The Lisa's successor, the Mac, would not get these features for another 19 years.

When you pressed the power button on the front of the machine, it saved all of your work in progress and then entered a deep, low-power Sleep mode. When you were ready to resume your work, it lit up instantly.

And you didn't have to worry about the difference between programs and documents; in fact, app icons were largely hidden. Instead, you used "stationery pad" icons. When you double-clicked one, you spawned a new, blank document.

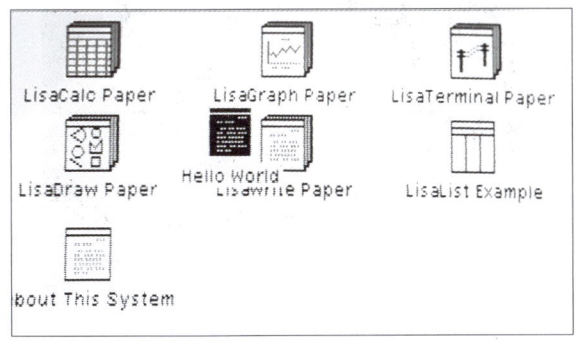

Stationery pads for creating new documents.

In the corporate world, data safety is everything, so Lisa's file system maintained duplicate sets of data about everything on your disk. If the system crashed, an app called Scavenger opened automatically to recover any corrupted files. And you could password-protect any document right at the desktop—a feature that never did come to the Macintosh.

Finally, for the first time in computerdom, you could give your files names up to 63

In 1983, every aspect of a graphic user interface was new to the public. Apple's ads often doubled as miniature instruction manuals.

characters long, including punctuation and symbols.

"Every 80-hour work week, every canceled vacation, every hot debate, and every wrenching management decision was motivated by one common driving force: we wanted our product to be the best," Tesler said.

In the end, after 200 person-years' worth of effort and $50 million in development, Steve Jobs unveiled the Lisa publicly at the Boston Computer Society in the spring of 1983. The reaction was euphoric. "Huge, huge ovations. Just incredible," says John Couch.

The Lisa ringleaders: John Couch, software manager Bruce Daniels, Wayne Rosing, and Larry Tesler.

"It surpasses anything available on the market in terms of ease of use," wrote the *New York Times*. Lisa could "revolutionize the personal computer industry and guarantee Apple's place in it."

When the machine went on sale in the summer of 1983, it was technically three years behind schedule—but the Lisa team considered the delay a feature, not a bug. The Apple III episode had taught the company a painful lesson: The mandate should be to get it *right*, not get it *soon*.

The Lisa Ad That Launched a Legend

Wistful music: piano and guitar. A handsome 28-year-old man rides his bike, silhouetted against the setting sun, sweater draped artfully around his neck. A faithful dog trots alongside.

"The way some business people spend their time has very little to do with the clock," says a gentle narrator. The handsome man and his dog ride an elevator to a huge, sunlit corner office. The man sits at his glass-topped desk. With his left hand, he moves the mouse to wake up his Lisa.

"At Apple, we make the most advanced personal computers in the world," says the voice. "Soon, there'll be just two kinds of people: Those who use computers..."

The handsome man answers the phone. He utters his only line in the 70-second ad.

"Hi. Yeah. I'll be home for breakfast," he says.

"... and those who use Apples," the narrator concludes.

As the scene fades to the Apple logo, we figure out who that handsome man is. It's Kevin Costner, still years away from starring in hit movies like *Field of Dreams*, *The Bodyguard*, and *Dances with Wolves* and winning two Oscars.

Steve Jobs always did have an eye for talent.

8. Crashes

Fortune did not smile on Apple during 1980 and 1981. The creative journey designing the Lisa was a rush of genius and excitement—but the *corporate* journey was regularly interrupted by bombshells.

As the Lisa was taking shape, Steve Jobs became more and more driven to make it perfect. He called Tesler at all hours of the day and night to share his ideas. "I loved it," Tesler said, "but it upset my bosses."

Those bosses included John Couch, whose job was to shepherd the Lisa to market. The 80 engineers he'd hired for the Lisa project were complaining. Jobs, they said, kept inserting himself into the process—and not in a friendly way. When he didn't like someone's work, he unleashed blistering barrages of insults.

Above all, Jobs saw the Lisa as a machine for the masses, and it made him crazy to watch the feature list, and the retail price, keep climbing. "There was a tug-of-war between people like me, who wanted a lean machine, and those from HP, like Couch, who were aiming for the corporate market," Jobs said.

Jobs's behavior irked CEO Mike Scott, too. "The leadership arrangement was a bit awkward," says Couch. "Many Apple employees considered Steve to be the real boss, and his product decisions as being the final word." Scotty was well-liked, but people didn't *really* consider him the source of power.

So in September 1980, Scotty and Markkula unveiled a corporate restructuring. Couch would now be solely in charge of the Lisa project; Jobs would be in charge of nothing. His new title was "nonexecutive chairman of the board," which meant very little.

Jobs felt deeply betrayed by Markkula. "He and Scotty felt I wasn't up to running the Lisa division," he said. "I was upset and felt abandoned."

Woz's Plane

Five months later, the company received another shock—one that shook its employees on a much more personal level.

Steve Wozniak hadn't been a big player on the Apple III or Lisa projects. His baby had always been the Apple II—and its June 1979 descendant, the Apple II Plus. It came with a full 48 KB of memory, Microsoft's Applesoft BASIC built in, auto-startup, and a few changes to the keyboard, all for a lower price: $1,200.

Woz's smaller role at Apple—and his IPO wealth—left him free to spend more time on other pursuits. He bought and renovated the Mayfair movie theater in San Jose. He and Alice divorced—and then, within a few weeks, Woz met and proposed to Candi Clark, an Apple employee and former Olympic kayaker.

Woz also learned to fly. In 1980, he got his pilot's license and bought a Beechcraft Bonanza A36TC, a six-seater propeller plane with a distinctive V-shaped split tail. He considered it "the most beautiful and unorthodox single-engine plane there is," and had it painted in gorgeous earth tones.

Woz never pretended that he was an expert pilot. "I knew the rules to follow," he says, "but still, I was a beginner pilot. I was still a pretty rough new trainee."

On February 7, 1981, he and Candi flew down to a small airport in Scotts Valley to pick up Candi's brother and his girlfriend. The plan was to fly on to San Diego, where Candi's uncle lived. (Woz had asked him to design a unique wedding ring for him: one with the diamond hidden *inside*.)

"I remember reaching for the throttle at the start of the runway, and that's it," Woz says.

The plane stalled during its initial climb, fell to earth, bounced along the runway, and broke through two fences. Finally, it plowed into an embankment at the edge of an ice-rink parking lot. All four occupants were injured, but Woz got the worst of it: severe head injuries, bruises, and

A Beechcraft Bonanza A36TC.

a lost front tooth. Most distressing of all, he had anterograde amnesia: He couldn't create any new memories. He didn't remember the crash, what happened afterward, or even, in the hospital, his own name.

His friends and colleagues flooded his hospital room with cards, gifts, and food, and visited him often, but he doesn't remember any of it. After two weeks, he returned home in a hazy state. After five weeks, he began to form new memories once again—but the hazy weeks are lost to him forever. Years later, he hired a hypnotist to see if he could recover some of the lost memories. He couldn't.

The accident shook Woz to his core. He treated it as a cue to make some changes, as a "lucky opportunity." He was 30 years old, and, ten years after dropping out of Berkeley, still had not finished his college degree. He reenrolled at Berkeley, set a wedding date with Candi, and told his colleagues at Apple that he wouldn't be seeing them for a while.

Black Wednesday

Only three weeks after Woz's plane crash, Apple got yet another gut punch—this time, from closer range.

Upon arrival at the office on February 25, 1981, Mike Scott began asking employees to visit him in his office. As each one showed up, he fired them. As the morning wore on, nobody knew who'd get the call next, or why. It was like a sniper firing random shots into a crowd. By 10:30 a.m., he had fired 40 people—including half of the Apple II engineering team. Some of them were superstars.

Survivors huddled in clumps, stunned. Apple had just pulled off one of the most successful IPOs in history. The company was growing in spectacular fashion. What possible reason could the CEO have to be unhappy?

Finally, at noon, Scotty convened an all-hands meeting in the basement of the Taco Towers building, where he stood with a beer in his hand and explained his actions.

Apple had grown too big, too fast, he said, and along the way had hired some mediocre people; they, in turn, had hired even worse people. Scott was especially convinced that the Apple II team had become complacent, that it had lost the old startup hustle. The place needed pruning, he said. "I used to say that when being CEO at Apple wasn't fun anymore, I'd quit," he told the team. "But now I've changed my mind—when it isn't fun anymore, I'll fire people until it's fun again."

Two weeks later, Scotty announced yet another reorganization of the company. Both times, he'd acted without the board's approval, and Markkula had had enough.

"He had some really serious personal issues, and it started affecting him at work," Markkula says. "It just became obvious that he couldn't run the company." After consulting with the board and Apple's top managers, Markkula removed Scotty as CEO, and offered to take on the role himself until he could find a permanent replacement.

Scotty's official post-demotion title was vice-chairman, which meant very little; shortly thereafter, Mike Scott quit Apple altogether.

Jobs himself became the chairman of Apple's board. He would not miss Scotty, with whom he'd clashed from day one. "I've never yelled at anybody more in my life," he said.

Welcome, IBM

The rapid corporate reshufflings, Woz's plane crash, and Black Wednesday weren't the end of the drama in 1981. Yet another crisis loomed: its first serious competitor.

Nobody in corporate America needed an introduction to IBM. Its mainframes and minicomputers dominated institutions. Apple had placed a losing bet that the Apple III would introduce *personal* computers to that deep-pocketed world.

On August 12, 1981, IBM introduced its own personal computer, called the PC. Jobs and his team quickly bought one, brought it to the office, and gave it a try. It was, as Jobs quickly decided, a "piece of shit."

On that day, Jobs began to see IBM not just as his enemy, but as "a force for evil." His David-vs.-Goliath view of the rivalry would drive him for years. "Once IBM gains control of a market sector, they almost always stop innovation. They prevent innovation from happening," he would say.

Working with Regis McKenna, Jobs came up with a now-immortal full-page ad, which appeared in the *Wall Street Journal* on August 24.

"Welcome, IBM. Seriously," said the headline. "Welcome to the most exciting and important marketplace since the computer revolution began 35 years ago. And congratulations on your first personal computer," it went on.

Here was Apple, this nest of young hippie types, welcoming a company 50 times its size. The ad was hilarious in its cockiness, and yet it reinforced that IBM was, in fact, the latecomer in personal computing.

The ad became an instant classic. But the IBM PC was a threat, no matter how much less elegant it was than Apple's efforts. First, its base model cost $1,656, about one-sixth the cost of a Lisa.

Second, IBM had built it with off-the-shelf parts and then published the specs so that other companies could make their own PC-compatible parts and add-ons, and customers could tinker. The Lisa's closed approach ensured smooth operation and a delightful experience for customers, but was less reassuring to potential customers who wanted the option of upgrading their machines down the line.

Apple's full-page ad in the Wall Street Journal.

Lisa Lands

When the Lisa went on sale in January 1983, everyone could see how advanced and superior it was. "Nothing on the market, including the IBM PC, compares with the new machine," said *Time* magazine. "Lisa is easy; you'd have to work at it to make a mistake," *Popular Science* said.

But when it came to *selling* the Lisa, Apple was in for a harsh lesson: Beauty and brilliance aren't the only factors that matter. Incredibly, or maybe inevitably, the Lisa crashed and burned.

It was much too expensive—$10,000, which is $33,000 in today's dollars. The Lisa's generous memory allotment, a key to its multitasking ability, had driven up the price; so had the external 5 MB hard drive, which cost Apple $1,000 each.

It was dog slow, too. When you dragged a window, only its outline moved with your mouse, not the image of its contents—and even the outline lagged about 20 pixels behind your mouse.

Jobs also regretted entrusting so much of the Lisa project to marketing and management experts. "We were 23, 24, and 25 years old. We had never done any of this before, so it seemed like a good thing to do," he said later. "But unfortunately, this was such a new business that the things the so-called professionals knew were almost detriments to their success in this new way of looking at business."

Jobs was less comfortable discussing the other factor that hurt the Lisa: rumors about *another* Apple computer that would offer Lisa goodness at a fraction of the price. Many of the rumors came from Steve Jobs himself.

Adrift and angry after being pulled off of the Lisa, he had heard about a tiny skunkworks project, little more than a concept, brewing in an Apple side office. It was, of course, the Macintosh, and its development overlapped with the Lisa's.

"Several times, there would be customers brought in to look at the Lisa," Bruce Daniels says. "And Steve would come running across the street and say, 'Come—you have to see the Mac!' It's kind of difficult to compete against the chairman of the board saying, 'Don't buy this—buy *this*, buy my thing.'"

For the second time, Apple had bet immense resources on its next revolutionary machine—and for the second time, it had bombed. Also for the second time, what saved the company was the Apple II.

"We virtually doubled in size in 1983," Jobs would say. "We went from $583 million in 1982 to something like $980 million in sales. It was almost all Apple II–related."

The Apple II was now six years old. Apple had spent five of those years—and $110 million—trying to design its successor. Twice, Apple thought it had a winner; twice, it failed.

Jobs considered the third project to be Apple's last chance.

9. Macintosh

Jef Raskin was the polymath's polymath. He was a programmer, writer, artist, photographer, author, inventor, composer, conductor, keyboard and recorder player, opera nut, bicycle racer, archer, target shooter, ham radio operator, toy collector, builder of remote-control airplanes. "People have often expressed surprise at the range of things I do," he'd say dryly.

He'd grown up in New York City, and studied math, physics, philosophy, and music at Stony Brook University. Then came a master's at Penn State and a PhD at the University of California San Diego, where he stayed as a professor of art, photography, and computer animation. When he eventually left that job, he rented a hot-air balloon and flew it over the chancellor's home, playing his sopranino recorder to get his boss's attention. "He came out, and I yelled down that I was resigning and floated off," Raskin said. "I was an art professor at the time, and it seemed arty to leave that way."

In 1977, he and Brian Howard, his friend, former housemate, and fellow Renaissance wind player and electrical engineer, started up a tiny company called Bannister & Crun (two characters on the BBC's *The Goon Show*) to write software manuals for hire.

Raskin and Howard had met Jobs and Woz at (where else?) the Homebrew Computer Club in 1976, and Jobs eventually invited them to write part of the Apple I manual.

Jef Raskin
1943–2005
Schooling: SUNY Stony Brook, Penn State
Before Apple: University of California San Diego
Apple: 1976–1982
After Apple: Author, inventor

Raskin's house was "practically one large playground," featuring secret doors and passageways, a 185-seat auditorium, and a model airplane room.

Raskin and Howard went on to write the BASIC manual for the Apple II—gorgeously laid out, in color, with humor, and, expensively, spiral-bound so that the book would lie flat. (Excerpt: "There is nothing you can do by typing at the keyboard that can cause any damage to the computer, unless you type with a hammer. So feel free to experiment. With your fingers.")

In January 1978, Jobs bought Raskin and Howard's little company to form Apple's new department of publications; they were employees #31 and #32. Raskin added a clause to his contract specifying that his Apple duties would have to work around his opera rehearsals.

In early 1979, the Apple III and the Lisa were well underway. Raskin, however, was philosophically opposed to both projects. He didn't think that the Apple III had enough technical pizzazz, and thought that the Lisa was going to be overpriced and too slow.

He'd been writing papers about creating something *much* simpler, smaller, and less expensive. He described it as the average person's amanuensis, meaning personal assistant.

Raskin wasn't the first person to think small, simple, and cheap. At Markkula's suggestion, Steve Wozniak had already begun designing a $500 game machine code-named Annie, to compete with Atari's products. But Woz never finished designing Annie. The plane crash took care of that.

So when Raskin described his idea for a cheap, simple machine, Markkula's first thought was: Maybe the Annie machine could be a starting point. Raskin's first thought was: No thanks.

"A game machine, although a good idea, was not something that I'd feel comfortable doing. So I counter-proposed a general-purpose, low-cost computer based on my own ideas—and dreams—for an interface," Raskin said.

Markkula approved the little project. As his first official act, Raskin announced that the new machine would *not* be called Annie. "I felt that the trend in the company to give new products feminine names was sexist," Raskin said. "I suggested Macintosh, naming it after my favorite kind of apple."

Of course, the apple variety is spelled "McIntosh," not Macintosh; Raskin says that he changed the spelling to avoid trademark trouble with McIntosh Laboratory, the maker of high-end stereos. (It didn't entirely work. When Jobs wrote to McIntosh Lab's president to seek permission to use the name, the answer was yes—for about $2 million.)

Over the course of a year, Raskin wrote a series of memos and brainstorms. The folder of neatly numbered idea documents came to be called the *Book of Macintosh*. The new machine would have a built-in screen (256 pixels square), priced at $500, built around the slow but very cheap Motorola 6809 processor. The much faster 68000 chip in the Lisa cost 20 times as much, and would have driven the Mac's cost well north of $500, or even $1,000.

Raskin's first hire was obvious: the pragmatic and soft-spoken Brian Howard, his Bannister & Crun partner. The next team member arrived on a silver platter via Bill Atkinson, who'd been one of Raskin's students at UC San Diego. "I've got someone you ought to meet," Atkinson said.

It was the awkward, funny, blond, 23-year-old technician Burrell Smith, whom Apple had hired right out of high school as an Apple II repair tech. (He pronounced his name name "BURrell," not "burRELL.") Atkinson had been floored by Smith's technical chops. "This is Burrell," Atkinson said to Raskin. "He's the guy who's going to design your Macintosh for you."

Smith dove into the project with all guns blazing. Working straight through the Christmas 1979 break, rarely leaving Apple, Smith built a prototype Macintosh in three weeks. It was only a circuit board—an add-on card for an Apple II, in fact. But it was decked out to Raskin's specs: 6809 chip, 64 KB of memory, and circuitry to drive a tiny 7-inch black-and-white screen.

Soon, the Macintosh team was a quartet: Raskin, Smith, Howard, and Bud Tribble, another colleague from UC San Diego. Tribble was in his fifth year of a seven-year MD/PhD program at the University of Washington. Joining the Macintosh project meant taking a leave of absence and moving into a spare bedroom at Atkinson's house in Santa Clara.

Bud Tribble

Born: 1953, Ann Arbor, MI
Schooling: University of California San Diego; U. of Washington (MD/PhD)
Between Apple stints: medical residency, UCSF Mount Zion
Apple: 1980–1985, 2002–2021
After Apple: NeXT, Sun Microsystems

In the early 1980s, Tribble played keyboards in Moit Moit, a local band in the San Francisco New Wave scene.

In the fall of 1980, following a lecture at Xerox PARC, Raskin found himself in a heated discussion about the future of computers with a Polish American grad student named Joanna Hoffman. She was entranced by his ideas. "Having this assistant that is an extension of your mind was just such an attractive vision," she says. Hoffman became the fifth member of the team. Her title was "researcher."

Joanna Hoffman

Born: 1955, Warsaw, Poland
Schooling: MIT, University of Chicago
Apple: 1980–1985
After Apple: NeXT, Lucid, Frox, General Magic

The Mac team gave out a fake award each year to whichever employee did the best job of standing up to Jobs. Hoffman won it two years in a row.

The little team set up shop in the Good Earth offices on Stevens Creek Boulevard—Apple's first home in 1977, and home of the early Lisa project.

Raskin's passions dictated the dominant decor elements: beanbag chairs, model airplanes, radio-controlled cars, musical instruments, Nerf guns, and cardboard barriers against the Nerf balls. It looked "more like a day care center than an engineering lab," says Andy Hertzfeld, who joined Apple in August 1979 as a system-software programmer.

"They were all very quirky personalities," Hoffman says. Raskin "loved people to be thinking big thoughts." Burrell Smith was "like the true heir to Wozniak." Brian Howard was an "almost Buddhist presence who would jump in and help anybody in anything." Bud Tribble was so low-key, you didn't even realize he'd said something profound. Hoffman herself dove into keyboard design. How could it be used internationally? She designed a scheme of keyboard shortcuts that could produce accent and diacritical markings.

There were games every day, lunches together at Cicero's Pizza (and endless rounds of Defender on its game machine), and tremendous creative freedom; in true skunkworks fashion, everyone did everything. Several times, alarmed at the lack of concrete progress, Markkula and Scotty proposed canceling the whole thing; each time, Raskin assured them that something promising was "forthcoming."

When Raskin finally showed him something physical, "It was just a cardboard mockup,"

Burrell's 'Stache

Andy Hertzfeld's book *Revolution in the Valley* tells the story of the Mac's creation through vivid vignettes. And the story of Burrell Smith's mustache is among the best.

When Smith signed on with Apple in early 1979, he was a low-paid service technician. Jef Raskin scooped him up—but even after nine months as a key engineer on the Macintosh project, he was still paid his old salary. Smith requested a promotion to engineer, but nothing changed. He was frustrated. Why wasn't he getting that promotion?

"It obviously wasn't a matter of talent or technical skill, because he was already far more accomplished in that regard than most of the other hardware engineers," Hertzfeld writes. "And it wasn't a matter of working harder, because Burrell already worked harder and was more productive than most of the others.

"Finally, he noticed something that most of the other engineers had in common that he was lacking: They all had fairly prominent mustaches. And the engineering managers tended to have even bigger mustaches. Tom Whitney, the engineering VP, had the largest mustache of all."

Pleased with his observation, Smith began letting his own mustache grow. After a month, he decided that it was complete.

"And sure enough," Hertzfeld says. "That very afternoon, he was called into Tom Whitney's office and promoted to 'member of technical staff' as a full-fledged engineer."

Markkula says. It had a low-slung, horizontal case like the Apple II's, but with a built-in screen. The keyboard could fold into the front for easy carrying.

But Jef Raskin was about to lose his baby—and his job.

Pirates

When Steve Jobs lost the Lisa project in 1980, he felt insulted, betrayed, and unmoored. Here he was, with no product responsibilities, at the company that *he* had cofounded.

He loved the fundamental promise of the Macintosh: simple, small, beautiful. And he deeply believed in the *Book of Macintosh* credo: a closed system. A case you could open, parts you could plug into sockets, extraneous keys on the keyboard, thick manuals—forbidden. "Ten points if you can eliminate the power cord," Raskin wrote. "It is better to offer a variety of case colors than to have variable amounts of memory."

In late 1979, Jobs encouraged Raskin to make the Mac "insanely great" and not get bogged down on cost. "Don't worry about price, just specify the computer's abilities," he said.

Raskin was annoyed. For months, he'd been making tough choices to keep the price down: black-and-white instead of color, slow processor, cassette storage instead of floppy.

He fired off a sarcastic response to Jobs. Don't worry about the price? In that case, why don't we design a color printer that weighs "only a fraction of a pound, and never needs a ribbon or mechanical adjustment . . . and costs $50"? Why not include speech recognition and speech synthesis? Why not make it capable of generating music? Heck, maybe it could "simulate Caruso singing with the Mormon tabernacle choir, with variable reverberation!"

Jobs was not amused.

He began dropping by the Good Earth office with increasing frequency. He'd become unhappy with the Lisa's direction anyway—too corporate, too expensive—and thought that the Macintosh could be a "more affordable, more accessible, consumer version of Lisa."

In December 1980, he moved the little group to an upstairs set of four office suites about half a mile away, at Stevens Creek Boulevard and Saratoga-Sunnyvale Road. The furniture was secondhand, and the desks were old and funky—a far cry from the sleek Herman Miller cubicles in other Apple offices. The office, on the second floor of the building, looked out over a gas station. With dry engineer wit, they named their new home Texaco Towers.

In the real world, January 20, 1981, was Ronald Reagan's inauguration day; inside Apple, it was the day Jobs announced that he would take over the project completely. "I just decided that I was going to go off and do [it] myself with a small group—sort of go back to the garage, to design the Macintosh," Jobs said.

Raskin's Amanuensis

After Apple, Jef Raskin did get his chance to build the small, cheap, mouseless, non-expandable computer that he'd envisioned in the *Book of Macintosh*—at Canon. In 1987, it brought to market a $1,500 machine called the Canon Cat.

True to Raskin's vision, it had no mouse, icons, or graphics. But when it came to editing text, the Cat flew. When you held down a special key labeled USE FRONT, you could trigger the commands whose names appeared on the *fronts* of the keyboard keys.

Below the space bar were two new wide keys, each labeled LEAP, to be pressed with your thumbs. (Raskin believed that our thumbs are underused in the typing process.)

If you noticed a word high up on the screen that needed editing, for example, you didn't need a cursor, a mouse, or a Find command. While pressing a LEAP key, you could type a couple of letters of the word—and the Cat would highlight it in real time. At that point, you could move, copy, or edit the text without having to choose commands from a menu.

Maybe it was too esoteric for the masses, or maybe (as Raskin insisted) Canon didn't do enough to market the Cat. Either way, only 20,000 Cats ever wound up on human desks. Soon enough, as Raskin noted, the machine was "a dead Cat."

In the end, the Apple Macintosh wasn't anything like the Raskin Macintosh. And yet its spirit—a simple, small, easy-to-use computer for the rest of us—never wavered. So what does that make Raskin to the Macintosh?

According to him, he was its inventor. According to Andy Hertzfeld, he was more like its eccentric great-uncle. According to Bill Atkinson, Jef Raskin was the father of the Macintosh *project*, but Steve Jobs was the father of the Macintosh itself.

In March 1987, Apple identified Raskin's role in its own way by presenting him with a special gift: the one millionth Macintosh computer. It was a Mac Plus. It bore an engraved plaque beneath the Apple logo:

THE MILLIONTH MACINTOSH
Presented by APPLE COMPUTER INC. to JEF RASKIN
ORIGINATOR OF THE MACINTOSH

Right away, Jobs could see that Raskin's concepts would have to change. Raskin, for example, hated the mouse. He didn't like taking his hands off the keyboard. You shouldn't have to live, as he put it, a "hand to mouse existence."

Raskin also despised icons. "An icon is a symbol equally incomprehensible in all human languages," he would say.

On Raskin's Macintosh, the screen was minuscule—only 256 pixels square—and there was no disk drive at all. There was no desktop, no icons, no Trash—and no apps. You were always typing, either word processing or programming. To Jobs, none of this was as elegant or limitless as what the Lisa team was creating.

The processor was another problem. Jobs wanted a graphic interface like the Lisa's—mouse, icons, folders, windows—and that would require the Motorola 68000, like the Lisa's, and it would need twice as much memory—128 KB instead of 64.

At Jobs's request, Burrell Smith sequestered himself in the lab for a month, working straight through the Christmas–New Year's break for the second year in a row, to create a new logic board that incorporated the 68000. The resulting design used half as many chips as the Lisa, and ran at 8 MHz instead of 5. It would be twice as fast as the Lisa, for less than a quarter of the price.

On the other hand, the new chip blew Raskin's $500 price goal to smithereens. The Macintosh would be more likely to cost $1,500. "What the Macintosh metamorphosed into," Hoffman says, "was exactly what he [Jef] didn't want in a computer."

Raskin fell into a bitter funk. He fired off memos to Markkula, informing him that Jobs had gone "from being a major asset to a significant company liability."

But the conflict went beyond interface philosophy. "Jef and Steve were oil and water, fire and ice," says Chris Espinosa. "You know: Raskin, the PhD, and Jobs, the dropout. Raskin, the deep-in-the-details, systems approach, and Jobs, the play-it-by-ear. The clashes were hideous."

Jobs, for his part, considered Raskin "really pompous." Raskin's Macintosh would have been a "piece of junk."

The last straw was Raskin's brown-bag lunchtime talk about the Macintosh on February 17, 1981. That morning, Jobs told Raskin that the talk was canceled. But when Raskin dropped by the auditorium just to confirm, a hundred employees were there, waiting. Jobs had told only *Raskin* that the talk was off—and nobody else. Livid, Raskin fired off a four-page letter to Markkula, documenting Jobs's failings: He missed appointments, he yelled at people, he didn't give people credit, he minimized how long tasks would take.

The clash had reached a boiling point. One of them would have to go.

To Markkula and Scotty, the Macintosh project seemed like a low-profile side hustle—a perfect place for Jobs to indulge his passions. "They didn't take us very seriously," Jobs said. "I think Scotty was just sort of humoring me."

Markkula put Jobs in charge of the Macintosh project, and put Raskin on a leave of absence. Soon thereafter, Raskin left Apple for good.

Staffing Up

With Raskin out of his way, Jobs wasted no time. He set a ship date for the Macintosh—January 1982, a year away. He needed a team of superstars—a *small* team.

He started with some heroes of the Apple II project, including Rod Holt, Dan Kottke, Randy Wigginton, industrial designer Jerry Manock, and programmer and writer Chris Espinosa.

Joanna Hoffman had been working on the Macintosh interface—but as far as Jobs was concerned, the Raskin interface was garbage. "You know what you're going to do now? Marketing," he told her. She was taken aback; she had zero experience with marketing. At the first opportunity, she visited the library and checked out a book on the subject.

Jobs also brought over programmer Andy Hertzfeld, who'd been a grad student in computer science at Berkeley. In 1979, Hertzfeld had become so enthralled by the Apple II he'd bought ("$1,295 plus tax, most of my life savings at the time") that he quit school to join Apple.

Andy Hertzfeld
Born: Philadelphia, PA
Schooling: Brown, University of California, Berkeley
Apple: 1979–1984
After Apple: Radius, General Magic, Eazel, Google

As a grad student, Hertzfeld wrote a program that gave the Apple II lowercase letters. Apple nearly bought it—and then hired him instead.

But Black Wednesday, CEO Mike Scott's unexpected round of firings, so rattled Hertzfeld that he considered quitting. Hoping Hertzfeld would stay, Scotty offered him a spot on the Macintosh team. That very afternoon, as Hertzfeld was working on DOS 4.0 for the Apple II, Steve Jobs approached him with his usual icebreaker: "Are you any good?"

When Hertzfeld asked for a day or two to finish up his project, Jobs yanked his Apple II's power cord, deleting everything he'd been writing. "The Apple II will be dead in a few years," he said. "The Macintosh is the future of Apple, and you're going to start on it now!"

Jobs picked up Hertzfeld's Apple II and started walking away. "I had no choice: I had to follow my computer. I couldn't work without it," Hertzfeld says.

Jobs drove the computer—and Hertzfeld—over to Texaco Towers in his Mercedes. "Here's your desk. Welcome to the Mac team!" he said. Hertzfeld was surprised to discover someone else's stuff in the drawers: camera gear, model airplane parts.

It was Jef Raskin's desk. He hadn't even had time to clear it out.

The Apple Bicycle

In the March 1973 issue of *Scientific American*, Steve Jobs had learned of a Duke University study about efficiency of motion: how much energy each animal species consumes to travel one mile. The study pitted humans against such competitors as mice, lemmings, fruit flies, locusts, hummingbirds, sheep, horses, pigeons, and seagulls.

The winning critter, in Jobs's frequent retelling, was the condor. The human on foot wasn't even close. Even if you put a man in a car or a jet plane, he still didn't win. He was now traveling much faster, but using far more energy.

But if you put a man on a *bicycle*, everything changed. "With the aid of a bicycle," the article said, "the cyclist improves his efficiency rating to No. 1 among moving creatures and machines."

Jobs loved this principle. He quoted it for years. (The *Scientific American* article identified the salmon as the winner, not the condor, but people got the point.)

"That's what a computer is!" he'd say. "It's the most remarkable tool mankind has ever come up with, equivalent to a bicycle for our minds. In the same way a bicycle amplifies our physical ability, technology can amplify our intellectual potential."

Jobs was so fond of this notion, in fact, that he and Rod Holt decided to change the code name of the new computer taking shape in Texaco Towers. From now on, it would be known as "Bicycle."

The charm of naming something technical and intimidating after something familiar and attractive had certainly worked in naming the co*mpany*. But the Macintosh team pushed back hard. "We simply refused to use the new name," Hertzfeld says.

After a month or so, Holt gave up. "It's only a code name, anyway," he said.

In late 1982, Apple hired a marketing firm to come up with a final, real name for the new computer. They generated a long list of ideas, including Allegro and Apple 40. None of them seemed as good as Macintosh.

And so it was that in January 1983, at a company off-site meeting in Carmel, Jobs swung a bottle of champagne onto one of the prototype machines and declared: "I christen thee Macintosh!"

Reality Distortion

Jobs had an incredible eye for talent—and the charisma to attract it. And as he put together his crack Mac team in Texaco Towers, he had no trouble wooing whomever he wanted. He had, in the phrase that Bud Tribble borrowed from the "Menagerie" episode of *Star Trek*, a reality distortion field.

The reality distortion field, in Hertzfeld's view, was the combination of "a charismatic rhetorical style, an indomitable will, and an eagerness to bend any fact to fit the purpose at hand. If one line of argument failed to persuade, he would deftly switch to another. Sometimes, he would throw you off balance by suddenly adopting your position as his own, without acknowledging that he ever thought differently. Amazingly, the reality distortion field seemed to be effective even if you were acutely aware of it."

It was like a return to the thrill of the old Jobs garage. He was leading a scruffy, small band of geniuses to make a dent in the universe. "This is the best team in the world at what they do," he'd say. "And being a pirate means really going beyond what anyone thought possible, with a small band of people, doing some great work. Really some great work that will go down in history." In a bigger organization, there are more layers, "almost a guarantee that it won't be great."

Most of the engineers and designers were night owls, mostly under 30. Most had no obligations outside of Apple—no spouses, partners, kids, churches, sports—and put in insanely long hours to build their insanely great computer. In early 1983, *Time* magazine quoted Jobs as saying that the team was working "90 hours a week." To honor that slightly inflated estimate, the Apple finance team ordered up gray hoodies for the Macintosh team that said "90 Hours a Week and Loving It."

"We were on a mission," says Guy Kawasaki, who was hired in 1983 to persuade software companies to write for the Mac. "We wanted to increase people's creativity and productivity, we wanted to make the coolest computer in the history of man, and we wanted to prevent totalitarianism of IBM and Microsoft. I mean, how can you argue with any of those three?"

Jobs considered them artists. He took the whole team to San Francisco's Fine Arts Museum to view an exhibit of stained-glass masterpieces by Louis Comfort Tiffany. "Tiffany was an artist who could mass-produce his work, just like we will be," he told them. And he installed, in the building lobby, inspiring examples of consumerized art. An $80,000 Bösendorfer piano. Ansel Adams photo enlargements. A BMW motorcycle.

On February 10, 1982, Jobs called the team together for a little ceremony. There was cake and champagne, and Jerry Manock laid out a big sheet of white drafting paper. "You're creating something beautiful," Jobs announced. "This is art. And real artists sign their work."

One by one, Jobs invited each of the 35 team members to step forward and sign their name on

The Story of the Signatures

At the Mac case-signing party, 35 current "artists" signed. Later, Jobs added the signatures of a few important figures who'd left the project—Woz, Jef Raskin, and Bud Tribble—as well as a few who joined after the signing party. In all, 48 names appear. Some are familiar to Mac fans: Bill Atkinson, Andy Hertzfeld, Burrell Smith, Chris Espinosa, Joanna Hoffman, Rod Holt, George Crow (engineer), Bill Fernandez, Bruce Horn, Daniel Kottke, Larry Kenyon, Debi Coleman, Jerry Manock, Mike Murray, Randy Wigginton, and Steve Jobs. Some are less familiar (Peggy Alexio, secretary; Colette Askeland, board layout; Bill Bull, keyboard/mouse; Matt Carter, factory setup).

These signatures appeared inside every early Mac model: the original, the Mac 512K, the Mac Plus (1986), the Mac SE (1987), and the SE/30 (1989). In fact, the names might have survived forever if not for one small problem: "Every time we added more ports, we had to cut out somebody's name," says Espinosa. "It was just attrition by expansion. 'If we're gonna put the processor-direct slot expansion card in there—well, there goes Bud Tribble!'"

By the time the Macintosh Classic came along in 1990, there were no names left.

But just inside the main entrance of Apple's current headquarters—the one-mile circular building known as Apple Park—there's a small, beautifully appointed room where an original Mac's open case is on display. The signatures, illuminated by a ring light, are clearly visible—a reminder that real artists ship.

---

the drafting paper—Burrell Smith first, then the other engineers. Jobs added his own signature near the upper center.

Jobs had those signatures embossed into the case interior, so that they would be part of every Macintosh that rolled off the line. "Most customers would never see them," Andy Hertzfeld says, "but we would take pride in knowing that our names were in there."

Jobs fostered an us-versus-them spirit. He'd point out that IBM's research and development budget was at least a hundred times larger than theirs. "There's something that makes a job a little more fun to work on when the odds are against you," Hertzfeld says.

He also cultivated a rivalry with the Lisa team. In fact, in April 1980, Jobs bet Lisa project leader John Couch $5,000 that the Macintosh would ship first. The Lisa, at that point, had a two-year head start. It also had ten times the head count, and its managers were paid twice as much as the

Mac team's stars. Couch took the bet. At the very least, he figured that the wager would motivate *both* teams to meet their deadlines.

The plan worked—at first. Soon, however, "Steve's Macintosh team started taking the friendly competition much more seriously than my Lisa team," he says. "Suddenly, I watched as important resources got redirected from Lisa to Macintosh."

Jobs began to poach key Lisa team members, including Bruce Horn, who'd come from Xerox PARC, and Bill Atkinson.

The Lisa did indeed finally beat the Mac to market—by a year. Jobs, man of his word, threw a triumphant celebration party for the Lisa team, and presented Couch with an oversized check for $5,000.

Defining the Mac

When Jobs became captain of the Mac, few at Apple considered the project much of a threat. He fostered a rebellious pride in his little band. "It's better to be a pirate than join the Navy," he'd say.

But in the second year of development, when actual prototypes began to emerge, tensions arose. The Lisa group thought it was crazy for Apple to cultivate two different machines, with identical 68000 chips, that couldn't run each other's software. Meanwhile, the Apple II team had its own reasons to feel disgruntled. After all, their division was *funding* the other two!

In the spring of 1982, the Macintosh team was no longer quite so scrappy; it was now approaching 50 people, and Texaco Towers was about to split open. "My desk consisted of the *eraser tray of a whiteboard*," says marketing manager Mike Murray. "The whiteboard was already claimed by someone else, but I got the tray."

The Mac group moved back to the main Apple campus and took up residence in the less colorfully named Bandley 4 building.

By August 1983, with 80 people, they had to move again, this time across the street to Bandley 3. The walls were white, the floors were Mexican tile, and the cubicle partitions were low enough that you could see everyone from one spot in the room. The software team nicknamed their area the Fishbowl because its glass doors let anyone passing by look in.

Jobs approved the acquisitions of a Ping-Pong table, a free soda machine, and two arcade game machines for the lobby: Defender (donated by Andy Hertzfeld) and Joust (from Randy Wigginton). He bought a high-end stereo system and almost every CD available. He had Odwalla fruit juices delivered daily to the fridge in the Bandley kitchen.

The central Macintosh idea was to capture all of the Lisa's genius—its black-type-on-white screen, its icons and windows, its powerful 68000 processor—while bringing down its size, weight, complexity, and cost. Instead of a $10,000 monster for well-heeled institutions, it would be a friendly $1,000 appliance for the rest of us.

"I loved the idea and didn't view it as a competition, because we were targeting two different markets," Couch says. "It would be good strategy to leverage Lisa's application software, adapt it for Macintosh, and let users run the same applications at home on a Mac that they did at work on a Lisa."

But Jobs didn't *want* the Mac to run the Lisa's programs. His Macintosh would have its own operating system, and run its own flavor of programs. One reason may have been his bitterness about being removed from the Lisa team. But from a technical perspective, his options were limited. To bring the Mac down to one-tenth of the Lisa's price, he had already decided to go with a smaller screen (9 inches instead of 12), one floppy drive instead of two, and no hard drive.

The Lisa's 1 MB of memory was one of its most expensive components. That's laughably little RAM by today's standards—the Apple *Watch* comes with 1,000 times as much—but in the early eighties, a meg was a luxurious helping. It was 64 times what you'd get on the IBM PC.

The Pirate Flag

Just before the Macintosh team moved into Bandley 3, programmer Steve Capps, who'd come to the Mac team from the Lisa group, decided that the building needed some adornment to celebrate its new occupants. If the Macintosh team were pirates, then they should have a flag!

He bought some black cloth, sewed it into a three-by-five-foot flag, and asked Macintosh artist Susan Kare to paint a white skull and crossbones on it. As the cherry on top, she added an eye patch: a multicolored Apple logo.

Finally, at ten o'clock the night before the move into the new building, Capps climbed to the roof, with other team members standing guard. He found a thin metal pole to use as a flagpole and some rusty nails to attach the flag. When Apple returned to work the next morning, the pirate flag was proudly waving to greet them.

Jobs loved it. Markkula tolerated it. The Lisa team stole it.

A crew from the Mac team blitzed the Lisa offices to steal it back; Capps discovered one of the secretaries hiding the flag under her desk, yanked it away, and returned triumphantly to Bandley 3.

The flag flew proudly for another year, even after the Macintosh made its debut. Finally, one day in 1984, it disappeared for good.

If Jobs hoped to bring in the Mac under $1,000, a megabyte was out of the question. 128 KB was more like it—one-eighth as much. And with so little RAM, no computer could run the Lisa's advanced, multitasking, protected-memory operating system. The Mac would have to run something much less powerful.

The Case

In Jobs's mind, the everyperson's computer should be small and friendly. Cases took a long time to get tooled and fabricated. Getting the Mac's housing done was critical—and urgent.

Jobs asked Jerry Manock, designer of the Apple II and III cases, to design the Mac. Jobs kept saying he wanted something enduring and sleek, like a Volkswagen Beetle or a Ferrari. After a week of debate, they had a breakthrough: They'd rotate the layout of standard computers (and of the Lisa) 90 degrees. Instead of a horizontal box, with the screen next to the disk slot, the Macintosh would be vertical, with the screen *above* the slot. Vents near the bottom created a natural cooling convection channel. This design cut the Mac's footprint in half, making it feel less like a suitcase on your desk than a lamp.

Designer Terry Oyama drew up sketches and made a plaster model for the team to inspect. Most of the team loved it. They found it clean, charming, and nonthreatening.

But Jobs saw a hundred things that needed revision. Oyama produced another mockup every few weeks, responding to the tweaks and ideas. Most of them were subtle, but one of the revisions added the now-famous handle. At 16 pounds, the Mac wouldn't exactly be portable, but that was only a third as heavy as the Lisa. Lifting it briefly—to put it into its carrying case, or to change desks—was now plausible. To accommodate Jobs's obsession with seamlessness, Manock had to hide the main screws inside the handle.

Because office workers found technology threatening, Manock beveled all of the Mac's edges, softening them. He also knew—or hoped—that the computer would be sitting on CEOs' desks. Anyone who came to visit would see the *back* of the Mac; it had to look just as good as the front.

Manock's Macintosh pencil sketch.

The jacks on the back would be labeled with a row of simple icons instead of words so that the Macintosh was ready to go international. "To localize Mac, all you do is change the keyboard, manuals, and the disks," Jobs pointed out. "Nothing in the box has to change."

The finished product shone with attention to detail that customers might register only unconsciously. Manock put the on/off switch on the back, where it wouldn't catch the eye of a curious toddler. Yet, groping around the back of an electrified box could be intimidating. And so, at the back left edge of the case, Manock interrupted the textured plastic with a 2-inch smooth, shiny stretch, so you could feel for the switch without looking.

The finished design was revolutionary in its simplicity. The vertical orientation elevated the screen, putting it closer to eye level. A front-panel overhang created a sweet little garage for the keyboard.

Some aspects of the design reflected Jobs's own philosophies. It would have no cursor (arrow) keys, so customers would have to learn to use the mouse. And the Macintosh would have no parts that could be expanded or upgraded. "We don't have slots. Slots cost a lot of money. They make the box much bigger. And you need a much bigger power supply, because you never know who's going to plug in what," Jobs said.

Apple service technicians would need to open the Mac, but Jobs didn't want customers to have that option. So Manock designed the four case screws to require a technician's six-spoked Torx screwdriver. In fact, if a consumer did attempt to open the case, they'd void their warranty.

Finally, the design was done, and ready to send to the tooling company. In February 1982, the team held a champagne toast to celebrate the first completed piece of the Mac.

The front tipped backward for a better viewing angle.

Twiggy and Sony

Back in 1978, Woz's controller for the Disk II floppy drive blew away everything else on the market and made a fortune for Apple. No wonder, then, that two years later, Jobs wanted Apple to design its own drive for the Lisa, in hopes of leapfrogging existing floppy technologies once again. The new drive earned the code name Twiggy, after the rail-thin 1960s British supermodel.

Right off the bat, this drive would feed data to the Lisa twice as fast as other 5¼-inch disks. And because the motor could vary the speed of the disk, slowing the rotation to pack more data onto the outer edges, a Twiggy disk could store four times as much data as other disks.

In practice, though, the Twiggy drive immediately ran into trouble. A floppy drive works like a record player, with a read/write head (the "needle") hovering a few millionths of an inch above the surface of a spinning disk of magnetized Mylar (the "record"). But the Twiggy drive had *two* heads, at opposite ends of the disk, requiring two holes in the protective case of each floppy to permit the heads to access the disk. No matter how you grabbed the diskette to pull it out of the drive, you put your finger on the Mylar. Once the disk got contaminated by skin oil, it often failed when storing or reading data.

The promised speed didn't materialize, either, thanks to that business of slowing down the disk's rotation to pack in more data.

Troubleshooting the drive held up the Lisa for months, and it wasn't flawless even when the Lisa hit the market. Eventually, Apple stopped selling floppy-only Lisas altogether. Every Lisa came with a hard drive, adding $1,000 to the price, because the Twiggy system just wasn't reliable.

Jobs intended to incorporate the Twiggy drive into the Macintosh, too. In fact, Wendell Sander had even managed to improve on Woz's brilliant Disk II controller card. To save space and components, Sander consolidated the eight chips of the Woz design into a single chip, a patented circuit he called IWM: the Integrated Woz Machine.

But the Mac engineers were terrified by the Twiggy drive. There was no hard drive to serve as a plan B for the Mac. If the Twiggy failed, you'd have no way to store data at all.

In the spring of 1983, a white knight appeared: a new floppy-drive system made by Sony. Analog board designer George Crow had worked with it at his previous job at HP, and loved it.

In truth, the Sony floppy disk was neither floppy nor a disk. There *was* a spinning Mylar disk inside, but a rigid, square plastic case protected it. You could handle it, drop it, mail it, write on it, toss it to a friend. And the whole thing was only 3.5 inches square—a better fit for the space-saving Mac, and small enough for a shirt pocket. Best of all, even a single-sided Sony disk held 400 KB of data. That was 25 percent more than even a *double*-sided 5¼-inch disk.

Jobs loved the new disk immediately. There was only one hitch: He didn't want to buy the drives from Sony. He wanted to buy them from Alps, the company that had been making the Apple II drives, and which had licensed Sony's 3.5-inch drive technology.

It seemed like a risky bet; the Macintosh's ship date was only seven months away. But Jobs flew with Holt, Crow, and engineering manager Bob Belleville to Japan to meet with both Alps and Sony.

Alps had only a crude prototype to show them. Sony's drive was ready to go, but would add $50 to the cost of every Macintosh. Jobs decided to go with Alps, and told Belleville to drop his work with Sony.

Belleville ignored him. He was so pessimistic about Alps creating the drive on such a short timeline that he hammered out a deal with Sony as a backup—behind Jobs's back.

When Holt, Crow, and Belleville were in Japan, keeping Jobs in the dark wasn't difficult. It was trickier in the office, where Jobs might stop by at any time. The absurdity of the clandestine drive program reached a peak the day Sony sent one of its engineers to Cupertino to assist, a young man named Hidetoshi Kamoto. As he sat in a cubicle with software engineer Larry Kenyon and Crow, they suddenly heard Jobs approaching.

The presence of a young Japanese engineer would have blown their cover. In a panic, Crow pointed to a janitorial closet and ordered Kamoto to get inside. "Hide in this closet! Please! Now!"

The young man, baffled, did as he was told—and Jobs never saw him. Kamoto wound up standing in the dark closet, motionless and befuddled, for five minutes before Jobs moved on.

Crow and Kenyon rescued Kamoto and apologized. Kamoto wasn't offended—only bewildered. "American business practices, they are very strange," he said.

In the end, Alps conceded that it would need 18 months to produce a Sony-like drive—nearly a year later than Apple needed it.

It was time for Belleville to reveal to Jobs his secret: that his stealthy deal with Sony would deliver a working drive on schedule. His disobedience had saved the Macintosh. "You son of a bitch!" exclaimed Jobs, grinning.

The Sony drives did indeed become part of the Mac, as great as they'd seemed—and nobody boasted about them more proudly than Steve Jobs.

Disobeying Steve, Part II

Secretly engaging Sony to make the Mac's floppy drive was the first time the Mac team explicitly ignored Steve Jobs's orders—but it wouldn't be the last.

Jobs was emphatic that the Mac should not be expandable. It would be born with 128 KB of memory, and it would die with 128 KB of memory. But to Burrell Smith, that edict was nonsensical. Even in his short time at Apple, he'd seen the standard computer's memory allotment expand. The Apple II came with 4 KB; the IBM PC had 16; the Macintosh would come with 128—and that, obviously, was not the end point. As the Mac's ship date kept slipping, Smith foresaw a nightmare scenario where it was outdated before it even hit the stores.

With a simple rewiring of the circuit board, Burrell could make it possible for tinkerers to snip one trace, yank out the original chips, and replace them with the quadruple-capacity ones—and they'd have themselves a 512 KB Mac. It would be capable of running more programs with less waiting.

Smith never did tell Jobs about his treachery. The Mac shipped with his sneaky expansion option, and technically proficient Mac owners all over the world were the beneficiaries.

Invisible Aesthetics

By the spring of 1981, the Mac's design was stable enough that Apple could order printed circuit boards. At one of the weekly team meetings, Burrell Smith laid out an enlarged blueprint of the board. Jobs studied it—and his critiques had nothing to do with the engineering.

"Look at the memory chips! That's ugly. The lines are too close together!"

Board designer George Crow was baffled. "Who cares what the PC [printed circuit] board looks like?" he said. "Nobody is going to see the PC board!"

"I want it to be as beautiful as possible, even if it's inside the box," Jobs replied.

He quoted his father, whose craftsmanship in his carpentry work had left a deep impression on him: "A great carpenter isn't going to use lousy wood for the back of a cabinet, even though nobody's going to see it."

Jobs ordered a redesign, at a cost of $5,000.

Ultimately, the better-looking layout didn't work as well electrically; Jobs reluctantly accepted the redesign as a failed experiment. But that principle—that the hidden guts of something should be just as beautiful as the parts you can see—became an Apple hallmark.

RoundRects

As 1981 wore on, the Macintosh team was working late into the night to write the Mac's software. At its heart was Bill Atkinson's masterpiece: a set of image-drawing algorithms called QuickDraw.

The Mac's black-and-white screen was a 512 x 342 grid of tiny square pixels, 72 of them per inch. QuickDraw's programming commands could form these dots into graphics—shapes, patterns, lettering, pictures—with maximum speed and minimum memory.

In May, Atkinson added the ability to draw ovals and circles. But Jobs wasn't impressed. "How about drawing rectangles with rounded corners? Can we do that now, too?" Jobs asked.

Atkinson was dubious. When would anyone need round-cornered rectangles? Did a shape like that really belong in the pantheon of basic lines, squares, and circles?

Jobs was appalled. "Rectangles with rounded corners are *everywhere*!" he said. "Just look around this room!" He pointed to the whiteboard; the desks; the tables, the computer screens. He even pulled Atkinson out of the building. "And he showed me rounded corner after rounded corner after rounded corner," Atkinson said. "Finally, when he got to a 'No Parking' sign, I said, 'OK! I'll figure it out!' "

Rounded rectangles—or, as the algorithm is called in programming, RoundRects—became an Apple hardware- and software-design hallmark that lives on to this day, in the shape of every laptop, phone, tablet, and watch.

> ### Atkinson's Accident
>
> One of QuickDraw's finest achievements was its ability to display what's inside of several overlapping windows—a trick that relied on compact representations of screen areas that Atkinson called *regions*. "It was really, at that point, Apple's jewel," says Atkinson.
>
> But one day in April 1982, Atkinson was driving to work in his Mazda RX-7 sports car. "A big truck with a tail was sticking out into the road, and somehow I drove under it," he says. "I sheared off the top of the RX-7, and I got bonked in the head. My head was pushed down into the passenger wheel well." Worse, his knee wound up on the accelerator, speeding the car down Los Robles Way and across Los Gatos Boulevard. After mowing down a row of hedges, the car finally came to a stop.
>
> When the police and ambulance arrived, they were horrified by the crushed-looking Mazda; the first officer on the scene radioed in a *decapitation* before discovering that Atkinson's body was actually mashed into the two wheel wells: legs in the driver's side, head in the passenger side.
>
> An ambulance took Atkinson to Los Gatos Community Hospital. Eventually, he regained consciousness, but—much like Steve Wozniak the previous year—he developed amnesia. "So I'm inside this brain, and I'm feeling retrograde amnesia come on," he says. "You know the Sherwin-Williams logo of paint pouring over the globe? Yeah, I felt like a wave of sadness pouring over my head. And I was feeling, 'Oh God, I fucked up my brain!'".
>
> When word of the accident reached Jobs, he raced to the hospital. He ran over to Atkinson's bedside. "Are you okay?" he asked. "We were worried about you!"
>
> "Don't worry, Steve," Atkinson said groggily. "I still remember how regions work."

10. Sculley

Mike Markkula's name was never familiar to the public, and that's just the way he liked it. Even when he wrote some early popular apps for the Apple II—Color Math and a checkbook program—he released them under a pseudonym: Johnny Appleseed.

Now, as Apple's acting CEO, his first act would be seeking a permanent one.

The board interviewed one candidate after another. Steve Jobs was too young and inexperienced. Floyd Kvamme, VP of sales and marketing, didn't seem like the kind of broad-strategy guy the board wanted.

Finally, Apple found the perfect candidate. A man with experience in the industry. A leader with a national profile. An innovator with drive.

His name was Don Estridge, the head of IBM's PC division. He turned Apple down.

Markkula brought in a headhunter: Gerry Roche, of the executive-placement firm Heidrick & Struggles. Most of the candidates he approached declined to leave their stable, well-funded jobs to gamble on a scrappy outfit like Apple.

John Sculley's preppy attire was at the polar opposite end of Apple's dress code.

As the options narrowed, Roche encouraged Markkula and Jobs to give serious consideration to one man in particular: the president of Pepsi-Cola, the soft-drink division of the beverage-and-snacks giant PepsiCo. His name was John Sculley.

Culturally, he was a strange choice. He was an East Coast establishment type—introverted, shy, self-described as scrawny and pale—who'd never stepped foot in Silicon Valley and had zero experience in technology. He'd attended Brown University and earned his MBA at Penn's Wharton School. His mother was a Manhattan socialite, his father was a Wall Street lawyer, and his first wife was the Pepsi CEO's daughter. Sculley's art collection specialized in eighteenth-century English landscapes and portraits.

But the corporate world saw Sculley as a marketing genius. When he joined Pepsi in the late sixties as a young marketing executive, Pepsi's market share had been dropping for four years in a row; Coke was number one. Sculley helped to engineer a turnaround. He'd launched a hugely successful "Pepsi Generation" TV ad campaign that depicted young, hip people drinking Pepsi.

Above all, Sculley had recognized the potential of a local Pepsi TV ad in Texas, where real people took blind taste tests and, more often than not, preferred the taste of Pepsi. "I can't believe it—I've drank Coke all my life," a grandmother says. Sculley took it national, calling it the Pepsi Challenge.

(When Sculley himself participated in a Pepsi Challenge, at Florida's Daytona 500 car race, he inadvertently chose Coke as his preference. Fortunately, it wasn't caught on camera. "Pepsi people were terrified that someone would find out," he says.)

What Sculley offered Apple, in other words, was a track record of expert marketing, international experience, and a natural appeal to Wall Street. "I thought: 'Good, this guy understands marketing,'" Markkula says. "That gives me an opportunity to retire again and not have to worry about this thing. So, hell, I was just hoping he would be hilariously successful!"

The wooing of Sculley began in New York, when Jobs invited him to put his hand on a mouse for the first time. He drew some shapes and printed them on the spot. He was amazed.

Still, taking the Apple job would mean parting with Sculley's longtime boss at Pepsi. It would be a huge gamble on a company that had had only one hit product. And it would mean persuading his new wife to uproot from their mansion in Greenwich, Connecticut.

There would be an enormous culture shift, too. When Sculley visited Apple, he was taken aback by the narrow hallways, which "would have violated most fire codes," and the employees' scruffy outfits, "less formally dressed than PepsiCo's maintenance staff."

But Sculley was dazzled by a Macintosh demo—which included a Hertzfeld program that bounced animated Pepsi cans and bottle caps around the screen—and by the Mac's creators. They were, he said, "universally young, passionate, idealistic, and brilliant."

After weeks of working on Sculley, Jobs flew to New York one more time, his reality distortion field set to high. "He didn't care that I didn't have a technical background," Sculley says. "He didn't have a technical background, either!"

Jobs and Sculley spent the entire day walking the city. They crossed Central Park, talking about nature; perused the Metropolitan Museum of Art, talking about design; and visited Colony Records, talking about music. "You're the best person I've ever met," Jobs said, according to Sculley. "I know you're perfect for Apple, and Apple deserves the best."

As the sun set, Jobs took Sculley to the thirtieth-story penthouse he was thinking of buying in the San Remo apartment building. And there, on a terrace overlooking the Hudson River and Central Park, Jobs made the pitch to Sculley that would become famous:

"Do you want to spend the rest of your life selling sugared water, or do you want a chance to change the world?"

It was, Sculley says, a "monstrous" question. "It simply knocked the wind out of me."

That pitch, and a formal offer from Markkula, sealed the deal. Sculley would get a $1 million salary, a $1 million sign-on bonus, 350,000 shares of stock, and $1 million if it didn't work out.

When Sculley arrived in Cupertino in May 1983, there was jubilation. "It was like a miracle for everyone, because people saw him as adult supervision," says Andy Cunningham, who was working for Regis McKenna to help launch the Mac.

There was also a cultural adjustment. Sculley was 44 years old; the average age at Apple was 27. After serving in the lily-white halls of PepsiCo, he now found himself surrounded by "people from every nationality and race, varying from Indians in turbans to scruffy, bearded kids from New Jersey." To his amazement, there were as many women as men in management positions. At Pepsi, conversations incorporated terms like "win," "competition," and "share points." At Apple, it was "building things," "vision," and "values." At Pepsi, the offices were hung with expensive, track-lit works of art. At Apple, the art was on T-shirts and posters.

But Sculley had barely settled into his clean, modern office when he confronted his first business crisis: a computer-industry collapse.

Only 3 percent of American homes had computers, but the market was flooded with machines from Commodore, Atari, Texas Instruments, HP, Apple, and others. The torrent triggered a price war that nobody could win. By the end of 1983, the industry had lost $1.2 billion. Apple's fourth-quarter profits fell 80 percent, and its stock fell from $63 a share to $23. Sculley was terrified. Even as the industry was crumpling, Apple was pouring money into three massive endeavors: a multimillion-dollar ad campaign for the just-released Lisa; the Mac, which would need marketing money soon; and Apple's new, state-of-the-art, automated factory in nearby Fremont.

Meanwhile, the company was hiring 250 people a month. Sculley slammed the brakes on hiring, and shut down the company's popular profit-sharing program.

The Plant in Fremont

Jobs had always been a fan of Henry Ford, who mastered mass-producing cars. And he'd been blown away by the spotless, zero-defect, automated assembly lines he'd seen in Japan. He wanted that kind of fabrication magic for his own company—so he built a plant in Fremont, 45 minutes from Cupertino.

It had over 120,000 square feet of floor space. Macs suspended on tracks, controlled by other Macs, moved from station to station, where workers installed the next components. As designed, the factory was capable of sending out one new Macintosh every 27 seconds. Machines did 90 percent of the assembly.

Jobs ordered the factory's walls painted bright white, and the machines painted bright blue, yellow, and red. It looked like "an Alexander Calder showcase," said plant manager Debi Coleman.

Jobs was constantly unhappy with the cleanliness of the place. "I'd go out to the factory, and I'd put on a white glove to check for dust. I'd find it everywhere—on machines, on the tops of the racks, on the floor." He'd ask Coleman to get it cleaned. "We should be able to eat off the floor of the factory," he'd say. She was baffled by the request.

By the time it was finally finished, the Fremont factory truly was a showpiece—at a cost of $20 million. In the end, it ran for only six years. By that point, director of operations Del Yocam had set up factories in Dallas, Ireland, and Singapore, with lower labor and operating costs—and much more traditional designs.

Pepsi's CEO called to ask if he wanted to come back. *BusinessWeek* declared IBM the winner of the computer wars. Pundits pressured Sculley to make Apple's computers compatible with IBM PCs. But one aspect of the new job made him feel confident and admired: Steve Jobs.

"We discovered incredible similarities in the way we thought and in the respect we had for ideas," Sculley says. "He would call me five or six times during the day. He'd think nothing of coming in and saying, 'Sorry to interrupt you, but I've just got to tell you what's on my mind. You're the only one who will understand.'"

Jobs and Sculley were the corporate power couple.

They took hours-long walks. "Steve did not like working in an office," Sculley says, "so when he had something that he really wanted to talk about, he liked to do it by walking around." They'd walk to the Stanford campus, or out and around Cupertino.

They'd meet on Sunday mornings for breakfast, or after work for pizza or sushi. "I had a terrible marriage at the time, so I spent most of my time with Steve and Apple," Sculley says. Sculley declared Jobs to be his "soulmate and near-constant companion."

Sculley was 16 years older than Jobs, but marveled at all he was learning from him: about art, music, technology, food, design, beauty. And yet Sculley also considered the ways that he could mentor Jobs. "I could give him something he couldn't get anywhere else," Sculley says. "I became a teacher to him."

In particular, Sculley hoped to teach Jobs to rein in his explosive temper.

Volatility

For 23 years of Apple's existence, Steve Jobs was its driving force. And anyone who worked with him will tell you that "driven" and "force" described him very well.

His penchant for rule-breaking has become famous—he routinely parked in handicap parking spaces, for example, and refused to put license plates on his Mercedes. But he became best known for his volatility.

He was quick to laughter, and quick to tears. He could be exhilarated by a development, or enraged. He'd put you on a pedestal, or he'd rip you to shreds.

Apple veterans have developed two different belief systems about his temperament.

The first philosophy holds that Jobs was simply capable of cruelty. "He was truly great," says Andy Hertzfeld, but he also had "a real mean streak that wasn't necessary. He could really reduce someone to tears."

In 1978, representatives from a potential peripherals partner flew to California for a meeting with Jobs. He examined their compact thermal printer for a few seconds, declared, "This is a piece of shit," and walked out.

In 1985, when Sculley opened discussions for Apple to acquire Xerox, Jobs torpedoed the meeting by telling its top executives, "You guys don't have any idea of what you're doing."

In 1998, when iMac team leader Doug Satzger—at Jobs's request—filled a conference room with swatches of alternative colors for the iMac, Jobs took one look and snapped, "You guys suck. Get the fuck out of the room."

He was equally difficult to please when it came to restaurants he visited, hotels he booked, and craftsmen he hired for his home.

At Apple, some people responded by quitting. Some stayed, but declined to demo their work for Jobs. Some describe a form of PTSD even years later. And some learned techniques to minimize the likelihood of getting eviscerated.

"Early on, it became clear that when you went to a meeting, whoever sat to the immediate right of Steve got first attention. He would come in and say, 'Now, let's go around the room,' and he always started to his right, the first chair—and he rarely got *beyond* the first chair," recalls one former executive. "For 45 minutes, it's you under the heat lamp. 'Let's talk about *that*. Let's talk about *that*. Why do you think that? Why don't you think this? Does this person know what they're talking about?'"

When presenting design options to Steve, you didn't use slides; Steve hated slides. You also learned to present exactly three ideas. If you came in with one idea, he'd hate it. If you came in with two ideas, he'd come up with a third one. So you came in with three ideas, with the one you liked most presented last. Even then, says software engineer Ken Kocienda, "you could never know how he would react. Sometimes he'd say he loved or hated something but then reverse himself in midsentence."

Chief financial officer Fred Anderson learned not to challenge Jobs in front of others. "Whenever there was something significant I had to say where I didn't agree, we'd go behind closed doors, one-on-one, and talk about it."

"And he would bait you," says iMovie and iPhoto author Glenn Reid. He'd ask someone what they thought of some technology. "And whatever they said, like, 'Oh, I think it's amazing, and blah blah blah,' he'd say, 'Well *I* think it sucks.' If you did a 180 and like, 'Yeah, it really does,' you were immediately thrown out of the room. He just wants to know that you have thoughts of your own, and if you could back up what your opinion was."

Jobs's crew often found, to their surprise, that standing up to him earned his respect thereafter, as though his criticism were a test to see if you could back up your position.

"He would go away, and he would think in terms of your arguments," says Rod Holt. "At the next executive staff meeting, he would have your position better than you had your position. That was his way of operating."

Often, Jobs not only adopted your viewpoint, but claimed credit for it. "You've heard this standard joke, the evolution of an idea?" says Steve Capps. "The first time you show him something: 'That sucks! What an idiot you are!' And then the second time, it's like, 'Oh, keep working on it.' Third time . . . 'Hey, did you see what I came up with?'"

Could Jobs have achieved all that he did without the mean streak? "Yes, I think so," Hertzfeld says. "The meanness was more of a bug than a feature. And a lot of it comes from insecurity, from his childhood."

There's a second philosophy, though, that maintains that Jobs's tirades were in the service of excellence. When Jobs judged someone's work and announced, "This is shit," what he was really saying, according to Bill Fernandez, was "I have a dream, and this isn't it."

"He's not saying that you're a terrible person. Now, that's the way it comes across—that you're a failure. That's because he didn't know how to express himself in a socially acceptable and effective way," Fernandez says.

Frequently, the provocation worked: Dozens of ex-Apple people describe attaining engineering or design achievements they'd never thought they could.

Getting better work out of his team is certainly how *Jobs* justified his leadership style. "A lot of times, people don't do great things because great things aren't really expected of them, and because nobody ever really demands that they try," he'd say. "If you set that up, people will, a lot of times, do things that are greater than they ever thought they could do."

Guy Kawasaki emphasizes the difference between "ego-driven assholes"—all too common in Silicon Valley—and "*mission*-driven assholes" like Jobs. "Having been in Silicon Valley for about 40 or 50 years, I'm an expert in assholes, okay?" he says. "Steve is one of the very rare mission-driven assholes. He was driven by a mission to make the greatest computer, by the greatest company. And if you got in the way of that, he would run you over. Run you over, back up, and run you over again."

Jobs didn't just drive others; he was himself driven. He was eternally impatient, and couldn't stand it when other people didn't appreciate the urgency of the mission. "As hard as he was on

Jobs and Journalists

Steve Jobs was always sensitive to his depiction in the press. If you wrote something he didn't like, he had no hesitation about calling you to complain.

In 2007, Apple introduced a new version of iMovie, its easy-to-use video-editing program. The design was simpler and the code was cleaner—but it had far fewer features.

I was the technology columnist for the *New York Times*, and also an avid iMovie fan, so I went public with my unhappiness. "iMovie '08 is an utter bafflement," I wrote. "It's nothing like its predecessor and contains none of the same code or design. It's designed for an utterly different task: throwing together movies quickly. It lets you scan through a clip to see what's in it, isolate the good parts, rapidly drop them into a sequence, and send a completed video to YouTube with one menu command."

But there were so many limitations! "The new iMovie gets a D for audio editing. You can no longer export only part of a movie. All visual effects are gone—even basic options like slow motion, reverse motion, fast motion, and black-and-white." There wasn't even a standard scrolling timeline of your clips. "Call it FlyMovie, or ByeMovie, or WhyMovie," my column concluded. "But one thing's for sure: it sure isn't iMovie."

That night, my phone rang. Steve came on the line, furious. "You have no idea what the fuck we do here at Apple, do you?" he said icily.

He informed me that my use of iMovie to build lovely short films, with background music and crossfades, was an outdated relic. "We have data. We know how people are using iMovie. And they're not editing home movies they shot with their camcorders. They just want to whip together some clips and post them on YouTube."

"But, Steve," I complained. "I have a hundred tapes I was planning to edit into little movies—and now I can't do that anymore!"

"Have you edited any of them?" It seemed like he was summoning all his patience, like he was talking to a child.

"Well, no. I'm super busy these days."

"I've got news for you," he said. "You'll *never* have the time. They're going to sit in your drawer. You're *never* going to edit those videos."

I thought he had a lot of nerve telling me what I was going to do with my videotapes. "I guess we'll see, won't we?" I said.

Ten years later, I paid a company to digitize all those tapes, so that they'd be sitting on a hard drive, ready to edit when I had the time.

I still haven't edited them. So far, all I've done with them is find a cute clip or two of the kids, whip together a quick montage, and post them on YouTube.

everybody, he was much harder on himself than anybody else," Joanna Hoffman says. "It was very hard for him to see that not everyone could meet his stamina and resilience."

Then there's the question of *which* Jobs you're assessing: Steve 1.0 or Steve 2.0.

Most tales of Jobs's blistering temper date back to his early years at Apple. By age 22, after all, he was running a rapidly growing global business—without a degree, without a learning period, with no experience managing people.

"My belief is that he wanted interaction, but he was too young to really know how to ask for it," concludes Jerry Manock.

But in 1985, Jobs left Apple for 11 years. He started NeXT and bought the computer graphics division of Lucasfilm, which he named Pixar.

"Over the years, he learned better how to deal with a wide variety of people in a wide variety of ways," says Fernandez. "He started out basically incompetent; he ended up hugely competent." He'd become, that is, Steve 2.0.

Many of Jobs's lieutenants, meanwhile, *never* witnessed Jobs the tyrant. "That definitely wasn't my experience," says font and icon designer Susan Kare. "He was sometimes dramatic. And there's stories about him saying something inappropriate in an interview, like waltzing in and waltzing out. It's not all just made-up. But there was also so much positivity and encouragement."

Woz, Andy Hertzfeld, Rod Holt, iPhone/iPad software lead Scott Forstall, design star Jony Ive, and Apple Store creator Ron Johnson were also immune. "There were people like myself who always had a white hat. He would never, never get negative about those certain people that were important," Woz says.

Many of the people who worked with Jobs most closely regret the emphasis that history seems to put on his temper. As Hertzfeld puts it: "He could berate you, but also inspire you, sometimes even in the same sentence. Almost any adjective you could think of could apply to him at different times. He was a man of contradictions."

11. Software

Every six months, Jobs took the group to some gorgeous California resort within a couple hours' chartered bus ride from Cupertino for a two-day meeting. Each off-site was a combination of rally, party, and Q&A break. Team leaders gave updates; guest speakers gave presentations; and Jobs offered what he called "Quotations from Chairman Jobs." They included such classics as:

- "It's Not Done Until It Ships"
- "Don't Compromise!"
- "The Journey Is the Reward"
- "Mac in a Book by 1986"
- "It's Better to Miss than Turn Out the Wrong Thing"
- "Real Artists Ship"

That final quotation, which Jobs wrote on the whiteboard at the January 1983 off-site, was becoming especially relevant. The Mac's ship date had been postponed twice; it was now scheduled for the annual Apple shareholders' meeting on January 24, 1984. Finishing the Mac's ambitious software in time would be a colossal task.

Jobs needed more engineering power. He liked them young, tireless, idealistic, brilliant, and, preferably, artistic.

"Pretty much everybody in the whole team had an artistic side, independent of Apple," says Steve Capps, who had come to the Lisa project from Xerox PARC in September 1981. "*He's* into photography; *he's* into music." Apple, Jobs often said, sits at the intersection of liberal arts and

Steve Capps
Born: 1955, Fort Wayne, IN
Schooling: Rochester Institute of Technology
Before Apple: Xerox
Apple: 1981–1985, 1987–1996
After Apple: Microsoft, PayNearMe

Capps has been continuously paid to program since 1969, when he was hired to automate his high school library.

technology. "He could go out and hire the best software coder in the world, but he'd rather hire a great software coder who also did stand-up comedy, wrote his own music, baked cakes, did calligraphy," says mobile ads VP Andy Miller. Only an engineer like that "could inject everything he's doing with some kind of culture, and design, and art."

In January 1983, Jobs poached Capps from the Lisa division to join the Mac team. "I was fresh, and everybody else was pretty fried," he says, "so it worked out pretty well."

Capps wound up helping to complete the Finder, the home-base program that Bruce Horn had begun. This was the Mac's starting point, the desktop metaphor that captivated first-timers.

As a graphic computer, the Mac would need graphics. So in late 1982, Andy Hertzfeld called up an old artist friend from his high school in suburban Philadelphia. She'd studied typography, she'd gotten her master's and PhD in fine art—and, as the cherry on top, she now lived in San Francisco. Her name was Susan Kare.

Susan Kare
Born: 1954, Ithaca, NY
Schooling: Mount Holyoke (BA), New York University (MA, PhD)
Before Apple: Excalibur Bronze Casting Foundry, Fine Arts Museums of San Francisco
Apple: 1982–1986
After Apple: NeXT, General Magic, Pinterest, freelance (Microsoft, IBM, Facebook, Google)

Kare's work for Microsoft included the deck of cards in Windows Solitaire.

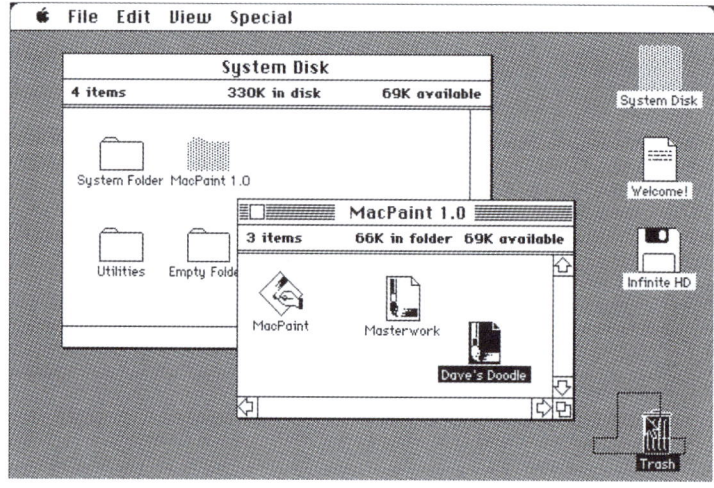

The original Finder—a home base where you could find and manage all of your documents—became the basis of every computer on earth.

Capps's Alice

When Jobs recruited Steve Capps from the Lisa team, it wasn't Capps's Lisa printer software that caught his eye. It was the game Capps had written on the side, called Alice, a 3D chess game where your job, as Alice, is to use legal moves to avoid the other players landing on you.

Capps had adapted Alice to run on the Mac, and the programmers in the Fishbowl were soon playing for hours at a time. "I want to nab you," Jobs told Capps. "Because I love Alice. We're gonna bundle Alice with every Mac!"

And sure enough, a few months later, standing at neighboring urinals in the men's room, Jobs offered Capps $100,000 for the rights to sell his game as an Apple product. "That was real money back then," Capps says.

Alice became the first Apple-branded Mac game ever sold. In fact, to this day, it's the *only* Apple-branded game ever sold.

Hertzfeld offered her an Apple II in exchange for drawing a few icons and fonts for the Mac. Despite her extremely limited canvas—32 pixels square—she managed to create charming, whimsical images. Soon, Jobs hired her. Apple's team of artists now had an *artist* artist.

At first, she drew the Mac's icons on sheets of graph paper, filling in the squares to represent dark pixels. Eventually, Hertzfeld wrote a program that made it easier to design icons at magnified scale; when reproduced on the screen at finished size, they looked crisp and clean.

Kare's first icons for the Mac included a happy Mac (which meant that the machine was starting up successfully) and a bomb (which meant that the machine had crashed).

City Fonts

Susan Kare designed not only the Mac's early icons, but also its first nine fonts—pixel by pixel, character by character. (Bill Atkinson contributed a tenth, a calligraphic font called Venice.) Her challenge was to build character shapes with graceful curves, descenders, and curlicues using only blocky, square, black-and-white pixels. As Larry Tesler put it, "Susan didn't know that making those fonts for the Macintosh was theoretically impossible, so she just went ahead and made them."

She and Hertzfeld named these fonts after stops on the commuter train in Philly, where they'd both grown up: Rosemont, Ardmore, Overbrook, and so on.

Jobs had no objection to using city names—but thought that her world-class fonts should be named for world-class cities. Kare obligingly renamed her typefaces after bigger cities—and added a mnemonic twist to each one. There was New York (a takeoff of Times, as in the newspaper); Geneva (a typeface reminiscent of Helvetica—a Swiss connection); Monaco (a *mono*spaced font); London (an Old English font); Cairo (a picture font, like Egyptian hieroglyphics); and so on.

The boldest font, the one that appeared in menus on the Mac (and, later, on the iPod), she'd named Elefont. Under the new scheme, that punny name was renamed Chicago.

And the font she'd made called Ransom, in which each character looked like it had been clipped from a headline in a different font—like a ransom note—became San Francisco.

But somewhere, on a Susan Kare floppy disk, Elefont lumbers on.

128K

To keep the price low, the Mac would have only 128 KB of RAM—already tight in 1984, and *impossibly* little by today's standards. (A single modern web page might require a hundred times that much memory just to appear on your screen.) Worse, about a third of the total was eaten up by the Mac's own operating software, leaving only about 85 KB for whatever program you were running. From the day it was conceived, the Mac was *gasping* for memory. And the less memory a computer has, the slower and less pleasant it is to use.

That's where Andy Hertzfeld came in. "The center of my contribution to the Mac was making things fly in the limited memory," he says.

His unavoidable programming contortions, however, would come back to haunt the Mac. "To get the low memory footprint, you had to do these wild side tricks that maybe weren't the most sustainable. If you thought of it as a platform that was going to evolve over ten years, you should make it as clean as you can, not as tight as you can."

One day in late 1981, Bud Tribble had an inspiration. The Macintosh didn't have enough memory

Mr. Macintosh

One night in February 1982, Jobs burst into Texaco Towers. "We've got to have Mr. Macintosh!" he announced.

"Who is Mr. Macintosh?" asked Hertzfeld.

"Mr. Macintosh is a mysterious little man who lives inside each Macintosh," Jobs said. "One out of every thousand or two thousand times that you pull down a menu, instead of the normal commands, you'll get Mr. Macintosh, leaning against the wall of the menu. He'll wave at you, and then quickly disappear. You'll try to get him to come back, but you won't be able to." He indicated that they could plant references to the legend of Mr. Macintosh in the Mac's manuals, sealing his mythology.

It would be a brilliant Easter egg—a hidden feature planted by programmers in their work to surprise and delight customers. Hertzfeld, charmed, promised to incorporate Mr. Macintosh into the Mac's code.

Shortly thereafter, Jobs met Jean-Michel Folon, a Belgian artist famous for his simple, playful style, especially his line drawings of top-hatted men. On a visit to Apple, Folon had fallen in love with MacPaint, the Mac's simple art program.

Jobs hired Folon to create some original art for the Macintosh marketing materials. Nothing much came of the collaboration, but Folon did draw Mr. Macintosh: a little top-hatted man wearing a trench coat (or possibly a raincoat—a Mackintosh).

In the end, Mr. Macintosh didn't make the cut. His image would occupy too much space in the already-crammed Macintosh ROM.

Only one trace of Mr. Macintosh still exists: On the very first prototype printed digital circuit board for the Mac, he appears next to the copyright notice.

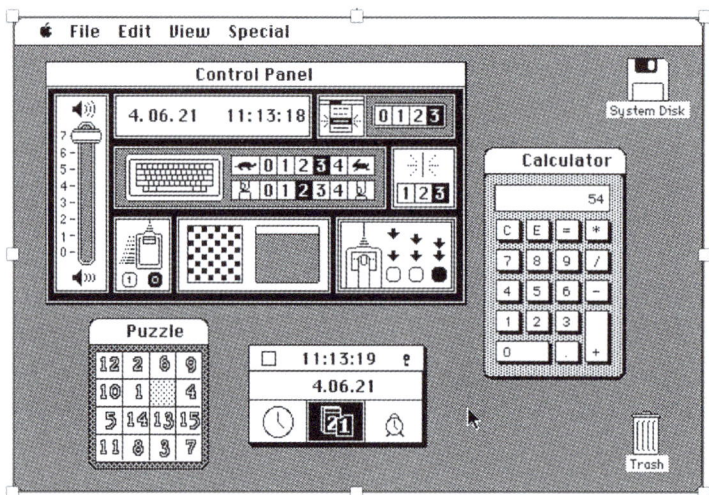

Desk accessories weren't exactly multitasking, but they let you perform simple functions without quitting your main program.

to keep two programs open at once. You couldn't pop into one program to make a chart and then paste the chart into your word processor, as you could on the Lisa. But what if there were *tiny* apps—useful one-trick ponies he called desk ornaments—that were so small, they could run simultaneously with the big ones?

Hertzfeld came up with Clock, Calculator, and Scrapbook—a storage app for bits of text or graphics you'd copied. Programmer Donn Denman contributed Notepad and Alarm Clock. Steve Capps wrote Key Caps, a keyboard map showing which symbols you'd get when you pressed which keys—an essential feature for typing diacritical marks for other languages. For example, if you were pressing the Option key while striking E, you'd get é, as in "résumé." (And no, Capps didn't name Key Caps after himself. "I mean, if I was gonna name it myself, I would have put two *p*'s in it," he notes.)

Finally, there was Control Panel: a purely visual app that, today, we'd call Settings or Preferences.

In the fall of 1982, the Apple publications team proposed renaming these handy little apps "desk accessories," maintaining that "ornaments" felt too insubstantial and frilly. DAs became a key part of the Mac for 17 years.

MacPaint

Once the Lisa was unveiled in January 1983, Bill Atkinson switched his focus to the Macintosh project. And one of his first acts was to adapt Sketchpad, his simple Lisa drawing program, into a Mac program he'd eventually call MacPaint.

MonkeyLives

In the fall of 1983, the engineers in the Fishbowl at Bandley 3 were burning all candles at all ends to finish the Mac. A huge part of the effort was hunting down bugs; the two primary apps, MacPaint and MacWrite, crashed constantly.

Steve Capps came up with an ingenious solution: an automated software monkey. The name referred to the infinite monkey theorem, which says that if you give a monkey a typewriter and an infinite amount of time, it will eventually produce the full works of William Shakespeare. "I just said, 'Why don't we have a desk accessory that just randomly does everything?'"

His little Monkey DA randomly banged on keys, moved the cursor, and clicked the mouse button at a furious pace, as though by a monkey on an infinite amount of coffee. The idea was to rapidly simulate event sequences that might crash your program.

It was an excellent debugging tool. When he first unleashed it, "MacPaint would survive for a minute and then die," he says. "MacWrite would survive for a millisecond." It was also fun to watch, especially when it actually created some "art" in MacPaint or typed an actual word in MacWrite.

Unfortunately, the Monkey often randomly hit the Quit command, thus shutting down the experiment—or, worse, the Print command, potentially printing hundreds of pages of gibberish.

Bill Atkinson came up with a fix: an option, a system flag, to turn off the Print and Quit commands while the monkey was running. He named this on/off switch MonkeyLives. (That's the verb "lives," as in "Elvis Lives!") If you set it to True, the Monkey could run uninterrupted forever—or until it triggered a crash.

"It became a measure of robustness for Macintosh applications: How many minutes or hours could it survive the Monkey?" remembers Hertzfeld. "MacPaint went two weeks once," points out Atkinson proudly.

Years later, Apple's testing teams created a much more structured, formal testing app called Virtual User. But in principle, it did exactly the same thing as Capps's little program.

Put another way, MonkeyLives lives.

MacPaint brought to consumers many of the computer-art conventions that live on in Photoshop and similar programs, starting with the tool palette hugging the left edge of the screen. The Pencil tool created a freehand line, the Brush made a fatter swath, the Paint Can filled in an entire enclosed area with one click, and so on. For greater precision, Atkinson also created a mode called FatBits, which magnified the pixels you were painting eight times. (Atkinson soon ordered a new California license plate: FAT BITS.)

Most important of all, he built in an Undo command. The knowledge that you could undo any

Jobs, Art, and Japan

The MacPaint version of the Japanese woman brushing her hair appeared in Jobs's Mac presentations, in magazine ads, and on the cover of Mac manuals. Susan Kare had drawn her from a scan of *Woman Combing Her Hair*, a *shin-hanga* made by artist Hashiguchi Goyo in 1920. (The *shin-hanga* technique begins as a woodcut, followed by colors painstakingly pressed onto the print one at a time.)

Jobs's first exposure to Japanese art was in middle school, when he loved Bill Fernandez's mom's collection of *shin-hanga* prints. In the seventies, Jobs spent years hanging out at the Zen Center in Los Altos, where he was deeply affected by the simplicity of Japanese furnishings, prints, and ceramics.

Jobs would make dozens of trips to Japan, often with his wife, Laurene, and their children. He maintained an obsession with Sony and its cofounder Akio Morita. Jobs's famous black turtleneck was designed by Issey Miyake. He sent Apple's cafeteria chef to study at the Tsukiji Soba Academy in Tokyo.

But the most visible Japanese influence on Jobs was in his design sense. His home was as sparsely furnished as a Japanese *ryokan* (inn). Only three pieces of art hung on his bedroom wall: a photo of Einstein, a photo of Gandhi, and a Japanese *shin-hanga* print of a woman called *Morning Hair*. (He would eventually collect 40 *shin-hanga* pieces.)

mistake made experimentation risk-free and judgment-free—essential elements of any creative endeavor.

MacPaint emerged as the poster child for the Mac's simplicity and joy of creation. Thousands of people's introduction to the mouse (if not computers in general) was the Mac, and MacPaint was the app that made them fall in love.

Cmd Key

The former HP engineers on the Mac team liked to keep their fingers on the keys for maximum efficiency. For people like them, the Mac team equipped the keyboard with a dedicated command key, labeled with Apple's logo. With the key pressed, you could tap the C key for Copy, Z for Undo, and so on. To learn these combinations, you could simply consult your menus. Next to the Copy command, for example, you'd see C.

One day in the summer of 1983, Jobs burst into Bandley 3. He'd just seen an early demo of a new program called MacDraw, whose long menus displayed correspondingly long vertical stacks of symbols. "You can't do that," Jobs said. "It's taking the Apple logo in vain! You have to come up with something else!"

But if they replaced the symbol in the menus, they'd also have to replace it on the keyboard itself, plus all references to it in the user manual, which was only days from going to press. But replace it with what?

Susan Kare flipped through a book of international symbols, hunting for a replacement. It would have to be distinctive, easily represented in very few pixels—

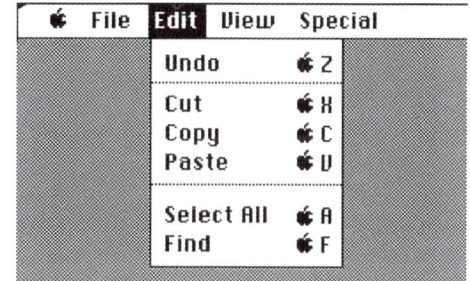

The column of Apple logos in the menus.

and it couldn't come with any cultural baggage. Finally, she stumbled onto the symbol that fans today refer to as a cloverleaf, propeller, or flower: ⌘. She found it in, of all things, a book of Swedish campground and trail markers. (It meant "remarkable feature.") It sits on every Mac keyboard to this day.

Soon thereafter, however, Apple began including the same keyboard model with Apple II's. The Apple IIe already had an key—too late to change course now—so Apple now had no choice but to label the Mac's command key with *both* the Apple logo *and* the ⌘ symbol.

Apple wound up taking the logo in vain after all.

"Business" Cards

In early 1982, the Macintosh project was getting real. The team was about to get business cards.

Andy Hertzfeld, whose title when he was working on the Apple II was "senior member of technical staff," had never liked his title. He declined a card altogether.

Steve Jobs swung by his desk the next day and urged him to change his mind; he could pick any title he liked. Hertzfeld liked the sound of that.

Anything?

He came up with "Software Wizard." He liked that name, he says, "because you couldn't tell where it fit in the corporate hierarchy, and because it seemed a suitable metaphor to reflect the practical magic of software innovation." (He also ordered MAC WIZ vanity plates for his car.)

His unconventional choice set off a domino effect that lasted for years. Burrell Smith chose "Hardware Wizard" for his cards. Bruce Horn went with "Trailblazer." Developer wooer Mike Boich was "Software Evangelist." Product manager John Medica went with "Big Shot." John Sculley chose "Chief Listener." Programmers Scott Knaster, Darin Adler, and Bill Dawson picked "Boy Guru," "Cheese Host," and "I Have No Pants On."

And programmer Ed Tecot honored the beginning of the whole thing by choosing, as his title, "Not Andy Hertzfeld."

Consistency

Software consistency was a hallmark of the Mac's concept. As you moved from one Mac program to another—word processor, spreadsheet, art, music—you didn't have to relearn anything. Every program would work the same way. The OK button would be in the same spot in every dialog box; the scroll bars in every program would look and work identically; the Cut, Copy, and Paste commands would be in the same menu, and have the same keyboard shortcuts, in every app. All of this was a radical departure from previous computer systems, where learning a new program was like learning a *new program*.

Joanna Hoffman drafted what would become a "how a good Mac program should work" bible for developers—the *Macintosh User Interface Guidelines*. "I'm not a technical writer, so, you know, it had flaws," she says. "But at least we had something that we could hand to people."

Saving Lives

In August 1983, the Mac was still taking two minutes just to start up, and Steve Jobs was unhappy. He approached Larry Kenyon, who was working on some of the startup routines. "You've got to make it faster," he said.

Kenyon tried to explain that the Mac startup process currently included a memory-chip test, which was a temporary measure; it wouldn't be in the final version.

But according to a famous story in Mac lore, Jobs uncapped a marker and began doing some math on a whiteboard. "In a few years, I bet five million people will be booting up their Macintoshes at least once a day," he said. "Well, let's say you can shave ten seconds off the boot time. Multiply that by five million users, and that's fifty million seconds every single day. Over a year, that's probably dozens of lifetimes! Just think about it: If you could make it boot ten seconds faster, you'll save a *dozen lives*! That's really worth it, don't you think?"

In the popular version of this story, Jobs's math was so persuasive that Kenyon doubled down and shaved another ten seconds from the startup time.

The truth isn't quite as much fun. In reality, Kenyon found the encounter "more irritating than motivational; as a software engineer writing assembly code at the lowest level of the system, I was already focused on speed."

Still, he realized that a faster startup time would help out Jobs when giving demos of the Mac for reporters, analysts, and investors. He removed the memory-diagnostic sequence.

"It did speed up the boot time considerably," Kenyon said, but "I got more credit for the 'speed up' work than I was really due . . . and Steve got more credit for his 'motivational' speech than was really warranted."

History may have gotten the story wrong, but Kenyon does think that Jobs had a point. "We do need to consider the time burden we place on people, even if it's minor," he says. "It's just that in this case, Steve gave us something more amusing than motivational."

Manuals

The Apple II's lucid, friendly manuals had helped it win a place in its owners' hearts. The Lisa manuals, on the other hand, felt dead, opaque, corporate. Jobs was adamant that the Mac would return to friendly, beautiful documentation. He hired Chris Espinosa, now 20, to put the Mac's manuals together, and authorized top-flight photography and printing. "It was lavish," Espinosa says. "Nobody ever got to spend that much money on manuals, but Steve wanted the whole effect to be crisp and professional."

Of course, the whole concept of a mouse was new to the public in 1984, so the Mac team also

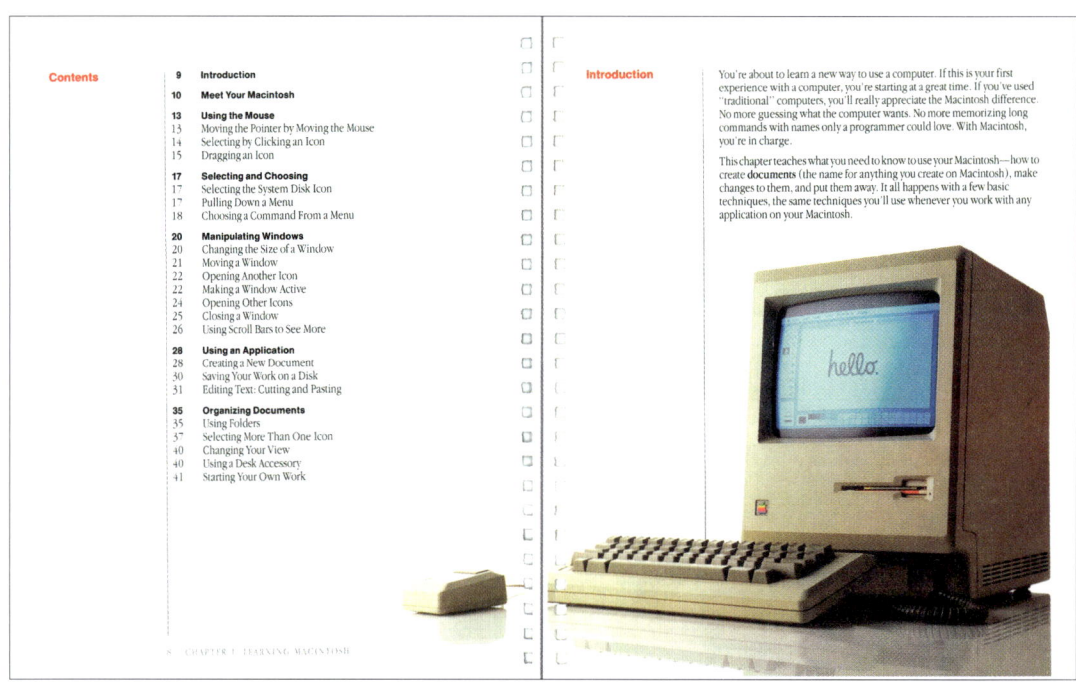

The Mac manuals used six-color printing.

created a tutorial app on the Guided Tour floppy called Mousing Around. It offered a series of simple exercises that taught you to move the mouse (a connect-the-dots game), click the button (an on-screen piano keyboard), hold the button down (a row of magicians' hats), double-click to open things (an animated wrapped gift), and drag objects on the screen (a tiny mouse in a maze).

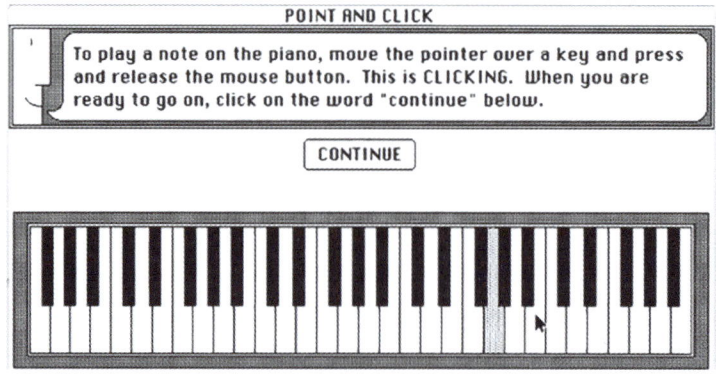

The Mousing Around app gamified the learning of a strange new skill: using a mouse.

12. Marketing the Mac

From the beginning, Steve Jobs had grand visions for the Mac's unveiling. "It's got to have the best introduction that any product has ever had," he'd say.

The objective was to build a global ecosystem for the Mac, but the number of moving parts, channels, and sales vectors seemed nearly infinite. It would require building a network of computer stores, training, promotions, advertising, marketing, and public relations. It would involve wooing hundreds of software companies and hardware makers to develop stuff for the Mac.

Stanford business school grad student Mike Murray had interned at Apple in the summer of 1981. In March 1982, Jobs hired him away from HP to build the Mac assault. He would wind up engineering a $15 million, 100-day launch campaign that incorporated a handful of initiatives, some straightforward and some unorthodox.

Mike Murray
Born: 1955, Klamath Falls, OR
Schooling: Stanford (BS and MBA)
Before Apple: HP
Apple: 1982–1986
After Apple: Microsoft, Unitus

In 1983, Jobs asked Murray to judge a love song he'd recorded for a girlfriend: piano, vocals, harmonica. Murray's verdict: "Great!" Murray never learned if Jobs actually sent it.

Software Evangelism

Without software, a computer is just an expensive box. The Mac had no chance unless there was an ecosystem of programs to go with it.

To cultivate one, Murray hired his old HP buddy Mike Boich—Harvard MBA, white Porsche Carrera, jet pilot's license—to become what Murray called a software evangelist. The job entailed traveling the world with a top secret prototype Mac—at this point, still containing a 5¼-inch Twiggy disk drive—showing developers how cool it would be.

Boich, in turn, hired his college buddy Guy Kawasaki. "Evangelism comes from a Greek word meaning 'bringing the good news.' Mike and I brought the good news of Macintosh," says Kawasaki.

The IBM Challenge

In late 1983, in hopes of gauging the Mac's potential appeal to business buyers, Mike Murray put together a focus group in New York. The subjects—corporate computer buyers—sat at a long table. Murray and his team watched from behind a one-way mirror.

An IBM PC and a prototype Mac sat on the table. A facilitator asked the participants to try various office tasks: moving a section of text, for example. On the PC, they had to type out commands; on the Mac, they'd just highlight the text with the mouse, copy, and then paste.

The difference was immediately obvious. "Suddenly, these business people in their forties and fifties are acting like kids! They're giggling, and they're just loving it," Murray says. "It was like going from a backyard birthday party to Disneyland."

When asked a series of comparative questions—"Which computer is easier to use?" "Which is more pleasant to use?" "Which one would you be more productive with?"—the subjects declared the Mac the winner every time. "It was much easier to use than the PC," they'd say. "It was intuitive." "It was fun." "I wish I could have one both at work and at home." Behind the mirror, the Apple team was exultant.

But once they settled back down into their metal folding chairs, they heard the moderator ask one final question: "If you were responsible for purchasing computers for your company, which of these two computers would you recommend?"

Every single participant pointed to the IBM PC. "That one."

It didn't matter how slow, frustrating, and clunky IBM's computers were; they were the *safe* bet. As the saying went, "Nobody ever got fired for buying IBM." Its salesmen locked customers in with long-term service contracts, and made it clear to customers that a different standard would be asking for trouble.

"We knew at that moment that we were hosed. Our task had suddenly become a hundred times more difficult," Murray says. He flew back to Cupertino, his mood dark. Maybe the Mac could succeed in higher education, or the creative arts, or among scientists.

But based on his little focus group, it would not be a player in the business world.

They figured that the job wouldn't be especially challenging; in the tech world, the hype about the Macintosh was already boiling over. Meanwhile, these road shows had a secondary benefit: At the time, the only way to write software for the not-yet-existent Mac was to use a *Lisa*. "I probably sold more Lisas than anybody in the world," says Kawasaki.

Then there was Microsoft. In 1983, of course, it wasn't the behemoth it would become. There was no Windows, no Microsoft Office. In fact, Apple's revenue in 1982 was $1 billion; Microsoft, with little to sell besides DOS, Flight Simulator, and some programming languages, barely cleared $32 million.

Jobs and Bill Gates, both 27, could not have been more different. Gates looked down his nose at the mercurial, passionate Jobs, who didn't know how to code. Jobs found Gates to be smug, passionless, and incapable of appreciating the Mac's elegance.

Guy Kawasaki
Born: 1954, Honolulu, HI
Schooling: Stanford, UCLA (MBA)
Before Apple: Edu-Ware
Apple: 1983–1987, 1995–1997
After Apple: ACIUS, Garage.com, Google, Mercedes, Canva

A fine-jewelry manufacturer hired Kawasaki while he was at UCLA. His first paid job was counting diamonds.

Still, they had a substantial mutual interest in collaborating. Jobs needed core apps for the Mac. Gates knew that if the Mac took off, its customers would buy a lot of software. Gates put 20 programmers on the project—more people writing Mac software than *Apple* had writing Mac software. Eventually, they hammered out a deal: Microsoft would write Excel, Chart, and File (a database) for the Mac. Apple would include them with every Mac, and pay Microsoft $10 per program.

But when other software makers objected to the exclusive arrangement—and when it appeared that Microsoft would miss the Mac's ship date—Jobs rescinded the offer. Microsoft was free to sell the programs on its own.

In the end, more than 100 software companies said they were excited to start writing Mac software. But the problem wasn't getting them to start; it was getting them to finish. Apple hadn't yet provided the software tools and documentation that programmers needed. Worse, the Mac's limited memory made it ill-equipped to handle sophisticated programs, especially business software. Four months before the Mac's launch date, only 20 software companies planned to have Mac apps ready.

Influencers

In the fall of 1984, Regis McKenna conducted 60 "sneaks" in the Picasso conference room in Bandley 3: seven-hour private Mac demos for key industry analysts and reporters. (The Bandley conference rooms were named for artists—Picasso, Matisse, Le Corbusier.) Meanwhile, three marketing teams fanned out in specially equipped trucks to visit dealers in 36 cities. By the time of the Mac's introduction, they had trained over 4,000 sales reps at computer dealers.

Sculley and Jobs also presented free, brand-new Macs to a group of 50 creative and corporate celebrities. Eleven of them wound up in the *1984 Apple Annual Report*, photographed using their Macs: entrepreneur Ted Turner, novelist Kurt Vonnegut, designer Maya Lin, Muppets creator Jim Henson, Broadway legend Stephen Sondheim, and so on. Michael Jackson, Andy Warhol, Beatle son Sean Lennon, and Mick Jagger got free Macs, too. (Jobs delivered Jagger's Mac himself, and didn't get much of a reaction. "I think he was on drugs," Jobs told the Mac team. "Either that, or he's brain-damaged.")

Publishing

Murray wanted to create the perception that the Mac was already a *thing* from the day it went on sale. In the IBM universe, *PC World* magazine gave computer fans a sense of context about their machines. What if the Mac had something similar?

Publisher David Bunnell loved Murray's idea of starting a magazine called *Macworld*. But how would it make money? In the barren landscape of Mac products, who would the advertisers be?

And so it was that Apple Computer got into the magazine business. Murray paid for the entire first year of *Macworld*—$200,000—to get it on its feet. The only requirement was that it had to be beautifully designed and written. The first issue appeared in January 1984, simultaneous with the release of the Mac itself, and continued in print for the next 30 years. (In 2014, it went online-only.)

The first issue of Macworld *ran stories about MacWrite, MacPaint, and the Mac itself.*

Undergrads

The Mac wasn't going to own corporate America, and the K–12 education market was already dominated by the Apple II. But what about colleges? In 1984, there *were* no personal computers on college campuses. Murray worked up a higher-ed deal: If a school committed to deploying 2,000 Macs within three years—to give away or resell to students—Apple would offer a 60 percent discount. Former Sony salesman Dan'l Lewin, who'd been working on marketing the Lisa, gave the program a vaguely academic-sounding name: Apple University Consortium.

By the time the Mac launched, Lewin had commitments from 24 schools to buy more than 50,000 Macs. Among them: Stanford, the University of Chicago, the entire Ivy League—and, of course, Jobs's beloved Reed College. Drexel University actually *required* every incoming student to buy a Mac.

In time, this program had a decades-long payoff: Thousands of college students whose first computers were Macs became lovingly locked into the Apple ecosystem for the rest of their lives—and, often, their children's.

"1984"

On Sunday, January 22, the L.A. Raiders beat the Washington Redskins in Super Bowl XVIII. But for millions of Americans, the most memorable moment was the 60-second ad that aired just after the second-half kickoff.

We're inside a dismal, gray sci-fi chamber—a spaceship or a prison, maybe. Endless lines of bald men march in sync, eyes dead. An authoritarian voice blasts from screens along the tunnel.

The "1984" ad generated more attention than the Super Bowl itself.

The drone men sit mutely on long benches while Big Brother (wearing *two* sets of glasses) indoctrinates them. "Our Unification of Thoughts is more powerful a weapon than any fleet or army on earth!"

But what's this? A blond woman is running into the hall, chased by an army of masked goons, carrying a sledgehammer. Her red track shorts and Macintosh T-shirt are the only *color* in this ad. She spins like an Olympian and flings the hammer at the big screen.

"Our enemies shall talk themselves to death, and we will bury them with their own confusion. We . . . shall . . . prevail!"—and BOOM. The screen explodes in white light. The men on the benches stare forward, their mouths open. "On January 24th, Apple Computer will introduce Macintosh," a narrator says. "And you'll see why 1984 won't be like *1984*."

The "1984" ad, we know now, became an instant classic, and its execution seemed flawless. But in fact, few aspects of its creation went as intended.

- **The line "Why 1984 won't be like *1984*" wasn't written for Apple.** It was a general-purpose tagline idea someone at Chiat/Day had written a couple of years earlier. The agency had tried unsuccessfully to find a client that might want it. Even *Apple* had passed on it when Chiat/Day pitched it as an Apple II ad. And so it sat, unused—just a printed-out headline, until head copywriter Steve Hayden and art director Brent Thomas found it in the spring of 1983.

  Jobs had recently approached Chiat/Day creative director Lee Clow with a tall order: "It's got to be dramatic. It's got to be famous," he said. Hayden thought that the *1984* idea might fit. Sculley swallowed hard and approved the unheard-of budget: $750,000 for the one-minute ad, and hired Ridley Scott, newly famous for directing *Alien* and *Blade Runner*, to direct it.

- **Big Brother wasn't supposed to be IBM.** Clow maintains that Big Brother wasn't a metaphor for IBM. "Our idea was that Big Brother represented the control of technology by the few."

- **It wasn't supposed to be a sledgehammer.** Hayden and Thomas's storyboard showed a young woman smashing the Big Brother screen with . . . a baseball bat. Ridley Scott figured that the hammer would be more memorable.

  Finding someone to *throw* the sledgehammer, however, was a trickier problem. To play the dead-eyed marchers, Scott hired 200 London skinheads, plus a few dozen fully haired extras who were willing to shave their heads for $250. But few of the female models who auditioned could handle the hammer.

  Finally, Scott discovered 18-year-old Anya Major, who had that rarest combination of talents: a model with a background in discus throwing.

- **There weren't supposed to be any words.** In Hayden's original concept, Big Brother didn't speak. But the production team in London needed something for aspiring performers to read in their auditions. So, after looking up some speeches by Mao and Mussolini, Hayden typed up Big Brother's oppressive speech and faxed it to London. Scott loved it enough to make it part of the ad.

- **The ad actually aired many times.** The usual story goes that Apple aired the commercial only once, during the Super Bowl—a stroke of genius that instantly made it rare, mythical, and talked about.

  But Chiat/Day wanted to enter the ad in various awards competitions for 1983, which meant that it had to have aired before the end of the year. So the ad actually premiered three weeks before the Super Bowl—once, quietly, at 1 a.m., on KMVT in Twin Falls, Idaho. It did wind up winning 35 awards, including the Oscar of advertising: the Grand Prix at the annual Cannes Lions ad conference.

  But after the Super Bowl, Apple also played the ad in movie theaters during the trailers, and on TV in the ten biggest U.S. cities. Apple also ran it in Boca Raton, home of IBM's PC division, just for spite.

  "It was designed to have a media life beyond the Super Bowl," says Clow. "But the board of directors at Apple decided it was irresponsible to continue running it [nationally], since the product wasn't even available yet. And that becomes almost part of the legend: 'Oh, the genius of just running it once on the Super Bowl!'"

When Jobs saw the finished ad, he was ecstatic. "That is *incredible*!" he shouted. "It's just awesome!"

But Sculley had concerns. The ad's punch line didn't work unless you knew the plot of *1984*, George Orwell's 1949 novel about a totalitarian government that brainwashes the masses. (The phrase "Big Brother is watching you" comes from the book.) Worse, the ad didn't show a mouse, or MacPaint, or even a computer. It mentioned Apple only once. Until the final seconds, you had no idea what was being advertised. "Are we going to blow the opportunity by getting so caught up in the theatrics of the commercial that people will forget the product?" he worried.

In October 1983, Jobs showed it at the company's Honolulu sales conference—its first public screening—earning a full one-minute, full-throated standing ovation. "The sales crowd went berserk after seeing it," says Mike Murray. "It was like feeding raw meat to hungry lions."

Jobs was confident that he was sitting on something spectacular. He wanted to air it during the Super Bowl—twice, in fact. Never mind the $1.6 million cost; it was the kind of splash the Mac deserved.

The Macintosh Dating Game

The purpose of a corporate sales conference is to fire up a company's salespeople about whatever they're supposed to be selling. But as Apple's 1983 Honolulu conference approached, there wasn't much to boast about. The Apple II was aging, and the III and Lisa weren't selling. The star of the show would have to be the Mac.

Mike Murray and his team proposed a parody of *The Dating Game*, the old game show where a bachelorette interviews three guys she can't see, and then chooses one as her date based on their answers. In Apple's version, the three "bachelors" would be the CEOs of the three biggest software companies, which were then making software only for IBM PCs: Microsoft, Lotus, and Software Publishing. In the role of the bachelorette: Steve Jobs.

All three bloodthirsty CEO competitors—Bill Gates, Mitch Kapor, and Fred Gibbons—agreed to fly to Honolulu, put on Macintosh polo shirts, and try to outdo one another in praising the Mac. They were, after all, young men who weren't averse to a free vacation.

As Jobs took the podium, the three "bachelors" entered and sat on folding chairs onstage. Jobs: "Software CEOs, could I please ask you to introduce yourselves?"

Gates: "My name is Bill Gates. I'm the chairman of Microsoft. And during 1984, Microsoft expects to get half of its revenues from Macintosh software." *[Crowd loses its mind for 20 seconds.]*

Gibbons: "If you were to put machine X on a table, and a Macintosh on the table beside it . . . we think Macintosh's benefits would be pretty obvious."

Kapor: "Macintosh sets a whole new standard."

As the game show ends, Jobs pretends to consult his notes, and then announces his choice among the suitors, in the cheesy way actual *Dating Game* contestants often did:

"Apples are red, IBM's blue; if Mac's going to be the third milestone—I need all of you!"

Murray had nailed both of his goals for the skit. First, it blasted the sales team's spirits into hyperspace. Second, the three CEOs inadvertently produced a year's worth of testimonials and promotional photos for Apple's ads, public relations, and dealer training.

A three-minute clip of the skit survives to this day on YouTube.

It's a good bet Bill Gates hasn't clicked on it recently.

In December, just before the big game, he and Mike Murray met with the Apple board to get their sign-off on buying Super Bowl airtime. Murray rolled the VCR into the meeting and hit Play. "And halfway through," he says, "I clearly remember one of the board members had his head down on the desk, pounding on the desk with his fist. And I thought: 'He really likes it!'"

Murray had misread the reaction.

"You will *never* show that commercial. *Ever*," one member said.

"That's the worst ad I've ever seen," said another.

"Can I get a motion to fire the ad agency?" Mike Markkula asked.

Sculley ordered Mike Murray to sell off Apple's Super Bowl slots.

Jobs and Murray were shattered; this ad was spectacular! Woz loved it so much that he offered to split the cost of the Super Bowl airtime with Jobs. Of course, for the board, it wasn't about the money. It was about the humiliation of airing a terrible ad.

As ordered, Chiat/Day set to work selling its Super Bowl ad slots: one to Heinz, one to Hertz. But as the sun set in New York, with two days before the big game, they couldn't find a buyer for the 60-second slot. (According to legend, Chiat/Day may not have tried especially hard.)

Apple was now the proud owner of one minute-long chunk of Super Bowl airtime. The full ad would air.

Today, everyone knows what happened. The ad was an instant cultural phenomenon.

"The following night on ABC, CBS, and NBC—those were the three networks then—it was

The ads and brochures introduced the notion that the Mac was the computer "for the rest of us."

Marketing the Mac • 133

> ### Test Drive a Macintosh
>
> Most people in 1984 had never used a computer at all, and almost nobody had used a mouse. Based on the success of the Pepsi Challenge, and taking a cue from the noble history of car test drives, Sculley proposed a trial program called "Test Drive a Macintosh": You could borrow a brand-new Mac from your local dealer, bring it home, and play with it for 24 hours.
>
>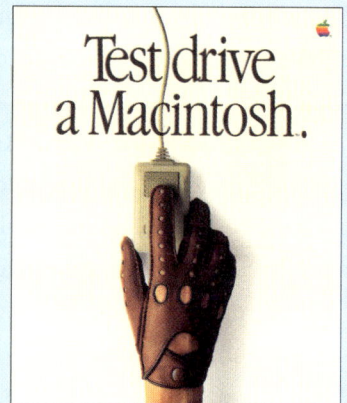
>
> Apple spent millions advertising the offer. And for the 200,000 people who signed up, it was incredibly effective. Many of them bought Macs. Inside Apple, the program was nicknamed the Puppy Sale. "If you borrow a puppy for a few days," explains engineer Scott Knaster, "you fall in love and you keep it!"
>
> For the dealers, though, the program was a living nightmare. Each loaned-out Mac involved as much paperwork as an actual purchase, without any guarantee of a sale—and meanwhile, now they had fewer Macs in the store that they *could* sell. Worse, many of the Macs came back with visible wear and tear.
>
> Officially, that was Apple's last attempt at a try-before-you-buy program. But in 2014, the company instituted a 14-day no-strings return policy on *anything* you buy directly from Apple.
>
> There was, however, no ad campaign. "Test Drive a HomePod," anyone?

shown as *news*. Not as advertising, not as PR," Murray says. "And they showed this outrageous ad, and they would show it in its entirety." It was, in essence, $150 million in free airtime.

There was a lasting societal effect, too: "It created a phenomenon where people started designing advertising specifically for the Super Bowl, and keeping it secret, and having it be a surprise," says Clow.

Jobs conceded that the ad didn't say much about the Macintosh. To fill in the informational hole, Apple bought all 39 available ad pages in a November *Newsweek* issue, for $2.5 million, to run a gorgeous, illustrated show-and-tell about this revolutionary machine. "It's unclear whether Apple has an advertising insert in *Newsweek*, or whether *Newsweek* has an insert in an Apple brochure," Sculley joked.

Computer of the Year

In 1983, with the internet a decade away, most people read about current events only in newspapers and magazines. Magazine coverage, and especially magazine *covers*, were deeply important to Steve Jobs.

He'd been on the cover of *Time* in February 1982. But nine months later, word reached Apple of something even more exciting: Jobs was being considered for *Time*'s annual "Man of the Year" cover story.

Over the summer, Jobs had invited *Time* editor Michael Moritz to observe the creation of the Mac, sitting in on meetings and conducting interviews. It was a surprising twist for the Mac team, which had otherwise been working in secrecy. But Jobs told them that Moritz was writing a book about Apple. "You can tell him everything," Jobs told them, "because he's going to write our story for us."

But when the "Man of the Year" issue appeared at the end of the year, Jobs was in for an upsetting double shock.

First, he was not on the cover. "*TIME*'s Man of the Year for 1982, the greatest influence for good or evil, is not a man at all," the article went. "It is a machine: the computer."

Worse, inside the magazine, there *was* a story about Jobs—a vicious hatchet job, in his view. Jobs, it said, was a "shill for a new gold rush." Woz looked like a "teddy bear on a maintenance dose of marshmallows." Jobs owed his success to "his smooth sales pitch and a blind faith that would have been the envy of the early Christian martyrs."

One anonymous source said that Jobs suffered from a "technical ignorance he's not willing to admit." Another: "Something is happening to Steve that's sad and not pretty, something related to money and power and loneliness." And Jef Raskin, recently exiled from the Macintosh project, quipped, "He would have made an excellent King of France."

Most upsetting to Jobs, the article revealed that he had a daughter, for whom he denied paternity and refused to pay child support.

"It was so awful that I actually cried," Jobs said about the article. "But it was a good lesson. It taught me to never get too excited about things like that, since the media is a circus anyway." He told the Mac team that if they ever spoke to Michael Moritz again, he'd fire them.

In the aftermath of the episode, Moritz, now a venture capitalist, said that he was as shocked as anyone by the story's tone; he'd supplied reporting to the article's author, Jay Cocks, but didn't write the story itself. *Time*'s editor, for that matter, said that Jobs had never actually been a candidate for Man of the Year.

Jobs didn't participate in interviews for at least another year.

$2,500

The Mac's mission had always been to change the world. To do that, it had to be inexpensive enough for the masses. But with each improved component—68000 processor, 128 KB of memory, Sony's 3.5-inch floppy drive—the price had inched upward: to $1,000, to $1,500, to $2,000. The team was deeply disappointed, but had little choice.

And then Sculley happened.

Sculley liked Jobs's dream of a splashy launch; he was, after all, a marketing guy. He shared Jobs's enthusiasm for hiring Chiat/Day (which had since bought Regis McKenna's PR agency), for placing ten million 20-page inserts in *Time*, *Newsweek*, and *Rolling Stone*, for buying a two-page spread in the *Wall Street Journal*, and for hiring talk show host Dick Cavett for TV and print ads. But where would the money for all of this come from?

To him, the simplest solution was to bake the costs into the Mac's price. He suggested raising the price to $2,500.

The team was gut-punched. $2,500 felt like a betrayal of everything they'd tried to accomplish. "We resented the idea that it was being artificially inflated to cover a glitzy ad campaign," Hertzfeld says.

Sculley argued that the early adopters would happily pay that price; Apple could always drop it later on. "Steve and I debated the issue for weeks and weeks," Sculley says. "I don't think a day went by when we didn't talk about it."

Jobs anticipated criticism from Apple's most loyal fans, who would mock the Mac's tagline "the computer for the rest of us." Who, exactly, was "the rest of us" supposed to be, when the price is $2,500 (about $7,600 in today's dollars)?

"You can't have it both ways," Sculley told Jobs. "Without the money, there is no advertising and no event." To the shock of the Mac team, Jobs finally gave in. The Mac would go on sale for $2,500, a price that would come back to haunt Apple in more ways than one.

13. Insanely Great

By today's standards, the size of the team that created the Mac was tiny: 15 engineers, plus a few dozen people in manufacturing, finance, and so on. They didn't have much of a Christmas in 1983.

Steve Jobs threw a lavish black-tie holiday party for them in the ballroom of San Francisco's St. Francis Hotel, with music supplied by members of the San Francisco Symphony Orchestra. Then it was back to work.

The Mac's grand unveiling would take place on Tuesday, January 24, 1984, at the annual Apple shareholders' meeting. But for the Mac software team, the real deadline was Monday, January 16, at 6 a.m. That's when the factory would begin duplicating the software disks that would come with every Mac.

Standing, from left: Bill Bull (analog board), Colette Askeland (printed board design), Jerry Manock, Mary Ellen McCammon (area associate), Andy Hertzfeld, Brian Robertson (manufacturing purchasing), Vicki Milledge (human resources), Debi Coleman, George Crow, Joanna Hoffman. Front: Chris Espinosa, Burrell Smith, Brian Howard.

By January 6, the Mac team was fried to a crisp. Their all-nighters had taken a tremendous toll in sleep, health, and hygiene—and yet MacWrite, the word processor, and Finder, the desktop, were still riddled with bugs. There was little chance that they'd have shippable software in the remaining week.

They decided to beg Steve Jobs for another couple of weeks. Maybe the first Macs could come with unfinished software; they could FedEx disks to the dealerships once the software was ready. "We all convinced ourselves that that was a supersmart idea," says Capps.

So on Sunday morning, January 8, the team got Jobs on the speakerphone from the Grand Hyatt hotel in New York, where he was promoting the Mac. They presented their solution, and then held their breaths for his response.

"No way," Jobs said. "There's no way we're slipping." He insisted that if they worked hard enough, they could finish on time. Ready or not, "I'm going to ship the code a week from Monday, with your names on it," he said.

The team returned to Bandley 3, knowing that they would not leave the building again for a week. Huge quantities of caffeine helped them stay awake continuously for the last two days, in the form of coffee, soda, and chocolate-covered espresso beans.

They thought they were making steady progress. But early Monday morning, two hours before the factory duplication was to begin, every app was crashing, even the usually reliable MacPaint. "I just lost it. I went totally hysterical. I just started laughing and couldn't stop," says Randy Wigginton. "I had to walk out of the building and literally walk around the block just to get a grip, because it was not looking good."

At 5:30 a.m., they rolled back to the previous version, which had worked better, and generated another release. This one seemed sound.

The Macintosh would ship on time—16 months behind its original schedule.

Flint

At 10 a.m. on January 24, 1984, about 2,700 people crammed into the auditorium of the Flint Center for the Performing Arts, on the campus of De Anza College, half a mile from Apple's headquarters. An overflow crowd of 800 was directed to a local movie theater to watch a live feed. 40 TV crews set up cameras.

The occasion was Apple's annual shareholder meeting, but everyone knew very well the real attraction: After months of buzz, hype, and secrecy, Steve Jobs was about to unveil the Macintosh.

Rehearsals hadn't gone well. Apple had rented a powerful projector, but the Mac had no video-output jack. Burrell Smith had hacked together a special circuit board to allow the Mac to connect, but the projector itself was temperamental, sometimes spontaneously shutting off.

28-year-old Steve Jobs, meanwhile, was nothing like the diligent preparer he would become. He hadn't managed a single full run-through of the show. At dress rehearsal, he was a nervous wreck. He asked Sculley to move from one seat to another, assessing the lighting. "He was driving people insane, getting mad at the stagehands for every glitch in his presentation," Sculley remembers. "I thought there was no way we were going to get it done for the show the next morning."

The Mac team wore Mac T-shirts and sat at the front of the auditorium. More stress was not what they needed. "It was the culmination of over three years of my life. We thought it would be a milestone for the industry, even for the human race," Hertzfeld says. "But we were also super nervous, because our demo could have crashed."

When Jobs finally took the stage, he was uncharacteristically well-dressed in a tailored dark suit and bow tie. The crowd went nuts.

"I'd like to open the meeting," he said, "with a 20-year-old poem by Dylan. That's *Bob* Dylan." He welcomed them by reading lyrics aloud:

> *Come writers and critics, who prophesize with your pen;*
> *And keep your eyes wide; the chance won't come again*
> *And don't speak too soon, for the wheel's still in spin*
> *And there's no tellin' who that it's namin'*
> *For the loser now will be later to win*
> *For the times, they are a-changin'.*

After opening remarks by chief counsel Al Eisenstat and John Sculley, Jobs returned to the podium and delivered a dramatic summary of the computer industry so far.

"It is 1958. IBM passes up a chance to buy a young fledgling company that has invented a new technology called xerography. Two years later, Xerox was born, and IBM has been kicking themselves ever since," he began.

He was setting up the Mac as the last hope for humanity against the mind-numbing corporate mediocrity of IBM. For him, it wasn't just personal. In November, Microsoft had announced its own graphic interface software for IBMs and compatibles, something called Windows. IBM PCs and clones were greatly outselling even the Apple II. *BusinessWeek*'s October 1983 cover bore the devastating headline "And the Winner Is . . . IBM."

"It is now 1984," Jobs continued. "IBM wants it all, and is aiming its guns at its last obstacle to industry control: Apple. Will Big Blue dominate the entire computer industry? The entire information age? Was George Orwell right?"

Flint Center, January 1984: Jobs slips a demo disk into a specially beefed-up Mac prototype and basks in the euphoric reaction.

"No!" screamed the crowd, losing its mind.

The lights dimmed. The "1984" ad played. The entire audience was on its feet.

Jobs approached a small table. He unzipped a cube-shaped carrying case and lifted out the Mac with one hand. That, in itself, was an amazing feat; the Mac was a third the size and weight of the IBM PC.

Jobs plugged it in, switched it on, and inserted a disk he pulled from his shirt pocket. The theme from *Chariots of Fire* played from the auditorium speakers.

As the enraptured audience watched, the word MACINTOSH scrolled magnificently by in a

System 1.0

Most modern Mac fans can easily guess at a few features that the very first Mac lacked: color, multitasking, and video playback, for example. It may come as a surprise, however, to learn that there was no Shut Down command. (That came in System 2 in 1985.)

The Trash didn't bulge when you put something in it, either. (System 5 introduced that feature in 1987.) List views showed only file names, without icons, until System 2.

For the Mac's first ten years, no clock appeared on the menu bar. (It debuted in 1994.) You couldn't put folders within folders until System 3. There was no New Folder command. Instead, a folder called Empty Folder was always present. The instant you renamed it, another Empty Folder spawned.

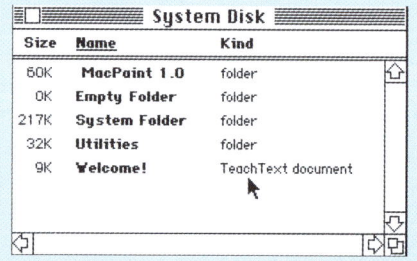

There was no online help of any kind until System 7. You couldn't eject a disk by dragging its icon to the Trash. When you copied files, no progress bar appeared—only a tiny dialog box that said, "Files remaining to copy: 3" (or whatever).

Macintosh: Revolutionizing computing, one empty folder at a time.

huge font—a demo Steve Capps had created. ("I pulled an all-nighter getting the intro going," he says. "The hair on the back of my neck still stands up when I hear 'Chariots of Fire.'")

Against a starry night sky, beneath the word "Macintosh," an invisible hand wrote out "*insanely great*" in cursive, courtesy of Bruce Horn.

Then a slideshow of programs appeared on the Mac's screen, prepared by Susan Kare: MacWrite. MacPaint. A programming environment. Steve Capps's Alice chess game. Even Multiplan, from Microsoft. No audience had ever seen images like these on a personal computer: typefaces, certificates, architectural renderings.

Then, the pièce de résistance. "Today, for the first time ever," said Jobs, "I'd like to let Macintosh speak for itself."

Mike Boich, in his visits to independent programmers, had come across a simple speech synthesis program. The moment Jobs heard it, he bought the program (later called MacinTalk) from its developer and incorporated it into the unveiling.

"Hello! I am Macintosh," the computer said in a nasal but clear voice. "It sure is great to get out of that bag!" The audience roared. "Unaccustomed as I am to public speaking," the computer went on, "I'd like to share with you a maxim I thought of the first time I met an IBM mainframe: Never trust a computer that you can't lift."

The audience came unglued. They were on their feet, fists in the air, cheering, shouting, applauding for five straight minutes. Jobs could only stand, beaming, holding back tears.

The first Mac was expensive, underpowered, and revolutionary.

It was an "I was there" moment. All three movies about Steve Jobs—*Steve Jobs*, *Jobs*, and *Pirates of Silicon Valley*—re-created the scene.

"He had created a church," Sculley says. "I was in the second or third row, cheering, with tears streaming down my face," remembers Jerry Manock. "It was on a scale of 'Jesus came and visited us,'" says Guy Kawasaki. "It was fricking religious. It was like: Show me the Kool-Aid!"

Years later, the world learned about one small deception: The Mac's 128 KB of memory would never have been able to handle the sound and animations in that demo. The Mac team had rigged up a special Mac for the unveiling with four times as much RAM.

The ebullient Mac team returned to Bandley 3 just in time to see a tractor-trailer truck pulling into the parking lot—and Steve Jobs stepping forward to greet it.

He had set up a surprise ceremony. He asked each of the 100 Mac team members to step forward to receive a handshake, a round of applause—and a factory-fresh Mac from the truck, with a tiny engraved plaque bearing that person's name.

Aftermath

For a while, a tidal wave of glory and rave reviews for the Mac kept the team aloft. "It is better than anything else of its kind," said the *New York Times* critic, "by a factor of 10." It "changes people's expectations of what a computer should be," said the *Los Angeles Times*. "Hands down, the best piece of hardware for its price," *Money* said.

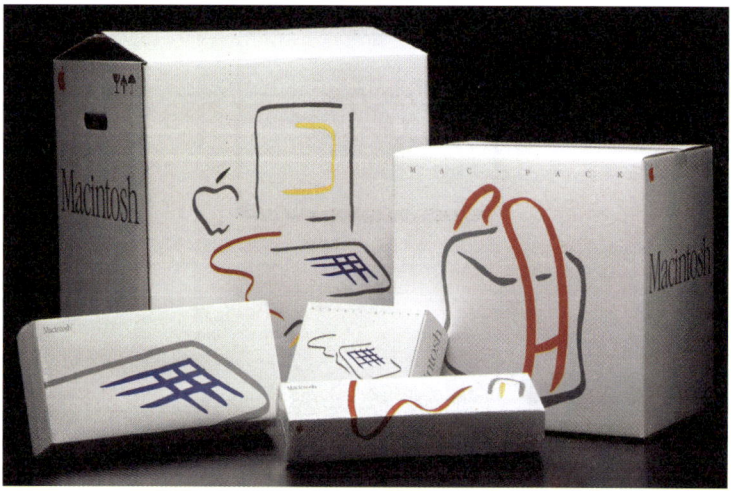

The Mac and its accessories were packaged as though something special was inside.

"The Macintosh is far easier to use than anything we've seen before," said Bill Gates, of all people. "I'm so excited about it. There's no question that I'll let my mom try it out."

The only people happier than the critics were the customers. The Mac created superfans, and the superfans formed Mac user groups.

The Mac was beautiful. Its software was beautiful. Its screen, with double the resolution (sharpness) of other machines, was beautiful. Even the *packaging* was beautiful. Opening the top flaps revealed the computer—and when you lifted it out, you then discovered the keyboard, mouse, manuals, disks, and cords sequentially from individual compartments. Jobs wanted to create a ritual in which you got to know each part of the machine.

Its carton wasn't some brown shipping box. It was white, bearing a colorful, Picassoesque drawing of the Mac itself.

Jobs had driven his box designers to distraction. "He got the guys to redo it 50 times," says Mac software evangelist Alain Rossmann. "It was going to be thrown in the trash as soon as the consumer opened it, but he was obsessed by how it looked."

As it turns out, people *didn't* throw the box away. To this day, closets all over the world are full of Apple boxes whose owners can't bring themselves to toss something so beautiful.

As Sculley had predicted, early superfans leapt for the Mac, no matter the price. They lined up outside computer stores across the U.S.; Apple employees were there, too, for the thrill of watching their baby meet the public. Apple sold $7.5 million worth of Macs on day one.

Jobs had estimated that Apple would sell 50,000 Macs in its first 100 days—an eyebrow-popping number, given the computer's low power and lack of available programs. But he was right—more than right; reaching the 50,000 milestone took only 74 days. (By contrast, it had taken two and a half years for the Apple II to sell that many.)

The numbers kept climbing; in June 1984 alone, 60,000 Macs flew out the door. The University Consortium program accounted for tens of thousands of sales to college students, bringing in $53 million.

Sculley bought ten million inserts in magazines, and two TV ads a night during the 1984 Olympics, making Apple one of the biggest advertisers in the United States. ComputerLand, Businessland, and Sears became dealers, adding hundreds of new places to buy the Mac. When Apple sought to hire 350 reps to service these chain stores, its ad in the *Wall Street Journal* generated 12,000 applications. "The deluge swamped the Cupertino post office for a week," Sculley says.

Fan mail poured in. Jobs liked to read this one, from a six-year-old boy, at presentations: "Dear Mr. Jobs: I was doing a crossword puzzle, and a clue was 'As American as Apple _____.' I thought the answer was computer, but my mom said 'pie.'"

Everything Apple touched was turning to gold, at a dizzying pace, and with no end in sight.

For the 1984 fiscal year, revenue jumped 54 percent (more than $1.5 billion). Jobs compared the feeling of disorientation to walking forward in a completely dark room; you just had no idea when you'd hit the wall.

Postpartum

Still, for Jobs, finishing the Mac was bittersweet.

"It's a wonderful, ecstatic feeling to create something and put it into the pool of human experience and knowledge. We just have this incredible chance to do that in the next five years, and then it'll be over," he told *Newsweek*. "Computers will be everywhere. Burrell will be off in Oregon playing his guitar, Andy will be writing the next great American novel, Joanna will have married some Soviet general. We'll be scattered all over the globe, doing other amazing stuff."

There would be a second Mac model, of course, with more memory. The Mac needed add-ons like an external floppy drive and a hard drive. Burrell Smith wanted to build a new model, Turbo Mac, with a faster processor, a screen that could display shades of gray, and a built-in hard drive.

But the Mac team couldn't muster the enthusiasm that they'd had for the first Macintosh. They'd just sprinted a marathon, and the letdown at the finish line was profound. Hertzfeld describes it as "a classic case of massive postpartum depression."

It wasn't only the work that was changing; it was the workplace. Only two weeks after the Mac's launch, Sculley reorganized the company as two divisions: an Apple II (and III) group, led by Del Yocam, and a merged Macintosh+Lisa group, led by Jobs. Each would have its own marketing, engineering, and manufacturing teams. They'd share sales, finance, and distribution.

The Mac team would fill all of the leadership positions in the Mac+Lisa group. They'd no longer be an elite skunkworks; they'd be cogs in a corporate division with more than 300 people.

To explain the reorganization, Jobs summoned the Lisa group to a meeting that would become legendary. "You're a B team. B players," he told them. "A players attract A players. B players attract C players." It was one of his favorite lines. Top talent won't tolerate working with lesser talent.

"So today, we are releasing some of your fellow employees to give them the opportunity to work at our sister companies here in the Valley."

He fired a quarter of them.

"The idealistic version of the Macintosh team that I yearned for had apparently vanished," says Hertzfeld, "subsumed by a large organization of the type that we used to make fun of, rife with bureaucratic obstacles and petty turf wars."

Some of the Mac team, utterly burned out, left Apple; a few of them never worked in tech again.

Bruce Horn quit. Steve Capps moved to Paris. Andy Hertzfeld took a six-month leave of absence, but never returned. "For a month after I left," he says, "I cried myself to sleep."

Randy Wigginton, who'd been hired to write MacWrite, spent most of the year lying on his couch watching TV. One manager became depressed, behaved erratically, and began carrying a pistol to work; he quit a few months later. Within a year, Burrell Smith, Susan Kare, and Joanna Hoffman had all left, too.

A few hung on, addicted to the mission, ready to dive into whatever came next.

Slump

In 1984, the FDA approved ibuprofen, Prince released *Purple Rain*, and Oprah Winfrey started a talk show. But on May 3, at a high-end Silicon Valley restaurant called Le Mouton Noir—the Black Sheep—Steve Jobs threw John Sculley a surprise first-anniversary party. Before the board, the executive staff, and their spouses, Jobs spoke of Sculley's accomplishments and presented him with a magnificent, five-foot-wide, acrylic-enclosed exhibit of artifacts from their first year together: newspaper headlines, magazine covers, the Mac's mouse, and even half of an actual, just-released Apple IIc.

Sculley was deeply moved. "We were still in love at that point," Sculley says. "Everything was working between us."

The story of Jobs, Sculley, and the Mac *wanted* to have a happy ending, it really did. What a tale! A tiny band of pirates, sacrificing everything on a quest to bring the power of computing to the masses. To change the world. To make a dent in the universe.

And for a few wonderful months, that cheerful fate seemed assured. But after the early-adopter wave, Mac sales began to droop.

Businesses had little interest in a machine with no cursor keys, no color, and no expandability; some called it a toy. There was so little memory, you couldn't write a document longer than about ten pages in MacWrite. Mac owners lived with frequent crashing, usually when their programs ran out of memory. "It's a miracle that it sold anything at all," Joanna Hoffman says.

And then there was the issue of Disk Swapper's Elbow.

The Mac had only one disk drive. Making a copy of a floppy, therefore, required swapping two disks—source and destination—in and out of the slot, over and over again, because the Mac's limited memory could store only a few kilobytes per swap. Duplicating one full disk could require as many as 20 swaps. (The Sony floppy drive spun at four different speeds, according to the data's location on the disk; each speed produced a different musical pitch. To this day, early Mac fans can sing to you the peculiar notes and rhythms of the Song of the Mac.)

What early adopters fell in love with, author Douglas Adams would say, "was not the machine

Apple Fellows

In 1983, Apple's leadership was seeking a way to honor Steve Wozniak, who'd returned to Apple after two years away. They made him an Apple Fellow: a prestigious title to acknowledge important work for the company. A Fellow receives stock options, a generous paycheck, and the freedom to do whatever kind of work he finds inspiring. In 50 years, only 12 people have ever earned the title.

The first inductees were Woz, Rod Holt, Rich Page, and Bill Atkinson. In 1984, Kay became a Fellow, too. Atari Pong creator and early game pioneer Al Alcorn became a Fellow in 1986.

"One of the rules initially was that you had to have made your notable accomplishment at Apple. Well, that rule didn't last even a year," remembered Larry Tesler. "But nobody really

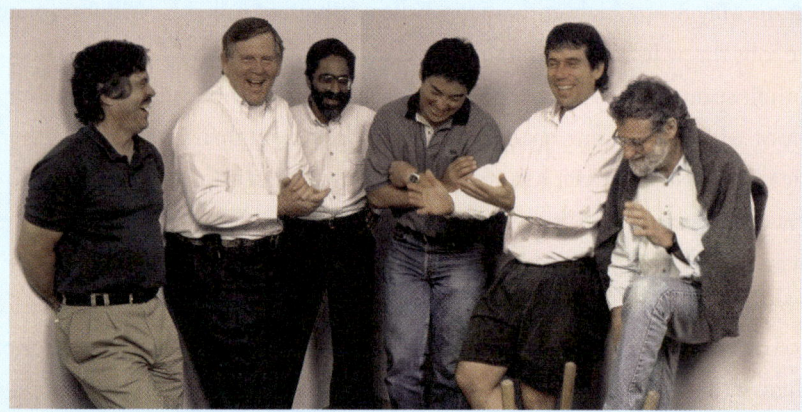

cared, because they were all excited about having these people in the company." He wound up in charge of the program, organizing quarterly meetups for all the Fellows.

Over the years, the list of Fellows grew only slowly. As Apple turns 50, they also include Gary Starkweather (inventor of the laser printer; 1988), Gursharan Sidhu (AppleTalk creator; 1993), Don Norman (user interface architect; 1993); Steve Capps (Finder, Newton OS; 1994); Guy Kawasaki (Mac evangelist; 1995); and Phil Schiller (marketing; 2020). (Shown here: Kay, Starkweather, Sidhu, Kawasaki, Capps, and Norman.)

Eventually, Apple also instituted a sort of second-tier Fellow program, called the Distinguished Engineer, Scientist, and Technologist level—the DESTs, as they became known. Over the years, this much longer list has included Brian Howard, Wendell Sander, handwriting-recognition pioneer Larry Yaeger, Final Cut Pro creator Randy Ubillos, and others.

itself, which was ridiculously slow and underpowered, but a romantic idea of the machine. And that romantic idea had to sustain me through the realities of actually working on the 128K Mac."

PARC alumnus Alan Kay, a pioneering computer scientist who'd become an Apple Fellow just after the Mac debuted, loved the Macintosh concept. But he told Sculley that it was "a Honda with a one-quart gas tank. It can only take you to the corner store for celery and back."

Apple was learning, for the second time, that an early-eighties computer could be inexpensive or decently powered, but not both. The Lisa had the power, but not the price; the Mac had the price, but not the power.

"If we had just waited eight months, it would have been a 512K Mac to start with. There would have been less user frustration," says Andy Hertzfeld. "But Steve was sort of cracking the whip to get it out as soon as possible."

It would be a constant theme for the early Apple: great technology too early. Lisa, Macintosh, Newton. "Apple was willing to take big risks and do ambitious things, but often way ahead of the curve in terms of what was possible with the physics," said Bill Atkinson.

Even after the Mac went on sale, its ecosystem was still empty. The external floppy drive was delayed to May. The Mac security kit, for locking down the eminently stealable Mac, July. The numeric keypad, September. Few software titles were available. Important apps like MacProject, MacDraw, and MacBASIC were delayed until the fall.

Sales slowed sharply. Jobs had anticipated selling 50,000 Macs a month; some months, dealers moved only a tenth as many. Even in the holiday season 1984, when Apple had expected to sell 75,000 Macs a month, it didn't even crack 20,000. The back rooms of dealerships nationwide were stuffed with unsold Macs.

Apple II sales *still* accounted for 70 percent of Apple's revenue.

It didn't help that the Mac arrived just in time to greet an industry-wide slump that lasted well into 1985. Consumers were realizing that computers were actually harder to use, and less useful, than the ads made them look. In 1984 and 1985, more than half of the computing industry's 300 magazines and newsletters shut down.

The mood was grim in the Macintosh division, which now employed over 700 people. In September 1984, they released a Mac with a much healthier memory allotment (512K), nicknamed the Fat Mac, for $3,200, but otherwise, there hadn't been much progress. There was still no hard drive and not enough software.

Snow White, Red Book

In 1982, Jobs had become alarmed at the state of design at Apple. There was no coordination among the designers in the Macintosh, Apple II, Lisa, and disk divisions. He wanted all of Apple's products to have a common design language: a unified set of defining shapes, patterns, colors, materials, and textures. "I want Apple's design to be not just the best in the computer industry, but the best in the entire world," he'd say.

Jerry Manock, Terry Oyama, and Apple II design head Rob Gemmell put together a design competition, open both to Apple's own designers and handpicked European invitees.

Manock outlined eight upcoming product lines, and—because he'd just read "Snow White" to his daughter the night before—nicknamed them according to the seven dwarfs. There would be a next-generation Lisa (Doc), a new Apple II model (Sneezy), a new Mac model (Happy), a laptop (Bashful), a new mouse (Sleepy), a line of printers (Grumpy), and an external floppy drive (Dopey). There actually was an eighth product, an external hard drive, but Manock had run out of dwarfs. Borrowing a name from *Bambi*, he called it Flower.

Apple would pay each competitor $50,000, but according to the Red Book—the documents that explained the contest—Apple would own their designs, and was under no obligation to hire the winner.

By March 1983, Apple had winnowed the field down to two designers. One of them, Hartmut Esslinger, had gone miles beyond the call of duty. At his studio in Germany, he had made full-scale, fully painted foam models—not just the eight required by the Red Book, but *40* Macs, printers, connectors, computer+phone systems, laptops, touchscreen tablets, and so on.

Esslinger had spent years designing the Walkman, Trinitron TV sets, and other personal electronics for Sony, a company Jobs dearly admired. After the crumbling of both his marriage and his firm in Germany, Esslinger had moved to California and set up a new company called frogdesign ("frog," he says, stands for Federal Republic of Germany).

Jobs offered Esslinger $2 million a year to take charge of all Apple design. For the first time, an Apple designer would report directly to Jobs, rather than to engineering.

Apple's own designers weren't universally thrilled by Esslinger's arrival, but Esslinger and Jobs bonded. They took long, conversational walks that covered history, philosophy, and culture.

The characteristics of Esslinger's Snow White design language would define the look of all Apple products from 1984 to 1990:

- **White.** Esslinger replaced Apple beige with an off-white or light gray. (But the term "Snow White" derived from the competition name, not from the plastic color.)

Esslinger produced dozens of design concepts in 1982 and 1983. (That's Steve Jobs writing on the prototype phone at left.)

- **Small and symmetrical.** "Smaller is friendlier," Esslinger says.

- **Smooth and grooved.** Esslinger eliminated the textured plastic of existing models; the smooth, reflective plastic of Snow White machines was interrupted only by a series of horizontal or vertical grooves, 2 millimeters wide and deep, 10 millimeters apart. Esslinger pointed out that they weren't purely decorative; they could also conceal ventilation slots.

- **Embedded logo.** On each Snow White product, the Apple logo appeared as a slightly bulging, 3D, inlaid shape.

Esslinger was also aghast at the state of Apple's manufacturing. "A car shop was painting the Apple II's. A body shop!" he says.

The paint itself was poisonous, and the resulting plastic could not be recycled. From now on, Esslinger said, the pigment would be incorporated into the plastic before molding.

Esslinger also wanted to eliminate the metallic paint that coated the insides of Apple's cases to prevent radio interference. "I mean, it was Poison 101," he says. Using a metal cage instead would release fewer toxins and, again, provide better recyclability.

Finally, Esslinger urged Apple to adopt a new method for molding its plastic. Typically, the side walls of a plastic shell can't be perfectly straight and rectangular—they must taper outward by at

Hartmut Esslinger on a 1985 talk show.

Insanely Great • 149

least 3 degrees toward the open side. Without this slight angle, known as draft, the plastic rips or scratches when it's removed from the metal molds.

But in designing Sony's TV sets, Esslinger had refined a technique called *zero-draft* molding. In this system, the inner mold is collapsible, and the outer mold is made of separate pieces that pull away in different directions. As a result, the finished plastic part can be removed cleanly, and its walls can be perfectly square, with flat, perpendicular sides. It's a cleaner, more modern look.

Eventually, a long line of important Apple products bore the Snow White look, including the Apple IIc (1984), ImageWriter II (1985), Apple IIgs (1986), Macintosh SE (1987), Macintosh II (1987), LaserWriter II (1987), Apple Scanner (1987), Macintosh IIx (1988), Macintosh Portable (1989), Macintosh SE/30 (1989), and the Macintosh IIfx (1990).

The ImageWriter II's cantilevered design left room for fanfold paper underneath.

And because Esslinger had set up frogdesign in California, every Apple computer's case would now bear the words "Designed by Apple in California."

Apple IIc

On April 24, 1984—three months after the Mac's launch party—another audience settled in to watch another Apple unveiling. This time, the event took place in San Francisco's Moscone Center, for an audience of dealers, developers, analysts, and press.

The lights dimmed. To the pearly sounds of a Fender Rhodes electric piano, a pop singer's voice entered familiar "Flashdance" territory:

> *Apple II forever!*
> *Making life better and better*
> *Apple II forever and ever,*
> *bringing the rainbow to you!*

The event name, "Apple II Forever," wasn't accidental. After all the hype about the Macintosh, Apple II fans had wondered if Apple was losing interest in their beloved machine.

Onstage, John Sculley held up what looked like a white book, about a foot square and only a couple of inches thick. It couldn't possibly be a computer, could it?

"Anyone want a closer look?" he said. On cue, the houselights came up. Strategically placed Apple employees reached beneath their seats. Each one pulled out an Apple IIc, fourth in the line of Apple II machines, the first Apple computer that wasn't beige. The Apple staffers proudly handed the computers to the dealers in attendance.

The *c* stood for *compact*, and yet it had, for the first time, a 5¼-inch disk drive built right into the side. The whole thing weighed 7.5 pounds, not including monitor or power brick. The IIc was Apple's first portable computer; it didn't have a battery, but you could flip out the handle on the back and carry it home from work, or chuck in your carry-on bag.

The IIc cost about half as much as the Mac. For another $200, you could get a compact color monitor, perfectly matched, that seemed to float, thanks to a cantilevered stand.

The Moscone event was a spare-no-expense, daylong extravaganza. Over dinner, dealers were

The First Flat Screen

Today, when every screen is a flat panel, it's hard to conceive of the era when every screen was a heavy, bulky cathode-ray tube (CRT) like early TV sets. As long as CRTs were the only way to display a computer's image, there would be no such thing as a laptop.

Incredibly, though, Apple sold its first flat-panel screen in 1984.

It was a $600 option called the Apple Flat Panel Display. It was black-and-white, of course, but it showed a full 80 columns of characters, 24 rows. It had no backlight, so it was hard to read in low light. It could display graphics, although because it was an oddball rectangle, they wound up distorted.

Apple created the Flat Panel Display for an obvious reason: It couldn't very well describe the Apple IIc as a portable machine if you also had to haul around a CRT monitor. If portability was more important to you than money, you could snap the Flat Panel Display right onto the IIc.

Very few people fell into that category. Apple soon discontinued the Flat Panel. Today, it's long lost to history—but it deserves a nod as the first flat-panel screen ever sold for mass-market computers.

The First "One More Thing"

After the Macintosh launch, Andy Hertzfeld, on a leave of absence, dreamed up a software utility that would address one of the Mac's most infuriating limitations. It would let you keep more than one program open at once, and hop between them—a baby step toward multitasking called Switcher.

After turning down a lowball offer from Microsoft, Hertzfeld accepted Steve Jobs's offer: $100,000, plus a royalty if Apple ever sold Switcher separately.

Jobs demoed Switcher live onstage at the 1985 shareholders' meeting. The audience gasped as they saw programs' windows switch like TV channels.

Jobs didn't give Switcher the ceremony Hertzfeld had hoped for. "It was fleeting in a low-key way, almost an afterthought, and only lasting a couple of minutes," he says.

But one aspect of the Switcher demo was historic. Steve Jobs presented it as a little delightful dessert. In fact, the way he worded it was: "Oh—and one more thing!"

It was the first time he'd ever used that phrase in a keynote presentation—but it certainly would not be the last.

Jobs, Sculley, and Woz present the Apple IIc.

treated to the music of Herbie Hancock—played *by* Herbie Hancock. At one point, a 6.2-magnitude earthquake struck 70 miles away; there was no damage at Moscone, but the floor shook and the hanging lights swayed. "How did Steve arrange *that*?" went the joke. But the Moscone event was only the beginning of the $20 million Sculley spent to market the IIc—25 percent more than he'd spent for the Mac. There were eight-page magazine inserts, newspaper ads, and TV ads during the Olympics.

Apple IIc
Sold: April 1984–August 1988
Price: $1,300 ($1,500 with monitor)
Processor: MOS 65C02 at 1.023 MHz
Memory: 128 KB, expandable to 1.25 MB
Equipment: Built-in floppy and keyboard; built-in circuits for mouse, floppy, 80-column text, printer, and modem

The handle flipped down into a stand, for ventilation and better typing angles.

Apple certainly expected the IIc to be a home run. After all, *Time* awarded it the 1984 Design of the Year, and New York's Whitney Museum of American Art acquired one as an exhibit.

But many consumers thought that *small* must mean *underpowered*; that had certainly been true of IBM's low-cost spin-off, the PCjr, and the Commodore 64. And in some ways, they were right. "This is not a technological improvement over the Apple II," sniffed *InfoWorld*'s reviewer. "The processor here is a toy."

Other customers shied away because there was no room for expansion cards. For most things that people bought cards for—modem, printer, external disk, 80-column display—Apple had built the circuitry right in, but that didn't seem to reassure people. All they could see was that expandability was far more limited than it was on the older IIe, which cost $300 less.

The company had hoped to sell 400,000 Apple IIc's in nine months—but the number was closer to 100,000 for the entire year. For much of 1984, Apple faced a mismatch for both Apple II machines: too many IIc's piling up in warehouses, not enough IIe's to meet demand.

In truth, the entire industry was still struggling. IBM's PCjr was a gigantic flop, and Wall Street was skittish about computers in general. Jobs, trying to cut costs, eliminated the perks he'd lavished on his team, like fruit juices, catered lunches, and overtime pay.

But this was an economic headwind too strong to dodge without bigger changes. And it would eventually bring the Two Steves era to a heartbreaking end.

14. Macintosh Office

By the end of 1984, 56 percent of companies with computers were using IBM PCs; a lowly 16 percent used anything from Apple. For the hundreds of millions Apple had spent on breaking into the corporate world, it had made little progress.

But Steve Jobs had heard through the grapevine about a new technology that might finally break the logjam. It wasn't a computer. It wasn't a printer. It was a technology that, for Apple, would change everything.

LaserWriter

From the day the Mac arrived, printing was a problem. The only option Apple offered was a $675 dot matrix printer: the ImageWriter. (It was essentially a rebranded printer from Japan's C. Itoh Electronics.)

ImageWriter printing was simple: In general, each dot on the screen produced a corresponding dot on the page. You got what-you-see-is-what-you-get printouts.

The chief virtue of the original ImageWriter was that it could print graphics.

When it came to typography, though, what you saw wasn't what you *wanted*. The sharpness of a screen image or a printout is measured by its resolution—how many dots it can display per inch (dpi). The Mac's screen resolution was only 72 dots per inch, so its ImageWriter printouts had jaggies—pixelated curves instead of smooth ones. It was a far cry from the professional, "letter-quality" printing that businesses expected.

Office daisy wheel printers produced text like a typewriter: by slamming metal letter shapes against an inked ribbon onto the paper. But a daisy wheel couldn't reproduce graphics or combine fonts. Steve Jobs hated daisy wheel printers.

In mid-1982, engineers at Apple's accessories office in Garden Grove, California, had been investigating a radical new printing approach. They'd gotten their hands on an early prototype of a Canon printing system, the LBP-CX10, intended for photocopiers.

In the Canon system, a laser "draws" the image of the page onto a rotating cylinder, creating patterns of static electricity. The cylinder rolls through black dust (toner), picking it up only where the surface was charged, and then rotates against paper, where heat fuses the blackness to the page. The result is clean, crisp, fast, nearly silent, very black printing. If Apple's engineers could use Canon's copier-engine technology in a printer, it would wipe away all of the drawbacks of dot matrix *and* daisy wheel printers. A laser printer could print text *and* graphics, with no jaggies.

In the end, their instinct was correct. But what saved Apple wasn't how the toner reached the paper. It was how typography reached the printer.

PostScript

The original Mac fonts were *bitmapped* fonts. Susan Kare had painstakingly drawn each dot of each letter, pixel by pixel, for each font, for each *size*. Her fonts looked good, therefore, only at the sizes she'd designed: 9-, 10-, 12-, 14-, 18-, and 24-point sizes, for example. If you tried to print a font in any other size, the Mac attempted to shrink or enlarge the original bitmaps, usually with hideous, jagged results. People learned to avoid nonstandard point sizes. The world was filled with newsletters featuring 24-point Helvetica and 12-point New York.

But change was coming.

In 1978, two bearded mathematicians at Xerox PARC discovered that they had a lot in common. John Warnock and Chuck Geschke both loved publishing and graphics; each had been educated with grants from DARPA (the military's advanced research wing); each was married, with two boys and a girl. Both felt that early Silicon Valley itch to change the world.

They came up with what they called a *page-description language*. That is, its computer code could describe the characteristics of a page, including text and pictures: "Draw a straight line, starting an inch from the top of the page. Turn left. Draw a curved line with radius .5 inches," and so on. They named it Interpress.

The beauty of this system was that it offered *device independence*: The same document could print at the highest quality on any printer: on an ImageWriter, 72 dots per inch; on a laser printer, 300 dpi; on a commercial printing press, 2,400 dpi.

When Xerox, in keeping with its long history of missing the boat, declined to commercialize Interpress, Warnock and Geschke quit PARC to start up their own company. They named it Adobe, after the creek behind Warnock's home in Los Altos. Another two-man, garage-based, Silicon Valley startup was born.

A bitmapped font (left) and a PostScript font.

In the spring of 1983, Steve Jobs heard about the project, now called PostScript, from yet another Xerox PARC alum, Mac manager Bob Belleville. The Lisa was about to ship, the Mac was coming along, and the Canon-based laser printer was underway in Garden Grove. Maybe, Jobs thought, Adobe's software and Apple's hardware could work together.

"John and Chuck were in their garage thinking about making laser printers, and we were in our garage working on ours," Jobs said. "We always felt that Apple should stand at the intersection of art and technology, and John [Warnock] felt the same way about Adobe."

To show Jobs what PostScript could do, the Adobe guys printed out a complex dental-office form, with tiny type and fine drawings of teeth. Jobs was blown away—and he wanted in. "I don't need the computer. I don't need the printer," Jobs told them. "I need the software."

Over breakfast at the Good Earth restaurant two days later, Jobs persuaded them to license their software to Apple. On the spot, he offered to buy 20 percent of Adobe for $2.5 million. To sweeten the deal, he promised to pay Adobe a $350 royalty on every laser printer Apple sold, with a $1.5 million advance on those royalties.

Sculley was shocked that Jobs would make such an offer to a scruffy startup with no product to show. (Six years later, Sculley would sell the company's Adobe stock that Jobs had bought for $2.5 million—for $87 million.)

The Mac itself had far too little memory and power to process the PostScript code. The solution was to pack the power into the printer itself, now called the Apple LaserWriter.

The LaserWriter wound up being a more powerful computer than the Mac; in fact, it was the most powerful computer Apple had ever made. It contained the same Motorola 68000 processor as the Mac—but it had 1.5 MB of memory, about 12 times as much as the Mac. It needed that much to store the image of a full page before printing it.

Fonts

In a PostScript font, letter shapes aren't stored as fixed maps of pixels. Instead, each character is a set of mathematical equations that describe the curves of each letter's outline. The PostScript printer fills in this outline with black.

Better yet, a PostScript printer can easily enlarge or reduce text; it just multiplies its equations by 0.3 or by 3 or by 33, for example, to create text that's much smaller or larger. The first LaserWriter could print type as big as 720 points (banner size) or as small as 3 points (prescription side-effects size). And because PostScript is a graphics *language*, it permits all kinds of wild text manipulations. You can stretch your type, print at an angle, distort it.

The original LaserWriter had more computing power than the Macintosh itself.

Adobe licensed a set of core font families from Allied Linotype and International Typeface Corporation—and in the summer of 1984, it all came together: the Macintosh, a LaserWriter prototype, and PostScript fonts. Jobs and the Adobe team watched the first page emerge, still warm, from the printer. It looked like it had been professionally typeset, with astonishingly crisp, smooth, black type—at 300 dots per inch.

The potential hit Jobs like lightning. It would change all the rules for businesses; now, ordinary workers could produce documents that looked almost typeset—right from their desks. "We knew we were going to hit it out of the park," Jobs said. "It was just a matter of how long it would take. You could sense the inevitability."

But not everyone *could* sense it. Apple's marketing managers confronted him about the printer's $7,000 price. It was almost three times the price of the Mac itself, and twice as much as the HP LaserJet, which contained the same Canon engine. Nobody would pay seven grand.

Jobs insisted that the LaserWriter would sell. The HP printer didn't have PostScript, and it wasn't networkable. Everybody in a small office could share a single LaserWriter. It wasn't expensive when you looked at it as a shared resource. Sculley and Jobs had dreamed up the Macintosh Office: a set of Macs, networked with cables, sharing a file server (a central shared hard drive), with a LaserWriter as the cornerstone.

At the January 1985 shareholders' meeting, they wouldn't just unveil the LaserWriter; they'd unveil the Macintosh Office. This time, surely, Apple would break into corporate America.

AppleTalk

In 1984, setting up an office network was infuriatingly complex and expensive. You'd need a file server—a computer that controlled the sharing of files and printers. You'd need a networking card for each computer, and hardware at each end of the chain of cables—a *terminator*—to prevent signals from reflecting off the end. All told, the price of connecting office computers was about $1,000 per machine—not counting the salary of the nerd you'd have to hire to configure them.

Stanford electrical-engineering PhD Gursharan Sidhu joined Apple in 1982, with an itch to design something radically different, a truly plug-and-play networking system: no central server, no terminators, no network administrator. You'd be able to add or remove computers and printers freely.

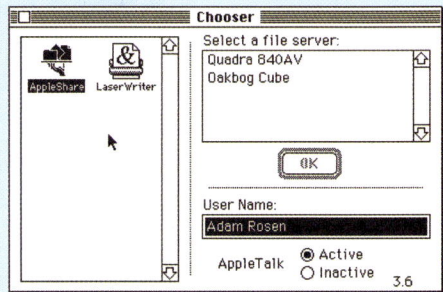

You wouldn't need networking cards for your Macs, either. You'd just connect a little $50 adapter to the printer port of each one, and then connect those adapters with cables. The icons of other Macs would appear in a little desk accessory called the Chooser, bearing plain-English names.

The new system was called AppleTalk. With a maximum cable distance of 1,000 feet and a maximum of 32 computers and printers, it was ideal for small offices, schools, and homes.

Eventually, hundreds of thousands of AppleTalk networks connected Macs all over the world. "The AppleTalk Personal Network is a breakthrough in price and value that brings Macintosh radical ease of use to work groups," raved the *Los Angeles Times*.

In 1993, Apple honored Sidhu by making him an Apple Fellow.

Office Unveiled

Sculley and Jobs did everything they could to make the 1985 Apple shareholders' meeting as triumphant, important, and ecstatic as the 1984 meeting had been. "Large corporations are telling us that they're going to use both Apple and IBM workstations, or that they want to use Macintoshes to talk with their IBM mainframes," Jobs said. "So for 1985, Apple proposes détente with IBM!"

This time, there were *two* Macs onstage, projecting onto *two* giant screens. In 1984, the enormous word MACINTOSH had scrolled horizontally across the Mac's screen; this time, the phrase was THE MACINTOSH OFFICE, and it spanned *both* screens. The point was clear: These Macs were, for the first time, connected.

Next, Jobs revealed the LaserWriter. The audience reception was rapturous.

Finally, he described the file server: a 20 or 40 MB hard drive with software that would handle various workers' access to shared files, folders, and printers. Eventually, it would accommodate email, multiuser databases, and shared calendars, too. The file server was coming in the fall, he said. "Today will be the beginning of an alternative to IBM's vision of the office—an alternative that starts with people rather than mainframes."

Soon, he went on, add-on AppleTalk products would come from other companies. Multiuser apps, file servers, connections to other kinds of computers. Apple would even be offering a little expansion card for IBM PCs, so that *they* could print on the LaserWriter, too. Détente!

After the keynote, attendees grabbed for the sheets coming out of the demo LaserWriters, astounded at the quality of its printouts.

Sculley bought yet another round of double-spread pages in magazines and newspapers, another round of TV ads. "AppleTalk is as simple to hook up as an extension cord," said one, "and almost as cheap." But today, if they remember the Macintosh Office campaign at all, people are most likely to remember only one ad: the one that aired on the Super Bowl.

Lemmings

The "1984" ad had become historic and insanely effective. Jobs hoped that Chiat/Day could make lightning strike a second time. Once again, copywriter Steve Hayden repurposed an ad from an earlier, unused campaign; again, it was high-concept; again, the implied target was an oppressive IBM. Ridley Scott was unavailable, so his brother, Tony, stepped in to direct the ad.

The ad was called "Lemmings," named for the rodents whose herd instinct is so strong, according to legend, that if one walks off a cliff, the others blindly follow. (It turns out not to be true.) In Apple's version, though, the lemmings are businesspeople carrying briefcases. Against a barren

*The "Lemmings" ad seemed to have all the ingredients of the "1984" ad—
but the result was a disaster.*

landscape, they march in lockstep, single file, blindfolded. They're whistling an under-tempo, mournful version of "Heigh-ho, heigh-ho, it's off to work we go."

"On January 23rd, Apple will introduce the Macintosh Office," a voice says. Now we see the front of the line, where we get the point: They're following each other off a literal cliff. They're *dying*.

Suddenly, one handsome guy in his thirties stops at the brink and removes his blindfold. "You can look into it," the voice-over continues, "or you can go on with business as usual."

Over the preceding months, Chiat/Day's ads had become increasingly explicit in their depiction of IBM as an evil empire. One ad showed people using chain saws to cut their PCs apart; another showed a businessman smashing his PC against the wall.

But this ad was extreme. Where the "1984" ad had been playful, rebellious, and optimistic, this one was depressing. Alienating. Arrogant.

Worst of all, it directly mocked the very people Apple was trying to win over: business owners currently using IBM equipment. It depicted them as mindless drones.

Neither Jobs nor Sculley loved the original storyboard, but Jay Chiat and Lee Clow pushed back hard. "I will put my whole reputation, *everything*, on this commercial," Clow said. He pointed out that Apple had had concerns about the "1984" ad, too—and it turned out to be a classic!

Sculley left the decision to Mike Murray, who pulled the trigger: the ad would air.

The decision made, Apple left nothing to chance. Newspaper ads teased: "If you go to the bathroom during the fourth quarter, you'll be sorry." A huge temporary screen stood at one end of the stadium, so that the Super Bowl crowd could watch the ad live. And 86,000 white cushions, bearing the multicolored Apple logo, padded the seats of the Stanford University stadium—a free ad in every TV camera shot. (Today, those cushions fetch a pretty penny on eBay.)

When the ad finally aired, in the final seconds of the game, Jobs and Sculley—along with Jobs's girlfriend and Sculley's wife—watched from their Apple-padded seats. But when it ended, there was no ecstatic cheering. "It must have been the only completely silent moment in Super Bowl history," Sculley says.

Apple's phone lines were suddenly flooded. Irate callers, disgusted by the ad, shouted that they would never buy another Apple product. Parent groups thought it depicted violence; school boards thought it glorified suicide. An *Advertising Age* magazine study found that fewer than 20 percent of viewers even got what it was about.

"We bit it, big-time," Hayden says.

Whereas the "1984" ad has been called one of the best ads ever aired, the 1985 ad would be called one of the worst.

Apple did not air another Super Bowl ad for 14 years.

What Happened to the Lisa

In hopes of reviving the dying Lisa, in January 1984, Apple relaunched it as the Lisa 2/5, with a 5 MB hard drive and a nicer price: $3,500.

Then, in January 1985, as part of the Macintosh Office effort, they renamed it the Macintosh XL, featuring a startup disk that ran the Mac's OS and most Mac programs.

Apple figured that its remaining stock of this Macintosh XL, priced at $4,000, would take 18 months to sell. Then, at last, the Lisa would, as Joanna Hoffman puts it, just "peter out into the sunset with nobody noticing." Instead, to Apple's astonishment, the Mac XL sold out in under three months. But Apple had no more to sell, and no more parts to make more. "It was a major, major scandal," Hoffman says. Mortified, she left Apple shortly thereafter.

For a time, the Lisa enjoyed a resurrection at the hands of Sun Remarketing, a company run by Apple dealer Bob Cook. He bought, on consignment, 7,000 older, broken, or incomplete Lisas, fixed them up, upgraded them, and sold them.

But one day in 1989, an Apple lawyer called. "We want to exercise our clause in the contract to pick up the computers that we own," he said. A line of trucks carried Cook's last 2,700 Lisa computers to a landfill in Logan, Utah, dumped them, and crushed them with bulldozers, ensuring that none of them could ever be recovered and sold.

And why destroy thousands of brand-new machines? For the tax write-off. Apple would be eligible for a tax break of 34 cents on the dollar—more than it would have netted by donating the machines.

In 2014, a Lisa fan in Alabama named Joseph Irvine came up with a thrilling plan: On the 25th anniversary of the Lisa's discontinuation, he'd excavate the landfill, live on TV, and recover those buried Lisas. He started a crowdfunding campaign on Kickstarter to raise $400,000.

It raised $85.

Celebration

In February 1985, Steve Jobs celebrated his 30th birthday with a black-tie party at the St. Francis Hotel. The San Francisco Symphony played; Ella Fitzgerald sang. (Bob Dylan had been unavailable.) Close to a thousand people attended, including original Mac team members who'd left Apple. Woz gave Jobs a special gift: a framed copy of the Zaltair brochure hoax he'd made for the 1977 West Coast Computer Faire. "As soon as he saw it, Steve broke up laughing," Woz says. "He'd never even suspected I'd done it!"

John Sculley raised his glass, and hailed his friend as "technology's foremost visionary."

It was the last time he would ever toast Steve Jobs.

Ghosts of the Missing Macs

For every product that winds up in the Apple Store, dozens never make it out of the lab. Apple may kill a prototype for any number of reasons—technological, economic, political. The internet is filled with cool-looking photos of canceled products.

In 1984, Apple began developing two new Macs. One, nicknamed BabyMac, featured a flat CRT screen, a faster chip (Motorola 68020), a handle, and a wireless keyboard and mouse. It was Hartmut Esslinger's "best design never to be produced."

But the most promising computer was the Big Mac: a powerful workstation with a vertical 15-inch screen, perfect for documents. It would run Unix, but the software would feel like a Mac (much like Mac OS X, which came along 17 years later).

For a time, Big Mac was intended to be part of the Macintosh Office initiative; Apple made six prototypes. But Jobs's successor in the Macintosh division, Jean-Louis Gassée, canceled the project. Neither the BabyMac nor the Big Mac made it past the frogdesign model stage. But echoes of their designs wound up in the iMac and Apple's later keyboards and mice.

15. Rift

As he surveyed the wreckage of the "Lemmings" ad, John Sculley longed for a return to something like Apple's ascendance in the summer of 1984. He'd had great hopes for Macintosh Office. It had every hallmark of a grand slam, with one tiny footnote: It didn't really exist.

There was no file server. There was a design for one, about the size of a paperback book, but its file- and printer-sharing software was not even close to being ready. The expansion card for IBM PCs, intended to let them join a Mac network, was also vaporware. To the business world, the spectacular rollout of Macintosh Office began to look like an exercise in hype.

The shareholders' meeting had left some bad feelings internally, too. The Apple II was responsible for 70 percent of Apple's revenue, but it didn't even get a *mention* in the all-day presentation. The Mac team was given front-row seats in the Flint auditorium; the Apple II team had to watch the feed in the overflow theater. They called the event the "Macholders' meeting."

Woz called Sculley directly to complain about Apple's corporate disregard for his masterpiece for months. "If you worked in the Apple II division, you couldn't get the money you needed or the parts you needed," he says. "I thought that wasn't fair."

Meanwhile, he missed the startup days. "I want to go back to the garage, where you don't have to fill out a dozen forms to get something done," he told a reporter.

Shortly thereafter, he left Apple for good.

Woz wasn't the only one. Dozens of executives, managers, and engineers left Apple. The organization chart was full of "to be hired" notations. In the Apple II division, the role of marketing manager turned over so many times that someone posted a sign: "If your boss calls, be sure to get his name."

To outsiders, Sculley could point to a record profit for the fourth quarter of 1984: $46.1 million, on record sales of almost $700 million. But in fact, those figures were far lower than Apple's own projections. Expecting better sales, Apple had built $100 million worth of more Macs and Apple II's than it could sell. They were piling up against the walls of the Fremont factory, threatening to block the hallways; Sculley ordered all four factories to shut down for a week.

Sculley was realizing that his history of marketing a consumer product—Pepsi—left him unprepared for four critical principles and lessons.

First, during the months of enormous ad spending, he had cut Apple's research and development

Woz After Apple

After his plane crash in 1981, Woz finished his college degree at UC Berkeley—under the pseudonym Rocky Raccoon Clark (his dog's name and wife's maiden name). Then he dove into a new venture just for fun: He conceived, funded, and helped to produce a music festival.

"I had so much more money than I could ever dream of spending. I was 30 at the time and probably worth a hundred million dollars or more," he says. "I thought: My God, why not put on a big progressive country concert with these groups I loved?"

It was called the US Festival (pronounced "us," not "U.S."); the bands included Fleetwood Mac, Tom Petty and the Heartbreakers, the Police, and Talking Heads. Woz lost millions of dollars on the venture, but enjoyed the experience enough that he produced a second US Festival in 1983.

And then he went back to Apple. He just wanted to do what he loved best: dreaming up cool engineering designs.

Apple offered him a small cubicle in the Apple II division. He got a thunderous ovation when he arrived, especially because he announced that he'd just bought $7 million in Apple stock. But at the same time, he felt strangely nonessential and out of place. He also continued to resent the Apple II's treatment as a second-class sidekick to the Macintosh. "Apple's direction has been horrendously wrong for five years," he told the *Wall Street Journal*.

And he began to long for the thrill of creating something entirely new.

When Woz left Apple for the second and final time in February 1985, he started up a new company, CL9, where he engineered and sold the first programmable universal TV remote.

Still, for years, he appeared at events on Apple's behalf and consulted on various Apple II projects. To this day, Woz receives a small salary from Apple and remains a shareholder.

Mostly, though, Woz has filled his post-Apple life with whatever he thinks is fun. He taught computer classes to middle schoolers and joined technology startups. He appeared as a contestant on *Dancing with the Stars*, started a Woz cryptocurrency, played Segway polo, had some kids, settled down with his fourth wife, Janet, and won boxcars' worth of awards and honorary degrees. He's given away a huge amount of money—he was a founding sponsor of the Tech Museum, Silicon Valley Ballet, and the Children's Discovery Museum of San Jose—and he travels the world giving talks.

As he writes at the end of his memoir, *iWoz*, "I hope you'll be as lucky as I am."

budget. "Robbing research to pay for advertising turned out to be a big, big mistake," he says. "We lost our new-product edge."

Second, in consumer goods like soda, the value of your product remains constant while it sits on the store shelf. In the computer business, unsold inventory loses value rapidly.

Third, the home and education computer markets were ruthlessly seasonal. Apple sold almost 40 percent of its computers during the holiday season. If you guessed wrong about how much product to build, you were hosed.

Finally, Sculley was learning that tech companies amortize their R&D costs into the price of a product from day one. His competitors *expected* to lower their prices over time—not just because of competition, but because they had front-loaded R&D expenses into early pricing. That never happened at Pepsi. "Our perspective had been hopelessly wrong. High tech could not be designed and sold as a consumer product," Sculley concluded.

Of course, the 1985 slump was hitting the entire industry. But Apple's struggles, gleefully reported by the press, seemed to be intensifying in every department. As the stock sank, Sculley got 50 calls a day from screaming investors: "How are you going to pay me back for the money I've lost?"

The French Connection

In late 1984, Jobs and Sculley tried to arm Apple for the war with IBM. They discussed buying Compaq and getting into the PC business. They held discussions about co-ventures or buyouts with Wang, AT&T, General Electric, General Motors, and Ross Perot's company, EDS.

In February 1985, a glimmer of hope shone from across the pond. French president François Mitterrand had ambitious plans to leapfrog the U.S. and Japan in computer literacy. He wanted to set up 40,000 computer centers to prepare the French people for the digital age. The charming, easy-to-use Mac seemed like the perfect machine.

Sculley and Jobs spent months in talks with French diplomats, tantalized by a deal that could be worth $500 million. If it went through, Apple's excess-inventory problem would disappear in a heartbeat.

Finally, they sat down at Élysée Palace, Mitterrand's official residence, over an "exotic meal of some kind of innards," as Sculley remembers it. There was much discussion of a new Mac factory in Marseille, and very little discussion of how France would pay for all of this.

In the end, it was all for nothing. The potential deal leaked, the press dumped on the idea of an American factory on French soil, and critics predicted a devastating blow to France's own computer industry. The deal was dead.

Power Shift

One aspect of Apple's organizational chart had always been awkward. Steve Jobs had had two roles: chairman of the board, above Sculley, and vice president of the Macintosh division, below him.

But when Jobs took over the newly combined Lisa/Mac division in the spring of 1984, he was now, at 29, an executive vice president of a $1.5-billion-a-year company, with operational power that he'd never had before. Sculley was fond of saying, "Apple has one leader: Steve and me," but this was different. The two-man power structure made other executives feel shut out of the decision-making.

Worse, in Sculley's view, the promotion seemed to make Jobs even more sharply demanding of others. "Not just the Macintosh division," Sculley says. "He began to dominate many of the conversations and discussions within the executive staff." For example, Jobs continued to show disdain for the Apple II team, calling them "bozos." The tension between the Mac and Apple II divisions, Sculley says, "could have cracked steel."

In early 1985, some board members were growing concerned. They told Sculley that he didn't seem to be running the company; he seemed to be *sharing* it. "I was startled," Sculley says. "I had always thought that part of my role was to help Steve grow, so that someday the board would have the option of allowing him to run his own company."

The Macintosh group, meanwhile, wasn't performing. A year after the first Mac's debut, there still wasn't a design for a successor. Critics were complaining that the much-vaunted LaserWriter, even with all of its horsepower, took an eternity to print graphics.

And the infernal file server software was still nowhere.

Sculley's relationship with Jobs was fraying. They once were like brothers and could agree on anything; now, under the strain of the company's misfortunes, they fought about everything.

Sculley saw Jobs as a toxic force that was burning people out. Jobs saw Sculley as an empty suit with no passion for products. "He had no idea how products are created," Jobs said. "After a while, it just turned into arguments."

Every computer company's sales were crashing; sales nationwide dropped 20 percent in March 1985 alone. But Apple's March figures were especially shocking: It sold only 10 percent as many Macs as it had projected.

The stress drove Sculley to his breaking point. He confronted Jobs. "I changed my entire life to come and work with you, Steve. But this is really not going to work," he said. He told Jobs that he planned to recommend to the board that Jobs step away from the Mac division.

"If you do that, you're going to destroy this company! I'm the only one who understands enough

> ### Apple Labs
>
> In December 1984, Steve Jobs met Steve Kitchen, an enthusiastic engineer and entrepreneur. His latest invention was a 4 x 5–inch flat-panel screen: a vacuum fluorescent display, clearer, bigger, and more power efficient than the half-height thing Apple was selling for the Apple IIc.
>
> A small, flat panel was a big deal. It could be the key to a thin, book-sized, mobile Mac. A Flat Mac!
>
> According to a subsequent lawsuit, Jobs offered Kitchen's company, Woodside Design, a $10 million deal to develop the screen. Jobs even found a building in Menlo Park, 20 minutes away, that could house the new R&D enterprise. It would be called Apple Labs.
>
> Apple Labs would solve everyone's problems. Sculley could run Apple without interference. Jobs could get back to developing insanely great products. Peace and progress could resume at the Macintosh division.
>
> The board, however, shut the pitch down. The two Steves (Jobs and Kitchen) were asking for $20 million to build a screen-making factory. The cost and time overruns of the existing Mac factory in Fremont gave them very little confidence in the Steves' estimates.
>
> Throughout the spring of 1985, Jobs agonized over pursuing the Apple Labs thing. In the end, of course, it never came to pass—and neither did the big Apple payday for Steve Kitchen. The project's collapse devastated Kitchen and nearly wiped out Woodside, which had already transitioned most of its staff to the Apple project. (They had even ordered an ice cream cake shaped like a flat-panel screen to celebrate what they assumed would be a triumphant board presentation.)
>
> Kitchen went on to sell his screen to the military for use in aircraft cockpits—and to sue Apple, in 1987, for $1 billion, for reneging on the deal. He did not prevail.

around here about manufacturing and operations," Jobs fumed. "If I'm not overseeing this, we're not going to get any new products out, and we're not going to succeed."

Board Meeting

The Mac's sluggish sales sent Steve Jobs into a depressive funk. "I don't understand it," he'd say. "Why isn't it selling?"

"I never saw Steve like that before, and I don't think we ever saw him that way since," remembers Bob Belleville. "He lost his confidence."

As his relationship with Sculley foundered, Jobs pulled the chairman card. He ordered Sculley to cut the Apple II's marketing budget, and to use that money to promote the Mac instead.

Sculley refused. "The Apple II is our only positive cash flow for the next several years," he argued. "If we do what you want, we could bankrupt Apple!"

They were at an impasse.

As Sculley had promised, he called a special board meeting. He and Jobs would each make their arguments, and the board would decide how to proceed.

Apple's board consisted of seven men, mostly powerful executives at other companies. Arthur Rock had become inconceivably rich by investing early in Fairchild Semiconductor, Intel, and Apple. Henry Singleton, the oldest member, had built the engineering conglomerate Teledyne from scratch. Peter Crisp was the managing partner of Venrock, another venture capital outfit. Philip Schlein was the CEO of Macy's California, chosen for his retail expertise. The internal directors were Jobs, Markkula, and Sculley.

On April 10, they gathered in the boardroom on the third floor of De Anza 2, home of the top executives. Sculley summarized his standoff with Jobs, and offered the board a choice. Option 1: Jobs leaves the Macintosh division, but remains Apple chairman and chief visionary, and runs a new new-products group—Apple Labs, or whatever.

Option 2: Sculley quits.

To the board, Sculley's ultimatum was a shocker. The meeting ran until 9 p.m., broke for the night, resumed the next morning, and dragged on until late in the afternoon. At one point, the board asked Sculley to leave the room; at another point, Jobs.

Finally, the board voted. It was unanimous. Jobs would step down as head of the Macintosh division.

Jobs, shattered, bolted from the conference room.

Del Yocam was just outside, waiting to make a presentation about the Apple II. "Steve came out and came up to me and put his arms around me—just crying, crying, crying, crying," Yocam says. "He was broken."

Sculley, also gutted, grabbed his briefcase and strode out of the office without a word.

In the weeks that followed, Jobs wrestled with the board's decision. Sometimes he was furious, insistent that Sculley had betrayed him. Other times, he sought a truce, reminding Sculley of the better days and promising that they could mend their relationship.

Jobs and his closest allies—Mike Murray, Debi Coleman, Susan Barnes (Coleman's successor as Mac financial controller), and Belleville—spent the days on calls, long walks, and house visits, discussing the next move.

Murray, Coleman, Belleville, and Barnes.

Apple's executive staff had a meeting scheduled for May 24, kicking off the Memorial Day weekend. Sculley would have to miss it. He was supposed to fly to Beijing to meet with the vice premier of China, to discuss the possibility of an enormous deal: supplying China's 980,000 schools with computers.

To Jobs, Sculley's trip was an opportunity: He intended to spend the week of Sculley's absence selling each board member, one at a time, on a new plan. Bob Belleville would lead engineering, Debi Coleman would run manufacturing, sales head (and former Columbia University football coach) Bill Campbell would take sales. Susan Barnes would be controller. And Jobs would become president and CEO of Apple.

But the night before the staff meeting, Apple lawyer Al Eisenstat and his wife hosted a barbecue at their home, attended by Sculley, Sculley's wife, and Jean-Louis Gassée, the colorful, outspoken head of Apple France, who was in Cupertino to discuss a possible new role. Gassée cornered Sculley in Eisenstat's den.

"If you get on that plane and go to China, you'll never come back an employee of Apple," he said. "Steve's planning a coup."

Sculley canceled his trip.

E-Staff Meeting

The executive staff meeting on May 24, 1985, has become legendary. Thanks to its inherent drama, colorful characters, and enormous stakes, it became the climactic scene of many an article, book, and movie about Apple.

Jobs showed up in a suit, late. He and Sculley sat at opposite ends of the long table.

"Steve, I heard that you were planning a coup," said Sculley, ignoring the printed agenda.

Jobs hadn't known that Sculley knew about the plan. ("I made the mistake of telling Gassée," he would say.)

But he didn't miss a beat. "I think you're bad for Apple and I think you're the wrong person to run this company," he said. "You have no understanding of the product-development process. You don't know how manufacturing works. You're not close to the company. The middle managers don't respect you."

"I made a mistake in treating you with high esteem," Sculley said, his childhood stammer returning.

Jobs reiterated that Sculley should step down. "I could run the company," he said.

"Well, obviously I can't run Apple if the executive team doesn't want to work for me," Sculley said. "Let's just go around the table and ask each person what they think."

Sculley maintains that he had no idea what they might say. Jobs, until his dying day, believed that the whole thing was a setup, an ambush.

Del Yocam had to go first. "I twisted my chair towards Steve," he says, "and told him how much our friendship had meant to me, how much I loved him, and that I wanted him to stay at Apple." But then he swiveled his chair to Sculley and handed him his support.

One by one, the others—Eisenstat, Campbell, Regis McKenna, CFO Dave Barram—offered variations of the same speech: "Steve, you're the visionary of this company. We love you. But you're not ready; we support John." HR director Jay Elliot cast the lone vote for Steve.

Jobs felt gut-punched, unable to breathe. He hadn't actually been fired—but maybe being ripped from power in the company he'd built was even worse. He stood shakily, tears in his eyes. "I guess I know where things stand," he managed, and then left the room.

It was a defining moment of his life, his first major humiliation, the first reminder of his professional fallibility. He never got over it.

Reorg

In the aftermath of the May 24 meeting, both Sculley and Jobs felt unmoored. Jobs contemplated leaving Apple completely; so, ironically, did Sculley. "I don't know whether I can go through with this," he told Eisenstat. "I'm going to resign."

The next day, Jobs called Sculley to propose a long walk in the hills around Stanford University. He'd spent 24 hours dreaming up alternatives to losing his operational role. Maybe *Sculley* should take the role of chairman, and *Jobs* should become CEO. Maybe they should split the company: Sculley would handle marketing, Jobs would handle product.

Sculley shot down every idea. The only way forward was for Jobs to surrender any management role.

In the following four days, Jobs and Sculley each sought private audiences with the final authority, Mike Markkula. Each made his own pitch for remaining in power. Sculley spent four hours at Markkula's home; Markkula visited Jobs's home in Woodside to meet with him and his four loyalists (Murray, Barnes, Coleman, and Belleville).

But once he'd heard the arguments for Jobs's plan, Markkula wasted no time. "I said I wouldn't support his plan, and that was the end of that," he says. "Sculley was the boss."

On May 31, Apple's middle managers crammed into the De Anza 2 conference room to hear an update from Sculley. It was all horrible. He broke the news of an after-tax loss of $17.2 million—the first quarterly loss in Apple's history—and then announced a new organizational chart.

No longer would there be separate Apple II and Mac divisions, each with its own sales, marketing, manufacturing, and product-development infrastructures. Now it would be two divisions: a newly merged marketing and sales operation, led by Bill Campbell; and operations (manufacturing, product development, distribution) headed by Del Yocam.

Jobs stood in the back of the room, staring at Sculley with fury. His name appeared on the new org chart, too—in a box labeled "Chairman," without any lines connected to any other part of the company.

Some Apple II employees enjoyed the schadenfreude of seeing Jobs fall. But the rest of the company felt "somber, depressed, and uncertain about the future," says Hertzfeld. It all reminded him of Black Wednesday.

Two weeks later, Sculley announced that he would lay off 1,200 people—more than 20 percent of the company. He also announced that the company would shut down its plants in Carrollton, Texas (Apple IIe and IIc); Millstreet, Ireland (peripherals), and Garden Grove, California (keyboards). The remaining factories in Fremont, Singapore, and Cork, Ireland, would have to pick up the slack.

Summer of '85

After his demotion, Jobs spent a few days at home with his girlfriend Tina Redse, listening to music with the shades drawn, ignoring the ringing phone. The three men Jobs admired most as father figures—Sculley, Markkula, and Arthur Rock—had all turned on him. "What had been the focus of my entire adult life was gone, and it was devastating," he'd say.

Sculley gave Jobs a new office in Bandley 6, a small, sparsely occupied building across the street from Bandley 3. Jobs called it "Siberia."

"I'd get there, and I would have one or two phone calls to perform, a little bit of mail to look at," he said, "and I would get depressed and go home in three or four hours, really depressed. I did that a few times and I decided that was mentally unhealthy. So I just stopped going in."

Mike Murray joined him in Bandley 6. He, too, had been left without a mandate. "I sat there for two or three weeks with absolutely nothing to do. I received no phone calls, no meeting requests, no nothing. Finally, Bill Campbell and I negotiated a small severance package, and I left. I was greatly depressed. No one ever said goodbye or thanks or anything. It was a personal train wreck."

Wayne Rosing and many of his engineers quit Apple. Soon, Bob Belleville quit, too. "I was so tired I couldn't see straight," he says. "I went on leave, and I never came back."

The company's center of gravity was shifting.

Eventually, Jobs picked himself off the floor. "I went for a lot of long walks in the woods, and didn't really talk to a lot of people, and gradually my spirits started to come back."

He spent the summer traveling with Redse: France, Italy, Russia. At some of the stops, he represented Apple; the rest he devoted to healing, thinking, and planning.

He received three offers to teach at colleges. Representatives from both political parties encouraged him to run. There was an option to become a passenger on an upcoming space shuttle launch. He turned all of it down.

But by the time he returned to Cupertino, Jobs had an idea for his next chapter. He was still young, passionate, and fantastically wealthy; he wanted to start again.

"I was writing down what were the things that I cared most about, that I was most proud of personally, about my ten years at Apple," he said. The top two items on his list were creating products and education.

Jobs asked the Apple board if he could speak at their September 12 meeting.

"I've been thinking a lot, and it's time for me to get on with my life," he began. He planned to start a new computer company, he said, but they shouldn't worry; his new product wouldn't compete with Apple's machines. It would focus on powerful workstations for college kids.

He also mentioned that he'd be taking a handful of people with him, "very low-level people that you won't miss."

The board wished him well, offered to buy 10 percent of the new venture, and even said he could remain on Apple's board.

But soon thereafter, Jobs presented Sculley with a list of the five people who intended to join his new venture. They were Apple Fellow Rich Page; Mac software head Bud Tribble; Mac engineering manager George Crow; financial controller Susan Barnes; and Dan'l Lewin, the architect of the University Consortium, and Apple's gateway to the education market.

These were *not* "low-level" employees; these were people who knew Apple's internal budgets, schedules, and product plans.

Gassée was outraged. Campbell was "fucking furious." "He said he was going to take a few middle-level people," railed Arthur Rock. "It turned out to be five *senior* people." Even Markkula, who rarely rose to anger, was offended. "He took some top executives he had secretly lined up before he left. That's not the way you do things."

Apple sued Jobs for $5 million, accusing him of "secretly scheming" to walk away with confidential details about the canceled Big Mac computer, the Unix-based workstation that Page had worked on. The new product Jobs intended to build sounded an awful lot like the Big Mac.

Jobs had thought he'd been transparent with the board. Deeply offended by the legal action, he launched a counteroffensive in the press. "When someone calls you a thief in public, there's just a time to say something," he said.

He invited reporters he'd met over the years to his house in Woodside. "Barely furnished, his home has the spartan air of a guerrilla camp," went the *Wall Street Journal*'s account the next day. "There are no curtains in the dining room and little more than a black baby grand piano and a bookcase in the living room. But an Apple Macintosh personal computer is perched on a shelf in the pantry."

Jobs no longer had any interest in the figurehead "chairman" title. On September 17, he drove to Mike Markkula's house and handed him a resignation letter that he'd printed on the LaserWriter.

> The Company appears to be adopting a hostile posture toward me and the new venture. Accordingly, I must insist upon the immediate acceptance of my resignation. As you know, the company's recent reorganization left me with no work to do and no access even to regular management reports. I am but 30 and want still to contribute and achieve. After what we have accomplished together, I would wish our parting to be both amicable and dignified.
>
> Yours sincerely, steven p. jobs

Jobs owned 11 percent of the company's stock, worth over $100 million. In a matter of months, he'd sold all of it except one share, which he held on to so that he'd get the Apple annual report and be allowed to attend shareholder meetings.

He gave a couple of interviews about his new venture, which was to be called NeXT. He mocked Apple's lawsuit. "It's hard to think that a $2 billion company with 4,300 employees couldn't compete with six people in blue jeans."

And he never spoke to John Sculley again.

> **Apple vs. Jobs**
>
> After four months, Jobs and Apple settled their lawsuit. The January 1986 agreement gave Apple the right to inspect prototypes of NeXT's computers to make sure they didn't incorporate Apple's "trade secrets." Jobs also had to promise that NeXT's computers would be more powerful than anything Apple made (and therefore wouldn't compete with Apple's home-and-student market), and that he wouldn't hire anyone else from Apple for six months.
>
> In short, the agreement put on paper the same restrictions Jobs had offered the board in the first place.

PART 2
Interregnum

16. New Ideas

It's not unusual for the founder of a startup to step aside as the company grows. In fact, it's very rare for one to remain at the helm.

"The evidence is strong that, in general, people with startup skills will not have the skills necessary to run a fast-growing enterprise," said Stanford management professor Steven Brandt at the time of Jobs's departure. "They can adapt, get out of the way, or fail. You can see all of those in the Apple situation."

Some at Apple demanded his return—there were "We Want Our Jobs Back" T-shirts—and some, weary of his erratic behavior, were relieved to be rid of him. Apple's stock jumped up by 7 percent when Jobs departed, and a few analysts put Apple back on their "buy" lists.

John Sculley began convening his executive team—Yocam, Campbell, Gassée—each morning. He told investors that 1985 had been "the year Apple grew up," and that it was now a "new Apple."

But internally, Apple was a shambles. To reach his office, Sculley had to pass clumps of distraught employees, some sobbing. Laid-off workers carried their stuff out to their cars. Reporters walked among them, looking for good quotes. "It was a mess, just an absolute disaster," Sculley says.

Part of the problem was that Sculley had reorganized the company along functional lines (sales, marketing, product) instead of product lines (Apple II, Macintosh), and the changes affected every single employee. Everyone was being moved into new buildings, reporting to new bosses.

In 12 months, Apple had experienced its first quarterly loss, failed a third time to capture the business market, laid off a fifth of its workforce, and lost both of its founders. Software developers were switching their efforts to write for IBM PCs, and even Apple dealers were changing their allegiances. Bill Gates called to suggest that Apple license its technology, to foster a marketplace of Mac clones. Rumors swirled that Apple was ripe for taking over.

"It was as if people were standing a deathwatch, waiting for the final collapse," Sculley says.

Sculley himself was miserable. Morale had flatlined. No new Mac models were in the pipeline. His wife, dismayed by his 18-hour workdays, had retreated to their house in Maine.

Whatever you thought of Steve Jobs and his management style, one thing was for sure: He always seemed to know exactly where technology was going, and what the public would want from it. To the employees, Apple had always been his company. Who could rival his taste, his eye for talent, his vision of the future?

With Jobs gone, the world wondered where Apple's new ideas would come from. Sculley wondered, too, and answering that question would preoccupy him for the next two years.

PageMaker

By the end of the 1985 summer, there were signs of progress. Yocam and Coleman managed to unload thousands of deeply discounted, unsold Apple IIc's in a European deal. Coleman consolidated Apple's 1,500 parts suppliers down to 250, whittling down the time to order from 15 weeks to 7. Bill Campbell set up a system that would track how many of the computers sold to dealers were actually selling through to *customers*, so that Apple would never again have a holiday season like 1984–85.

And the LaserWriter was a huge hit. "The Mac-cum-LaserWriter may well become, besides a corporate fixture in typesetting departments, the central facility of a new breed of printing centers springing up across the land," predicted the *New York Times*. " 'Just bring in your disk and we'll do the rest.' "

A market of graphic designers and printing shops, however, would not be big enough to save Apple. The company needed something more like a miracle—and a former newspaperman named Paul Brainerd supplied it.

At the University of Minnesota, where he'd earned a master's degree in journalism, Brainerd had edited and produced the school newspaper. That's where he learned the antiquated, messy method of laying out newspaper pages: with X-Acto knives, razor blades, and wax on the back

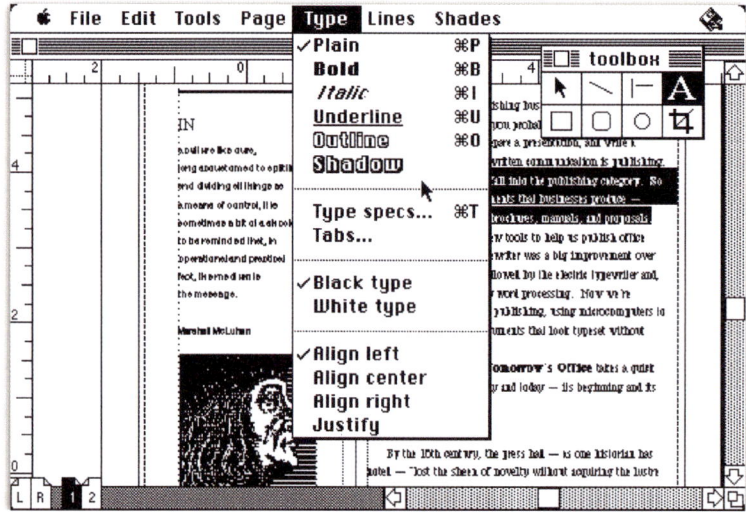

PageMaker introduced the world to the floating, movable tool panel.

of printouts. Nine years later, after jobs at a newspaper and a publishing software company, he rounded up five engineer buddies to start a software company of his own.

He called it Aldus, after Aldus Manutius, the Renaissance-era printer who invented italics. Brainerd's big idea was to write a program that would computerize the grisly layout process. You could paste your articles into columns, drop in photos and graphics, choose your fonts and sizes, all on the screen—and then print it on the spot. No razor blades, no wax.

Brainerd called the software PageMaker, and he called the process desktop publishing.

Brainerd knew about the Macintosh. In early 1984, a salesman from Apple's Bellevue, Washington, office had visited Aldus as part of Apple's developer evangelism efforts. "He didn't know us from Adam," Brainerd says. "He got two Macs out of his trunk, brought them into the office and said, 'Here. You take them for the next six months.'"

Brainerd saw instantly that PageMaker's simple layout tools, together with Adobe's PostScript graphics and Apple's LaserWriter, would make a $10,000 system that could replace an entire universe of ink, sweat, and printing presses. "Freedom of the press is guaranteed only to those who own one," once wrote journalist A. J. Liebling—and now, almost anyone could. The Apple-Adobe-Aldus trio would democratize publishing.

Sculley loved Brainerd's vision and instantly embraced the potential. He began pumping $1 million a month into advertising the system—Macintosh, LaserWriter, PageMaker. Unsolicited testimonials poured in from small businesses, offices, churches, schools, and small publishers. Dan Putnam, Adobe's second employee, remembers opening the first envelopes that came in: a pornographic lesbian newsletter and a fundamentalist Christian sect. "It wasn't exactly what we had in mind, but we gave them the voice to present their point of view," he says.

In Providence, a guy named Tom Oat produced his business newspaper entirely on four Macs and a LaserWriter. "Two writers, myself, and a copy editor are putting the paper out," he said, "for around $13,000 capital investment, which is very very small."

"It's what saved Apple from going completely out of business," says Bob Belleville. "Had the laser printer not been there, the desktop publishing business wouldn't have evolved, and nobody would have bought those Macs in the quantities that they were bought. We did not forecast that."

The desktop publishing revolution was the rare stroke of good luck for Apple in 1985. Wall Street analysts had predicted that Apple would post a loss for its fourth quarter; instead, Apple eked out a $22.4 million profit.

On the other hand, there was *still* no file server, still no networking card for PCs. Marketing head Bill Campbell admitted to the press that the whole "Macintosh Office" marketing concept was a dud. "It made people concentrate on what we didn't have, rather than on what we had," he said. "It was a mistake, we admit it."

ATG

In 1984, just after the Mac started shipping, Wayne Rosing had started up what he called the Education Research Group—a skunkworks with a mission to cultivate new technologies for an inexpensive color Mac. When Rosing left Apple in 1985, Larry Tesler took it over, and, in 1986, gave it a new name: the Advanced Technology Group (ATG).

It was a well-funded, pure-research playground that reminded Tesler of PARC. Its 100 engineers had no deadlines and no executive mandates. They could pursue ideas that might or might not make it into a product. The offices were in Apple's City Center, far from regular executive scrutiny.

It was also a perfect place for burned-out engineers to heal. "Sometimes it was 'This guy's fried,'" said Tesler. "And we'd give them a year in ATG to get a refresh break."

ATG groups looked into speech recognition, educational tech, human-computer interaction, networking, collaborative computing, and multimedia. Over the years, ATG became a natural home for Apple Fellows, those engineering celebrities with no particular assignment.

City Center is two connected eight-story buildings.

Many ATG efforts never went anywhere. It started up an artificial intelligence group, for example—"and then we realized we were about 40 years ahead of ourselves and maybe we should come back to that later," Tesler said. Some of its developments surfaced in Apple products years later.

But a few ATG projects led to technologies that changed Apple—and, often, the world—for good. They included ColorSync (matching screen colors and printed colors), AppleScript (simple programming of Mac functions), HyperCard (software construction kit for nonprogrammers), PlainTalk (speech recognition), and the Newton's handwriting software (another story).

Apple IIgs

By 1986, three million Apple II's occupied desks in homes and schools all over the world. Its design was now ten years old.

Sculley had learned his lesson about letting the Apple II languish, but trying to update it was dangerous business. How do you incorporate all the advances in processors, color, and sound without breaking the quirky circuitry that 10,000 existing Apple II programs relied on? How do you move forward while maintaining backward compatibility?

The answer turned out to be the Mega II.

Apple IIgs
Sold: September 1986–December 1992
Price: $1,000 ($1,900 with monitor and drives)
Processor: WDC 65C816 at 2.8 MHz
Memory: 256 KB, expandable to 8 MB
Equipment: 7 slots; color at 320 × 200 or 640 × 200; Ensoniq ES5500 sound-synthesis chip

The IIgs was the first Apple II to come with a mouse—and Apple's first color OS.

It was an Apple-II-on-a-chip, a single integrated circuit that combined all the functions of an Apple IIe's *30* chips. When engineers Dan Hillman and Jay Rickard created the Mega II, it was just a side project; nobody had any particular plans for it. "The Mega II just seemed like a way to make the Apple IIc smaller," says Hillman.

But late in Woz's two-year stint back at Apple, he'd become interested in creating a *16-bit* Apple II. (A 16-bit computer is faster than an 8-bit one because, other factors being equal, it can move data in bigger chunks.) Of course, a 16-bit machine would be totally incompatible with the Apple II's traditional 8-bit circuitry and all those thousands of programs.

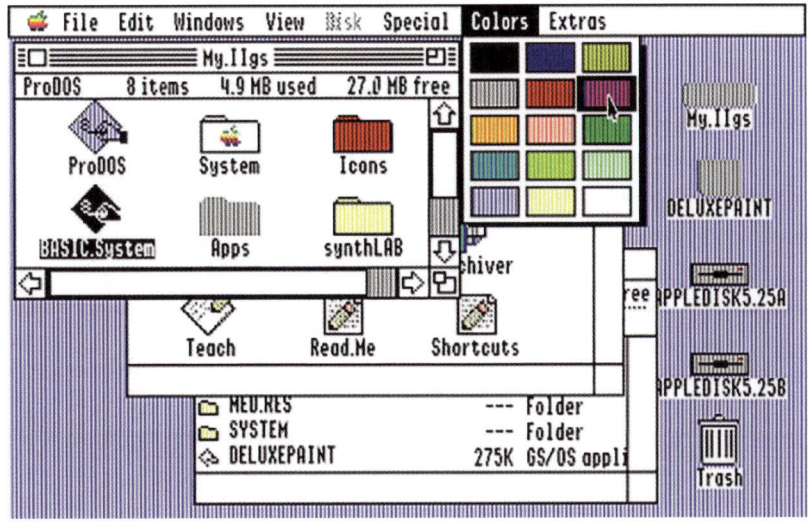

GS/OS, as the Apple IIgs's OS was known, was Apple's first color OS.

New Ideas • 181

The Apple Jonathan

The most radical computer Apple never made was the Jonathan, named either after the apple variety or its creator, Apple II hardware engineer Jonathan Fitch. Or both.

It was to be a *modular* computer, containing Motorola's powerful new 68030 chip. You could buy a $500 base, and then snap additional book-sized modules into a horizontal backbone, like a high-tech bookshelf. Some modules would contain software—operating systems like DOS, Windows, or Unix. Others held hardware: storage, memory, modems, and so on, made by Apple and others.

"A basic system would have a short shelf with one or two books. A business setup would have three or four books. And a power system would have seven or eight books on a wider shelf," says Fitch.

Hartmut Esslinger loved the idea: an expandable, evolving machine, a computer that looked more impressive the more power it had. His team designed a sleek, dark gray, sharp-edged machine like nothing Apple had ever made.

By June 1985, Jonathan's budget, marketing plan, and even packaging were ready to go. It was time to show it to Apple's executive staff.

After a slideshow, Esslinger whipped away the cloth that had been covering the prototype.

The executives were shocked by the design and the color. "It's not that they didn't like the idea," says frogdesign designer Tony Guido. "They were *afraid* of it."

They showered concerns on the project: It competed with the Mac II, now in development. It wouldn't be profitable enough. Sculley even worried that Jonathan's multiple OS feature would be a Trojan horse in reverse. Instead of giving DOS users their first taste of the Mac OS's elegance, he said, the machine might lure Mac users to *DOS*. "That reasoning floored us," says Fitch. "Apparently, Sculley had less faith in the Mac than we did."

Maybe Sculley actually believed that people would prefer DOS, or maybe the Jonathan smacked too much of a Steve Jobs project. Jobs had left only a month before.

In any case, Sculley and Gassée canceled the project a week later.

But what if this new machine came equipped with a Mega II chip? It could be a computer with two personalities. It would run *new*, 16-bit programs with spectacular speed. But when you ran an *old* Apple II app, the 8-bit Mega II chip stepped in, impersonating a standard Apple II.

This computer, Project Phoenix, could be the bridge between the Apple II and Macintosh worlds. Its code names were Brooklyn and Golden Gate—but the world would know it as the Apple IIgs, short for graphics and sound.

It was the first Apple II to come with a mouse and a full-blown Finder like the Mac's. The fonts weren't as nice, but the familiar landmarks were there: menus, windows, file and folder icons, the Trash, desk accessories, control panel, and so on. In fact, the IIgs Finder had something important that the Mac *didn't* have: color.

To the exhilaration of music nerds, the IIgs came with an Ensoniq sound chip. It could not only play 15 notes simultaneously but could also record and play back audio with excellent quality. With the right add-ons, the IIgs could be the world's first digital answering machine.

Woz left Apple before the IIgs was finished. But he couldn't have missed the double homage to him: First, Apple had finally invested in a supercharged Apple II. Second, the first 10,000 Apple IIgs machines bore his signature, "Woz," silkscreened onto the front panel.

When it went on sale in October 1986, the IIgs got mixed reviews. It was pricey, especially when compared with the similarly equipped Atari ST and Commodore Amiga. Then there was the awkward fact that the IIgs was only *mostly* compatible with existing Apple II software (95 percent) and hardware (80 percent). Bill Gates, after seeing a demo, announced that Microsoft would be making no new programs for it.

In the end, the IIgs was a moderate success; Apple sold 1.5 million during its six-year lifetime.

It wasn't enough, though, to hold back the Macintosh tide. Apple was about to give the Mac the power boost and expandability it desperately needed—and by the end of 1986, the Apple II's reign as Apple's biggest cash cow would be over.

17. Heresies

Jean-Louis Gassée was the brash, uninhibited head of Apple France, who'd previously worked at HP and (during his shaved-head years) Data General. Under his watch, France had become Apple's most profitable territory outside the U.S.; in fact, Gassée's outfit sold more Apples than IBM sold PCs.

In the spring of 1985, he had offered John Sculley a modest idea: bring Gassée to Cupertino to start a software division. The lack of available programs was killing the Mac, and maybe, with his charisma and ingenuity, he could persuade developers to get their butts in gear.

In the end, the discussions went nowhere. "Jobs was adamant that I should report to him; I was equally determined not to work for Steve," Gassée says.

But when Jobs left Apple, Sculley brought Gassée back—not to start a software division, but to take over Jobs's role as the visionary guy: VP of product development.

Gassée, of course, was not Steve Jobs, but it was impossible to miss some similarities: a belief in elegance and simplicity, charm as a speaker, and intense charisma.

"He ran hot, opinionated, passionate, profane, shoot-from-the-hip, intellectual, and witty," says engineering manager Mike Potel. Gassée usually dressed in black and wore a diamond stud earring. In his French-accented English, he was equally adept at spontaneous hour-long talks and colorful sex metaphors, which became known as Gassée-isms.

"We must always give our user pure sex," Gassée might say. "It's like a rendezvous in the back seat of an automobile with a beautiful girl." Or: "Looking to Wall Street to measure a company is like trying to measure Marilyn Monroe by chest size. It doesn't give you the gestalt."

Also like Jobs, Gassée was 90 percent marketer, 10 percent engineer—yet he'd now been handed the reins to Apple's engineering organization. He found himself stepping in between two warring divisions: Apple II and Macintosh. The street that separated their buildings earned the nickname "the DMZ."

"You ventured across it only at your peril," Sculley remembers. "The anger was poisonous."

If Gassée's mission was to unite the engineers, he got off to a terrible start. In Paris, where he'd learned conversational engineer, people didn't mince words. "We don't say, 'I love what you're doing, but maybe we can tweak it?'" he says. "The French don't give 'feedback'; they perform live dissection."

When he tried that direct approach at Apple, he was met with hostility—and no action to address his concerns. Within days, a human resources rep was fielding complaints and suggesting that Gassée lighten his tone.

He began asking questions. As he met each engineer, he asked what they were working on and how it fit into the larger Apple mission. Little by little, the engineers warmed to him, recognizing that he could be their champion. It helped that he chose an office right there in engineering, instead of in the Mariani building with the other top executives.

They especially liked that Gassée was willing to violate a key tenet in the religion of Steve Jobs: He wanted to open up the Mac.

Designing the Mac to be closed-off and non-customizable had never made sense to Gassée; after all, the slots in the Apple II had made it a global success. So Gassée let it be known that he would be authorizing the creation of at least two new Macs *with slots*. And to hammer home the point, he ordered new license plates for his Mercedes: OPEN MAC.

Gassée credits surviving a 1993 stroke and couples therapy with making him a "recovering assoholic."

Mac Plus

First, Gassée directed the Mac team to add an expansion jack to the back of the next Mac, to be called the Mac Plus—a SCSI port, which the engineers happily pronounced "scuzzy." It stood for small computer system interface, and it let you connect, in a chain, up to seven hard drives, scanners, tape drives, CD-ROM drives, and so on. It wasn't quite a slot, but it was expandability.

Macintosh Plus

Sold: January 1986–October 1990
Price: $2,600
Processor: Motorola 68000
Memory: 1 MB, expandable to 4 MB
Equipment: 9-inch monochrome, 512 x 342 pixels; 800 KB 3.5-inch floppy drive; SCSI port; numeric keypad and cursor keys

In January 1987, Apple retired its classic tan color; the case was now platinum gray.

Heresies

The new machine, code-named Mr. T, looked identical to the existing Mac. But inside was, as Apple's press release put it, "One Full Megabyte of Memory"—eight times what the original Mac had had. Its new, double-sided floppy drive could store twice as much data. The keyboard had cursor keys, too—another sin in the book of Jobs—and a number keypad off to the right, the better to use with the new spreadsheet program, Microsoft Excel.

On January 15, 1986, Sculley unveiled the Plus at a special two-day event, produced by Apple itself, called AppleWorld. The Plus remained part of Apple's product line for nearly five years—longer than any other Mac model, before or since.

Macintosh SE

On March 2, 1987, Apple hosted its second annual two-day AppleWorld conference, this time in L.A.'s Universal Amphitheatre. The audience was about to witness the births of the first two wholly new machines of the Gassée era.

First, a stagehand rolled out the Macintosh SE, for "system expansion." Except for the gray, grooved plastic of the Snow White design era, it looked at first like all previous Macs. But as the first Mac that had had no input from Steve Jobs, you could open it without voiding your warranty. It contained, yes, a slot. It could accommodate accelerator cards, networking cards (with ethernet jacks), video-output cards, DOS cards, and so on.

Nor was that the only characteristic that would have appalled Jobs: It made noise. It was the first Mac with a fan, ending an era of silent Mac computing.

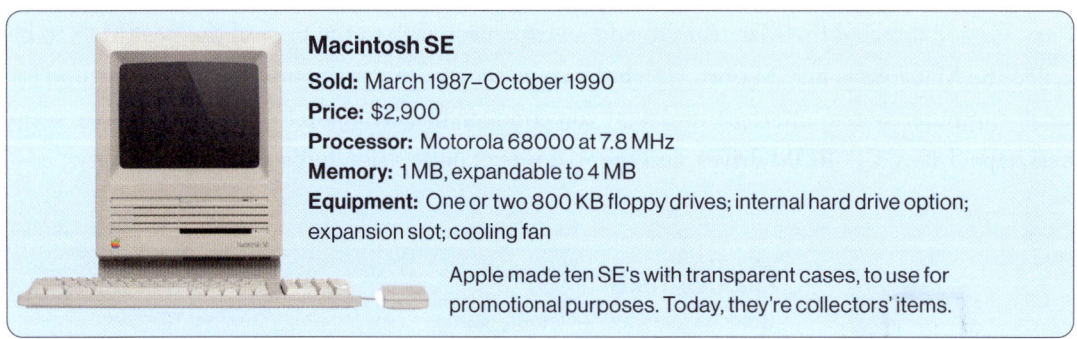

Macintosh SE
Sold: March 1987–October 1990
Price: $2,900
Processor: Motorola 68000 at 7.8 MHz
Memory: 1 MB, expandable to 4 MB
Equipment: One or two 800 KB floppy drives; internal hard drive option; expansion slot; cooling fan

Apple made ten SE's with transparent cases, to use for promotional purposes. Today, they're collectors' items.

At $2,900, the SE was no Apple II. Apple hoped that it would be adopted by—what else—the business market. In that regard, it did well enough. Better yet, once the Mac had a slot, the nightmare of unsupervised user meddling that Jobs had feared never came to pass.

The First Apple Easter Eggs

An Easter egg is a little surprise hidden in software, buried by the programmers in such a way that only the savvy know how to trigger it.

Since the day Steve Wozniak buried "Woz" in the Apple II's first ROMs, Apple's engineers have been brilliant Easter eggers. In 1982, after a company called Franklin Computer had illegally produced an Apple II clone (Apple sued and won), Steve Jobs asked the Mac team to hide a watermark in the Mac's ROMs to make future lawsuits easier to prove. Using astonishingly few pixels, Susan Kare drew an adorable software thief behind bars.

But the first time the Mac team gave photographic evidence of its own existence was in the Mac SE. And you really had to work to get it.

Early Macs came with a strange accessory in the box: the programmer's switch. It was a small plastic panel that you could snap into the vents on the side of the Mac. It offered two big square buttons—an Interrupt switch and a Restart switch—intended for use by programmers.

If you pressed the Interrupt switch, the Mac's debugger screen appeared. If you now typed *G 41D89A* and hit Enter, you'd trigger a four-frame slideshow, featuring the faces of the Mac SE team.

When Steve Jobs returned to Apple, he issued a decree: No more credits in About boxes or Easter eggs. In an industry where top talent is often poached by predator rivals, a credits box like the one in the SE amounted to a billboard for headhunters.

Mac II

The next table rolled onto the AppleWorld stage shattered the link to past leadership unmistakably. It wasn't a graceful wedge like the Apple II. It wasn't a compact all-in-one like the Macintosh. It was an enormous, rectangular box, "something cooked up by the geometry police from IBM-land," as *Newsweek*'s Steven Levy put it.

And when Gassée played a slideshow on its screen, the audience gasped: It was in color.

Macintosh II

Sold: March 1987–January 1990
Price: $3,770
Processor: Motorola 68020 at 16 MHz
Memory: 1 MB, expandable to 8 MB
Equipment: One or two 800 KB floppy drives; internal 40 MB hard drive option; 6 NuBus slots

George Lucas's Industrial Light & Magic used Mac II machines to create special effects for *The Abyss* (1989).

This was the Macintosh II: the first color Mac, the first without a built-in screen, the first to use Motorola's 68020 chip (four times the speed of the Mac Plus), the first 32-bit Mac, and the first Mac with multiple slots. If you filled them all with graphics cards, you could connect up to six screens—and even drag windows from one monitor to the next, your cursor crossing the air between them. "It was the largest product we've ever done at Apple Computer," Gassée said. And it was as far from Jobs's Mac ideal as it was possible to get.

The Mac II's NuBus slots, a format developed at MIT, required no configuration; you could put any card into any slot without having to flip any so-called DIP switches just to tell your clueless computer what you'd done. Better yet, these cards could work independently of the processor. Apple foresaw a day when you'd insert a card containing a faster, next-generation processor. It would simply bypass the one your Mac II already had.

(In fact, a year later, Apple introduced a new operating system called A/UX: Unix with the Mac's own system software running on top of it. A truly committed geek could add to that a Mac286 card, and presto: a single machine could now be *all three* of the world's most popular computers—DOS, Unix, and Mac.)

The Mac II team had struggled to answer a key question: Exactly how colorful should the Finder be? On the Apple IIgs, windows, menus, and icons exploded with color. For a business machine, it felt like too much.

In the end, the Mac II desktop was still entirely black-and-white except for the tiny, colorful Apple logo at the corner of the screen.

At $7,000 with a color screen and 40 MB hard drive, the Mac II certainly wasn't priced for "the rest of us." (An 80 MB drive was also available, an inconceivably huge amount of storage for 1987. As one Apple magazine pointed out: "It's hard to imagine one person filling 80 megabytes, and it's unlikely that any one person will.")

But the Mac II was an engineering marvel, and everyone knew it. Apple couldn't keep up with demand, and its stock leapt up 11 percent.

The Mac SE and Mac II took 18 months and $150 million to develop. But they proved two things: First, that expansion options would not ruin the simplicity and joy of using a Macintosh. And second, that a traditional structured, corporate development path didn't have to produce a soulless product.

By July 1987, it seemed as though Apple had turned a corner. Quarterly profits had risen 65 percent from a year earlier, to $53.5 million. Morale was surging. The product pipeline was full once again: Sculley told reporters that Apple would be announcing more new products in 1987 than it had in its entire first decade.

According to Sculley's calculations, Apple's revenues would reach $20 billion by the year 2000. It was an optimistic forecast—but, as it turned out, not nearly optimistic enough.

HyperCard

Bill Atkinson had done plenty of essential work for Apple. He'd solved the overlapping-windows problem; written QuickDraw, which made possible instantaneous graphics on the Mac; and created MacPaint, the art app that captivated thousands of fans.

Then, one night in 1985, after swallowing a medium dose of LSD, he dreamed up what would be his last great Mac software project.

"I spent most of the night sitting on a concrete park bench outside my home in Los Gatos, California. I gazed up at a hundred billion galaxies, each with a hundred billion stars," he said. "I thought if we could encourage sharing of ideas between different areas of knowledge, perhaps more of the bigger picture would emerge, and eventually more wisdom might develop. This was the underlying inspiration for HyperCard."

HyperCard resembled a stack of note cards. You could type onto a card or paint onto it with MacPaint-style drawing tools, and then—this was the magical part—add buttons that *did* something. Usually, clicking a button took you to a different card. In that regard, Atkinson said, "HyperCard was a precursor to the first web browser, except chained to a hard drive." But in the larger sense, HyperCard was, as Atkinson liked to say, a software Erector Set. It let anyone snap together parts to build useful programs.

At the August 1987 Macworld conference in Boston, HyperCard was an instant megahit. Some people just used HyperCard "stacks" as they came. Some made little changes to existing stacks, or went on to build their own. Some used it to prototype new programs (the party trivia game You Don't Know Jack); some used it to build *finished* ones (the megahit puzzle game Myst).

And some mastered the simple scripting language (HyperTalk) to write their own complete new programs: databases, notepads, address books, calendars, interactive adventure games,

Claris

Apple included three programs with every original Mac: MacWrite, MacPaint, and, starting in the fall of 1984, MacDraw. (A painting program changes the colors of individual pixels. A *drawing* program creates fundamental shapes—circles, rectangles, lines—that remain editable objects.)

Steve Jobs thought he was doing the world a favor by including three beautiful apps with every Mac, but independent software companies were deeply annoyed. How could their own word processors and art programs hope to compete with free?

And so, in 1987, Sculley created a new subsidiary called Claris Corp. in Santa Clara. Claris became the new owner and developer of MacWrite, MacPaint, and MacDraw, plus a project-planning app called MacProject and an all-in-one suite called MacWorks.

Claris sold enhanced versions called MacWrite II ($200), MacPaint 2.0 ($125), and so on. In the 1990s, it had hits with Claris Emailer (email), Claris Home Page (website design), Claris Organizer (calendar, contacts, notes), and ClarisWorks (spreadsheet, word processor, database, presentation, and graphics). The company also produced some flops—for example, a Windows slideshow program called Hollywood.

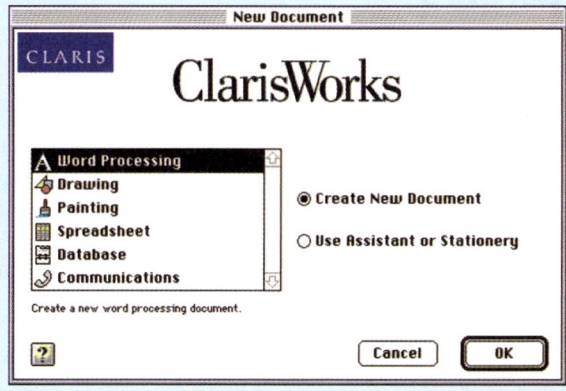

Claris's biggest hit of all, though, was the FileMaker database program that it bought in 1988. FileMaker combined the Mac's elegant graphics with powerful list- and form-making tools, and became the heart of thousands of small businesses.

In fact, by 1998, FileMaker sales were just about the only thing keeping Claris going. Apple shut down development of the other apps and changed Claris's name to FileMaker.

At least, that is, until 2019, when FileMaker Inc. began to branch out again into other kinds of programs. Now that the company was no longer an all-FileMaker company, it needed a new name.

After some thought, Apple's leadership came up with a really good one. The newly rebranded company would be called . . . Claris.

interactive tutorials. They'd distribute them on the CDs that came with computer magazines, or through user groups.

And because stacks could control external equipment, HyperCard became the front-end browser for multimedia CDs, DVDs, and laser discs. Software companies put entire art galleries, anatomy textbooks, or musical works onto these discs; you'd navigate them with HyperCard stacks on your Mac. There was a HyperCard version of the 10,000-page counterculture magazine/catalog *Whole Earth Catalog*. There were deep, self-guided dives into Beethoven's Ninth, the Beatles' *A Hard Day's Night*, and *Macbeth*.

Beethoven's Ninth Symphony, from the Voyager company, controlled playback of an off-the-shelf Beethoven CD connected to your Mac.

"The examples that I was most proud of were where someone who wasn't a programmer was able to communicate a passion that they had pent up for years and years, and finally had a way to let it out," Atkinson said.

Atkinson had made a deal with Apple: HyperCard would come with every Mac for at least two years, and Atkinson would get $1 per copy, up to 5 million copies. ("There was a year when I literally paid a million dollars to the IRS," he said. "I wish I'd had a better accountant.")

The last version of HyperCard came out in 1998. But its core ideas live on: in JavaScript (the programming language that tells web pages how to respond when you click buttons, fill out forms, or scroll); in slideshow programs like PowerPoint and Keynote; and, of course, in the World Wide Web itself. Tim Berners-Lee, inventor of the web, cites HyperCard in his original proposal for it.

Knowledge Navigator

It's early morning in a lovely, two-story, open-atrium home office. Georg Benda's peaceful Harpsichord Concerto in C Minor fills the air. A man in his forties enters his study and takes off his jacket. On the desk, he opens what looks like a folding, double-width iPad.

In a small video window at the corner of the tablet's screen, an AI-generated bow-tied butler appears. "Your graduate research team in Guatemala just checking in," he says. "Today, you have a faculty lunch at twelve o'clock." Wait . . . So this rich guy in his drool-worthy two-level house and his futuristic folding iPad is a *college professor*?

Yes, because we're watching a fantasy: a six-minute video that Apple released in 1987.

Sculley had grown weary of hearing that without Steve Jobs, Apple had lost its vision for the future. He was slated to give the keynote talk at EDUCOM, the biggest academic computing conference.

Bud Colligan, who led Apple's college marketing group, had just read Sculley's memoir, *Odyssey*. He was inspired by the epilogue, in which Sculley describes a futuristic Mac with a flat color screen, voice control, and joysticks on each side. In the book, Sculley called it the Knowledge Navigator. He wanted to make the point that "we weren't at the middle of an industry or the end of an industry; we were at the beginning of an industry."

Colligan had an idea: What if, as the heart of Sculley's conference speech, he showed an actual video of this thing?

For Sculley, the AI butler was the most important part. "It will wander around throughout dozens of databases," he wrote. "You won't have to search through the stacks of libraries—the world's largest library will exist on your desktop or your lap." (The actual public internet was still six years away.)

"All of the technology that it takes to build a Knowledge Navigator is today incubating in our universities or in the laboratories of our companies," Sculley said to wrap up his EDUCOM speech.

Today, the 1987 Knowledge Navigator video is a little cringey—the acting is stilted, the AI implausibly perfect—and yet Sculley was right. Almost all of the technologies that the video depicted have now become commonplace: tablet computers (like the iPad in 2010) with a touchscreen interface (iPhone, 2007), wireless networking (Wi-Fi, 1999), video playing on a computer screen (QuickTime, 1991), a front-facing camera (Connectix QuickCam, 1994), real-time document collaboration (Google Docs, 2006), clickable links (World Wide Web, 1993), video chatting (CU-SeeMe, 1992), removable memory cards (PC cards, 1990), a folding flat screen (Royole FlexPai phone, 2018), and spoken interaction with an AI agent (ChatGPT, 2022).

The Knowledge Navigator video was a special effects fantasy of a Mac 20 years in the future.

The educators loved it. Sculley used the video as the centerpiece of his subsequent presentations for employees, reporters, analysts, and investors. It triggered conversations throughout the industry, including within Apple, about how computers would and should evolve.

But all of the fuss amused Hugh Dubberly, the project's creative director. "These pieces were marketing materials," he says. "They were not inventing new interface ideas. They were about visualizing existing ideas."

The project's real significance, he argues, was its contribution to video prototyping. Apple produced sequels of the video in 1988, and HP, Sun, and AT&T would soon make similar videos of their own. "By releasing Knowledge Navigator, Apple gave permission to others to make videos about software that did not yet exist," Dubberly says.

  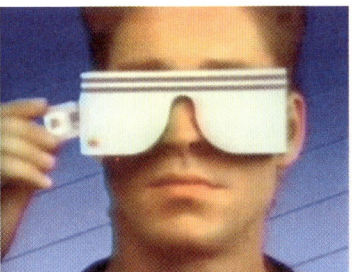

Apple's sequel videos showed a micro-laptop for kids (left); a thin proto-iMac; and a virtual-reality headset, 36 years before the Vision Pro—that still accepts floppy disks.

Apple vs. Microsoft

As the Mac was taking shape in 1982, Microsoft agreed to create programs for the Mac. Naturally, Apple would have to give Microsoft some of its prerelease Lisa machines, which were required at the time to write Mac software.

"Steve, you just sold the corporate jewels," John Couch told Jobs. What would stop Microsoft from borrowing Apple's graphic interface, bypassing the sleepless nights and passionate design debates that had consumed Apple for three years?

And sure enough: In November 1983, Microsoft announced Windows 1.0, a $100 program that would run on top of MS-DOS. It took two years to finish, and it was crude, slow, and limited, but the writing was on the wall: Soon, anyone with an IBM PC or PC clone could use a mouse, open multiple windows, and choose commands from menus.

The first time he learned of Windows, Jobs tore Gates apart. "You're ripping us off!" he shouted. "I trusted you, and now you're stealing from us!"

"Well, Steve," Gates supposedly replied, "it's more like we both had this rich neighbor named Xerox, and I broke into his house to steal the TV set, and found out that you had already stolen it."

Apple's larger problem, however, was the bargain *Sculley* made with Microsoft.

Back in 1977, Apple had paid Microsoft $31,000 for the right to incorporate Microsoft's Applesoft BASIC programming kit into every single Apple II—for eight years. Nobody imagined that the Apple II would still be a hot seller eight years later.

So when the deal was about to expire in 1985, Apple was nervous. Most Apple II programs were either written in Applesoft BASIC or relied on its existence. Without it, many programs wouldn't even run. Apple desperately needed to renew the Applesoft deal.

Since Bill Gates now held all the cards, he made two aggressive demands. First, he wanted Apple to kill off MacBASIC, its own elegant, friendly BASIC language, so that Microsoft could have exclusivity for its own BASIC.

Second, Gates wanted a perpetual license to use the Mac interface.

Sculley had no room to maneuver; Apple *needed* Microsoft. Without Word, Excel, and PowerPoint, the Mac was as good as dead in the business world.

The Loma Prieta Earthquake

On October 17, 1989, a magnitude 6.9 earthquake shook the Bay Area. Almost 15,000 buildings were damaged; 63 people died, most in the collapse of a freeway overpass, and almost 4,000 were injured. The one lucky break: Because the earthquake struck as Game 3 of the 1989 World Series was about to begin at Candlestick Park, the freeways were largely empty.

"I was in a meeting on the second floor of De Anza 3," remembers Apple engineer Paul Mercer. "We'd all been trained; we all went under the table, then we ran outside. We all milled around the parking lot for a couple hours wondering if they would let us back in the building."

As it turned out, the engineering and software floors were sloshing in water from burst sprinklers; the fire marshals sent everybody home. For the next month, the engineers worked from Mariani 1 a few blocks away while repairs were underway.

When he returned to De Anza 3, Mercer spotted the old Lisa computer that he'd found early in his time at Apple. "My desk had tipped at an angle, and my Lisa was hanging off the desk, held on by just the mouse cord," he says. "I kid you not!"

Overall, Apple faced only temporary disruptions in power, communications, and commutes. Today, the earthquake's primary role in Apple history is as a metaphor for the internal chaos of that time.

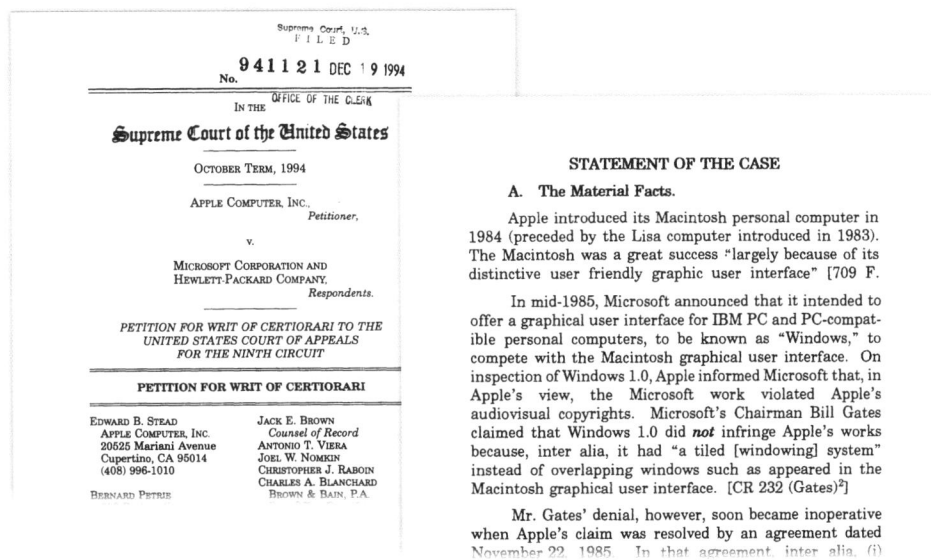

Apple's final attempt to win the case was to appeal to the U.S. Supreme Court.

He agreed to Gates's terms. "Apple hereby grants to Microsoft a non-exclusive, worldwide, royalty-free, perpetual, non-transferable license to use these derivative works in present and future software programs," the agreement said. It even specified *which* interface elements Microsoft could use: icons, menus, scroll bars, overlapping windows, dialog boxes, a Trash icon, and resizable windows with title bars.

According to Andy Hertzfeld, it was "the single worst deal in Apple's history."

Still, in 1987, when Microsoft introduced Windows 2.0, now with a far more Mac-like desktop, Apple felt that Microsoft had gone too far.

On March 17, Apple sued Microsoft for copyright and patent infringement in federal district court in San Jose, seeking $5.5 billion in damages. Microsoft's defense rested on two key points. First, there was Sculley's 1985 agreement, giving Microsoft perpetual rights to use overlapping windows, icons, and so on; second, Microsoft had licensed some of its Windows interface elements from Xerox PARC, which had been Apple's inspiration in the first place. Apple couldn't claim authorship of ideas that it hadn't originated.

Experts on both sides analyzed every tiny aspect of the Mac and Windows. Was the animation the same when you double-clicked an icon to open it? How many of the 189 interface elements in question had Microsoft actually stolen? The case dragged on for four years, cost both companies millions, and fascinated the legal and tech worlds.

In April 1992, district court judge Vaughn Walker handed down his decision: Microsoft had not infringed. The 1985 agreement granted Microsoft permission to use all but ten of those 189 items—and the remaining ten weren't protected by copyright at all, because they were so obvious.

Apple appealed, lost, appealed to the U.S. Supreme Court, lost again.

The aftereffects of the ruling were titanic. First, it handed much of the personal computer industry to Microsoft. As the primary supplier of software for the world's IBM PC clones, its offering of a Mac-like desktop was irresistible. By 1994, Windows had a 90 percent market share, a dominance it maintained for the next 20 years.

Second, the case shifted how companies thought about intellectual property. Now they're more likely to patent specific underlying technologies than interface designs.

Finally, the ruling began the modern era of tech companies imitating tech companies. Windows looked more Mac-like with every edition; the Android phone's interface looks just like the iPhone's; and modern Apple, Google, and Samsung smartphones offer almost identical lists of features.

All of these companies now know the golden rule: You can't copyright an idea.

18. High/Right, Low/Left

The success of the Mac II gave Jean-Louis Gassée the confidence to embrace a new motto: "55 or die." That was the profit margin he thought Apple's machines could generate.

Dell's profit margins were 30 percent; Compaq's were 40. Of course, Apple would not get there by making junky PC clones. Gassée's plan was to take the product line "high and right." That is, if you plotted Apple's offerings on a graph of price and power, they'd appear high and to the right: expensive, but fast.

The II Family

The Mac II, with its color, its slots, and its choose-your-own-monitor scheme, was a critical and commercial hit. Its high/right sequels arrived thick and fast.

- **Macintosh IIx.** The fastest computer Apple had ever made—and the fastest most people had ever encountered. Its same enormous Mac II case contained the new Motorola 68030 chip. The IIx also introduced a built-in SuperDrive, as Apple called it: a floppy drive capable of handling "high-density" disks—not just double-sided, but double-packed, so that each could hold 1.44 MB of data. For many professionals who worked on graphics, video, modeling, statistics, and so on, the IIx's nosebleed price was worth it.

    Its 1990 successor, the Macintosh IIfx, was the most expensive computer in Apple's history—and at the time, the fastest personal computer ever sold. If you were the special effects company Industrial Light & Magic, and you were trying to finish *Terminator 2: Judgment Day*, you didn't let the price tag stop you ($9,000 to $12,000).

Macintosh IIx
Sold: September 1988–October 1990
Price: $7,700
Processor: Motorola 68030 at 16 MHz
Memory: 1 MB, expandable to 128 MB
Equipment: Apple SuperDrive; internal 40 MB hard drive option; 6 NuBus slots

The *x* in Mac II names (IIx, IIcx, IIvx) came to mean "includes the 68030 chip."

- **Macintosh IIcx.** The IIx may have been a technological success, but it was the size of an aircraft carrier. Six months later, Apple offered the same guts in a smaller, lighter box. (*C*, as usual, meant "compact.") It achieved its smaller demands of your desk space in two ways. First, it had three expansion slots instead of six. Second, it could sit *either* flat on your desk—or turned on its edge, thanks to an included set of four little rubber feet.

Gassée unveiled the IIcx onstage at the January 1989 Macworld Expo in San Francisco in an unforgettable way: Rather than hauling out a finished machine, he began with an empty case—and then *built* the computer onstage, component by component, snapping them together: circuit board, speaker, hard drive, power supply, video card. It powered up flawlessly: "Welcome to Macintosh," the screen said. ("I can confess to being a ham on occasion," Gassée says.)

Gassée unveiled the Mac IIcx by snapping all of its components together, live onstage.

It may have been the strangest hardware demo in Apple history—a reverse teardown—but it illustrated just how far Gassée had taken his "Open Mac" philosophy. This machine was so modular, you could Lego its pieces together without tools: breezy upgrading for customers, lower labor costs for manufacturing for Apple. Mac fans loved the IIcx; many called it the best-designed Mac yet.

Macintosh IIcx
Sold: March 1989–March 1991
Price: $5,370
Processor: Motorola 68030 at 16 MHz
Memory: 1 MB, expandable to 128 MB
Equipment: Apple SuperDrive; 40 MB hard drive; 3 NuBus slots

Apple design apprentice Gavin Ivester designed the IIcx and IIci while he was still in college at San José State University.

In September 1989, Apple offered a IIcx twin: the Macintosh IIci, at the time the fastest Macintosh ever made ($6,270). Its processor ran at 25 MHz instead of 16, giving it enough horsepower for color graphics and movie playback. The *i* in the name stood for *integrated video*; it was the first Mac II machine that didn't require you to buy a $400 video card to connect a monitor. It was among the bestselling, longest-lived Macs ever.

The Compact Family

Not everyone needed a massive, modular color Mac. What would happen, Gassée wondered, if Apple engineers could pack the speedy guts of the Mac IIx into the body of a Mac SE?

The answer: the Macintosh SE/30. It was the first compact Mac with a 68030 processor, the first with an internal SuperDrive, and the first to include color QuickDraw, so that you could connect it to a color monitor. This thing blazed at four times the speed of the original SE.

"The SE/30 wasn't just a terrific system just when it debuted," blogger John Gruber would write 20 years later. "It remained eminently usable for years to come. When I think of the original Mac era, the machine in my mind is the SE/30."

Macintosh SE/30
Sold: January 1989–October 1991
Price: $4,370
Processor: Motorola 68030 at 16 MHz
Memory: 4 MB, expandable to 128 MB
Equipment: 9-inch monochrome screen; 40 or 80 MB hard drive; Processor Direct slot.

A Mac SE/30 appeared on Jerry Seinfeld's desk in *Seinfeld*—the first of many Macs that appeared there over the years.

Mac Portable

In the early eighties, Seiko, Tandy, HP, Toshiba, Sharp, Grid, and others made valiant attempts to create laptops. Battery life was terrible—if there *was* a battery; some machines could run only when plugged in. The screens were black-and-white, tiny, and dim. Hard drives and CDs were years away. And for all this misery, you might pay $8,000 or more.

In 1985, Toshiba introduced the T1100, a PC clone that actually managed to be workable. At nine pounds, it was no MacBook Air. But with an "eight-hour" battery and a $1,900 price, it got the industry's attention—and Apple's.

Of course, laptops had been on Apple's radar for years; as early as 1984, Steve Jobs would chant the mantra "Mac in a book by 1986." The Apple IIc, easy to carry in a briefcase, was a start.

But on September 20, 1989, Apple unveiled a true, battery-powered, full-blown portable Mac. Well, *mostly* blown.

"When we set out to design the Macintosh Portable, we had three sets of challenges in mind: portability, consistency, and battery paranoia," said Gassée on a Universal City stage before 5,000 attendees. "No subset of applications. No 'Mac Junior.' No compromise."

The audience got its first glimpse of the Portable in a video. They saw it being carried by carefree, happy businesspeople, beachgoers, construction foremen, and even mountain climbers. It looked like a wedge-shaped suitcase, with a built-in handle that doubled as a screen-hinge lock.

As he had done with the IIcx earlier in the year, Gassée snapped together a Portable from its parts onstage. When it turned on, its screen displayed the words "Ta Da!"

In this case, though, the snappability meant convenience. If you were left-handed, you could snap the trackball to the *left* side of the keyboard. If you had a mouse, you could replace the trackball with a number keypad. If you wanted to install more memory, you could pop in a RAM board. If your battery died halfway through a presentation, you could swap in a freshly charged one without skipping a beat.

But in other ways, it meant *inconvenience*. The snap-in feature, designed in part to accommodate the Fremont factory's robots, required extra space around every component, making the whole thing bigger.

The Portable didn't use the NiCad (nickel-cadmium) batteries used in rival laptops, for one key reason: They gave you no warning as they approached depletion. You might be in the middle of something important, and then *boom*—those laptops simply shut off, and you'd lose your work.

Macintosh Portable
Sold: September 1989–October 1991
Price: $7,300 with hard drive
Processor: Motorola 68000 at 16 MHz
Memory: 1 MB, expandable to 9 MB
Equipment: SuperDrive; 40 MB hard drive option; 10-inch active-matrix LCD screen; video-output jack

The signatures of the Portable's designers are embossed inside the case—the last time any Mac would include this style of Easter egg.

For the Portable, Apple introduced a new kind of battery: sealed lead-acid gel. It weighed more than a NiCad, but it could run for ten hours and had a smoother exhaustion curve. The Portable, for the first time in laptop history, displayed a battery-charge graph in a desk accessory, showing when your charge was getting low. Nor did this battery exhibit the "memory effect" of NiCads, where the battery would "remember" a lower capacity if you didn't deplete it completely before recharging.

Even if the worst should come to pass, you wouldn't lose your work. For this Mac, Apple invented sleep mode: a low-power state where the machine would be dark and quiet, but preserved everything you'd been working on.

The real star of the show, though, was the screen. It had no backlighting, so it was hard to read in dim light. But it was the first-ever *active*-matrix LCD screen in a consumer machine. Every one of the 256,000 pixels had its own transistor, so that it could be individually switched on or off. The resulting display was stunningly crisp and visible from wide angles. (It was also fiendishly difficult to manufacture. In 1988, nobody could make an active-matrix screen with fewer than ten dead pixels. To meet Apple's requirements—fewer than six—its supplier, Hosiden, had to build a whole new factory, which delayed the Portable by over a year.)

Despite all of its advancements, the Portable was a sales dud. Apple sold only 10,000 of them in the first year, about a fifth of what it had predicted.

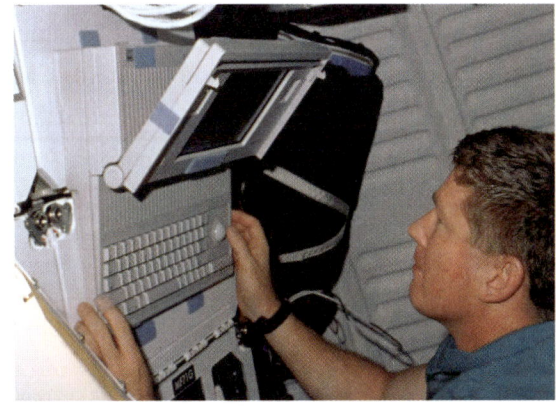

On the space shuttle, nobody cared how much the Portable weighed.

The reasons were obvious to everyone: At the modern equivalent of $18,000, it was incredibly expensive. It was too big—bigger than an airline tray table. And it was absurdly heavy: 16 pounds, roughly the weight of a Mac SE. (Critics renamed it the Mac Luggable.) When Compaq released its seven-pound LTE laptop only six weeks later, "it was just embarrassing," said Apple director of hardware engineering John Medica.

After seven months, Apple lowered the Portable's price by $1,000. In 1991, it released a new Portable with a backlit screen (and half the battery life). It was all too late.

Developing the Portable, however, wasn't a waste of effort. It pioneered dozens of power-saving technologies and software innovations that, two years later, would turn Apple's *next* laptop attempts into world-changing triumphs.

Golden Age

Gassée was a polarizing figure. He was charismatic, witty, and eloquent in two languages—but also blunt, stubborn, and unfiltered. "I had developed this mentality of fighting," he says now, "which served me well sometimes, and other times gained me a reputation of being difficult."

But he got results. The Mac was growing up, the engineers felt supported, and the product pipeline was full. In a 1988 restructuring, Sculley promoted Gassée: As president of Apple products, he now had dominion over manufacturing and product marketing. Gassée's new office was next to Sculley's in Apple's sleek new De Anza 7 building, across the street from the De Anza 3 engineering building where he'd been. "And as I crossed De Anza, I thought, 'That's the end. I'm not going to do well,'" Gassée says. "In retrospect, I'm a gang leader, I'm not a corporate executive."

Overall, it was a thrilling time for Apple. The computer industry slump had ended, the desktop publishing movement fueled spectacular sales, the stock had risen almost 80 percent in three years, and the company's leadership was competent and stable.

For talent, Apple was the hottest place on earth. Engineers earned 20 percent more than other companies were offering; they got automatic 10 percent raises every six months, plus profit-sharing checks that added another 15 percent to their annual pay.

And the fans! Mac fans were a special breed. No matter how big Apple got, to them, it would always be David the underdog—and Goliath was the clunky, tasteless Wintel (Windows + Intel chip) empire. Macheads spread the word of Mac like it was a religion, to debate PC fans over OS superiority, interface design, and the moral failings of Microsoft. Plenty of them got Apple tattoos.

John Sculley was enjoying his first golden age at Apple.

Gassée Trouble

The thrill of firing on all cylinders lasted only until the end of 1989, when Apple experienced its first holiday season sales drop in years—and a 20 percent drop in the quarterly stock price. The cracks in the high/right philosophy were starting to show.

The issue was market share. Microsoft's strategy—selling Windows to hardware companies, collecting a royalty on every machine anybody sold—was unimaginably successful. The Wintel market exploded. Dozens of PC-clone makers competed with one another, which kept the hardware prices down; IBM's halo of corporate competence kept the business market interested; and the ubiquity of Windows ensured a tidal wave of software titles. In ten years, Apple's market share had dropped from 16 percent to below 8 percent.

Developers, press, and customers kept telling Sculley how obvious the lessons of Wintel's dominance were. First, Apple should make some inexpensive Macs to compete with those cheap PCs. Cheaper Macs meant more Macs sold—a larger installed base—which would make a more attractive platform for software companies.

Second, Apple should consider an idea that had been kicking around since 1985: License the Mac operating system to other computer makers, just as Microsoft licensed Windows.

Gassée forcefully disagreed. A low-cost Mac would be a terrible mistake. It would violate the very heart of Apple's mission: Make the finest machines available. In fact, when a shortage of memory chips increased Apple's cost of components, Gassée *raised* Mac prices—an unheard-of strategy in computerland. (Customers were livid; Apple undid the price hike three months later.)

By the same logic, he predicted that licensing the Mac OS would be a disaster. The analogy with Microsoft was flawed. Apple was not primarily a software company—it was a *hardware* company. Why would it deliberately cultivate a market of cheap competitors to eat its lunch?

Furthermore, Mac clones would erode Apple's ability to control the quality of the platform. If any old PC maker could throw together a junky Mac clone, the public might stop associating Macs with premium, beautiful, consistent products.

Sculley and Gassée could not find common ground; their clashes, personal and professional, intensified. More than ever, Gassée regretted moving to the office next to Sculley's. "Proximity to the executives had proven to be the diplomatic disaster I anticipated; my 'raise prices' advice was openly scorned," he says. "My behavior was considered strange, almost embarrassing."

Their antagonism came to a head in January 1990. HR director Kevin Sullivan had proposed that the two men sit down for a dinner to hash things out. ("He was a rather even-mannered individual," Gassée says, "so we used to call him Kevin the Bold.") Sculley, Gassée, and Sullivan met in a private dining room at Maddalena's, a high-end Palo Alto restaurant.

Over dessert, Gassée didn't hold back. He said that the engineers didn't respect Sculley, that *he* didn't respect Sculley, that Sculley didn't know enough about technology. "Gassée basically told me that I was the wrong person to be at Apple," Sculley remembers.

Two days after what Gassée calls "the last supper," Sullivan stepped into his office. "I've been asked to separate you from Apple, effective immediately," he said.

Gassée wasn't entirely shocked.

The news spread like wildfire. Dozens of engineers made signs ("JLG Don't Go!"; "JLG 4 CEO") and marched in front of the building. Gassée was touched, but Apple's executive staff was alarmed. What if some of those engineers followed Gassée out the door to some new venture, as had happened with Jobs?

Within hours, Sullivan returned with a new message from management: Could Gassée "stay around" until the end of the company's fiscal year?

For eight months, Gassée was "a minister without a portfolio." He took up Japanese calligraphy. He represented Apple at conferences. He answered the occasional question from Sculley. He joined Steve Jobs and Mike Murray in Apple's Fired by Depurposing club. "How does it feel to be fired?" Steve Jobs asked him in a phone call. It was "a smirking question I deserved," Gassée says, "given my role in his own dismissal."

Gassée finally left in the fall of 1990. "He and I didn't get along particularly well, but I always admired him," Sculley says now. "And in the years since we've left Apple, we've corresponded and said, 'Gee, I wish we had been friends when we worked together.'"

In the meantime, Sculley had to fill his role.

Michael Spindler, born in Germany, had been with Apple since 1980, when Mike Markkula asked him to run marketing for Apple's tiny European office. His strategy had been to tailor the approach for each country. "Apple beats with two hearts: our California heart and the heart of the local company," he would say.

His 18-hour-a-day work ethic earned him the nickname "the Diesel," and his rise was stratospheric. By 1984, he was running marketing and sales for Apple Europe. In 1985, he took over *all* international. By 1988, he was running Apple's 13 Europe divisions, the Middle East, and Africa, and he was spending a good deal of time in Cupertino. From 1988 to 1990, Spindler tripled Apple Europe's sales, making it a quarter of Apple's overall business.

Michael Spindler
1942–2016
Schooling: Technical University in Cologne
Before Apple: DEC, Intel
Apple: 1980–1993
After Apple: Upstart Capital, Daimler-Benz, Bertelsmann

After Apple, Spindler served as a trustee for the American Film Institute in Los Angeles.

In January 1990, Sculley invited him to take over Gassée's portfolio, and more. Spindler became Apple's chief operating officer.

Low and Left

Sculley had grown increasingly alarmed at the high prices of Gassée's new Macs. How many customers would pay $4,370 for a Mac SE/30, $7,700 for a Portable, or $12,000 for a Mac IIfx? Now that Gassée was decommissioned, Sculley could put his low/left strategy to the test. In October 1990, Apple released three low-end Macs.

- **Macintosh Classic** (October 1990). Steve Jobs's dream fulfilled: a $1,000 Mac. Of course, the modest price also meant modest equipment: the original 68000 chip and no hard drive. But Mac fans, encouraged by Sculley's $40 million ad campaign, snapped up 100,000 Classics a month—especially in schools, which enjoyed $800 educational pricing.

- **Macintosh LC** (October 1990). *LC* stood, without much subtlety, for "low-cost color." Its front foot lifted the front edge—not to make the LC look friendly, as it turns out, but to give you clearance over the keyboard to insert floppy disks.

How Amy Picked the LC

Robert Brunner's Industrial Design Group designed three different versions of what would become the low-cost Macintosh LC: One was a mini tower, one resembled a boom box, and the third was the low-slung "pizza box" design.

When the executive team inspected the three models, they couldn't decide which to launch. So they wound up asking John Sculley's assistant, Amy Bonetti, what she thought.

She pointed to the pizza box design. "Oh, that one. That one by far," she said. Her conviction was strong enough that the choice suddenly seemed obvious.

"You think of Apple, and everything's perfect, and really thought out," Brunner says. "But in any company, there's machination and politics and personalities and legacies and all this stuff—that just is the human condition. It's seldom linear and it's seldom pretty."

In this case, Bonetti's choice was a good one. Mac fans snapped up half a million LCs in its first year, and its successors—the LC II, LC III, LC 475, LC 580, and so on—remained a staple of Apple's product line for years.

- **Macintosh IIsi** (October 1990, $2,500). The LC was cheap, but it was a big drop from the expandability and speed of a Mac II. To fill the gap, the IIsi had the same 68030 chip as the Mac II and, with an adapter, could even accept one NuBus expansion card. And it came in a new, elegant, not-as-big-as-a-Mac-II case design, which Apple never used again.

Sculley had hoped that these new, low-end Macs would increase Apple's market share. He got his wish; they sold like crazy. Within six months, Apple's U.S. market share reversed its slide and inched back up to 8 percent. International sales boomed; low-income countries in Asia and South America placed huge orders of inexpensive Macs.

It was a disaster.

The Macintosh Classic (left), Macintosh LC, and Macintosh IIsi were Sculley's low-cost Macs.

The first problem was supply. Within a month of the unveilings, Apple had $525 million worth of orders that it couldn't fulfill. The Fremont and Ireland plants were maxed out, the Singapore factory began running 24/7, and Apple started shipping the Classics by air instead of by sea. None of it was enough; the Classic remained back-ordered for six months.

Wall Street was unhappy for a different reason, the one that Gassée had warned of all along: Cheap Macs meant lower profit margins. Apple was bringing *in* money by the boxcar—revenues were up 19 percent in a year—but making far lower profits.

To keep the wheels from coming off, Sculley engineered another corporate reorganization in May 1991—Apple's fifth reorganization in five years. "Obviously, we don't like having to do this," he said in a company-wide voicemail. "It will be hard for all of us to see some wonderful and talented employees leave the company." He laid off 10 percent of the company's 15,600 people—the first major layoffs since 1985. The cost of the restructuring: $224 million, which resulted in Apple's first quarterly loss ($53 million) since 1985.

Sculley took a 15 percent pay cut himself (from a reported $16.7 million). He named himself chief technology officer—an act that dismayed the engineers. He also closed two of Apple's five U.S. sales offices, and cut budgets for conferences, travel, bonuses, research projects, and even marketing. No more free luxury cars for Apple's top 100 executives. From now on, you'd be eligible for a raise once a year, not every six months.

There would be marketing cuts, too. In 1990, Apple had spent $750 in marketing for each Mac it sold. That, clearly, wasn't sustainable.

"Apple has a tough few years ahead," concluded a *Macworld* analysis. "If the company can throw away old perceptions, catch up on the technology, and fix its organizational problems, it can buy the time it needs to come up with a clear plan for the next decade. If not, Apple may well become a bit player in tomorrow's technology market."

19. System 7

The 1984 Mac team had written the Mac's original system software on a desperately short timeline, for a machine with absurdly little memory.

"These guys were brilliant—the Hertzfelds and the Steve Cappses and Atkinsons," says software engineer Paul Mercer, who worked on the Finder in the late eighties. "They were brilliant in what they did; I mean, that code is *revered*. But the problem is Steve made them run this whole thing in 64K of memory! So they had to torture this code. It was just a lot of hackery."

By today's definition, what they put together was not even an operating system. It lacked key features on "real" computers, including multitasking, a kernel (a traffic cop software chunk between the hardware and your apps), and, in particular, memory protection (separate memory bubbles for each program). "Everything ran in one big-ass memory space. So one bug took down the entire system," Mercer says.

Macs in the early years crashed *a lot*.

Over time, memory became less expensive, competition began to intensify, and Apple's engineers began to discuss ways to shore up the Mac's OS. But a complete overhaul seemed to be fundamentally impossible. How do you replace the hackery with something more solid—without breaking compatibility with existing Mac programs?

Of course, there had been improvements. There was System 2 (April 1985),

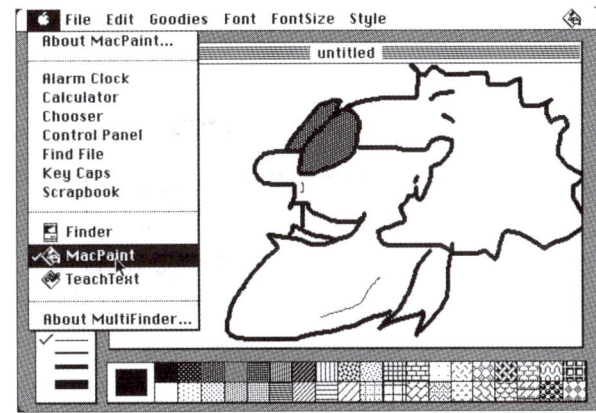

In MultiFinder, you could switch between open programs using a menu.

which introduced the New Folder and Shut Down commands. System 3 (January 1986) brought the Hierarchical File System (HFS), meaning that you could, for the first time, put folders inside other folders. System 5 (October 1987) offered MultiFinder, which let you keep multiple programs open simultaneously. It was something of a hack—your background programs weren't actually *running* unless the foreground app voluntarily relinquished control, and one crashed app still took them all down—but it was better than nothing.

But by 1988, Apple's technical staff got serious about fixing the Mac's OS. Parts of it were starting to lag behind *Windows*, for heaven's sake.

In March, five senior engineers, led by Gifford Calenda and Sheila Brady, convened at the Sonoma Mission Inn & Spa, a high-end resort an hour from San Francisco, to brainstorm. Calenda instructed the engineers to write down their wish-list features for a new OS on index cards: quick and easy ones on blue cards, harder ones (like advanced multitasking) on pink cards. The really ambitious ideas, proposals that would require rewriting the whole damn thing, they wrote on red cards.

A group of about a hundred—the Blue team—dove into the blue-card ideas. Getting the new release ready took $100 million and three years of effort—interrupted only by pizza, Friday beer busts, Ultimate Frisbee, and volleyball at a local apartment complex. But in May 1991, Apple unveiled the biggest advance in Mac system software history.

System 7

This upgrade wasn't the major overhaul the OS desperately needed; it added some delicious features, but did nothing for the rickety plumbing underneath. And the upgrade wasn't for sissies: It required a Mac with at least 2 MB of memory and a hard drive, and it came on 15 floppy disks. (Months later, Apple would offer System 7 on, for the first time, a CD.)

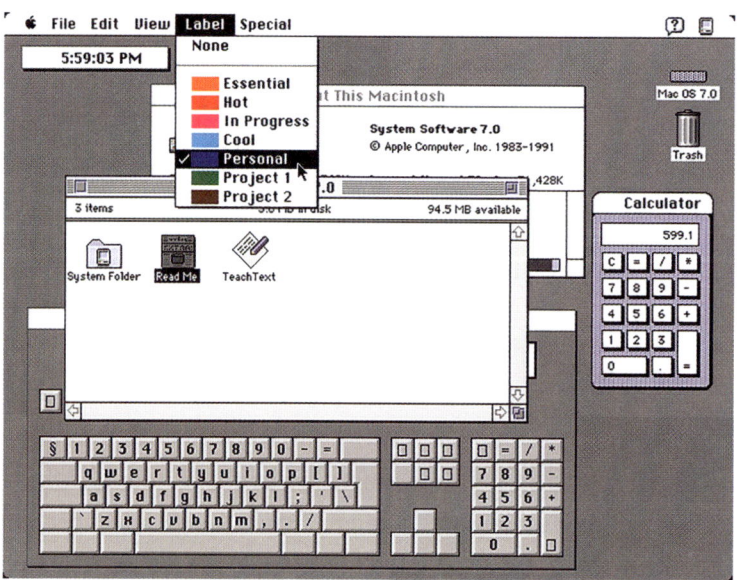

In System 7, you could choose colors for your desktop, text highlighting, individual icons, and window borders.

Sosumi

System 7 came with a sweet little customization feature: a choice of sounds to use as the Mac's error beep. Most of them were onomatopoetic, spelled like they sound: "Wild Eep," "Quack," "Droplet," and so on. For that, you can thank Jim Reekes, who wrote most of Apple's audio software in the 1990s.

To assemble some good sounds, he'd conducted an internal competition. One engineer did a flawless duck quack impersonation, so that made it in. Another engineer's wife did a great monkey screech. One guy contributed a recording of himself just singing, "Ta-da!" which Reekes used as the reward when you solved the Puzzle desk accessory.

The new Mac LC and IIsi models were the first to include microphones, and Apple's legal department was not thrilled. They still remembered 1978, when the Beatles sued over the name "Apple," and Steve Jobs had agreed that Apple would stay out of the recording business. "I had all these lawyers asking me all these questions," Reekes remembers. "They got really paranoid about anything that hinted at Apple doing anything musical."

Most of Reekes's error beeps were sound effects: monkey, duck, and so on. But one of them was a staccato, E-flat, diminished triad, which he named Chime.

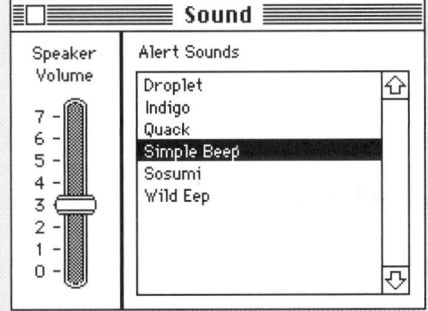

"You can't use this sound," the lawyers said. They considered "Chime" a musical term, and therefore in violation of the 1978 settlement.

"So I'm like, 'Are you kidding me?'" Reekes says. Rolling his eyes, he sat with his buddies on a conference room sofa, drinking beers, trying to come up with a new name for the sound. Maybe, he suggested, he'd just call it Beep.

But "beep" was probably a musical term, too. "Can't do that," one of his buddies said.

"So sue me!" Reekes shot back.

No sooner was the phrase out of his mouth than it hit him: "Holy shit, that's the name! I just have to spell it funny so that no one realizes what I'm doing!"

The sound's new name was Sosumi. Reekes had no further hassle from the lawyers.

Did Jim Reekes put one over on the top legal minds in Silicon Valley? "I don't know. In the back of my mind, I think they actually did get it, but they were willing to let it go."

But if your Mac could handle it, the rewards were rich. Right off the bat, System 7 was in color.

The new features included balloon help (point to something, and a cartoon balloon pops out to explain it); a customizable menu (you could stock it with your favorite applications, folders, and documents); built-in file sharing, for collaborating over a network; the option to paste custom icons onto files; 32-bit addressing (which finally allowed newer Macs—or older ones with a software patch—to access more than 8 MB of memory); aliases (which let a file's icon seem to be in more than one folder); and virtual memory, a scheme that fools the Mac into thinking that a hunk of disk is RAM so that you can keep more programs open. A new smooth-font technology called TrueType debuted, too—a shot across the bow of Adobe and its smooth-font monopoly.

There was also a new Easter egg: If you opened the System file with a text editor, you'd find a little note: "Help! Help! We're being held prisoner in a system software factory!" It was signed by the Blue Meanies, a team of senior engineers who troubleshot and integrated System 7's various features.

But the System 7 feature that changed the world in the biggest way was a little software piece called QuickTime.

QuickTime

Today, more people watch YouTube than television. Today, people routinely shoot and edit video on their phones. So it may seem hard to believe that in 1990, nobody had ever *seen* digital video playing on a computer screen.

It was possible to pump the output of a laser disc player straight through to a window. But that wasn't *digital* video; the computer wasn't involved in storing or playing it. It was just analog video using your monitor as a glorified TV.

For years, John Sculley had been intrigued with the notion of multimedia—any combination of images, text, animation, sound, video, and interactivity—especially its promise for education. As a burgeoning field, multimedia was a perfect fit for Apple, whose computers had been advancing graphics and sound since 1977.

In early 1990, Dave Nagel, who'd taken over the Advanced Technology Group, pitched Sculley on a drive to develop multimedia features. "We could get years ahead of Microsoft," he said.

He pointed out that in May, Apple would be holding its Worldwide Developers Conference, an annual, weeklong gathering of thousands of programmers. WWDC was a place to get programming tips straight from Apple engineers, to meet other developers, and, best of all, to hear what Apple was planning for the coming year. It would be the perfect place to announce video on the Mac.

Pencil Test

At the 1988 SIGGRAPH computer graphics conference, Apple played a three-minute video called "Pencil Test." It was a charming, wordless comedy created by the Advanced Technology Group (ATG). In it, the pencil in the MacPaint tool palette comes to life, jumps off the screen into the 3D real world, explores other items on the desk, and then tries to return to the Mac's screen.

Pixar, the 3D animation studio, had already made a few simple computer-generated 3D movies (and Steve Jobs was already its majority shareholder). But Pixar used specialty computers that cost hundreds of thousands of dollars. "Pencil Test" was created entirely on Mac II's.

Standard video creates the illusion of motion by flashing 30 pictures a second before your eyes. It took 18 MB of memory to hold a *single* frame of "Pencil Test," which is more memory

than any Mac on earth had—so the ATG team was forced to render each frame one horizontal strip at a time, and then digitally stitch the strips together.

Even then, the Mac II required 40 minutes to compute each frame of video. At that rate, the movie's 500 frames would require over three months to render—too long to finish in time for the conference. So the ATG team wrote custom software that could distribute the rendering work among 28 connected Mac II's. "We collected a bunch of machines from other groups, and a bunch of machines from our group," says ATG engineer Al Kossow, "and then brought in our machines from home. And we stacked them up three high."

The video blew people's minds at the conference, and lives on YouTube to this day. Attentive viewers may notice a few familiar names in the closing credits: Andrew Stanton (future director of *Finding Nemo* and *WALL-E*), Galyn Susman (future producer of *Ratatouille*), John Lasseter (future Pixar CEO), and Nancy Tague (future Mrs. Lasseter).

And so it was that at WWDC 1990, marketing manager Don Casey took to the stage of the San Jose Convention Center to announce a thrilling new initiative: QuickTime. It would be a system for controlling external devices like laser disc players and fax machines. QuickTime would be groundbreaking, he said. It would be cross-platform (Mac and Windows). It would arrive within a year.

There was only one problem: It didn't exist. Nobody in engineering had heard of it.

Back at the office, they were baffled. "Um, that sounds really good . . . But who's doing that?"

Bruce Leak was intrigued. He knew a little something about displaying images on a screen. In 1987, he'd worked on a color version of Bill Atkinson's QuickDraw software, the heart of the Mac's graphics displays, for the Mac II; in 1990, he'd expanded Color QuickDraw's palette from 256 colors to millions.

But to him, the real magic wasn't controlling external hardware. He wanted QuickTime to be a software feature so that you could have *video* anywhere you could currently have a still image: in presentations, in word processing documents, in copy/paste operations.

That idea, says QuickTime engineer Peter Hoddie, changed everything. "It took video from being something where you needed an expensive Mac with a few thousand dollars of expensive gear attached to it—to, you only needed an expensive Mac."

Leak set up a team of six engineers on a first-floor area of De Anza 3, across a busy intersection from the rest of the OS team in Mariani 1, where they could toil in relative privacy.

One of the biggest challenges was compression. ATG's three-minute "Pencil Test" video, for example, would have occupied 5 GB if it were a digital file—far bigger than the biggest hard drive on the planet. ("Pencil Test" was played from videotape.) Unless its file size could be radically shrunken, computer video would be impossible.

Eventually, the QuickTime team came up with five tricks that, together, reduced the data size of a video by up to 90 percent:

- **Lowering the frames per second.** Standard TV video is made up of 30 frames a second, but a QuickTime movie might get away with only 12, which uses far less data to process and store. The motion isn't as smooth, but it's bearable.

- **Shrinking the window size.** Early QuickTime movies didn't fill your screen. They occupied a window about a quarter screen size.

- **Storing only parts of the image**—the areas that *change* from frame to frame (someone's mouth moving, for example, while the background remains static).

- **Clumping pixels together.** The software studied each 4 x 4 clump of pixels, which the team called a cell. If those 16 pixels were all *roughly* the same color, the algorithm might say, "Eh, that's close enough to being a solid color"—and it would store the information for only *one* color instead of 16. In other situations, *two* colors might suffice to describe what was in that cell. "Then the trick got to be picking the best colors," Hoddie says. "What color, what two colors most accurately represent that cell?"

- **Multiple codecs.** QuickTime could accommodate different *kinds* of compression-decompression systems, known as codecs. One might be best for cartoons, full of solid blocks of color, while another might work best for action sequences. The original QuickTime 1.0 came with five codecs.

The other challenge was keeping audio and video from drifting out of sync. "We kind of got shocked fairly early on in the process," Leak says. "If you're only doing a short video, you never notice it. But once you get a 30-minute video, it's like, 'Wow, the timing is way off.'" To solve the problem, engineer Gary Davidian created a timing algorithm that was accurate down to a millionth of a second.

The human-interface team designed the standard scroll bar, coded by Hoddie, that now appears on every digital video. Today, we take that controller for granted—but at the time, computer playback interfaces were designed to look like VCR buttons: Rewind, Stop, Play, and Fast-Forward. Hoddie's controller, on the other hand, offered total random access to a video.

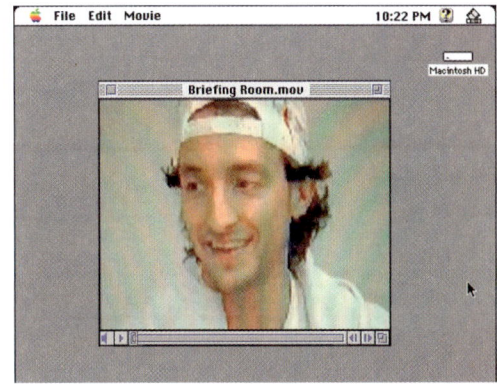

Bruce Leak appears in one of the first QuickTime movies.

By late 1991, it was all working. You could play, edit, and save videos, or even copy and paste videos into your documents, as easily as though they were graphics. If you bought a video capture card, you could also import it from a camcorder or another source.

The final challenge was making QuickTime the *standard* for digital video, worldwide—which meant creating a Windows version. "None of us on the QuickTime team were super excited about our stuff going to Windows," Hoddie says. "It's hard to understate how much that was seen as pure evil."

But plans were plans. QuickTime became the first software Apple ever wrote for Windows.

Leak himself introduced QuickTime at the 1991 Worldwide Developers Conference. It wasn't the first time Apple had ever played its "1984" ad at a conference—but it was the first time a *Mac* ever played that ad for an audience. Apple had just invented software-based video.

In these pre-internet days, there were still very few ways to get video *into* your computer. The

> **The QuickTime Logo Rule**
>
> In hopes of fostering its rapid adoption, Apple let any software company license QuickTime for free. The only requirement: the QuickTime logo must appear on the box. "So we had QuickTime logos in computer stores everywhere, which was spectacular," says Hoddie.
>
> At that point, the story took a surprising—or, perhaps, unsurprising—twist: The porn industry found QuickTime. For the first time in history, people could watch adult videos on their computers.
>
> All of which put Apple in something of a bind. Did it really want its name and QuickTime logo on porn CD boxes?
>
> That day, the QuickTime licensing agreement changed. Apple would still offer QuickTime software for free to anyone who requested it. But from now on, developers either *had* to include the logo on the box—or were *forbidden* from including it, at Apple's discretion.

obvious solution was a CD-ROM; each disc could hold 600 MB of data, far more than a hard drive. Starting in 1990, every Mac model came with a CD drive; soon, consumers and educators could choose from thousands of interactive CD-ROMs.

In the end, QuickTime's compression schemes got better, video sizes got bigger, and frame rates got higher. In 2001, QuickTime became a key component of MPEG-4, the video format of Blu-ray discs, video calls, Android phone videos, and streaming services like YouTube and Netflix.

In 2011, in Mac OS X Lion, Apple replaced QuickTime with a more invisible, more modern, more secure, smoother-playing technology called AV Foundation. But most of the video files themselves still work. Millions of QuickTime and MPEG-4 files play every day, making QuickTime among Apple's most important and lasting inventions.

KanjiTalk

"In the eighties, Japan ruled the world," says Satjiv Chahil, Apple's global marketing chief in the mid-nineties. "They took over industry after industry: motorcycles, cars, consumer electronics. They bought up Hollywood studios, golf courses, and the best real estate in the country. We referred to the nation as Japan, Inc."

What Japan did not have, though, was personal computers.

The Japanese take deep cultural pride in their writing system, which involves three different scripts. There's kanji, used for nouns and basic verb forms (over 2,000 characters); hiragana, for native Japanese words and verb endings (46 characters); and katakana, for foreign words (46 characters).

Satjiv Chahil
Born: 1950, Amritsar, India
Schooling: Punjab University, Thunderbird School of Global Management
Before Apple: IBM, Xerox
Apple: 1988–1997
After Apple: Sony, Palm, HP

Chahil's turban always color-matches his shirt, tie, sweater, or jacket.

There's a directional challenge, too. In novels, newspapers, and literature, you write vertically, in right-to-left columns; in emails, websites, and textbooks, you write as in English: horizontally, left to right.

No wonder, then, that as late as 1987, no personal computers ran in Japanese. NEC and seven smaller Japanese companies made proprietary, incompatible word processors. But if you wanted to use a real computer, with a full universe of apps, you worked in English.

The Mac, with its graphic display, should have had a natural advantage in displaying Japanese characters—but its presence in Japan was dismal. Apple had handed the job of selling Macs to Canon, who marked up Macs by 60 percent and left them to die. (The Apple brand was so obscure in Japan that at one point, a transport company, having seen the Apple logo on the shipping documents, sent a *refrigerated truck* to pick up boxes of Macintoshes.)

In 1988, Sculley decided to get serious about Apple's overseas operations. He split up the Apple International division into Apple Europe, led by Michael Spindler, and Apple Pacific, led by Ian Diery. (Apple used the term "Pacific" loosely: it covered Canada, Central and South America, Australia, Japan, and even the Caribbean.) Sculley gave Diery a challenge: Grow Japan into a market with $1 billion of sales a year within five years.

In early 1989, Diery and Chahil rented and staffed a headquarters for Apple Japan and canceled Canon's exclusivity. Chahil, in turn, hired Carlos Montalvo, a Xerox PARC veteran whose specialty was Unicode, a standard system of coding every character in every language system (including symbols and emojis) so that every computer type can display them correctly.

The first step was creating KanjiTalk, an invisible software layer above the Mac OS that let you type the Japanese words phonetically, using the standard Mac keyboard; it converted your words into kanji characters. Apple would now be able to market the very first Japanese-language personal computer.

Then, to let Japanese buyers know about its superior tools, Apple launched a visibility blitz.

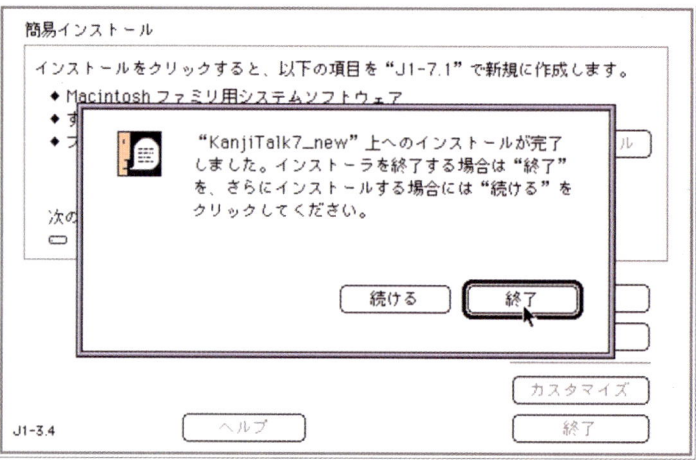

KanjiTalk brought the nuances of Japanese-language processing—typing direction, phonetic input interpretation, punctuation, line breaks, hyphens, and so on—to the Mac.

It sponsored the first Tokyo Macworld Expo; sponsored Janet Jackson's Tokyo tour and covered each of the 60,000 seats with Apple-logo seat cushions; opened AppleCenter franchises around Japan; and hosted a popular annual Pacific Forum conference in San Francisco, where potential developers could meet with Japanese distributors, software translators, and lawyers.

The initiative was a triumph. Within four years, Apple hit its $1 billion sales goal. Mac fans could choose from 500 Japanese-language programs and three Japanese Mac magazines. Crowds of 170,000 people attended Macworld Expo Tokyo. Apple's reputation, Montalvo says, "became almost mystical in Japan."

"In the process," *Time* wrote, "Apple has joined a select group of American companies that have debunked the myth of Japan as a fortress impenetrable to outside products."

Apple Pacific repeated the KanjiTalk playbook in other regions. There would be HangulTalk (Korean), HanziTalk (Chinese), HinduTalk (India), BhutanTalk, and ThaiTalk. (The king of Thailand, in fact, became such a Mac fan that he became a developer. He used Fontographer, a typeface-design program, to create Thailand's first digital fonts.) Those markets added another $1 billion a year to Apple's sales reports.

The Font Wars

Adobe produced an add-in card for Apple's $3,000 LaserWriter NTX that could print super-sharp Japanese characters. With this printer, a Japanese business could have a complete desktop publishing system for about $10,000. "Our sales went from $37 million a year to about $200 million," Montalvo says. "But it was still high-end."

And then, in a serendipitous encounter, Montalvo discovered a Boston software company called Bitstream. It had developed a competitor to PostScript: a system of outline fonts that produced crisp type on the screen *and* on any printer, right down to $200 inkjets. The LaserWriter NTX-J had been making money hand over fist for Apple in Japan, but desktop publishing on *inkjet* printers would blow the doors off. With these fonts, even noncorporate individuals could get great-looking type—and not just in Japan.

Montalvo called Chahil. "Satjiv, you gotta meet me in Boston! Buy the rights to this company. Just trust me!"

Apple now had its own smooth-font technology, soon renamed TrueType. Adobe's monopoly on smooth, crisp fonts was about to end.

In the beginning, of course, Adobe had been Apple's best friend. Its PostScript technology made the LaserWriter sing, and desktop publishing kept Apple alive.

But by 1988, John Sculley was resenting the cost of the relationship: $27 million in royalties paid to Adobe every year. The LaserWriter didn't even have an exclusive anymore; Adobe was now offering PostScript to other printer companies.

Adobe, meanwhile, had dreamed up a technology called Display PostScript, which brought its "smooth-fonts-at-any-size" magic to the *screen*. But to Adobe's astonishment, Apple declined to adopt it. Sculley wasn't about to start paying Adobe *another* set of royalties—and wouldn't have to, because he now had Bitstream's TrueType technology in hand. It offered smooth fonts on the screen *and* on any printer.

Furthermore, to give TrueType extra oomph in the marketplace, Apple had teamed up with its most bitter enemy: Microsoft. Apple supplied the font software, and Microsoft developed the printer software, called TrueImage. Neither company would need PostScript anymore.

When Adobe founder John Warnock learned of TrueType, he called Sculley, livid. "Your guys just committed a terrorist act!" he shouted. Adobe's stock fell by half. The font wars had begun.

To save its business, Adobe launched a frantic, sixth-month effort to create a new, free product called Adobe Type Manager (ATM). It would offer the same advantages as Apple's TrueType, but it would work with everyone's existing Adobe fonts—and it would beat TrueType to market by a year.

In the end, the two technologies coexisted. Most Mac fans did fine with TrueType fonts, and graphics professionals went with ATM and Adobe fonts. (Microsoft's TrueImage efforts went nowhere.)

The two companies finally buried the hatchet in 1991, when Apple incorporated Adobe's font technology into System 7. The font wars were over.

20. Newton

"Pen computing" was one of the buzziest of the industry buzzwords in the late eighties. Evolution had dictated the sizes of our hands, our hands dictated the size of our keyboards, and keyboards dictated the size of our computers. If we wanted smaller computers, shrinking our hands wasn't an option; we'd have to get rid of the keyboard.

Handheld computers weren't a new idea; Isaac Asimov's novels mentioned them in the 1950s. It was just a matter of waiting a few decades for batteries to last long enough, the screens to get flat enough, and the processors to be power-stingy enough.

By the mid-eighties, Japanese companies began producing the first handheld computers: crude, heavy, and barely usable, but real. Within Apple, Sculley was still feeling cornered by Wintel's growing dominance in computers. Maybe, he thought, a completely new technology platform could be a way out. Apple could win the war by leapfrogging the battlefield.

By the early nineties, pen-computing experiments were bubbling all over Apple.

Scribe

In May 1991, the USSR launched two cosmonauts to the Mir space station, *Dallas* wrapped up its 14-season run on TV, and Queen Elizabeth II became the first British monarch ever to address the U.S. Congress. But onstage at Apple's 1991 WWDC, the Advanced Technology Group's Laurie Vertelney was demonstrating a new prototype tablet code-named Scribe.

It didn't attempt to recognize handwriting; that feature, ATG interface designer Tom Erickson felt, was still too primitive. Instead, your handwriting remained handwriting—but became useful when you *tagged* it. That is, you could drag digital stamp icons onto one of your scrawls: Important, To Do, Calendar, Email, Remember, or Telephone. Later, you could round up all your To Dos, or Remembers, or Telephone notes with one tap.

The Scribe was the iPad's great-grandparent.

You could import documents and email from your Mac. If a document was too big for the screen, you could swipe with your finger to scroll it. The page continued moving until you tapped to stop. That's right: Apple demonstrated inertial scrolling, a universal smartphone feature today, in 1991.

Unfortunately, the prototype was incredibly heavy—"holding it in one hand and writing on the other was really difficult," remembers Erickson—and it would have cost far more than the PowerBook laptops that Apple was already shipping. "Costs more, does less" has never been a good slogan. The Scribe didn't move forward as a product.

"There were always little skunkworks projects going on," Sculley says, "but remember, these were not expensive projects. It was somebody writing some code and somebody getting some parts."

Sakoman

Hardware engineer Steve Sakoman had left HP in 1984 because he couldn't stomach a future of career boredom. "I saw myself building DOS clones for the rest of my life. I thought Apple might be more fun," he says.

He became Apple's director of hardware engineering. But after shipping the Mac Plus, Mac SE, IIgs, and the Mac II, he grew antsy again; he was still cranking out gray plastic boxes. In the summer of 1987, he told Jean-Louis Gassée that he was ready to move on. Together with former Lotus executives Mitch Kapor and Jerry Kaplan, he'd decided to form a startup dedicated to handheld computing.

But instead of wishing Sakoman well with his next venture, Gassée suggested that *he*, Gassée, should join the startup, too.

Sakoman replied that he didn't think it would work—too many CEOs in the room.

So Gassée pivoted. What if, instead, Sakoman created a startup *within Apple*? He'd have no corporate meddling—but he'd have all of the resources a huge company could provide.

Sakoman went for it. He named the project Newton, an homage to Apple's original logo. He found an abandoned warehouse on Bubb Road—train tracks, broken windows—and fixed it up, complete with black-and-orange newts spray-painted on the exterior walls.

One of Sakoman's first hires was Steve Capps, a key player on the original Macintosh software team. "We were all thinking about pen-based," Capps says. "A portable, slate-like computer—think of an iPad Pro–size computer, but much thicker."

As promised, Gassée protected the secretive group from corporate meddling—but that also left them undirected. "There was no feature that wasn't good enough to throw in this thing," Capps says.

It would be only three-quarters of an inch thick. It would run two hours on a charge. It would contain AT&T's new processors, nicknamed Hobbit. It would have a slot for PC cards so that people could add storage, a modem, ethernet, and so on. It would send faxes. It would beam data to other Newtons with infrared light. It would dial phone numbers through its speaker. It would recognize printed handwriting, cursive, or a mix of both. When you used the stylus to draw geometrical shapes, it would automatically neaten and straighten them.

All this for only $8,000.

The Touchscreen Mac

In early 1991, Paul Mercer, then one of five engineers writing the Finder for System 7, got his hands on a handheld computer called the Sony PTC-300. It was the size of a VHS cassette, with a black-and-white touchscreen, handwriting recognition, and a PC card slot. And it contained a Motorola 68000 processor— same as the Mac. He got to wondering: Could the Sony be the basis of a full-blown, handheld mini-Mac?

He put the Mac's ROM and RAM, plus MacPaint and a text editor, onto a memory card and, in short order, got it working. He called it Swatch.

Swatch was a part-time project, an "engineer to engineer" thing. One buddy got the palmtop onto an AppleTalk network; another pal offered handwriting software; a designer in the Industrial Design Group made a set of sleek, colorful housings.

In early 1992, Randy Battat, head of Apple portables, invited Mercer to show the executive staff what he'd been working on. "We hand it around," Mercer says. "People were like, 'Holy crap!'"

Sculley, in particular, adored the idea of a pen-driven Mac that cost under $400—and would fit in a coat pocket. Bob Ishida, Apple's liaison with Sony, invited Mercer to Japan to discuss a Sony collaboration. But no sooner had he bought his plane tickets to Tokyo than Sculley called him to his office.

"Listen, I just did this deal with Sharp to make Newtons, to be a manufacturing partner," Sculley said. "You can't go talk to Sony. With the Japanese, you just don't do that. You can only do one big deal with one Japanese company at a time."

And with that, the Swatch project was over. Paul Mercer joined the Newton team, left to wonder what might have been.

Tesler Era

In May 1988, Sculley sent Larry Tesler, then leading the Advanced Technology Group, over to Bubb Road to give the Newton group some direction.

Tesler's first directive was to get the price and the weight down. By using one processor instead of three, adopting a smaller screen, and incorporating a smaller battery, they brought the projected price down to $1,500.

For two years, the Newton team worked 12-hour days, with breaks only for Ping-Pong or basketball, convinced that they were onto something great. The screen would be like an endless paper tape (or roll of toilet paper)—no modes. An Assist button would parse whatever you'd written and move it to the right app (date book, address book, and so on).

The stylus permitted ingenious editing gestures. To correct a typed letter, you simply wrote a new letter on top of the wrong one. You could draw a carat (^) to insert something new. You could delete something you'd written or drawn by scribbling up and down; it vanished with a puff of animated cartoon smoke. (Steve Capps, who devised and even patented that charming bit, recorded the *whoosh* sound effect by blowing across a microphone.)

Capps also created the animation for deleting an entire page: It crumpled into a paper ball and

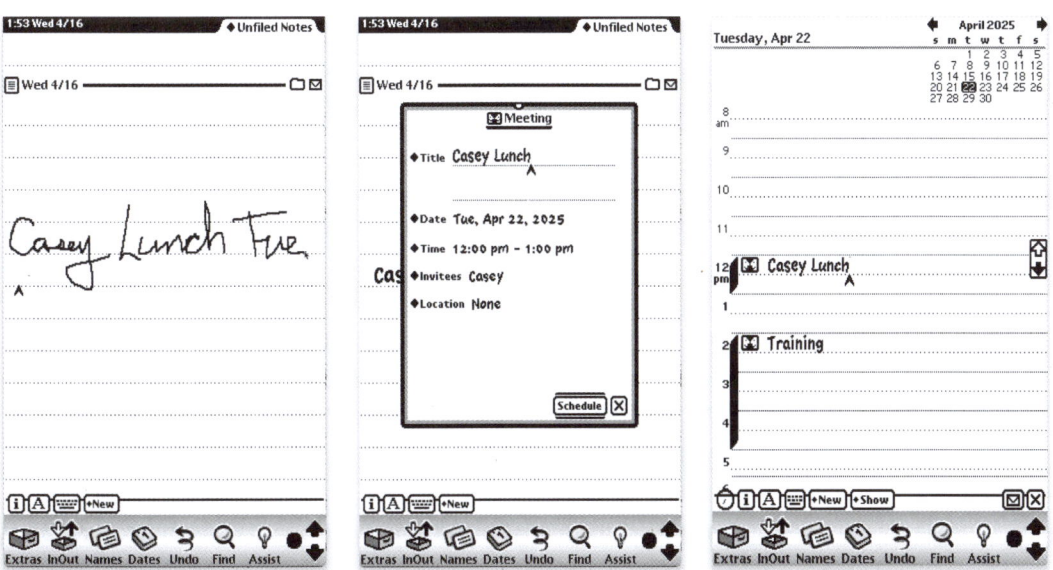

If you wrote "Casey Lunch Tue" and then tapped Assist, the Newton asked you to confirm its interpretation (center) and then added the appointment to the calendar.

flew into a wastebasket icon. "I felt it was too cute," he says, "but people really loved it. You'd give demos to people, and they would say, 'Do that again! Do that again!' And you'd be sitting there throwing stuff away."

Sculley loved the demo, too. He green-lit the Newton project, and gave the group two years to finish.

What made things awkward, however, was that *another* handheld touchscreen computer was underway at Apple, too. This one, code-named Paradigm, was the brainchild of Marc Porat, a consultant Apple had hired to help identify the next wave of computing.

So in May 1990, Porat suggested that Apple spin off the Paradigm project into its own little startup, which would be called General Magic. Apple would own a minority stake; Sculley would be on the board. Sculley agreed.

General Magic

When he founded General Magic, Marc Porat intended to create "a tiny computer, a phone, a very personal object," with "the comfort of a touchstone, the tactile satisfaction of a seashell, the enchantment of a crystal."

Many former Apple superstars fell under his spell. From the original Mac team, Bill Atkinson, Andy Hertzfeld, Susan Kare, and Joanna Hoffman. From the System 7 team, Phil Goldman, Bruce Leak, and Darin Adler. From the Apple II and III projects, Wendell Sander and Walt Broedner.

General Magic would create a prototype, then license the software, called Magic Cap, to other companies, much as Microsoft had licensed Windows. The charming visual interface was built of "rooms": games in the Living Room, work tools in the Office, and so on.

Sony, Motorola, Matsushita, Philips, and AT&T became investors and partners. Nine telecom partners contributed $6 million apiece for the right to offer cellular service for mail and messages. General Magic's IPO was a feeding frenzy, raising $96 million. It was the hottest, most secretive project in the Valley.

Then the hype faded. As years passed, General Magic still had no product, in part because it was custom-building every piece. "In many projects, you stand on the shoulders of giants," programmer Kevin Lynch says. "We were building the giant from the toes all the way up to his head."

When Sculley announced the Newton in 1992, General Magic's team was stunned by its similarities to Magic Cap. ("I thought they would coexist," says Sculley. "I wasn't really concerned that Newton would hurt General Magic.")

The Newton announcement put pressure on General Magic to reveal its *own* project, but Magic Cap wasn't anywhere near ready. The team set up bunk beds in the office and burned themselves out for weeks to complete it.

The fall of 1990 was a turbulent time for the Newton project. Gassée was Steve Sakoman's friend and protector; when he left Apple, Sakoman left, too.

At about the same time, Tesler started thinking about how to market the Newton. He brought in Michael Tchao, who'd worked on marketing for the Mac Portable and Apple's multimedia exploits. Tchao was alarmed by what he found.

"When I got there, the Newton was a collection of 'no one's ever done *blank* before.' It literally had three different wireless networking technologies built into it, and there wasn't a single one shipping in any Apple product. You know, they were all science projects," he says. "The AT&T Hobbit processor had never been shipped before. It had a digital signal processor chip in it, which had never been put in a Mac before. It had an all new power connector. It had a whole new card-slot architecture. It had a whole new software."

"AT&T was late; they could not get their network done. Sony was late; they could not get the consumer electronics piece done. We were late, because we were perfectionists," Porat says. "Everyone was late. Two years late."

The first device, Sony's Magic Link, finally arrived in September 1994 for $800 — and nobody bought it. Most Americans, it turned out, had no particular interest in *doing* email on the go. As engineer Tony Fadell put it, "Joe Sixpack had literally no reason to put a computer in his pocket."

Worse, in the new era of free internet, paying a subscription for AT&T's private network made no sense to anyone. AT&T bowed out. The stock crashed. Porat left the company, then got divorced. "It all fell apart," he says.

Some of General Magic's innovations live on — in animated emojis, push notifications, and apps — and many of its alumni went on to engineering stardom. Tony Fadell led the iPod project; Kevin Lynch led the Apple Watch; Pierre Omidyar founded eBay; Steve Perlman, Phil Goldman, and Bruce Leak founded WebTV; Andy Rubin created the Android phone's OS; and Megan Smith became the U.S. chief technology officer under President Obama.

General Magic was, Sculley says, "the most important company to come out of Silicon Valley that nobody's ever heard of."

The Eye of Sauron

The original full-size Newton, code-named Cadillac, was supposed to have a super-cool collaboration feature: diffuse, wide-area infrared communication.

The early tablet design, by Italian car designer Giorgetto Giugiaro, had a massive lens that the group nicknamed the Eye of Sauron, after the all-seeing evil eye in *The Lord of the Rings*. Its purpose was to blast out infrared light, visible only to other Newtons nearby.

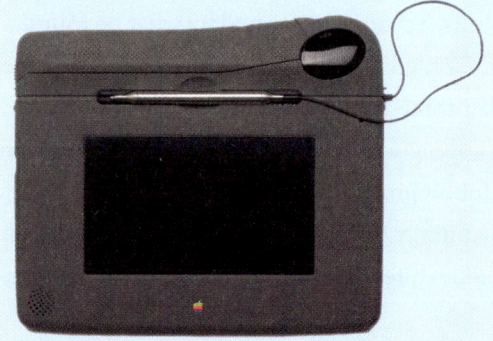

"If you walked into a conference room with three or four of these, the idea was you could scribble on your Newton, and the marks would show up on their Newtons," says Tchao.

But there was a hitch: It didn't work in rooms with fluorescent lighting.

Fluorescent light bulbs flicker rapidly—too fast for you to see, but disruptive to the Newton's own infrared transmissions.

Unfortunately, in 1982, most rooms in most offices all over the world used fluorescent lighting, making one of the Newton's most magical features a nonstarter.

It also had its own, new, object-oriented programming language called Dylan, supposedly short for *dy*namic *lan*guage. (And yes, Bob Dylan sued for trademark infringement. Apple settled.)

To Tchao, the whole thing was too ambitious and too expensive, but mostly too big. If something is the size of a thick clipboard, you have to rest it on something every time you want to write on it. Even the midsize prototype, code-named Vic, seemed too unwieldy to write on without sitting down or balancing it on your forearm.

Tchao, Capps, and hardware engineer Mike Culbert thought that something one-handable, like a notepad, made much more sense. You'd be able to pull it out, write something, and put it away, all while standing in line or even walking.

Tesler pushed back. A smaller model wouldn't accommodate many of the coolest features. He worried that what remained would be little more than an organizer, a Sharp Wizard thing.

But in February 1991, on a flight to Japan for the Tokyo Macworld Expo, Tchao pitched Sculley on the superiority of the smaller, less expensive Newton. It wouldn't have the infrared beaming, but it would still convert handwriting and do an amazing job of calendars, tasks, phone book, notes, and sketches—for under $1,000.

Sculley agreed to fund it, on one condition: It had to fit inside a man's suit pocket.

The Speech

On January 7, 1992, in a speech at the Consumer Electronics Show in Las Vegas, Sculley revealed his vision for handhelds. He hinted that Apple was developing a general new category of gadget, which he called a personal digital assistant. He predicted the birth of an entire PDA industry: wireless, handheld, smart computers. Within a decade, he said, it could generate $3.5 trillion in annual revenues.

His talk started a towering wave of hype—and started the countdown clock; there could be no more pursuing endless dream-machine ideas. Sculley told the Newton team to have a working prototype ready for the Chicago CES in May, four months away—and have the Newton ready to ship eight months after that. Until that moment, nobody on the Newton team could have said for sure that their project would even become a product.

Getting a demo ready for Chicago was a superhuman effort—a string of 18- and 24-hour days. As late as 2 a.m. the night before the keynote, the compact Newton prototype, "Junior," was still

The Newton-Sized Pocket

Sculley's goal wasn't just to demonstrate a working Newton at his presentation at the Chicago CES in 1992. He also dreamed of pulling it out of his inner coat pocket.

The problem was that the Newton model was a fraction of an inch too wide for the sport coat Sculley planned to wear. There wasn't much time. "We thought we had already reduced the product's size to the absolute minimum," says design head Robert Brunner. "The size issue became a crisis."

Brunner's team flattened the Newton's curves a bit, shaving off another millimeter or two. Back they went to Sculley's office closet—the Newton still didn't fit.

On another attempt, they flattened the stylus, in hopes of eking out another millimeter. Now the Newton was narrow enough to fit in *most* men's suit pockets. But for some reason, Sculley's pocket was a hair narrower than the standard. "I seriously thought about trying to get in there and opening up that pocket a little," says Brunner. "We were convinced that he had some fifth-percentile coat pocket."

Three times, Brunner and his team brought revised Newtons to Sculley's office; three times, the Newton wouldn't fit. Finally, they produced a version that could squeeze into the pocket fully.

A few weeks later, Sculley took to the stage in Chicago. As the audience watched, he reached inside his blazer for the Newton.

"And it fits in my coat pocket!" he said.

Michael Tchao and Steve Capps, barely awake, hold up Newtons at the Chicago CES.

crashing. The engineers were so exhausted that they fell asleep on the stage floor between rehearsals, or even between sentences in conversation.

But the Chicago demo was a smash hit. The audience marveled as Capps showed a future pizza-ordering app: He dragged icons of toppings onto a pizza and then tapped the Fax button. (His prototype Newton was secretly cabled to a Mac IIfx backstage.)

Still, the Chicago presentation created a double problem. First, Apple had announced a ship date—an immovable deadline. Second, rivals now had an entire year to gear up competitive products.

The challenge to finish the Newton was unimaginably huge. It had a brand-new, untested processor, containing a brand-new multithreaded OS, with an entirely new programming language. "From an engineering point of view, that's incredibly exciting," software engineer James Joaquin says. "But from a business point of view, it's completely crazy."

To make matters worse, AT&T's new Hobbit chip was riddled with bugs. Late in the game, Tesler made the difficult decision to replace it with a brilliantly engineered new processor from a tiny British company called Acorn Computers. The beauty of this chip was not just its speed, but its low power requirements. It was *born* for battery-powered devices, a natural for the Newton.

The Birth of ARM

At this moment, you probably have at least one ARM processor on you or near you. There's one inside every modern phone, tablet, and smartwatch on earth. And no matter whose logo is on the case, the chip was brought to you by Apple.

In 1983, Acorn Computers, the British computer startup, was seeking a more powerful chip for its home computers. The Motorola and Intel chips of the day offered either speed or power frugality, but not both. But a hot new chip technology was being developed at Stanford and Berkeley, something called RISC (reduced instruction set computer). Because these chips respond to a far shorter list of software commands, even if they have to crunch more *of* them, they were faster and required less power than traditional processors.

Acorn decided to design its own RISC processor. They nailed it on their first attempt: the ARM1, in 1985. It required half as many transistors as traditional chips, and therefore required much less power and generated much less heat.

Larry Tesler, on the quest for a better Newton chip, persuaded Sculley to look into Acorn. And in November 1990, Apple, Acorn, and chip-fabrication company VLSI formed a joint venture called Advanced RISC Machines Limited, or ARM. Twelve of Acorn's designers set up shop in a renovated turkey barn in Cambridgeshire, England.

Apple, having contributed $3 million, owned 43 percent of the new company. Tesler joined its board, to ensure that Apple got exactly the kind of processor it needed.

ARM began licensing its technology—to Nokia, Qualcomm, Samsung, Broadcom, and a thousand other companies. Today, ARM's licensees make 1.5 billion ARM chips a year, and Apple's investment became worth a towering fortune.

The real punch line of this story, however, is not how ARM saved the Newton; it's how, 16 years later, it saved Apple.

Now, however, the engineers would have to rewrite the million lines of code they'd already done for the AT&T chip—in eight months.

The Newton missed its April 1992 deadline and dropped the original large-format Newton project, known as Cadillac or Senior; Larry Tesler left the project and went back to ATG. "I'm a starter type of person," he said. "Somebody else comes in to clean up."

It missed its January 1993 deadline, too. The new target was the Boston Macworld Expo in August 1993.

Newton became a five-alarm fire. Nobody slept. Health and relationships suffered. For his wedding, Steve Capps drove down to Monterey wearing shorts, brought along Michael Tchao as the witness, exchanged vows, and returned directly home to keep coding.

Sculley had arranged for Sharp to manufacture the Newton; Sharp, Motorola, and a couple of other companies licensed the Newton, intending to make their own versions. On June 27, Apple finally shipped the ROM code off to Japan, full of bugs, magic, and Easter eggs.

When the first shipments of the Newton MessagePad arrived from Japan, they were unsellable. To the team's horror, the little rubber feet kept falling off, and the tutorial PC card didn't work. A team in Sacramento spent an entire day gluing feet and manually patching PC cards. Another team fanned out to update the software in the Newtons that were already in stores. One Apple rep would distract the store's owner while another slipped into the stockroom to upgrade the Newtons. "It was not pretty," says Tchao.

Finally, the moment arrived. The fans packed into Boston's Symphony Hall, where banners bore the Newton's double-meaning slogan: "Now taking orders." Steve Sakoman's dream machine was finally on sale—for $800.

The Newton was ahead of its time in both good and bad ways.

Handwriting

The Newton had an on-screen keyboard that you could peck with your stylus, but the real magic would be handwriting recognition. No consumer device had ever offered such a feature.

The technology behind it came from a tiny company called ParaGraph, founded by a Russian programmer named Stepan Pachikov. (He would go on to create the popular Evernote organizing app.)

ParaGraph was remarkable in its ability to recognize both printed lettering and cursive, no matter your writing style. It worked by comparing what you'd written with a word list stored in the Newton's ROM. But as Newton development approached the finish line, there was very little space left in the ROM; it had room for a list only 10,000 words long. The ParaGraph engine would always find a match for the word you'd written—but it wasn't always the word you wrote.

When the recognition worked, it was magical. When it didn't, which was often, it was terrible. "People started to make jokes about the handwriting recognition," Sculley says. "Here you've got some of the most brilliant computer scientists ever, doing breakthrough stuff that no one has ever seen before, and we're getting slammed in the press because the handwriting recognition didn't work."

The peak of Newton-mocking arrived on August 26, 1993, in Garry Trudeau's *Doonesbury* comic strip. In it, Mike Doonesbury is trying out Newton's handwriting recognition. He writes, "Catching on?" And the recognizer types out, "Egg freckles?"

To produce the custom Easter egg, you'd write "egg freckles" and then tap the Assist button.

The strip became what today we'd call a meme. Along with a similar gag on *The Simpsons*, it turned the Newton's poor handwriting recognition into a national joke.

But Steve Capps had an even funnier idea: He'd immortalize Trudeau's joke *in the Newton*—as an Easter egg.

Getting the rights to the published cartoon, however, would have alerted Apple's lawyers to what Capps was up to. Instead, he decided to perform an end run. He'd ask Trudeau himself to draw a *custom* version of that same comic frame, exclusively for use in the Newton's 2.0 software.

Improbably, Capps and his wife tracked down Trudeau's phone number and reached him at home. Trudeau had never heard of an Easter egg—but once he understood the idea, he agreed immediately.

And so it was that Garry Trudeau drew a custom version of his cartoon, satirizing the Newton, that Steve Capps immortalized *in* the Newton.

The Newton team eventually sent a MessagePad to Trudeau to thank him. He got the handwriting to work pretty well, he says, "but you have to write really slowly. I found myself printing more legibly than any time since the fifth grade."

After Newton

The reviews for the Newton were painful. "Apple promised too much and failed to deliver a useful device," said the *New York Times*. Newtons were "best suited as random non sequitur generators."

"By the time we shipped the first product," says Newton marketing exec James Joaquin, "it was more expensive than we thought. It was heavier than we thought. The battery life was shorter than we thought. The handwriting recognition wasn't quite as good as we thought. All of those forces of gravity hurt us." Early adopters snapped up 50,000 Newtons in August—and that was about it. During the holidays, Apple sold only about 7,500 a month.

eMate 300

Apple has never sold a touchscreen laptop—or so most people think.

Since the Newton's arrival, Apple's Industrial Design Group had experimented with spin-offs. There was a medical records version (white case, red cross); a cellular Newton; and a brightly colored Newton for kids.

But in 1996, when school-system IT buyers were complaining about the prices of Macs, a new idea emerged: Turn the Newton into a touchscreen mini-laptop that kids could take home with them. After all: The Newton had no spinning hard drive or floppy—no moving parts at all—so it could be incredibly rugged. And it had no color screen, so it could be cheap, and the battery could last all day.

Apple's industrial design team came up with a clamshell with—for the first time on a Newton—a built-in keyboard. It looked like something from Porsche: curved, sculpted, sensuous. Its nearly indestructible green plastic shell was translucent, to evoke a sense of accessibility and openness. Designer Thomas Meyerhöffer put no electronics beneath the keyboard, to prevent damage from spills. A carrying handle hid inside the hinge.

"I tried to make the design friendly and accommodating for the little ones, yet sophisticated enough so that older students will think it's cool," says Meyerhöffer. In focus groups, kids said it reminded them of gummy bears.

It was called the eMate 300 ($800). Its 480 x 320–pixel grayscale screen could open completely flat, so kids could draw on it. Its infrared lens let kids beam their work to each other or to the teacher.

Teachers, students, and critics adored the eMate, but it was never the bestseller it should have been. Apple, reaching financial and administrative rock bottom in 1997, spent almost nothing marketing it; few educators even knew of its existence. Then, only a year after the eMate's birth, Steve Jobs cut all Newton efforts. That was the end of the weirdest, most adorable laptop Apple ever made.

Apple moved quickly to fix the handwriting and address the shortcomings. The MessagePad 100 (March 1994) cut the price by $100 and added letter-by-letter recognition. The 110 featured a gorgeous, thinner body, featuring a protective cover that flipped around like a stenographer's notebook. It was designed by a young British designer, newly arrived at Apple, named Jony Ive.

The MessagePad 120 (January 1995 in the U.S.) let you rotate the screen, or attach a physical keyboard, or switch on backlighting in dim environments.

And yes, Apple fixed the handwriting recognition. Apple ATG's Larry Yaeger came up with Rosetta, a technology that offered superb recognition even of made-up words or weird last names. Rosetta was part of the 2.0 version of the Newton software.

There would be a MessagePad 2000 in March 1997, with a screen that could show shades of gray, two card slots instead of one, and a processor ten times faster. After having lost money consistently for four years, the Newton was finally breaking even.

And then Steve Jobs returned to Apple.

21. Moonshots

In the spring of 1990, the Hubble Space Telescope launched, *Pretty Woman* was a movie blockbuster, and Microsoft introduced Windows 3.0. It was buggy and slow, and required a huge amount of memory. But most businesspeople didn't seem to care. Microsoft sold 4 million copies in a matter of months.

90 percent of all new personal computers now had Intel inside. As Sculley was painfully aware, a vicious cycle was taking shape: The more people bought PCs, the more software companies wrote programs for them—which made the PCs even more attractive.

Meanwhile, Apple had a processor-pipeline problem. From the day the Mac was born, its processors had come from Motorola. But Moto's pace was slowing down just as Intel's was picking up.

And that is why Apple's 1990–1992 period became an era of moonshots: huge, expensive, ambitious attempts to avoid circling the drain. There was talk of Apple buying Sun. Of making Mac clones and selling them mail-order. And, always, of the nuclear option: licensing the Mac OS to other hardware companies.

The only one that went anywhere began with a phone call from IBM.

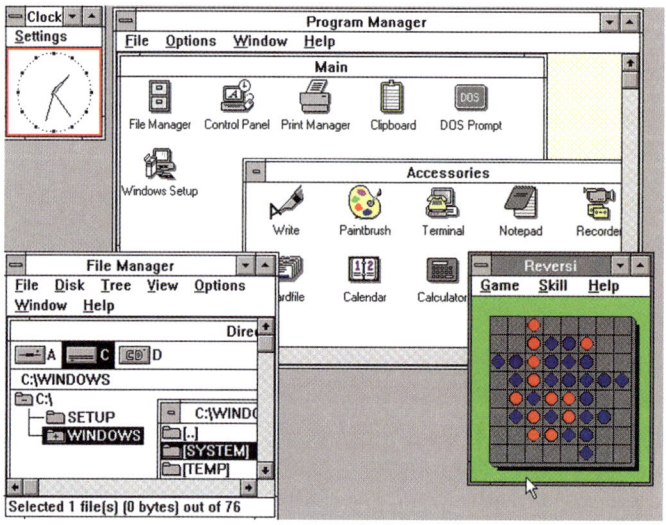

Windows 3.0 was the first commercially successful version.

Big Blue

Like everyone else in Silicon Valley, Apple had become tantalized by the prospect of RISC processors. These chips ran faster, cooler, and more efficiently. They could open up fantastic new uses for Macs, like video editing, 3D graphics and, someday, video chats.

RISC projects were percolating all over Apple. One was a brand-new, clean-slate supermachine called Jaguar. Other groups were testing new RISC processors from chipmakers AMD and MIPS. Sun's experience in RISC computing—it had been the first to market a RISC chip—is what made Apple investigate an acquisition in the fall of 1990.

But IBM had a RISC chip, too, and was looking for customers. So in 1990, IBM president Jack Kuehler called Apple about a possible collaboration. It seemed to make sense: Apple was fishing for a better processor supplier, and IBM had little expertise in consumer software design.

Sculley was intrigued, but didn't want Apple to be beholden to *another* single source of processors. He proposed inviting Motorola to join the discussions, too, to form an Apple-IBM-Motorola alliance. It even had a great acronym: AIM.

In June 1991, after two weeks of negotiations at an IBM research facility in Austin, the three companies announced three new collaborations.

- **PowerPC.** IBM and Moto would design the new RISC processor chip, with Apple's input. Motorola would shut down its own RISC project.

- **Kaleida** was a joint venture of Apple and IBM that would develop ScriptX, a programming language for making multimedia CD-ROMs.

Robert Brunner's team designed foam mockups of the Jaguar RISC Mac.

Project Aquarius

Apple's 1990 collaboration with IBM and Motorola wasn't its first attempt to take control over its own processor destiny.

In 1986, engineer Sam Holland presented Jean-Louise Gassée with an intriguing pitch. Why remain at the mercy of Motorola's pricing, designs, and timelines? Apple should design its own blazing-fast, four-core processor! Project Aquarius was born.

To simulate this new processor, Sculley authorized the purchase of a huge, purple, $15 million Cray X-MP/48 supercomputer, along with an electrical substation to power it and a special building to house it. It took six weeks for 24-hour crews to install the AC, removable floor, power lines, and cooling tower with high-pressure Freon plumbing for the Cray.

With all of that construction fuss, it was hard to keep Aquarius secret. The cover story was that Apple was using the Cray to model new Macs. What it was actually doing, of course, was developing Apple's first home-brewed silicon.

Over time, though, Aquarius became a money pit. "It was getting ugly," says former Atari game designer Al Alcorn. "And I was asked by the troops, would I take over the project? It was considered career suicide." Then again, he was already an Apple Fellow. "I had nothing to prove, so I could take on a case like this."

Eventually, the effort became mired in budgetary, design, and technical snarls. Alcorn shut the project down.

Apple would indeed make its own RISC chips—when it bought Palo Alto Semiconductor in 2008. Sam Holland had been right about the advantages of Apple controlling its own chip destiny. He'd just been 20 years early.

- **Taligent** was another Apple/IBM joint venture, this one dedicated to creating a new, universal, rock-solid operating system.

  Soon after Apple's 1988 off-site to plan its OS future, senior engineer Erich Ringewald had rounded up a couple dozen engineers to study the feasibility of the medium-hard ideas, the ones written on pink index cards. He'd set up shop in a former warehouse on Bubb Road, with plans to build one of the world's first *object-oriented* OSes: built of software modules that can easily be extended, reused, or changed, for better stability and consistency. Someday, developers could write individual *features*—a spelling checker here, a chart maker there—rather than entire monolithic programs. Customers would be able to buy and plug in only the modules they needed.

But after two years of work, the Pink OS was still only a wobbly idea. Taligent would see it to the finish line.

Three powerful tech companies, three promising endeavors. It seemed to be, as *BusinessWeek*'s headline put it, "An Alliance Made in PC Heaven."

AIM's three initiatives.

Project Star Trek

In February 1992, representatives from Novell contacted Apple with a tempting offer. If Apple could make a version of the Mac OS that would run on Intel chips, Novell would install it on the PCs that connected to its server software.

Apple perked right up. Novell made NetWare, the most popular OS for corporate servers. Here was the chance to introduce millions of corporate drones to the elegance of the Mac! It would be a corporate Trojan horse. Mac OS could run on ten times as many computers!

On the other hand . . . Mac on Intel? *Insane!* It was an engineering task that could break mere mortal programmers.

Apple set up the Star Trek project: a top secret team of 18 engineers in a Novell office five miles from the Bandley campus. Software VP Roger Heinen gave them three months to develop a prototype, encouraged by a promise of $25,000 bonuses.

It was a small, gifted team, racing against an impossible deadline, with huge ramifications for success – a rekindling of the spirit that drove the original Mac team in 1983.

They pulled it off. By November, they had the Finder, QuickDraw, and QuickTime ported to Intel, running on a generic PC. It was time to demo Star Trek for the top brass: Sculley, Spindler, Joe Graziano (who had returned to Apple for a second stint as CFO), and the rest of the executive staff. They were confident that they'd get an enthusiastic green light to proceed.

What they didn't consider was Apple's *other* corporate collaboration: the AIM partnership. The big bosses worried that Apple didn't have the bandwidth to pursue *both* massive, all-hands-on-deck projects, and that IBM and Motorola would frown on the distraction of another major side project.

By June 1993, Star Trek had boldly gone where many projects of the era wound up: nowhere.

In the end, two of the ventures collapsed by the end of 1995. The internet ended Kaleida; what use would anyone have for CD-ROMs in the age of downloading? And Taligent's mission, creating an object-oriented OS, proved to be crushingly difficult, especially as a collaboration of two companies. "The cultures were just too different between IBM and Motorola and Apple," Sculley says. Apple and IBM had spent $300 million on the two failed ventures.

Emulation

The plan to move Macs onto the PowerPC chip came with a footnote the size of Texas: Existing apps don't run on a new processor.

At this point, there were thousands of Mac programs. Every single one of them was written for the Motorola chip. Putting a new processor into a computer dooms the investments that your customers and developers have made in the old one. "It's a transition that no company has pulled off without trouble," wrote *InfoWorld*'s Stewart Alsop.

The obvious solution would be to write an *emulator* program: software that *impersonates* the older chip. In theory, existing programs think they're still running on the original processor, and operate just fine.

Unfortunately, just as it takes longer to carry on a human conversation through a translator, emulation makes software run slower.

Apple did employ, however, an experienced writer of microcode, the layer of software between pre-RISC processor chips and the code that programmers see. After nearly three years at Apple, he'd become intimately familiar with the Motorola chips and their interactions with the Mac. In fact, he was already writing an emulator—for Jaguar, Apple's effort to put a Motorola RISC chip into a Mac. His name was Gary Davidian.

Gary Davidian

Born: 1956, New York City, NY
Schooling: SUNY Buffalo
Before Apple: Nanodata, Data General, Rational, NEC (consulting)
Apple: 1987–1995 (employee), 1999–2013 (consulting)
After Apple: AMD, Power Computing, Fast and Small Software

Young Davidian learned how things work by taking apart vacuum cleaners, bikes, and toys. His favorite was the robot toy Mr. Machine.

His emulator, code-named Cognac, was looking good. It held the promise of bringing RISC to the Macintosh without sacrificing compatibility with existing software. But in the summer of

1991, the Jaguar team was called into a conference room. On the whiteboard, someone had written "DOTC."

"And I think: 'What the hell is DOTC?'" Davidian says. "It was 'Deal of the Century.' That was when we were told that Apple was going to switch to PowerPC."

He had just spent two years writing an emulator for the wrong RISC chip.

Davidian knew nothing about the PowerPC processor; it didn't exist. But Apple needed a Mac emulator for it, and he was the guy.

Two years later, Apple introduced its first RISC computer, the Power Macintosh 6100. Thanks to Davidian's emulator, existing Mac software ran at the delightful speed of a Mac IIci. And *new* programs, written with the RISC chip in mind, absolutely flew.

In the end, the move to RISC was the only one of Apple's moonshots that succeeded. And today, a RISC chip powers every Apple computer, phone, tablet, and watch on earth.

22. PowerBooks

When Steve Jobs chose frogdesign to become his design department, Apple's own in-house designers were largely sidelined. Some moved to other parts of Apple; some left.

Once Jobs was gone, however, frogdesign's star began to dim. Two forces made sure of that: Jean-Louise Gassée, who was still paying Hartmut Esslinger's firm more than $2 million a year, and Apple's remaining designers. The frogdesign era ended abruptly in 1985, when Apple bought out the remaining time on Esslinger's contract. (He rejoined Jobs at NeXT.)

Richard Jordan, VP of design, now faced a daunting challenge. He needed to repopulate Apple's own industrial design department (meaning physical design, as opposed to electrical or mechanical design). And he needed a star to lead it.

After visiting designers all over the world, he realized that the perfect candidate was right in his backyard: Robert Brunner (rhymes with sooner).

Over the years, Jordan had hired Brunner's little firm, Lunar Design, to design three machines that never shipped and one that did: the Macintosh LC, which became a colossal hit. Brunner, he'd always thought, was talented and chill.

Brunner started work on January 3, 1990. The era of the Apple Industrial Design Group was about to begin.

Robert Brunner
Born: 1958, San Jose, CA
Schooling: San José State University (BS)
Before Apple: Lunar Design
Apple: 1989–1997
After Apple: Pentagram, Ammunition Group

Brunner and his team have designed every Beats by Dr. Dre audio product since its founding in 2005. When Apple bought Beats, Brunner's Apple history came full circle.

Brunner

"My first day at the job, I thought I'd made a horrible mistake," Brunner says. During the frogdesign era, Apple had moved the scraps of the design group seven miles away, to a strip mall in

Santa Clara. "I had this ten-by-twelve-foot Herman Miller Action Office cubicle with no windows, right in the middle of the building, all warm gray. I'm mildly claustrophobic. I'm just like, 'What have I done?'"

It took over a year for Brunner to set up a better space for the new design department: a building near the Apple campus called Valley Green 2. It was far enough off the beaten path to give the designers some privacy, but close enough to feel part of the Apple action.

Now he had to build a team. To convey the message that Apple was once again a cool place for designers, he bought the back cover of *Industrial Design* magazine for a few months, filling each with photos of whimsical, attention-grabbing mock products, like a language-translating mask, a digital food taster, and a handheld currency exchanger.

But before he could settle in with his new team, Brunner would have to deal with the Giugiaro problem.

In mid-1989, Apple had offered famed Italian car designer Giorgetto Giugiaro $900,000 to come up with a new Apple design language. His concepts were fascinating, but it was obvious that Giugiaro had never designed a computer. His work focused solely on the chassis, and he provided very few measurements and dimensions.

"I'll never forget the day I knew that it was gonna end," Brunner says. He'd been awaiting a shipment of models from Giugiaro's studio for a new computer. "I walk in there expecting all these

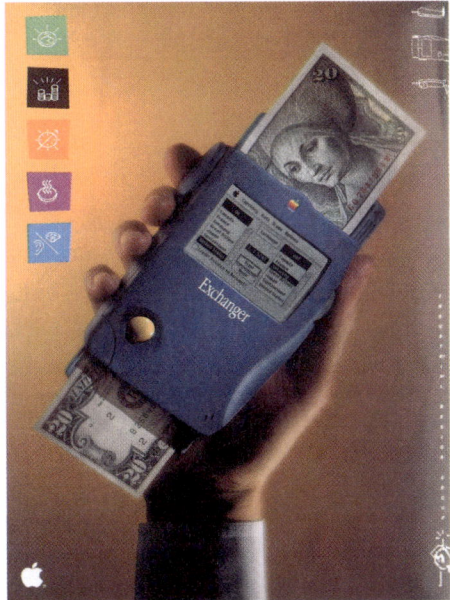 

Brunner's fanciful inventions graced the back cover of Industrial Design *magazine.*

boxes, and there's *one box*. And we open it up, and they're little, teeny, fifth-scale models—like that you would do with a car, right? I remember looking at these little toys going, 'Oh, no. This is just not gonna work.'"

Espresso

Brunner admired Esslinger's designs. But the Snow White ethos wouldn't fit Apple's portables and handhelds. The colors and geometric shapes worked well when you saw them together on a desk, but made less sense in products you carried around.

Then there were those long parallel grooves—the "Snow scars," as the new ID team called them. They were 2 millimeters deep, which required that the plastic be at least 4 millimeters thick—otherwise, the grooves would weaken the plastic. But 4 millimeters was much too thick for a laptop's case. So the grooves had to go—and with them, Snow White.

Brunner shifted away gradually. His Macintosh LC design, for example, featured the Snow White grooves and platinum color—but also adopted curves that gave it a hint of whimsy.

But in time, Brunner put his own stamp on the designs, which—in an homage to the coffee machine in the design studio—became known as Espresso.

Symmetry would be a key element. Colors would be darker. Subtle curves would take Espresso designs away from the rigid corners of Snow White. The idea would be to maintain "something that felt precise and technical, but had a much more humanistic feel to it," Brunner says.

PowerBook

On the day Brunner arrived at Apple, an urgent project was already on his desk. It was the product that would change Apple's fortunes for decades to come: the PowerBook.

Windows laptops were coming on strong. Apple needed a Mac laptop, and soon. Sculley was projecting that by the beginning of 1991, 30 percent of all computers sold would be laptops. He wanted an Apple laptop no bigger than 9 x 11 inches, no heavier than 7 pounds, and on the market by fall 1991.

Time to market was so essential, in fact, that product managers wrote "TTM" on whiteboards in planning meetings. At one point, an engineer wandered in, misread the handwriting, and asked, "What's this TIM?" A code name was born.

In January 1990, Brunner and principal TIM designer Gavin Ivester arrayed the components of the new machine before them on a table—circuit board, battery, hard drive, screen, keyboard, modem, trackball—and began seeking the best layout. The first step was to sculpt life-size models

*PC laptops of 1990 featured keyboards at the front—
and a trackball hanging off the edge.*

out of foam—"literally surfboard foam," says Brunner. "It's a really good density, like a really light gauge of wood, almost like a balsa wood."

On most laptops of the time, the keyboard hugged the front edge of the machine, like typewriters before them. Most laptops didn't have pointing devices, either. A few Windows laptops had detachable trackball units that hung off the side, just waiting to be smashed by a flight attendant's cart.

TIM would have a built-in, centered trackball, convenient for both lefties and righties, but there was no obvious place to put it. One quickly abandoned idea placed it *above* the keyboard. On most of Ivester's initial designs, though, it was below the keyboard, forcing the keyboard away from you by an inch. Even that gap, however, left room for only a tiny trackball, about 20 millimeters in diameter (three-quarters of an inch).

Jon Krakower didn't think that was big enough.

Krakower, a board designer, systems integrator, and puzzle aficionado, had spent three and a half years working on the Macintosh Portable. One night in September 1989, during a press tour for the Portable, he sat in his New York hotel room, deeply disappointed by the Portable's terrible initial reviews. In a burst of inspiration, he grabbed one of the glossy card-stock brochures for the Portable, cut out photos of its various elements, and began to rearrange them.

If you wanted any decent precision with the trackball, it would have to be bigger than the current proposal—say, 30 millimeters (1.2 inches). And if you wanted room for that, you'd have to push the keyboard even farther from the front edge. And if you're going to do *that*, why not push it *all* the way back—up against the laptop's hinge? "That idea just fell into my head," Krakower says.

Suddenly, you'd have *palm rests* on either side of the trackball. Your wrists wouldn't have to hover, and the sharp front edge of the machine wouldn't dig into them. Your hands would be stabilized when the laptop wasn't on a level surface. You'd have built-in support when using a tray table on a plane or train. And you'd be able to move the cursor with your thumb, without taking your fingers off the keys.

Krakower's functional TIM mockup.

Krakower was not an industrial designer; it wasn't even remotely his job to suggest a redesign of Apple's new laptop. So he didn't present his idea to Ivester and Brunner right away. "I had to keep it under wraps as long as possible, collect enough evidence to justify the idea," he says.

Over the Christmas break in 1989, Krakower built a working model. It had a real screen, keyboard, and trackball, but no circuitry; it was all a front end for a Mac Portable nearby.

He began orchestrating secret, after-hours focus groups in a conference room. In all, he videotaped 45 Apple employees as they assessed various trackballs, side buttons, and slopes. "How weird is it to have the keyboard back there and putting your thumbs on the track ball?" he'd say.

Finally, he was ready to present his radical redesign to his managers. Most weren't thrilled. "This is just too different," they'd say. "We already screwed up once, with the Portable; we can't afford taking a risk this size."

Brunner, however, immediately became its champion. "Wait a minute. This is perfect, 'cause it creates this consistent work surface!" he said. "And it looks really amazing." As a bonus, the heaviest components—the battery and drive—were now beneath your hands, away from the hinge, making the laptop less likely to backflip off your knees.

In that moment, Krakower redefined humanity's concept of a laptop. Within about a year, every laptop from every company had adopted the same layout—keyboard against the screen, palm rests for you—which prevails to this day.

Years later, his design won a patent, and, at a little ceremony, Apple presented him with a $1,000 check. But Krakower doubts that his rogue exploits would fly today. "They didn't have cameras everywhere," he says. "They didn't know what people were doing at night when people weren't around," he says. "I can't imagine anyone doing that again. Apple now is locked doors everywhere."

Brunner and Ivester made new TIM models that adopted Krakower's overhaul, and in March 1990, Sculley approved them. The palm-rest models would become Apple's new laptops.

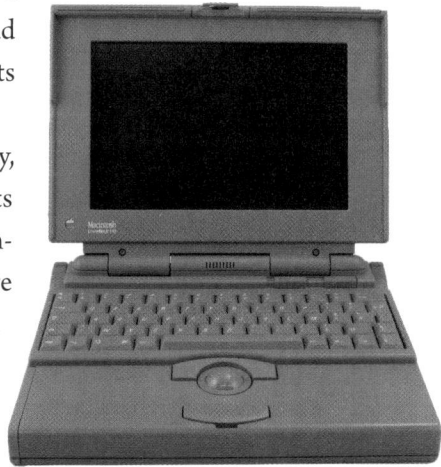

The first PowerBooks revolutionized the fundamental form of a laptop.

PowerBook 100

Sold: October 1991–September 1992
Price: $2,500
Processor: Motorola 68000 at 16 MHz
Memory: 2 MB, expandable to 8 MB
Equipment: Passive-matrix grayscale screen; 20 or 40 MB hard drive; internal connector for optional modem card. Weight: 5.1 pounds.

The raised lines on the lid are closer together than the 10-mm Snow White spec—a mistake that arose when Sony Xeroxed an Apple design fax at a sight reduction.

In the end, Apple created three models. The PowerBook 140 and 170 (same design, different specs) would have the floppy built in, but at the expense of size and weight. A little subnotebook, the PowerBook 100, would come with an *external* floppy drive, which you were free to leave behind.

Through the spring and summer of 1990, the design team conducted user tests and tweaked the design. They settled on two buttons, above and below the trackball, to accommodate different people's clicking styles.

To help meet the deadline, Apple contracted with Sony to engineer and manufacture the 100 model. Sony had considerable expertise in miniaturizing electronics, and was already making disk drives, monitors, and other peripherals for Apple, so the partnership made perfect sense—and went extraordinarily smoothly.

The PowerBook 140/170 was a rougher ride. "Everything happened so fast, we were under constant strain until the moment the products shipped," says Brunner. "At times, it got kind of ugly—this crucible of churn of emotion and fear and doubt."

Six months before the ship date, for example, the design group got the first samples of the cases back from the plastics company. They were a mess. Brunner counted 150 defects. Flaps were hard to open, hinges creaked, the Caps Lock key didn't lock, the gap around the trackball buttons was too big, and the icons that labeled the jacks on the back were stamped upside down.

Somehow, in October 1991, the PowerBooks shipped—on time—and were instantly hailed as masterpieces. The 140 cost $3,200 with a 40 MB drive and a passive-matrix grayscale screen; the 170, with a faster chip and an active-matrix (sharper) grayscale screen, went for $4,600. Both had flip-out feet at the back edge for comfortable typing, a ridge around the screen bezel to block crumbs from falling into the closed lid, and a beefy-looking hinge to give some attitude.

The PowerBooks were titanic hits. They were infinitely better-looking than any previous laptop. It took Apple only four months to sell 100,000 of them. Despite back orders and parts shortages, Apple sold over $1 billion worth of PowerBooks in the first year. Apple's market share popped up to 8.5 percent, its highest in four years.

Infinite Loop

In its first 50 years, one throughline has rarely flagged: Apple has always needed more employee space. With only a brief pause in the mid-nineties, the company has steadily grown, and steadily gobbled up more buildings. Local residents used to call Apple the "Pac-Man of Cupertino."

By 1990, its various acquisitions of the Bandley, Mariani, and other buildings weren't enough to contain the whole company. Sculley decided that it was time to build a proper campus for Apple's growing R&D efforts. It would be the biggest development in Santa Clara County in over a decade.

Apple's real estate team found a perfect spot: An old 564,000-square-foot Motorola campus—four buildings—on De Anza Boulevard and Interstate 280. Motorola had wanted about $40 million to vacate those buildings. But after the 1989 earthquake left three of them unusable, the property's value plummeted. Apple got it for $25 million.

Apple built a university-style campus: six buildings in an oval, surrounding a courtyard. The road that circled the whole thing came to be called Infinite Loop. (It was an

The PowerBooks won more design awards than any product in Apple's history. *Fortune*, *Time*, and *BusinessWeek* all hailed them as Products of the Year. Even a decade later, the honors would still be raining down: *PC World* named the PowerBook the tenth-greatest PC of all time, and *Mobile PC* hailed the PowerBook 100 as the greatest gadget in history.

Over the next four years, Apple introduced more PowerBooks: the 145, 145b, 150, 160, 165, 180, and 190. Along the way, the screens went color, the hard drives got bigger, and the expansion options expanded.

In 1994, the PowerBook 520/540 introduced a built-in microphone and speaker—and, for the first time on any laptop, a clickable track*pad* instead of a trackball. (No more having to pop the ball out to clean out the lint and dust!) In 1995, the PowerPC processor came to PowerBooks, which now sprouted *four*-digit model numbers to differentiate them from their older, Motorola predecessors.

engineering joke. An infinite loop is a programming sequence where the computer instructions call on each other recursively, intentionally or not, so that the code runs forever.)

Infinite Loop became a 32-acre, 785,000-square-foot complex—offices, shops, auditorium, gym, cafeteria, and outdoor seating—that could hold 2,600 employees. Each of the six buildings had its own character and scale, between two and four stories, to avoid looking mass-produced. A third of the parking spaces were underground; the rest were wrapped around the complex. Construction costs were around $200 million.

Building 1 held the executive suites and the Advanced Technology Group. IL 2 was software, IL 3 was developer relations, tech support, and product marketing. IL 4 was a glorious library and the cafeteria, IL 5 was dedicated to peripherals, and IL 6 was hardware.

There was also a bar across the parking lot, the Peppermill Lounge (later, BJ's), which Apple employees jokingly referred to as Infinite Loop 7. "I'll meet you in IL 7" was code for "Let's go drinking."

In the Bandley and Mariani buildings, everyone worked in open-plan cubicles; in IL, every person would have an office. Everybody wanted to move into Infinite Loop.

Infinite Loop was intended to house Apple's entire R&D organization. But by the time it was finished in late 1992, R&D had already outgrown it.

Only once did these buildings manage to contain the entire R&D operation: in the dark years of Jobs's absence from the company, when the company's steady stream of layoffs reduced its workforce by 40 percent.

PowerBook Duo

The wildest, weirdest laptop Apple ever brought to market, though, was the PowerBook Duo, in October 1992.

The idea was ultimate portability—three pounds lighter than the regular PowerBooks—by leaving behind everything but the bare minimum. No floppy drive, very few jacks. The screen was only 9 inches diagonal, the trackball was more of a trackmarble, and the keyboard keys were about 88 percent of the standard size. "These are not computers for the ham-handed," noted *Macworld*.

But the name Duo hinted at the coolest part. Once you were back at your desk, "small and light" didn't matter. You could slide the closed laptop like a VHS tape into the motorized opening of the Duo Dock, a sort of laptop garage. It connected you to the *rest* of your computer system: big color screen, floppy drive, full-size keyboard, real mouse, full set of connectors, expansion slots, optional second hard drive, more video memory. Now you had yourself a full-blown desktop system.

The Duo on the road (left) and as part of a full desktop system, in the Duo Dock.

When it was time to head back out on the road, an Eject button pushed out the laptop, and off you went. It was, Apple asserted, the best of both worlds, laptop and desktop. (The Duo's code name was BOB W.)

The Duo Dock was a two-toned affair: PowerBook dark on the bottom, platinum gray on top, to match Apple's monitors. It was, Brunner says, "this Frankenstein thing. In retrospect, it was a design failure." But for those who had smallish hands and biggish bank accounts, the Duos really were the best of both worlds.

Mac-Like Things

Through the early nineties, Sculley's commitment of money and talent to the Newton project seemed to suggest that consumer electronics might be Apple's next realm. Small engineering groups began exploring audio gear, cameras, touchscreens, and handheld CD-ROMs.

Marketing exec Satjiv Chahil felt that these motley projects ought to be more officially assessed and developed. The result was an informal group called Mac-Like Things, whose mission was to identify the next big product category and develop it before competitors did. Projects involving touchscreens, stylus interfaces, and CDs popped up in various combinations.

Some creations never saw the light of day. The Paladin project, for example, combined a Mac with a speakerphone, fax machine, answering machine, and inkjet printer in a single unit on a swiveling base, complete with a slide-out keyboard and trackball. (That is correct: Apple once built a fax machine.) It got as far as being designed and tested before being cut from the lineup. "I was told that the company would not fund it and that I should stay focused on driving the multimedia market," Chahil says.

There was also an Apple TV years before the Apple TV: the Interactive Television Box. It was

PenLite: The Tablet Mac

Apple has never produced a touchscreen Mac—but in 1993, it got very close. Tom Gilley, of Apple's ATG, became a champion of the PenLite project, as it was called. PenLite was essentially a PowerBook Duo with a touchscreen—flipped to face outward—so it didn't cost much to develop. In fact, it could slip neatly into the standard Duo Dock. "If you open up [the PenLite], there's a lot of air space in that unit," says Gilley. "We kept it thicker than it needed to be, so that it would fit into the Duo Dock!"

The Newton's Rosetta handwriting software worked brilliantly. "You could start writing anywhere on the screen, and wherever the cursor was blinking, it would insert the text."

Logitech did the manufacturing. Manuals and packaging were complete, FCC and UL approvals secured, optional keyboard and carrying case designed. The first 500 units were packaged up for shipping to Japan—when Larry Tesler found out about the project. His group had just spent years developing the Newton. Why, he wondered, would Apple need *two* pen-computing devices?

Gilley was asked to appear before the Apple board to explain how PenLite and Newton were different. "I was so young, I really didn't know how to present it to board-of-director-level people," he says now. "If I had presented it in a different way, more clearly and concisely, and disassociated itself from it being just a tablet and being more of an extension of your desktop, it would have seen the light of day."

Well, maybe; at $3,500, PenLite would have cost more than twice as much as a Duo, and five times as much as a Newton.

The PenLite team was devastated at the cancellation—especially when they learned that most of the 500 existing PenLites would be pulverized by a commercial disposal company. (As with the Lisa before it, Apple sought a tax write-off.)

In hopes of cheering up the team, Gilley proposed one last engineering assessment: a drop test. He climbed to the rooftop of Apple's eight-story City Center building. As the team watched from the windows, he let a sacrificial PenLite plummet to the pavement below. The screen cracked, but incredibly, the machine still started up.

"And yeah, I did get in trouble for that," he says.

a subscription TV service with fast-forward, pause, and rewind features—in 1993. It was nearly complete—FCC certification cleared, manuals written, test runs in 2,500 British households—before Michael Spindler shut it down.

A few Mac-Like Things, though, did become products. The battery-powered Apple PowerCD (1993), a rebranded Philips player of music and data CDs, made a perfect portable partner to PowerBooks, which didn't yet have CD drives.

The AppleDesign Powered Speakers (1993), Apple's first computer speakers, had a subwoofer jack and a headphone jack on the front. They were popular enough that Apple introduced a smaller, darker II version the following year.

To insert the paper into Apple's combo Mac/fax machine, you slipped it vertically behind the screen.

It's tempting to cluck at the time and effort spent on these go-nowhere projects of the early nineties. But experimentation has always been part of Apple's culture. Getting an invention to the manufacturing stage is always a long, difficult road—but building a prototype is always the first step. Without concepts and trials, the winners would never have a chance to rise to the top.

In other words, there are more Apple products you've never heard of than those you have.

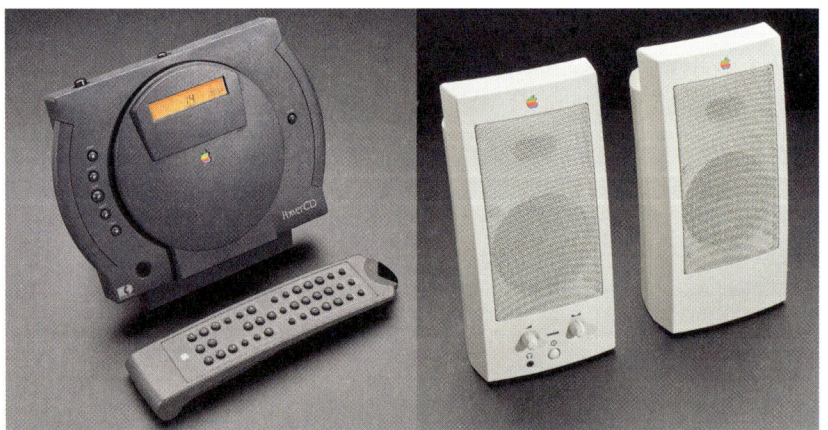

The PowerCD (left) and AppleDesign Powered Speakers (right) were actual Apple products.

23. Spindler

On June 17, 1993, the Apple board fired John Sculley.

From a historical perspective, his firing might seem surprising. In his ten years as CEO, Apple's stock valuation soared from $600 million to *$5.6 billion*; its revenues had ballooned from $1 billion a year to $8 billion. Sculley had successfully moved the Mac onto RISC processors. He'd halted the market-share slide, and brought it from 7.5 percent up to 8.5. The company had $2 billion in cash—a record. It celebrated selling its ten-millionth Mac.

And yet, in 1993, Sculley's departure surprised nobody. It was the climax of five simultaneous storylines.

First, there was his book. In August 1987, Sculley published *Odyssey: Pepsi to Apple*, a memoir about his first years at Apple. Even with a coauthor, it was a time-consuming project. Sculley appeared on talk shows and interviews, and spent time away from Cupertino for a book tour.

Second, his involvement with Bill Clinton's campaign for president made things even more complicated. "I am still a Republican, but I am voting for Bill Clinton," he told reporters, "because I don't believe America's industries can survive four more years of President Bush."

Sculley and Hillary Clinton (and Alan Greenspan) at the State of the Union address.

After Clinton won the election, Sculley made frequent visits to the White House, and was even offered a job as secretary of commerce. (He turned it down.)

But Apple's board members—all Republicans—were growing increasingly annoyed, especially when Sculley sat next to Hillary at Clinton's first State of the Union address. To them, that nationally televised moment seemed to symbolize Sculley's detachment from Apple. "He was running back and forth to New York and Washington, D.C.; he just wasn't paying attention to the company," Mike Markkula says.

The third factor in Sculley's ouster was his failure to orchestrate a merger. By 1992, the dominance of the Wintel alliance had become so extreme that the board couldn't see how Apple could survive by itself. Apple, they believed, would have to merge with another company.

The thought of Apple subsumed into some bland corporate Borg was shocking and depressing for Apple's fans. ("What do you get when you cross Apple with IBM?" went one joke. Answer: "IBM.") But Markkula was emphatic. "Every day and week and month that goes by without a challenge brings the folks in Redmond closer to controlling our fates," he said. "I don't think preserving the legal entity that is Apple is as important as preserving our organization, philosophies, brand name, employees, customer base, and technology."

Sculley and a small team spent months discussing merger or acquisition possibilities: British Telecom. Philips. Siemens. Kodak. AT&T. IBM. None of it went anywhere.

His meetings with IBM did, however, lead to a twist. IBM's transition from mainframes to personal computers had been rocky, and in 1992, it lost around $5 billion, then the largest annual loss in U.S. history. IBM was floundering, and looking for a new CEO—and *Sculley* was on the short list.

The Apple board wasn't pleased—not so much because Sculley might depart, but because he'd be leaving the merger project incomplete. "You can't leave us high and dry," Peter Crisp told him.

IBM told Sculley that the Apple-IBM merger might still work out, but it would have to wait a year while the company sorted out its leadership problems. When the Apple board emphasized to Sculley that those priorities were backward, Sculley took himself out of the running. "I am not available or interested in being CEO of IBM," he told the *Wall Street Journal* in March.

There was a fourth strike against Sculley, too: the Newton. For over a year, he'd given talks and interviews about how the MessagePad and its descendants would change education, business, and computing. He'd spent over $500 million to bring it about—and then it flopped.

As though anyone needed a last straw, Apple's June quarterly sales numbers came in at a measly $4.6 million. The popularity of PowerBooks had suddenly cratered. Apple had saturated its installed base, and there weren't enough new customers to keep up the wave.

The board's four outside members—Markkula, Crisp, Arthur Rock, and Bernard Goldstein, who ran a venture capital company called Broadview—had had enough. The outside commitments,

Splitting Apple

John Sculley was certainly throwing a lot of darts at the bull's-eye of Apple's recovery. But one of the strangest—yet most intriguing—was his initiative, in the early summer of 1993, to break Apple into two separate businesses.

One, called the Macintosh Company, would sell Macs directly to consumers, as Dell and Compaq were doing. The Mac business could be leaner, more efficient, and more responsive to changing trends. Then there would be the Apple Company, which would work on software and new technologies. It would be a Silicon Valley R&D lab that could license its innovations to other companies.

Separate hardware and software entities—wasn't that the essence of the Wintel dynasty? These two companies, Sculley felt sure, would be more valuable than one. Sculley would become the head of the Apple Company; Michael Spindler could run the Macintosh Company.

It was a dramatic and risky proposal, but Apple had no options left that weren't dramatic and risky. The board approved the plan, and Sculley hired Goldman Sachs to start working on the restructuring.

The executive team, however, was aghast and against it. In the end, nothing came of the idea—because Sculley, its principal architect, had left Cupertino.

Sculley's executive team. From left: Kevin Sullivan, Joe Graziano, Michael Spindler, Sculley, David Nagel, Al Eisenstat.

the flirting with IBM, the disastrous numbers—it was just too much. Michael Spindler, who was already running sales, marketing, product, and R&D, would become the new CEO.

Officially, Sculley wasn't fired; he was demoted to chairman. "This is not my swan song. I hope to be around for a while," Sculley told reporters. He thought maybe he'd set up an office in New York, closer to his home in Greenwich, Connecticut. He could focus on building alliances and thinking about future technologies.

But Sculley knew as well as anyone how the Apple pattern went. He'd seen it with Jobs, Mike Murray, and Gassée: They don't fire you. They just give you a job without any responsibilities, and wait until you leave on your own.

After a two-month sabbatical, he resigned. His replacement as chairman of the board was its longtime leadership understudy, Mike Markkula.

The truth was, Sculley had long contemplated life after Apple. His third wife, Leezy, unhappy in California, had already moved back east. In 1992, exhausted and stressed, he'd even told the board that he wanted to step down and join her. The board had asked him to stay on long enough to merge Apple with another company.

Still, the public nature of the split made it especially painful. "I had to fail in front of the whole world," Sculley says. "It really hurt me for a long, long time. And it wasn't till about 15 years later that I could actually get my head around and deal with it."

Michael Spindler

On paper, Michael Spindler was the perfect CEO. He was an engineer by training. In his brisk rise through Apple's ranks, he'd done marketing, sales, product management, and manufacturing. He knew the numbers, the players, the products.

But he arrived in the CEO's office at a delicate time. True, Apple had 11 million fans who, as the saying went, "bled in six colors." Apple had no debt and billions in the bank. Above all, the company had an ace in the hole: the PowerPC chips, coming in March 1994. Macs would beat Wintel to RISC processors, offering incredible speed from day one.

On the other hand, Spindler inherited a national computer price war, the end of the PowerBook sales joyride, and, of course, the ever-looming market-share problem.

He jumped in with the biggest layoffs in Apple history: 2,500 workers, 16 percent of Apple's workforce. The cost of the reorganization, $321 million, resulted in a loss of $188 million for the quarter ending in September. Even engineering, usually considered sacrosanct, lost 10 percent of its talent. (One engineer made an animated screen saver called "Spindler's List.")

Even as Spindler trimmed the workforce, he mourned the loss of talent that might have helped

Apple innovate its way out of trouble. "You can't win either way," he said. "You cut off your arm, and now you don't weigh as much. But then you don't have an arm."

Spindler also froze salaries for 18 months, and reorganized the company yet again. Now there would be four new divisions: the PC division (Macs); Applesoft (the effort to build a modern operating system); Personal Interactive Electronics, which would oversee the Newton and eWorld (Apple's version of America Online); and Business Systems, for developing corporate storage servers.

Butt-Head Astronomer

Apple people refer to the products they're working on by code names, both to maintain secrecy and because, often, nobody yet knows its final name.

The nickname for the midrange Power Mac was "Carl Sagan," an homage to the Cornell astronomer who, in books and TV shows, became famous for conveying the spectacle of space.

But in 1993, he became famous for a different reason. He interpreted the 7100's code name to mean that Apple was using his name for *marketing* purposes.

"My endorsement is not for sale," he wrote. "Through my attorneys, I have repeatedly requested Apple to make a public clarification that I knew nothing of its intention to capitalize on my reputation.... Apple has refused."

With an eye roll, Apple engineers changed the Power Macintosh 7100's code name to "BHA."

It stood for "Butt-Head Astronomer."

This time, Sagan was even less pleased. He sued Apple for unfair competition, infringement of right of publicity, invasion of privacy, libel, and intentional infliction of emotional distress.

(It can be hard to read the case documents without giggling: "Because a reasonable factfinder could not conclude that 'Butt-Head Astronomer' implied that Plaintiff was a less than able astronomer, or that Plaintiff was legally wrong in asking Defendant to cease using Plaintiff's name, the only remaining assertion is the bare statement that Plaintiff is a 'Butt-Head Astronomer.' Clearly this phrase cannot rest on a core of objective evidence.")

Sagan lost the case.

He appealed, and then settled with Apple in late 1995. "Apple has always had great respect for Dr. Sagan," the company wrote in a statement. "It was never Apple's intention to cause Dr. Sagan or his family any embarrassment or concern."

As part of the settlement, the engineers were required to change the 7100's code name one last time. It was now known as LAW.

It stood for "Lawyers Are Wimps."

Spindler also took the opportunity to update Apple's board of directors; apart from a brief stint served by space shuttle astronaut Sally Ride in 1988, it had been nothing but old white men for years.

Arthur Rock resigned, saying that he wanted to avoid a conflict of interest (he was a longtime board member of Intel). So Spindler welcomed four new directors: Katherine Hudson, CEO of W. H. Brady Company, a maker of industrial equipment; Delano Lewis, CEO of National Public Radio; Jürgen Hintz, a former marketing executive from Procter & Gamble; and Gil Amelio, CEO of National Semiconductor. Markkula, Peter Crisp, and Bernard Goldstein remained on the board, along with chief financial officer Joe Graziano.

It was a brutal time to be Apple's CEO, and Spindler did not handle stress well. He suffered from anxiety attacks, a bad back, and heart troubles. When the pressure really got to him, he lay under his desk in a fetal position. "Sort of like you would sleep in bed, with your legs pulled up and head on your arms," remembers Del Yocam. On one occasion, after helping Spindler out from under the desk and onto the couch, Yocam was so alarmed by Spindler's rapid pulse that he called 911.

Sculley had enjoyed being Apple's public face. Spindler, on the other hand, hated the spotlight. In his first four months on the job, he didn't make a single appearance.

Fortunately, Apple's new RISC Macs did much of their own publicity.

PowerPC

In the spring of 1994, *Schindler's List* won the Best Picture Oscar, the first planets beyond our solar system were confirmed, and Richard Nixon died. But at New York's Lincoln Center on March 14, 700 reporters and analysts settled in to watch a most unusual performance.

"Today, we make a quantum leap forward," Spindler told the crowd. He unveiled three new computers, each faster, bigger, and more expandable than the last: the Power Mac 6100, 7100, and 8100.

The first three Macs with PowerPC chips were the Power Mac 6100, 7100, and 8100.

OpenDoc

In 1991, the software buzzword was *component software*. You'd start with a single blank page and add only the features you needed: "Small, self-contained, reusable modules that can be easily assembled into seamless solutions," as the narrator of a 1991 Apple video explained.

Apple's version, created with partners like IBM, WordPerfect, and Oracle, was OpenDoc. In theory, it let you drop individual features—a chart editor, a spreadsheet, a text editor—into a container document. When you clicked on the inserted chunk, the menus changed accordingly: text-formatting commands, chart-making commands, or whatever.

Big software companies wouldn't have to create monolithic new versions of their programs every year; small ones would be able to sell individual *features*, not whole big apps. Customers would no longer have to pay for clunky apps with features they'd never use.

Microsoft had its own component software system called Object Linking and Embedding (OLE). Microsoft wanted Apple to adopt OLE; Apple wanted Microsoft to adopt OpenDoc. Spindler and Bill Gates met in 1993 to break the impasse, but neither budged.

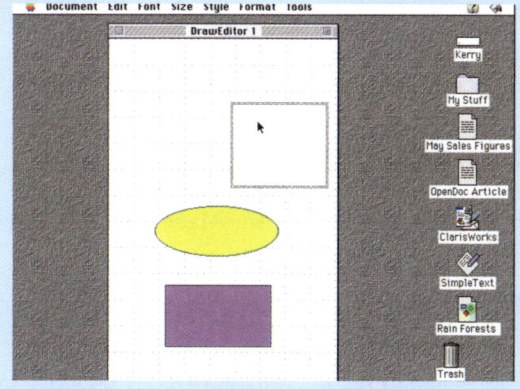

Apple was convinced that if it adopted OLE, Microsoft would always add new features to the Windows version first, just as it had been doing with Microsoft Office. "We would have a time-to-market deficit we would have to live with the rest of our lives," says Dave Nagel, head of software engineering in 1993. "Windows would always have it first."

The pinnacle of OpenDoc's career was Apple's 1996 WWDC, when an Apple engineer demonstrated an early OpenDoc-compatible version of ClarisWorks. In seconds, just by dragging and dropping OpenDoc parts, he assembled a dazzling multimedia document from graphics, links, scrollable maps, and animated buttons.

This oddball technology story, however, met a familiar end: When Steve Jobs returned to Apple in 1997, he shut the whole thing down.

For what it's worth, OLE didn't take off, either. It wound up letting you insert an Excel chart into a Word document while maintaining a link to the original data, but that was about it.

Product manager Jim Gable ran a set of side-by-side speed competitions between a Power Mac and a Windows PC equipped with Intel's most powerful new chip: the Pentium. The Power Mac won, every time, by a laughably big margin. "Just as Macintosh changed the course of computing ten years ago," Spindler said, "today we are defining a new era in personal computing."

Executives from Aldus, Adobe, and Microsoft spoke, too, promising to adapt their popular software to run on the new, faster chip.

The Power Macs were instant bestsellers; within three months, Mac fans snapped up 200,000 of them. After six months, their sales set a record for Apple: $9.2 billion. In the third quarter of 1994, Apple was the number one bestselling computer company, pushing Compaq into second place.

Gary Davidian's emulator allowed thousands of old apps to run unmodified on the new chip. But for developers who wanted to RISCify their apps for greater speed, a tiny Montreal startup called Metrowerks stepped in with a suite of conversion tools called CodeWarrior. In a matter of minutes, it could convert an existing Mac program to run at full speed on the new chip. CodeWarrior became the hottest programming tool on the market.

Model Explosion

In the beginning, there was the Macintosh. It had successors—the Mac Plus, Mac SE, and so on—but it was one machine. For the rest of us.

The Mac II series introduced a higher-end, color tier. It, too, evolved: the Mac IIx, Mac IIcx, and so on. For professionals.

1990 brought the Macintosh LC and Mac IIsi. For schools and homes.

But Sculley always worried about the dominance and variety of Windows machines. One way to fight back, he decided, was for Apple to fill every market niche by itself. Soon, Apple introduced the Quadra line. The PowerBooks. The Centris family. The Performa series.

And then the wheels *really* started to come off.

In the beginning, there was a naming logic. A Macintosh LC meant "low cost," for homes and schools. Centris Macs represented the center of the lineup. The high-end Quadras were named for the Motorola 68040 chips inside. And a Performa was a Mac sold in a complete package—monitor, printer, software—to make decisions easy for families buying their first computers.

But in the early nineties, every chain—Sears, Price Club, CompUSA—had its own price guarantee: "If you find the same model anywhere else, we'll beat that price!" Apple's sales team worked with them to ensure that that situation would never arise. You *couldn't* find the same model elsewhere. Every chain got its own unique Mac model numbers so that comparisons were impossible.

The QuickTake Camera

Apple may be well-known for commercializing the personal computer, the mouse, the graphic user interface, Wi-Fi, CD-ROM, the touchscreen smartphone, and tablets. But few remember that, in 1994, Apple also introduced the first digital camera.

Digital cameras at the time were hulking, $10,000 experiments, many of them black-and-white. But at $750, the QuickTake 100 was the first color camera for consumers.

Kodak had developed the light sensor (the CCD chip)—but, fearful that digital would cannibalize its film business, offered it to Apple, who contracted with Chinon to do the manufacturing. (Spoiler: Pretending that digital photography didn't exist did not save Kodak.)

The camera's shape, designed by Robert Brunner's team at the peak of his Espresso period, suggested futuristic binoculars.

There was no memory card, and its built-in memory held only eight photos at a time, with a resolution of 640 x 480 pixels (one-*third* of a megapixel). There was no screen, so you couldn't see your photos until you synced the camera with your Mac. You couldn't delete an individual photo. The camera had a flash, but no zoom; the camera was fixed at 50 mm. You had no control over focus or exposure.

Still, the photos were surprisingly good for 1994, especially in bright light. The colors were true, and the detail was as crisp as 640 x 480 would allow. And it was *digital*. For the first time, you could see your photos as soon as you returned to your Mac, without the time and expense of developing film.

The QuickTake 150 (1995) doubled the storage and added a snap-on close-up lens, which could focus a foot away from your subject. 1996 brought the QuickTake 200 ($600), made for Apple by Fujifilm. This model was shaped more traditionally, with a screen, focus and aperture controls, and a memory card.

Apple's digital cameras sold briskly to schools and small businesses. But by 1997, Apple had increasing competition from Canon, Nikon, Fuji, and even Kodak. When Steve Jobs returned to Apple, he quickly took down the QuickTake.

Under Spindler, that system devolved into an explosion of models. The Macintosh LC 475, the Quadra 605, and the Performa 476, for example, were all the same computer. At one point, Apple introduced *seven* different Performas in the 6200 family, differentiated only by details like modem speed or software bundle.

All told, in the five years from 1992 to 1997, Apple introduced 10 Quadras, 28 PowerBooks, and 45 Performas.

Not even the most rabid Apple supporters could tell you the difference between a Performa 6410 and a 6360, or between a PowerBook 540c and a 550c. The model multiplication was baffling even to computer store salespeople; as a result, they sometimes didn't bother displaying those machines in the stores.

Inevitably, the quality of the products suffered, despite the best efforts of the 600 members of Apple's quality-assurance group. "The number of configurations was just astronomical, and you had to test them all," says Ike Nassi, Apple's chief technology officer at the time.

The final victims of the Era of Model Chaos were Apple's own forecasters. Predicting how many of a given Mac model will sell is a massive challenge under the best of circumstances. Trying to figure out exactly how many of *60* different products to manufacture was hopeless.

Christmas

There wasn't much Christmas cheer at Apple in 1994. The exciting multimedia programs on CD-ROM that Sculley had predicted had finally arrived; they drove sales of Windows PCs to a historic record, up 32 percent from the previous Christmas.

But Mac sales didn't budge—and Apple's market share dropped almost 12 percent. It was increasingly apparent to Spindler that only a massive, desperate act could keep Apple in business.

24. Clones

From the moment the Macintosh was born, an existential question hung over the company: Should Apple allow other companies to make Macs? For eight years, the arguments pro and con had not shifted a millimeter.

On one hand, look at Microsoft! It had achieved world domination by selling Windows with every new PC. Apple made software—why couldn't it sell its OS the same way?

On the other hand, look at IBM! Its PC business had been decimated by sales of cheap clones from other companies. Apple made hardware—why wouldn't clones cannibalize its computer sales the same way?

Microsoft was a software company. IBM was a hardware company. Apple was *both*. It was impossible to predict what would happen if it licensed the Mac's OS.

While Apple executives, pundits, and consumers spent thousands of person-years debating the issue, Wintel's dominance only grew. By 1994, Apple had sold 25 million computers; PC makers had sold 210 million. It wasn't just about losing hardware sales; it was also about losing software companies. They were writing their programs first for Windows, and sometimes *only* for Windows.

Most terrifying of all, Chicago was coming—the code name for Windows 95. It would let you drag-and-drop files. It would let you give your files names up to 255 characters long. You could create file shortcuts—duplicate icons for one actual file, in effect putting the same file in multiple locations. The Mac had had all of this for years, of course. As the bumper stickers put it, "Windows 95 = Macintosh 89."

But the *Apple v. Microsoft* ruling had emboldened Microsoft to borrow from Apple's designs far more freely. Windows 95 would be the most polished, Mac-like Windows yet.

As far as Michael Spindler was concerned, Apple had already tried *not* licensing, and it wasn't going well. The only other option was to license.

Within months of becoming CEO, he hired consultants to help shape the program. He believed that such a program wouldn't succeed unless a big-name PC maker became a licensee, so he sent delegations to IBM, Compaq, and Dell to sound them out.

Finally, at the January 1993 shareholders' meeting in the Flint auditorium, on the very spot where the original Mac had been announced, Spindler dropped the bomb: Apple would clone.

"Yes, there will be cases where Apple loses sales to licensees," licensing head Don Strickland told reporters. "But the overall market, the pond, is growing larger."

By that, he meant that clone makers could build Macs for industries, locales, and market segments that were too nichey for Apple to bother with. All the advertising for their Mac OS machines would raise the public's awareness of *all* Macs. More Macs sold would mean more people buying software, printers, and accessories—often from Apple.

Finally, of course, Apple would be collecting licensing fees. Add it all up, and Strickland calculated that Apple would do better with clones than without them.

Little by little, he and Spindler ironed out the thorny questions involved in licensing. Who would serve as tech support for all those new Mac makers? (Apple.) Would licensees be allowed to call their machines Macs? (No.) Would Apple adopt new Mac OS versions first, releasing them only later to its licensees? (No.)

The potential clone makers—IBM, Dell, Compaq, Gateway, Motorola, Zenith, Toshiba, Goldstar—had a few questions of their own. Would licensees have to buy parts from Apple? (No.) Wouldn't Apple have a price advantage, since *it* wouldn't have to pay any licensing fees? (Well, no, because Apple would be shouldering all the R&D costs.) And above all: How much would Apple charge?

Answer: A per-machine licensing fee of around $50, plus a one-time fee ($50,000 to $100,000) for the circuit board design.

As Spindler was aware, one aspect of Apple's situation *wasn't* like IBM's: IBM had never *wanted* clones. It didn't have a cloning program. It didn't issue standards. That's why the market filled up with cheap, junky PC compatibles. Mac clones, on the other hand, would have to incorporate Apple's ROM chips, ensuring that every clone would behave like a "real" Mac.

The Mac clones were not design masterpieces.

Little by little, Apple signed up four companies willing to take the dive:

- **Power Computing.** Founder and CEO Stephen Kahng hired a posse of former Apple engineers. Kahng adopted Dell's model: not available in stores. Each computer would be built to order and shipped directly to the customer.

    The first two models rolled off the line in May 1995: clones of Apple's Power Mac 7100 and 8100. Because they were built of off-the-shelf parts, they looked like generic PC clones—but they cost less than Apple's machines and worked identically, which was the entire point. Power sold 100,000 in a year—half of them to people who'd been considering PCs. The pond, it appeared, was growing.

- **Radius, Umax.** In 1986, a handful of original Mac team members—Andy Hertzfeld, Burrell Smith, Alain Rossmann, Mike Boich, and Matt Carter—established a company of their own, called Radius, to make Mac enhancement cards, like accelerators and video capture cards. Radius was the second company to sign up to make clones, but, in March 1995, the first to ship one: the System 100, a competitor to Apple's Power Mac 8100. These machines were supercharged like rocket engines, costing as much as $12,500.

    But soon, in dire financial straits itself, Radius sold its Mac license to a scanner maker called Umax, which hired most of its engineers and carried on the clone business. It renamed the computers SuperMacs.

- **DayStar** began as a maker of accelerator boards for the Mac. So it wasn't surprising that its first Mac clone, the Genesis MP, would be built for speed—with *multiple* PowerPC processors. The first model cost a heart-stopping $14,300 and up. At 50 pounds and nearly two feet tall, it was the biggest and heaviest Mac ever made, thanks to its metal case, *seven* hard drive slots, six expansion slots, and twelve memory slots. Two fans made the Genesis sound like a jetliner taking off.

- **Motorola.** In September 1996, Apple finally got what it had always wanted: a big-name cloner. Motorola's StarMax machines ($1,600 and up) were clearly aimed at businesses. For example, Motorola offered an unheard-of five-year warranty—just the sort of thing a corporate buyer would appreciate.

"The floodgates will open soon," wrote *Macworld* in April 1995. "By next year, expect to see up to 30 Mac clone makers." But that wasn't quite the way things turned out.

In early 1995, Gateway expressed interest in becoming a licensee. Gateway was a huge company that would take a huge bite out of Apple's sales—and nobody dreaded that more than Apple USA president Jim Buckley.

"Look, if we give this thing to Gateway," he told Spindler, "there's no way I can go back to my channel and say, 'We want a big order at the end of the quarter—and oh, now you have new competition from Gateway.'"

Spindler finally had the big-fish licensee he'd dreamed of—and turned it down.

CHRP

The PowerPC chip was the shining star of the Apple-IBM-Motorola alliance. But since IBM planned to use it in its own machines, too, an ambitious idea surfaced in 1995: Why not design a single computer that could run Mac OS *or* any of IBM's operating systems?

The pitch was irresistible: A PC maker could sell one model across multiple markets. These Common Hardware Reference Platform (CHRP) machines would run Mac OS, OS/2, AIX, Solaris, NetWare, Windows NT—pretty much everything except standard Windows. Corporate IT teams could support a single type of computer across departments. Need fewer designers and more engineers? Just turn the Macs into PCs—no new hardware required. Apple, finally, might break into the corporate world.

Apple intended to phase out its own designs and rely on off-the-shelf parts like everyone else. Licensees like Motorola were thrilled—no more paying Apple for special ROMs. Maybe, at last, Apple would have a shot at challenging Intel-Microsoft.

But in the end, the stampede of PC makers adopting CHRP never showed up, even after the alliance chose the marginally friendlier name PowerPC Platform. The design was too complex, the benefits were unclear, and cheap PCs still ruled the day.

When Steve Jobs returned to Apple at the end of 1996, surprise!—he shut down the clone program and the CHRP platform along with it. But once again, the effort produced some useful tech that lived on in Jobs-era Macs, including the "New World" architecture (removing the ROM chip and putting its contents in RAM, where it's more easily updated); the PCI expansion-slot format that CHRP would have incorporated; and Open Firmware, a software switch that, at startup time, let you choose which OS you wanted it to run.

The IBM + Apple Plan

In the fall of 1993, after months of frustrating meetings with IBM about CHRP, Spindler rekindled a shocking but sensible idea: IBM and Apple should merge. IBM, the world's biggest computer company, could finally bring the Mac OS into businesses. Apple, meanwhile, was strongest where IBM was weakest: in homes and education.

It wouldn't just be a merger of markets; it would be a merger of software. Mac OS was gorgeous and fluid, but did not have a kernel (the OS core that sits between the hardware and software) to prevent crashes; IBM's OS/2 had a kernel, so it rarely bombed, but wasn't as elegant as the Mac.

"I thought it could be as helpful to IBM as it could be to Apple," says Mike Markkula. "We really had good technology and good product, and with the strength of IBM, I figured we'd no longer have that roadblock in the business market."

Months of meetings followed. Dallas, Chicago, Cupertino, San Francisco. Boards were alerted, lawyers hired, and investment bankers secured. The stakes were high; the combined company would have 17 percent market share and sales of $74 billion a year.

Finally, at the Westin hotel near the Chicago airport, IBM CEO Lou Gerstner made his offer: IBM would buy Apple for $40 a share. (Apple's share price at the time was $37.)

Markkula was taken aback. He'd been thinking more like $60. "His perception of what Apple was worth was down *here*, and our perception of what Apple was worth was up *here*," Markkula says. "We couldn't get anywhere close to an agreement."

To Spindler's dismay, just like that, the merger was dead.

It seemed evident that unless *somebody* bought Apple, Apple would soon cease to exist. The executive staff began to panic. Spindler got on the phone. He spoke to all the logical candidates: HP, Compaq, Sony, Toshiba, Philips.

The talks went nowhere. For better or worse, Apple was on its own.

Diesel Out

For Apple in the mid-nineties, even good news was bad news.

The top-of-the-line Power Mac 8100, Apple's most profitable model, was so popular that you couldn't find one to buy. Apple was sitting on $1 billion worth of orders for high-end Macs that it couldn't fill—and $1.95 billion of low-end Macs it couldn't *sell*.

The new PowerBook 5300 should have been cause for celebration, too: the first laptop with a PowerPC chip. Its new lithium-ion battery technology lasted four hours on a charge, triple the duration of the old nickel-metal hydride ones. And you could fill the swappable accessory bays on each side of the laptop with whatever you wanted: two batteries, a battery and a floppy, and so on.

But just as the marketing blitz was about to begin, two early 5300 units caught fire: one at an

PowerBook 5300

Sold: August 1995–October 1996
Price: $2,700–$7,000
Processor: PowerPC 603e at 100 or 117 MHz
Memory: 8 MB, expandable to 32 MB
Equipment: Grayscale or color screens (passive or active); two PC card slots; 500 MB hard drive; pop-out feet

The infrared port let you beam files wirelessly to other PowerBooks in the same family (PowerBook 190, for example).

Apple employee's home, and one at the factory. Sony's lithium-ion batteries had overheated. The computer press had a field day, mocking up pictures of exploding PowerBooks.

Morale was at rock bottom. Spindler and Markkula started selling their Apple stock. Great engineers were taking jobs at other companies. Fiefdoms had emerged among different departments. An exciting idea supported by a group of engineers could be shut down by one short-sighted middle manager—or, as the saying went, "At Apple, a vote can be 15,000 to 1 and still be a tie."

The running joke was that the media had renamed the company "Beleaguered Apple." "The beleaguered Apple announces layoffs." "The beleaguered Apple announces a decline in sales." "The beleaguered Apple loses a key executive."

CFO Joe Graziano was growing increasingly concerned about Apple's finances. In 1994, Apple had had $1.2 billion in the bank. But as the third quarter ended in 1995, the stash was down to $756 million. When he made an emphatic case to the board that Spindler was taking the company off a cliff, the directors were appalled by his lack of support—and fired him.

On top of everything else, Apple now had no chief financial officer.

Apple's 1995 holiday season was devastating. Analysts had forecast that Apple would earn $87 million for the quarter; it wound up *losing* $69 million. Apple had never before lost money during the holidays, the busiest time of year. The stock plunged 16 percent; another 1,300 people lost their jobs.

Spindler, crushed by stress, checked into Stanford Hospital with heart palpitations. The San Francisco Macworld Expo, the biggest gathering of Apple fans on earth, would begin the next day; he didn't attend.

The timing of the 1996 Apple shareholders' meeting two weeks later—January 23—could not have been worse. The stockholders practically carried pitchforks and torches. "You have brought a great company to its knees!" said one angry investor.

"Mr. Spindler, it is time to go," said another.

Snapple

But Spindler had one last idea up his sleeve. During the fall of 1995, he'd been meeting with Sun, maker of powerful corporate workstations. CEO Scott McNealy, hoping to exploit the rise of the internet, thought that Apple's technologies and brand recognition might help him with the effort. The rumor mill went into high gear over the "Snapple" deal.

After the shareholders' meeting, Apple's board convened in a City Center 3 conference room to hear McNealy's pitch to buy Apple. His offer: $23 a share, far below the current price of about $31. The board was shocked. You didn't buy a company for *less* than its current share price!

Apple was a company with $11 billion in annual sales and $300 million profit for the previous year, and 20 million superfans. Apple was one of the most recognizable brands on earth. Had it come to this?

The board told McNealy they'd think about it.

The lowball Sun offer cratered Apple's stock. Stock-rating companies like Moody's and Standard & Poor's downgraded AAPL to junk-bond status. This moment had to be either the end of something—or the beginning.

For the next few days, board members conferred, trying to figure out how to save the company they loved. On January 31, 1996, they flew to New York to meet in person at the St. Regis hotel to finalize their decision. They asked Spindler to meet them that afternoon in the fifty-fifth-floor Rockefeller Center office of Venrock Associates, where board member Peter Crisp was a managing partner.

And there, for the third time in Apple's history, Markkula removed its leader from power.

Spindler felt blindsided, but managed a few dignified words before leaving the room. "I tried the best I could," he said. He flew back to Cupertino, where he and his two assistants would spend the weekend packing up his office stuff into boxes.

For the rest of the night, the board turned to the matter of his successor.

They wanted somebody who knew Apple well. Somebody with a technical background. And, preferably, someone with experience not just running a big tech company, but turning it around.

Ordinarily, you'd hire a headhunter firm to search for such a person. But on this night, a guy like that was already sitting in the room.

25. Amelio

Gil Amelio was born in 1943 in the Bronx, the son of immigrants from Italy. His father had done truck maintenance for General Patton in World War II, and then moved his family to Miami after the war.

At the Georgia Institute of Technology, young Gil earned a master's and a PhD in physics. As a researcher at Bell Labs, he was part of the team that, in 1970, developed the first working CCD (charge-coupled device, the sensor in early digital cameras). His name wound up on 16 technology patents.

He spent about a decade each at Fairchild and Rockwell, and then in 1991, landed at a struggling National Semiconductor. He consolidated plants, laid people off, and focused on the company's chipmaking business; by 1994, Nat Semi posted record revenues of $2.3 billion, and Gil Amelio became known as a turnaround artist.

Amelio joined Apple's board in January 1995—and exactly one year later, near the end of its marathon meeting in New York, the board offered him the CEO job. He wasn't especially surprised; Crisp and Markkula had leaked the possibility with him earlier in the week.

Gil Amelio in the lobby of the City Center building.

Two days later—Friday, February 2, 1996—Spindler sent out a gracious note of farewell to the entire company. "The end of a long voyage. A page in my life has turned," it said. "I take personal responsibility for things that didn't work and should have worked. I tried to give it my best."

He concluded the memo—the last one of his 16-year Apple career—like this: "In fading away from the place which I loved and feared, I will become whole again—hopefully renew the father, husband, and self I am."

Amelio was well aware that he was Apple's best hope. He held out for impressive compensation: $2.5 million and 200,000 shares each year, a $5 million loan, and a $10 million golden parachute if he were ousted in a takeover or merger.

In addition to the CEO job, he also wanted to be chairman of the board, which had been Markkula's title since the day Apple was founded. Markkula agreed, becoming *vice*-chairman.

The announcement finally hit the airwaves at nine o'clock that night. Gil Amelio would be Apple's new CEO.

Dr. Amelio

Amelio's first act was to shut down the Sun negotiations for good. "No way did the numbers add up to a desperation sale," he says.

He spent the next two weeks on a listening tour of executives, managers, and employees. What he learned was terrifying. There were too many divisions, too many fiefdoms. The company had 22 different marketing organizations, by product groups and geographies.

And the software! Imagine, as a Mac fan, trying to figure out the difference between System 7.5.3 and System 7.5 Version 7.5.3. Or the difference between System 7.5.3 Revision 2 and System 7.5.3 Updated to Revision 2. (Answer: There was no difference. These were different names for exactly the same software.)

At one point, the lawyers from two different Apple divisions showed up in the Patent and Trademark Office, prepared to sue *each other*. Each wanted the same trademark for a different Apple business.

At a time when Apple's finances were in dire shape, the company had no chief financial officer. The job remained unfilled, five months after Graziano's departure.

Worst of all, a self-perpetuating sales cycle had set in: Nobody wanted to buy computers from a dying company, so sales declined, so the company died a little more. Traditional corporate Mac strongholds, like US West, Deloitte & Touche, and Electronic Data Systems, began phasing out their Macs. Apple may have chosen a new CEO, but the death spiral didn't seem to care.

Software companies were treating their Mac apps as second-class citizens, releasing them long after the Windows versions, or in cruddier forms. Microsoft's Office programs opened instantly on Windows, but took 20 seconds on a Mac. And that's if developers continued making Mac programs at all; Quicken, QuickBooks, and CorelDRAW for the Mac disappeared altogether.

Then there were the quality problems. Two Power Mac models and 12 Performa variations began exhibiting random freezes. On the PowerBook 190 and 5300, the hinge plastic was cracking and the AC adapter pin was snapping off, leaving people unable to charge their laptops.

Amelio ordered production stopped and the laptops recalled, which created shortages and killed sales. It took four months to resolve the PowerBook problems. "Everything at Apple felt like bench-pressing 500 pounds," he says.

"Years of overlooked opportunities, flip-flop strategies, and a mind-boggling disregard for market realities have caught up with Apple," went a *BusinessWeek* article called "The Fall of an American Icon." "Management is in near-meltdown: Out of 45 vice-presidents, 14 have been axed or left in the past year." It concluded that "Apple is rapidly becoming a minor player in the computer business."

Apple PR tallied up more than 1,000 articles a month about Apple and its troubles.

Time: "[Apple is] a chaotic mess without a strategic vision and certainly no future."

Fortune: "By the time you read this story, the quirky cult company . . . will end its wild ride as an independent enterprise."

Forrester Research: "Apple as we know it is cooked. It's so classic. It's so sad."

And then things got *really* bad.

BusinessWeek *left Apple for dead.*

$740 million

At an April 1996 board meeting, Amelio bore some gut-wrenching news. In a single quarter, the company had lost *$740 million*. It was the biggest shortfall in Apple's history, before or since.

At the rate the company was burning through cash—$50 million a month—it was six months away from bankruptcy. Or less: Apple subsidiaries in Japan and the Netherlands held $400 million in loans, with payments coming due within weeks. If they called in their loans, a great American brand would close its doors.

Amelio ripped the bandage off hard. He laid off another 1,500 people; Apple had now shed a fifth of its entire workforce. (Apple hired Chas Pfister, the master carpenter who'd built the Apple I cases 19 years earlier, to oversee the dismantling of thousands of now-empty office cubicles. "I never got paid for that, by the way," he says.)

Amelio cut the advertising budget by a third. He installed a manufacturing-quality czar; no product would ship without his blessing.

And most important of all, he hired a CFO.

The Anderson Plan

Fred Anderson seemed to have been born for this moment in Apple's story. He had an MBA from UCLA. He'd learned auditing (at Coopers & Lybrand); IPOs and turnarounds (at MAI, a minicomputer company); Wall Street and analysts (as CFO of computer services company ADP).

Fred Anderson
Born: 1944, Loma Linda, CA
Schooling: Whittier (BA), UCLA (MBA)
Before Apple: Air Force, Coopers & Lybrand, MAI, ADP
Apple: 1996–2004
After Apple: Elevation Partners, NextEquity Partners

Anderson and wife, Marilyn, own a 50-foot sports yacht called *Dream Weaver II*, which they regularly sail from Catalina Island, San Diego, and Santa Barbara.

As CFO, Anderson boarded the sinking Apple ship with his eyes wide open. Still, he had three reasons to accept the deal: He and his wife were both Mac fans. He'd felt under-challenged at his previous job. And the offer's stock options meant that if he could turn Apple around, he'd be *set*.

His first day of work was April 1, 1996—Apple's 20th anniversary. It was, he says, "like going into an emergency room to keep the patient alive." He tackled the cash crisis from every possible angle:

eWorld

By the early nineties, Apple had grown unhappy with AppleLink Personal Edition, a customer support service, which was being run for Apple by a company called Quantum Computer Services. Its CEO was a guy named Steve Case.

In 1991—two years before the internet became available to the public—Quantum relaunched its own service as America Online. Thanks to its easy-to-use, graphic interface, AOL quickly became the world's biggest online service, dwarfing CompuServe and Prodigy.

Apple saw its opportunity. It shut down AppleLink and hired AOL to help it develop its own online service, to be called eWorld. When you dialed into the service ($9 a month), a charming cartoon town appeared. To access eWorld's features, you clicked little buildings like Marketplace, Newsstand, eMail Center, Learning Center, and Community Center (chat rooms and message boards). When the internet came along in 1993, eWorld also offered access to early web pages.

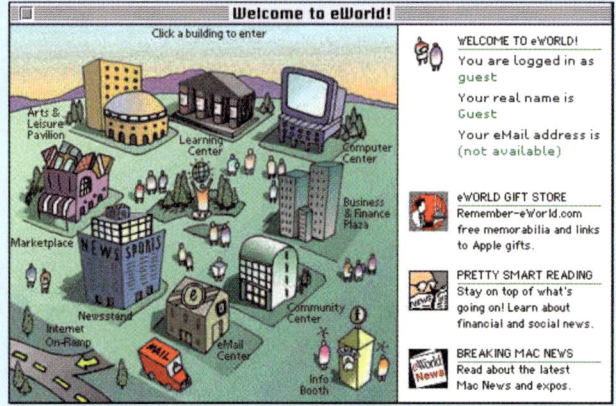

eWorld had its fans, but it was available only to Mac and Apple IIgs owners; a Windows version never made it out of testing. At its peak, it had only about 115,000 members; AOL was 30 times bigger.

No wonder, then, that when Gil Amelio became CEO, he shut eWorld down. Apple fans got their last glimpse of the cute cartoon town on March 31, 1996.

- **Renegotiate the loans.** Those $400 million in loans, mostly from Japanese banks, were the most imminent danger. Anderson flew to Tokyo to meet with the bankers. Through an interpreter, he described his plans to restore Apple's profitability, and assured them that if they could give the company more time to repay, they'd be made whole. "I had a hard time, because I didn't speak the language—I didn't know the culture," he says. "Are they buying in here? I didn't know!"

  A few days later, he heard the good news: The banks would give Apple more time.

- **Raise some cash.** In June, Anderson hired Goldman Sachs to help raise $500 million with a convertible bond: an investment that gives the backers regular interest—and the option to get repaid with stock along the way. If Apple did climb out of its hole, those shares would reward the investors handsomely. The bond was even more successful than Apple had hoped: It raised $661 million, which Anderson used to repay some of its loans.

- **Sell some assets**, including Apple's Colorado and Sacramento Macintosh plants. Apple would outsource more of its manufacturing.

- **Extend payment terms.** Apple had always paid its suppliers within 30 days, which was, for the industry, unusually prompt. Most agreed that Apple could pay them within 45 or 60 days, further easing the cash crunch.

- **Cut expenses.** "People were flying first class, going to seminars, using consultants that were unnecessary—it was crazy," Anderson says. He clamped down.

- **Cash in on ARM.** As he pored over Apple's assets, Anderson uncovered a very happy surprise: Six years earlier, Sculley had invested $3 million in ARM, the British RISC chipmaker, to create a low-power processor for the Newton. Now, in 1996, that stock was worth a staggering *$800 million*.

The Maglite Incident

In an effort to connect with the Apple workforce, Amelio held a series of what he called "coffee klatches" in early 1996: small, off-the-record rap sessions with a dozen employees, invited at random. At one such meeting, he brought two flashlights: a $2.98 cheapie, and a $20 Maglite.

"Their flashlights cost more, but you know it's going to be reliable and long-lasting," he told the group. "When the electricity goes out, you know the Maglite is going to work."

His analogy leaked—to Apple's own professional bearer of good news, Guy Kawasaki, who'd returned to Apple in 1995 as an Apple Fellow. He told the story in his daily *EvangeList* newsletter. And on its way to the mainstream media, the story became: "Apple intends to go premium." That, of course, had been Gassée's philosophy—and it had bombed.

"If the premium-brand direction is one that Mr. Amelio pursues," wrote John Markoff in the *New York Times*, "it will be a remarkable about-face for Apple. It was just such a strategy—attempting to differentiate the Macintosh from the PC pack, and charging higher prices—that led to the downfall of Apple's former chairman, John Sculley."

It took Apple PR days to calm the hysteria, explaining over and over again that "Mr. Amelio's remarks should not be construed as an all-encompassing strategy statement."

Amelio was learning that when it came to press coverage, Apple wasn't like any other company.

It was the best investment the company had ever made, and now was the perfect time to start cashing out. (If you think about it, as Sculley points out, "Newton was actually profitable!")

- **Stop paying dividends** on Apple stock, saving $60 million a year.
- **Shrink the head count.** Two months after Anderson's arrival, another 3,000 people were let go.

In 1995, Apple's revenue had been $11 billion; Anderson's goal was to make Apple small enough to be profitable even at $9 billion. And sure enough: In September 1996, the company eked out a $25 million profit, despite another 23 percent drop in sales. Anderson had restored the company to profitability six months ahead of schedule.

The Shortest Honeymoon

Amelio did what he could to cheer up Apple's depressed workforce. He sent out regular emails called ReachOut, where he spoke frankly about Apple's challenges and what he was doing about them. He directed Satjiv Chahil to send a daily all-company email bearing whatever good news there was. And at his first all-hands meeting, on April 19, he spoke so earnestly that the employees gave him a standing ovation.

It didn't take long for Apple's workforce to discover, however, that Gil Amelio was no Steve, John, or even Diesel. For one thing, he preferred to be called Dr. Amelio, a reference to his PhD. He also wore a suit and tie every day—a far cry from Sculley's polo shirts, let alone Jobs's jean cutoffs. And instead of eating lunch with the employees in Caffè Macs, the corporate cafeteria, he had his lunch brought to his office on china plates.

The press wasn't impressed. "I was criticized for my hairstyle, the way I dress, and because I'm not thin enough," Amelio says. He was also blasted for charging Apple $1,700 an hour for the use of his seven-passenger Citation II business jet, which carried top executives to meetings—even though, he pointed out, Apple had been paying Mike Markkula's charter company even more for *his* jets for the last 12 years.

For Amelio, the lack of employee respect was like nothing he'd ever experienced. "I would meet with one of the vice presidents, and we would discuss a particular problem and what needed to be done. We'd agree on a course of action. And nothing would happen. *Nothing.* It was as if the conversation had never taken place."

"He'd come up with some idea—to be honest, usually not great ideas, right?" says a product manager at the time. "So he tells his team: 'This is what we've got to do.' And we're like, 'That's a dumb idea. We're not doing that.' So that's where he would say, 'Apple can't be led.' Well, no—it can't be led with dumb ideas!"

Amelio didn't always score bedside-manner points, either. When, in January 1997, he announced *another* layoff of 2,700 people, he acknowledged the difficulty of the task. "I slept about five hours last night," he said. "I will step up and do it, but I'm not happy about it. Dammit! Don't ever put me in this position again!"

Mission: Tom Cruise

In the early 1990s, John Sculley challenged marketing VP Satjiv Chahil to sell a million CD-equipped Macs in a year. Thinking that these Macs would appeal to the film and music industries, Chahil met with the two most powerful men in Hollywood: superagents Michael Ovitz (Creative Artists Agency) and Jeff Berg (International Creative Management).

Apple wound up hiring Berg. For $1 million a year, he would supply a five-person team to orchestrate marketing partnerships and open the doors to the music and film worlds.

The pinnacles of the program were the two movie blockbusters of 1996. In *Mission: Impossible*, Tom Cruise's character uses a PowerBook 5300c to break into the central CIA vault. And in *Independence Day*, Jeff Goldblum uses a PowerBook to destroy an alien spaceship—and literally save the earth.

Chahil proposed, to the movie studios, a tie-in promotional campaign. Apple would run PowerBook ads that simultaneously plugged the movies.

In one of the resulting TV ads, we hear the famous *Mission: Impossible* theme. We see the most thrilling sequences from the film—and then a narrator says: "After you see the movie, you may want to pick up the 'Book.'"

In the end, the $8.5 million campaign didn't sell many PowerBooks, because there were very few PowerBooks to buy; it was the summer of quality-related recalls. But it worked wonders for Apple's image. For years to come, Macs were the darlings of product placement—in movies like *Batman & Robin* (1997), *Volcano* (1997), *Wag the Dog* (1997), and TV shows like *Friends* (1994–2004), *Seinfeld* (1990–1998), *NewsRadio* (1995–1999), and *Home Improvement* (1991–1999).

At least among TV, movie, and music people, Apple's market share rose to nearly 70 percent.

The employees were aghast. "He's blaming the employee base?!" says the product manager.

On May 13, 1996, Amelio took the stage in San Jose to deliver the opening address for the annual Worldwide Developers Conference. Many of the 4,000 software engineers believed that it might be the last WWDC ever.

But Amelio announced a reasonable plan. He'd halve the insane proliferation of Mac models. And to help mend Apple's relationship with them, the developers, he would allot $20 million to the efforts of former software company CEO Heidi Roizen, who had just become Apple's head of developer relations.

He concluded his speech by playing Apple's *Mission: Impossible* ad on the big screens.

In general, what Amelio said reassured the WWDC audience. What worried them, however, was what he didn't say: that they could now pick up an early copy of Copland.

Copland

As Gil Amelio took charge of Apple, the Mac OS was 12 years old—and still crashing.

Over the years, Apple's engineers had taken to dropping little patch files to it, each designed for compatibility with some new feature: 7.5.2 Printing Fix, 040 VM Update, 630 SCSI Update, Serial Update 406, and so on. The System folder became a sprawling mess.

These extensions and INITs, as they were known, interfered with each other so often that Conflict Catcher, a third-party program whose sole function was to help you figure out which extensions were causing your Mac's crashes, became a bestseller. (Its author, Jeff Robbin, would go on to write an even bigger bestseller: iTunes.)

Apple desperately needed to write a modern, clean, and stable new operating system. So, in March 1994, a massive project called Copland was born, named after composer Aaron Copland. Ordinary non-nerds may not know or care about the buzzwords of operating system plumbing. But inside Apple, everybody knew what Copland needed:

- **Real multitasking.** Since 1985, the Mac had had only *cooperative* multitasking: The program you were using got to decide when other apps could use the processor. If a program was greedy, it could hog the Mac's power; if it froze, everything froze, and you'd have to restart the computer.

    What the Mac needed instead was *preemptive* multitasking, where the *operating system* directs the use of the processor, and can take away control from an obnoxiously written (or frozen) app at any time.

    The Lisa had had preemptive multitasking in 1983. Windows 95 had preemptive multitasking in 1995. But a full decade after its release, the Macintosh was still crashing away.

- **A kernel.** At the core of a modern OS is the kernel, a sort of air traffic controller that oversees how apps share the processor and memory. It's the key to preemptive multitasking.

- **Multithreading.** A software program often has a lot going on at once. A web browser, for example, may have to detect your mouse clicks, download text, and play video, all at once. In a modern operating system, these various *threads* can be running simultaneously; no wristwatch or hourglass cursor appears to make you wait for one to finish.

- **Platform independence.** The Copland kernel, code-named NuKernel, would be independent of Apple hardware, making it ready to run on CHRP machines.

- **Native code.** Copland would be designed exclusively for the PowerPC chip, with almost every key component written in PowerPC code. Why not harness all that processor speed where possible?

- **Protected memory** is a scheme in which each app runs in its own carbon-fiber-lined memory compartment; one app bombing doesn't bring down the rest.

    But to preserve backward compatibility with existing apps, Copland would have only *partially* protected memory. The OS would live in one bubble, but *all other* apps would live in

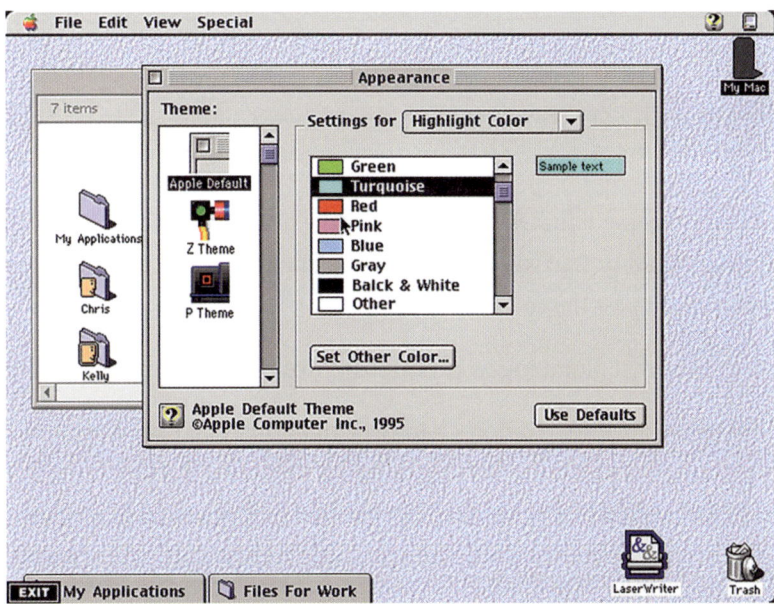

Copland could find words inside your files—a feature that arrived in Mac OS X nine years later.

Pippin

In April 1996, Apple introduced a video game console. "This isn't a misprint or an April Fools story," noted *Macworld*.

The Pippin ran Mac OS, contained a PowerPC chip, and played games on CD. The idea was to license it to other companies—a strategy to increase the Mac OS's market share.

Only one company took the bait: Bandai, the Japanese toy maker best known in the West for the Mighty Morphin Power Rangers franchise. The result, called the Bandai @World, cost $600, came with four games on CD, and could get onto the internet for $25 a month.

About 60 titles were produced for Pippins, including such immortal classics as Tetsuman Gaiden: Ambition of Great Game and Mr. Potato Head Saves Veggie Valley.

In the end, though, the Pippin bombed hard. Bandai sold about 25,000 in Japan and almost none in the U.S. It was too expensive for a game machine (a Sega cost $200 at the time) and too cheap to be a real computer. "Pippin could have been the first, best, and least expensive internet computer on the market," Amelio says, making its demise "surely the saddest of all the Apple stories."

---

another—an arrangement that Amelio says "left the problem basically uncorrected." Full memory protection would not arrive until the *next* planned OS version, nicknamed Gershwin—which never made it past the code-name stage.

In addition to all of that plumbing work, Copland would be bursting with more visible goodies:

- **Interior design.** Copland's windows had realistic-looking 3D shading. You could customize the desktop pattern, screen saver, text highlighting color, and the font for dialog boxes and window titles. What we'd call Dark Mode today (white type against dark windows) made its debut here. For kids: a choice of window designs, like one with wiggly yellow borders, and windows and menus that made sounds.
- **Window tabs.** If you wanted easy access to a frequently used window—but didn't want it taking up space on your screen—you could drag it to the bottom of the screen, where it turned

into a tab—an idea that had been kicking around since Bill Atkinson's work on the Lisa. You could pop them open or closed as needed, without cluttering up your screen.

- **Spring-loaded folders.** To move a file into a deeply nested folder, you didn't have to open a bunch of windows. In one continuous drag, you could move the icon onto the first folder, pause until its window opened, move onto the next nested folder, and so on. When you finally reached the destination folder, all of the windows you'd opened along the way closed automatically.

- **Global search.** A new Find window could search for your files by name—and could even search for the *words inside them*.

- **User profiles.** When you turned the computer on, you could choose your name from a list of people who used this Mac. You'd arrive at a desktop designed to match your level of technical

Iguana Iguana

Even in Apple's darkest times, the company's engineers never lost their sense of humor—well, not entirely. Thanks to the internet, it was more fun than ever for Mac fans to pass along the secrets of Mac Easter eggs they found.

In System 7.5, if you typed *secret about box* into the Note Pad app and dragged that phrase onto the desktop, the Note Pad turned into a full-fledged game of Breakout. Each brick bore the name of an Apple engineer.

In System 7.5.2, that same technique produced a video of the Infinite Loop courtyard, complete with an animated flag bearing the legend *iguana iguana powersurgius*.

It was a reference to Herman, System 7.5.2's mascot: a 3.5-foot iguana who lived in the cubicle of OS engineer Dave Evans. (Power Surge was the second-generation Power Mac code name.)

By moving the mouse, you could change the direction of the waving flag—or, if you whipped the mouse hard enough, you could break it off the flagpole entirely, fluttering it off the screen.

If you waited long enough, the picture on the flag changed to show the development team. The text beneath the flag scrolled by: "Watch out for low-flying iguanas."

The great Cupertino iguana flag flapped proudly all the way through System 7.6.1.

sophistication: big, simple icons if you were a newbie, longer menus and password protection if you were a power user.

- **A universal inbox** for email, faxes, and online messages. (The world *still* awaits this feature.)

In August 1995, Apple managers began giving Copland demos to the press. Mac fans watched in amazement as an Apple rep duplicated a bunch of files, emptied a very full Trash, and opened a program—*simultaneously*. It was a multitasking miracle.

Copland's 1996 budget, $250 million, funded an army of 500 engineers. But by WWDC in May, the whole thing was sinking under its own weight. As various departments bombarded software VP Dave Nagel with feature requests, Copland began to look unfinishable. "Copland was still just a collection of separate pieces, each being worked on by a different team, with what appeared to be an innocent expectation that it would all somehow miraculously come together," Amelio says. "But it wasn't, and it wouldn't."

Shortly after WWDC, Amelio announced another corporate restructuring. Now Apple would focus on four product groups: Macintosh; information appliance products; printers and monitors; and "alternative platforms" like CHRP.

In July, Amelio found his permanent CTO: Ellen Hancock, who'd worked with him at National Semiconductor and had spent 28 years at IBM before that. Her first assignment: solving Copland.

Product managers spent several hours showing her what they had. Hancock was stunned to discover that Copland wouldn't even offer the holy grail, full memory protection. Continuing with the project seemed nuts. It was time for a mercy killing.

Hancock and Amelio decided to harvest whatever features they could from Copland's corpse and weave them into the existing Mac OS.

And sure enough: In the summer of 1997, Mac fans delighted in a new OS release called Mac OS 8 (to the confusion of Apple watchers who thought that *Copland* was going to be OS 8). It was a thoroughly gussied-up System 7, now with Copland features like multithreading, spring-loaded folders, and pop-up window tabs. Apple sold 1.2 million copies in the first two weeks.

A year later, Mac OS 8.5 inherited more of Copland's features, including a desktop decor control panel and that supercharged Find window, now called Sherlock. In 1999, Mac OS 9 would get those user accounts, where you logged in with a name and password.

But in 1996, Hancock and Amelio were no closer to fixing the fundamental problem: The Mac didn't have a stable OS, and they were running out of time and options.

Starting over, writing a new OS from scratch, would take years. Licensing an existing OS and trying to make it run on the Mac—Sun's Solaris, for example—would be a technical nightmare.

Amelio even called Bill Gates about adapting Windows NT, Microsoft's corporate OS, for the Mac. Gates was thrilled, and vowed to put hundreds of engineers on the project. But Apple's engineers didn't just find the proposal distasteful; the technical contortions needed to get Windows to run on the Mac were too daunting to be practical.

That left Apple only one way out: *buying* an OS.

26. NeXT

When he was ousted from the company, Apple's charismatic leader took some of Apple's top engineering talent with him. He founded a startup with an offbeat, one-syllable name. The goal: to create a state-of-the-art workstation computer with an all-new, rock-solid OS. But even after years of effort, the new machines weren't selling. He abandoned the hardware idea and pivoted the company; now it would focus on selling only its OS.

"He," of course, is Jean-Louis Gassée. His post-Apple story mirrors Steve Jobs's almost exactly.

Steve Sakoman, originator of the Newton project, had left Apple at about the same time as his friend Gassée, in February 1990. Over the summer, they discussed how unhappy they'd become with "the complicated layers of hardware and software silt the Mac had accumulated," as Gassée puts it. "Surely, we could come up with a simpler, cleaner architecture, a more agile personal computer." They decided to go into business together.

Gassée proposed calling their new venture United Technoids, Inc.—UT for short. ("This was both cute and useless," he acknowledges, "as there was already a huge conglomerate called United Technologies.")

A few days later, Sakoman called to say that he had been reading the dictionary to find a better name, and had gotten as far as *B*. Gassée, misunderstanding, was delighted. He *loved* the name Be! Be, Inc. sounded "simple, resonant, romantic even—a call to action," Gassée says.

For four years, nobody knew what Be was working on—only that it had some powerhouse talent, including former top Apple OS engineers. Finally, in 1994, the plan was revealed: They were building a new, multiprocessor computer from scratch, called the BeBox. It would run a brand-new operating system, BeOS—with a kernel, preemptive multitasking, and protected memory.

By September 1995, the company had 30 employees (about half from Apple), $14 million in

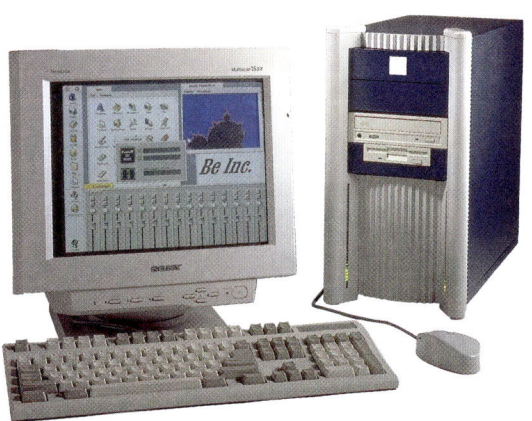

The BeBox's best feature was its OS.

investment capital, and a working prototype. The BeBox was a tall, plastic-cased, two-toned tower, with lights that blinked to represent the activity of its five PowerPC processors. When he demonstrated its speed and flexibility at a conference for computer executives, Gassée got a standing ovation.

In truth, though, Be was struggling. It had burned through cash and wasn't finding much interest in the BeBox. So in 1996, Gassée refocused the company on finishing BeOS.

Word was getting around that BeOS could run on PowerPC Macs. In August, Be's booth at the Macworld Expo in Boston overflowed with admirers; simultaneously, over at the Power Computing booth, crowds marveled at the BeOS running on one of Power's Mac clones.

Gil Amelio invited Gassée to his office in City Center 3, requesting a demo. The timing could not have been better for Be. "I was looking for an exit sign," Gassée recalls, "and here came Amelio."

By the end of the summer, Apple was in talks with Gassée about buying Be. BeOS was not complete—it still lacked printing and network file sharing, for example. But it held out the promise that the Mac, at long last, would get a modern operating system.

By October 1996, it was time to talk price. Gassée requested 15 percent of Apple's stock, worth about $500 million.

Amelio was stunned. "You have zero sales, you've got an operating system that's three years away from any reality, and you want 15 percent of the company!?" he said. "That's not even within the realm of possibilities."

Gassée's price dropped to $300 million; Amelio's offer went up to $125 million. Rumors swirled. The press went nuts.

Gassée countered with $275 million; Amelio came up to $200 million. And as December approached, that's where the deal stood.

NeXT Calls

When Steve Jobs founded NeXT in 1985, he had no baggage. He could operate without interference from executives who had no taste. He could handpick his engineers, executives, and designers.

He hired Hartmut Esslinger to design NeXT's first computer: a striking black cube, one foot on a side. He hired Paul Rand to design the logo—the guy who'd done the logos for IBM, UPS, Westinghouse, and ABC. Rand charged $100,000, no changes permitted.

But starting from scratch was a more massive project than anyone had imagined. When one reporter asked why the computer kept being delayed, Jobs quipped, "It's not late; it's five years ahead of its time."

When the NeXT Computer finally shipped in mid-1989, it was dazzling. Among other breakthroughs, it was the first computer to include a magneto-optical disc drive—like an erasable CD—and a digital signal-processor chip for spectacular audio features.

The crown jewel, though, was the operating system: NeXTSTEP. It looked like a futuristic version of the Mac OS—windows, icons, menus—but underneath was Unix, the beloved, battle-hardened OS that ran the world's banks, governments, and corporations.

NeXTSTEP was fully buzzword-compliant. It had memory protection, preemptive multitasking, and virtual memory. It was object-oriented, making it quick to build new programs. It was incredibly secure. It displayed any typeface gorgeously and crisply at any size, because it incorporated Display PostScript (an on-screen version of Adobe's PostScript printing technology).

But the Cube would never be the dorm room fixture Jobs had envisioned—not at $6,500.

In 1990, NeXT offered the slab-shaped NeXTstation ($5,000) and a second-generation cube, now called the NeXTcube, for $8,000.

NeXT machines had their fans; Tim Berners-Lee, who invented the web, used a NeXT machine to write the first web browser and web server. But the thing just didn't sell. After eight years, NeXT had sold only 50,000 computers.

In 1992, NeXT gave up on hardware altogether, changed its name to NeXT Software, and adapted NeXTSTEP to run on Intel PCs, and on RISC chips made by Sun and HP. Even so, NeXTSTEP just wasn't selling.

The Cube was stark and futuristic for its time.

Fortunately, NeXT had one last trick up its sleeve. It had also created WebObjects, a software suite for designing sophisticated interactive web apps. FedEx used it to build its package-tracking site. Dell used it to build its entire PC ordering website, with real-time pricing, configuration details, and online ordering.

In 1996, WebObjects was a hit, and revenues were beginning to turn around. The company was even planning an IPO.

Inside NeXT, employees were furious to learn that Apple was negotiating with Be. "Our feeling was that Be was an incredibly thinly veiled attempt to copy what NeXT had already built," says Garrett Rice, NeXT's director of development and developer relations. "It had a half-assed version of multitasking, an old-school user experience, and not-great developer tools. Plus, Jean-Louis Gassée was at least partially considered a villain in Steve's departure."

Rice was so offended that—without even consulting Jobs—he cold-called Apple's head of

The Mach Kernel

NeXTSTEP was a dazzling OS. On the surface, it was colorful and easy to use. But underneath, it ran Unix, one of world's most stable multitasking operating systems. And at the core of it all, NeXTSTEP had the Mach microkernel.

In the mid-eighties, two researchers at Carnegie Mellon began tinkering with one of the most popular Unix variations, Berkeley Unix. They rewrote its kernel—the beating heart of the OS, the part that controls access to the processor and memory—to be more stable, easier to debug, and easier to adapt for other processors and computers. It was years ahead of its time.

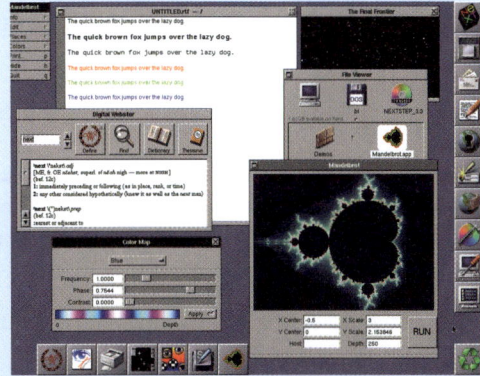

Those researchers were Richard Rashid, a professor, and Avie Tevanian, his PhD student. Eventually, Bill Gates hired Rashid to come to Microsoft—and Steve Jobs hired Tevanian to come to NeXT.

NeXT's engineers saw Tevanian's Mach presentation at a conference. Next thing he knew, one night in 1987, he was having dinner with Jobs and a few OS engineers. "Our guys have seen your work," Jobs announced. "They love it, and we want to use it."

Nobody had ever adopted Mach in a commercial OS, but Apple had come close. In the summer of 1995, software VP Ike Nassi ran a secret project to build a version of the Mac OS running on Mach. It looked and felt like the traditional Mac—but it had all the stability and security of Mach underneath. Nassi distributed this Mach-based Mac OS at WWDC in 1996, convinced that it might finally solve Apple's OS problems.

By that point, however, Gil Amelio was entertaining the notion of buying Be, believing it to have better multimedia features. "I thought that was the dumbest idea ever, because I already had this up and running," says Nassi. He was so disheartened that he wound up leaving Apple.

Mach did, of course, find its way into Apple's operating systems, through a more circuitous route. To this day, it's at the core of every Mac, iPhone, iPad, and Apple Watch.

---

software to suggest that Apple consider NeXTSTEP instead. "In any version of the multiverse other than this one," Rice says, "this all ends with me getting my ass unceremoniously fired."

But when Ellen Hancock returned his call, he pitched her hard. "NeXTSTEP is by far the best operating system on the market. It's multitasking, multi-user. It has a user experience built by ex-Apple people. It has a full object-oriented development environment that nobody else has. There's no way you can't be talking to us!"

"I hadn't even thought about that," Hancock said. She asked for a meeting.

Now Rice panicked; his spontaneous call was having real consequences. He confessed to his boss about making the call. Jobs and Tevanian were summoned, and Rice had to explain what he'd done.

Jobs did not fire Rice. "Let's set up the meetings," was all he said. (Months later, Jobs toasted Rice at an all-hands meeting—and sent him a $20,000 check.)

On December 2, for the first time in 11 years, Steve Jobs walked into Apple's headquarters to give Hancock and Amelio a demo of NeXTSTEP. They saw right away that it was rock-solid and complete.

"I'll structure any kind of deal you want—license the software, sell you the company, whatever you want," Jobs told them.

Now Apple had *two* modern OSes to choose from.

After weeks of meetings and engineering consultations, Amelio and Hancock decided to stage a bake-off. Jobs and Gassée would make consecutive presentations before the executive staff, and then Apple would choose which company to buy.

The date: December 10. The arena: two meeting rooms in Palo Alto's Garden Court Hotel. The judges: Amelio, Hancock, Fred Anderson, Satjiv Chahil, and mergers-and-acquisitions head Doug Solomon.

The Apple contingent sat around a U-shaped table. Tevanian ran the demo on a NeXTstation and provided technical details. Jobs, at the peak of his charismatic game, did most of the talking. Amelio remembers the demo as "dazzling."

Avie Tevanian
Born: 1961, Portland, ME
Schooling: University of Rochester, Carnegie Mellon (MS, PhD)
Apple: 1997–2006
After Apple: Elevation Partners, NextEquity Partners

Tevanian, a former motorcycle drag racer, can waterski barefoot. Probably still can, anyway.

"Steve has them completely mesmerized, speaking so humbly, thoughtfully," remembers Chahil. "They were in love with Steve by the end of that meeting."

The demo featured four interactive apps, all running simultaneously. One displayed a spinning 3D molecule. Two QuickTime videos were playing at once (or, as NeXT's version was called,

NeXTIME). Even with all of this background activity, Tevanian could still type in a word processor without any lag. "Their jaws were on the floor that we could do this," he says.

After that act, Gassée didn't stand a chance. He had no formal presentation, not even a demo. Apple's technical teams had already met with Be's, so Gassée didn't think there was much more to say.

At that moment, Jobs and Tevanian were taking a walk around the block. They ran into one of the Apple staffers. "That went really well," he told them.

"We kinda knew we had the deal," says Tevanian.

The Deal

Both BeOS and NeXTSTEP came with a catch the size of Texas: backward compatibility. These were brand-new OSes. No existing Mac programs would run on them.

Jobs had told Apple that Tevanian's team would need about a year to create an emulator. Existing apps would all run inside a single bubble; within that simulation, they wouldn't offer protected memory or preemptive multitasking. But *new* apps, and old apps updated for NeXTSTEP, would get all of the modern benefits.

Meanwhile, NeXTSTEP was only part of what Amelio was buying. He was also getting 300 brilliant, world-class engineers. He was getting WebObjects. Above all, he was getting Steve Jobs.

Jobs and Amelio were not what you'd call a mutual-admiration society. Jobs called Amelio a bozo. "All he knows is the chip business," he'd say, "where you can count your customers on one hand. They aren't people—they're companies."

For his part, Amelio considered Jobs a world-class schemer. But bringing Jobs back into Apple's orbit would have an enormous symbolic meaning to Apple's employees, fans, and shareholders.

Jobs invited Amelio to his house and introduced him to his wife, Laurene, whom he'd married in 1991. (In 1989, she'd attended a lecture he gave at Stanford's business school. He was so smitten that he blew off his next obligation so that he could invite her to dinner. "I was in the parking lot, with the key in the car, and I thought to myself: If this is my last night on earth, would I rather spend it at a business meeting, or with this woman?")

Amelio and Jobs worked out a deal over tea: Apple would buy NeXT for $378 million in cash and 1.5 million shares of Apple stock. Jobs's take would be about $157 million.

The Apple board found the price outrageous. $427 million for NeXT? They hadn't even wanted to pay $275 million for Be!

But at this point, they had no room to maneuver; they approved the deal. "In hindsight," Fred Anderson says, "it was one of the biggest bargains in the history of the technology industry."

Jobs, uncertain what his role should be in the company, settled on "advisor to the chairman," and offered to appear at Apple's keynote at the Macworld Expo, only a few weeks away.

Gassée took the defeat with grace. He emailed Amelio and Hancock: "Congratulations for all aspects of an extremely well-crafted deal. Clearly we were not in the same technical and financial league. Thanks for having considered us."

At a press conference on December 20, 1996, Amelio broke the news. "I can boil it down to: We've picked Plan A instead of Plan Be," he said.

Macmaker

In the May 1997 issue of *Macworld* magazine, a witty columnist offered new lyrics to the song "Matchmaker" (from *Fiddler on the Roof*), which summarized Apple's efforts to buy itself a new operating system.

Macmaker, Macmaker, find an OS,
Buy us a way out of this mess!
Macmaker, Macmaker, please show some guts—
Do something to save our butts!

For Tesler, make it Net-savvy;
For Hancock, make it simply not suck;
For Gil, well, he would be happy
For something below half a billion bucks!

Macmaker, Macmaker, you know the tune:
Time's running out, buy something soon!
BeOS, NeXT OS, who could care less?
So find us a find, try us a try,
Deal us a deal, buy us a buy—
Just get us a new OS!

(Okay, the columnist was me.)

27. Transition

When Steve Jobs came back to Apple in 1997, he didn't know if *he'd* be able to orchestrate a turnaround, either. Everything was still going the wrong way: sales, market share, profit margin, stock price, and employee retention. In fact, soon after the NeXT purchase, Apple posted an $816 million loss for the fiscal year—an unfortunate new company record.

Amelio had designed his cuts so that Apple would break even at $8 billion in revenue. But now it looked like the actual sales figure for 1997 would be $7 billion. Fred Anderson, alarmed, quietly drew up a what-if plan for a $6 billion outcome.

But even his most ambitious cash-crisis maneuvers could not fix the central problem: People weren't buying what Apple was making.

Keynote

Apple's presentation at the January 1997 Macworld Expo in San Francisco was supposed to be the greatest keynote of all time. Gil Amelio would celebrate some historic news: Apple now had a cutting-edge operating system, an army of brilliant engineers—and Steve Jobs.

Marketing VP Satjiv Chahil stocked the event with superstars: Muhammad Ali! *Independence Day* star Jeff Goldblum! Broadway legend Gregory Hines! Comedian Sinbad! Celebrity airplane designer Burt Rutan! Rock star Peter Gabriel! And then, at the finale: Steve Wozniak! The two Steves would be onstage together for the first time in years.

Chahil would then unveil the Twentieth Anniversary Macintosh, a stunningly thin, vertical, flat-panel Mac designed by Robert Brunner and Jony Ive. Chahil would present the first two machines—serial numbers 0000 and 0001, of course—to the Steves, to celebrate the dawn of a new era.

The only dark omen was Amelio's absence at the dress rehearsal the night before. "Don't worry—I've got it," he told the show's anxious producers.

The day was crisp and sunny as 4,000 Apple fans packed into the Yerba Buena Ballroom of the San Francisco Marriott. Jeff Goldblum took the stage to introduce Amelio. "In 'Jurassic Park,'" he joked, "I play Ian Malcolm, an expert on chaos theory. So I figure that qualifies me to speak at an Apple event."

Gil Amelio entered, to applause. Leaning with one elbow on the podium, he began to speak. And speak. And speak.

"Uh, you know—when the, when the, uh, first Mac came out in 1984, it only had a 128K, uh, OS. And, uh, you run, run—am—ran everything on floppies. Uh, so you'd have the 128K OS, your applications, your documents all on a single floppy. So now if you—in the—uh, if you can imagine the Macintosh being analogous to a, to a small Cessna, a small airplane . . ."

Chahil, watching from the wings, was horrified. "I didn't know what the fuck he's talking about! It was a horror of horrors. I wanted to die. I wanted the earth to open so that I would fall in!"

Amelio was clearly nervous. Worse, he was lost.

The Twentieth Anniversary Mac

The Twentieth Anniversary Macintosh, that thin slice of the future, had been kicking around in the Industrial Design Group for five years.

In the fall of 1992, Robert Brunner had invited his designers, plus five outside designers, to create concepts for a one-piece computer of the future. He sought "provocative forms, rich materials, unique configurations, and added functionality, using miniature components."

The various entries looked like wedges, vases, desk lamps. Brunner himself made a foam model of a curved, vertical, flat slab code-named Spartacus: flat screen above, vertical CD drive below, circuitry behind the screen. There were speakers on each side, and a semicircular foot holding it all up.

When focus groups declared it the winner, Brunner turned the project over to his newly hired young designer Jony Ive, who turned Spartacus into a more practical machine. The hoop foot could now fold up to become a handle; the exposed distraction of the spinning CD got hidden behind a trapdoor; a detachable "backpack" could accommodate an expansion card. He hid the Mac's power supply inside the subwoofer.

A texture consultancy produced over a dozen

"Uh . . . uh, while, uh, we're on this aero—aeronautic theme that I—uh, I did visit, uh, down there with, uh, with Burt and, uh, and got to look at some of these—uh, these tremendous things . . ."

The teleprompter operator scrolled frantically, trying to find Amelio's place. (Later, an audience member, leaving the show, would ask an Apple employee if Amelio had had a stroke.)

Amelio's speech continued into a second hour. Jobs, backstage, grew increasingly irritated.

Finally, in the third hour, Amelio welcomed Jobs to the stage. The room exploded in cheers. There were so many camera flashes that Jobs had to ask them to back off. "You guys have to stop making me blind, or I'm gonna fall off the stage here," he said.

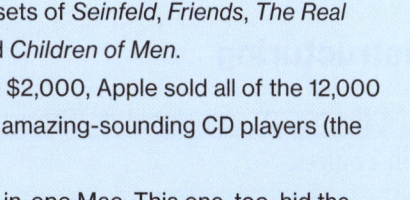

mockups, made of wood, fabric, metal, carpet, plastic, and leather. In the end, the Apple group chose a bronze-colored plastic—plus fabric (speakers), leather (palm rest), and rubber (subwoofer top). The keyboard could tuck away inside the foot when not in use; the centered trackpad could pop out and sit on either side of the keyboard.

Spartacus, now called the Twentieth Anniversary Macintosh (TAM), went on sale in March 1997. Apple positioned it as a deluxe, premium machine. The $7,500 price included delivery and setup by a "concierge."

Critics were not amused. They blasted the slow CD-ROM drive (a 4x speed model, because faster ones could not operate vertically) and, of course, the price.

Hollywood, however, loved it. TAMs popped up on the sets of *Seinfeld*, *Friends*, *The Real World*, and *Sabrina*, and in movies like *Batman & Robin* and *Children of Men*.

Eventually, after cutting its price to $3,500 and then to $2,000, Apple sold all of the 12,000 machines it had made. Today, collectors still use TAMs as amazing-sounding CD players (the sound system was designed by Bose).

Seven years later, Ive designed another flat-screen, all-in-one Mac. This one, too, hid the electronics behind the screen; it, too, had a vertically mounted CD and a keyboard that tucked away inside the foot when not in use.

It was called the iMac G5. It was a smash hit.

Woz, Jobs, and Amelio onstage.

He spoke frankly about the Mac's stalled progress, and Microsoft's success. With a few clear diagrams, he explained the virtues of Mach. He demonstrated NeXTSTEP playing five movies simultaneously without dropping a frame. To illustrate the virtues of an object-oriented OS, he built a simple app from scratch—in about a minute.

The crowd went crazy. It went even crazier when Woz came onstage, and the two Steves clapped each other on the back.

The audience gasped at the futuristic design of the Twentieth Anniversary Macintosh. But when Chahil tried to get CNN to play—"and with the click of a mouse . . ."—nothing happened. Apple had rented a cable connection for the underground ballroom. But the keynote had gone on so long that the window of TV service had timed out.

It was, Amelio says, "one of the most embarrassing and unforgiving experiences of my life."

Restructuring

When the NeXT deal closed on February 7, 1997, most NeXT employees began reporting to Apple's offices.

It was a strange time, full of transition and uncertainty. How would the two companies merge? What was Steve Jobs's role? Who was really in power?

Jobs told Amelio repeatedly that he didn't want a title or an office. He declined an offer to run the OS project. He left his first Amelio executive staff meeting halfway through and never attended another.

He sold all but one share of the Apple stock he'd won with the NeXT purchase—at a $13 million loss. He rarely appeared on campus. He rented an office in Palo Alto, and continued to drive up to Pixar in Richmond a day or two a week. "It was a nice life. I could slow down, spend time with my family," Jobs said.

He did, as promised, advise Amelio. "From September 1996 to the following July, I talked to Steve more frequently than to my children," Amelio says. They talked about whatever was on Amelio's list of burning issues: networking, Java, product road maps, the NeXTSTEP+Mac OS project.

For nine months, Amelio and Anderson had been working with a consultancy, McKinsey & Company, to devise a structure for a smaller, more focused Apple. At the time, the company was organized by business unit: Macintosh, peripherals, networking products, software, PDAs, and online services. Each division had its own engineering, legal, marketing, sales, finance, human resources, and administrative teams, and each was responsible for turning its own profit. It was an expensive arrangement, better suited for a big company than a small one, that led to a huge amount of duplication and infighting among divisions.

McKinsey recommended replacing that divisional structure with a *functional* one, more typical of smaller companies, where the primary divisions were marketing, manufacturing, engineering, and so on.

The new plan also accommodated another round of layoffs—3,000—and also shook even more of Amelio's lieutenants out of power, including COO Marco Landi and CTO Ellen Hancock, both demoted; manufacturing chief Fred Forsyth; developer-relations lead Heidi Roizen; and Satjiv Chahil, whom Amelio replaced with Guerrino De Luca from Apple's Claris software subsidiary.

Of the 47 executives named on Amelio's first organizational chart ten months earlier, only 18 now remained at Apple. The rest had quit or been fired.

To replace Hancock, Jobs encouraged Amelio to install two of his trusted former NeXT team members: Avie Tevanian as head of software, Jon Rubinstein to lead hardware. Rubinstein had once been NeXT's VP of hardware.

Jon Rubinstein
Born: 1956, New York City, NY
Schooling: Cornell (BS, MEng); Colorado State University (MS)
Before Apple: ComputerLand, HP, Ardent Computer, NeXT, FirePower Systems
Apple: 1997–2006
After Apple: Palm, HP, Bridgewater; boards of Qualcomm, Amazon, Robinhood Markets

Rubinstein's wine collection includes over 3,000 bottles.

But upon his arrival at Apple in February 1997, "it was an insane asylum," Rubinstein says. "Worse than I thought."

His friends could not fathom why he'd want to work for the dysfunctional, shriveling company. But Apple, Rubinstein maintains, was "the last innovative, high-volume computer company in the world." If it went under, it wouldn't be his fault. And if the new team succeeded, "it'll be the most amazing turnaround in the history of business."

When outside observers noticed the NeXT team taking these top jobs, many concluded that Jobs was sneakily insinuating himself into Apple. "To the Machiavellian eye, it looks as if Jobs . . . might be scheming to take over Apple for himself," wrote *Fortune*.

It certainly looked that way to Amelio, who grew alarmed by press reports that Jobs's friend Larry Ellison, CEO of Oracle, was considering a hostile takeover of Apple.

Did Jobs intend to retake Apple all along?

Jobs certainly denied it, internally and externally. "People keep trying to suck me in," said Jobs. "But I have no desire to run Apple Computer. I deny it at every turn, but nobody believes me."

Of course, he had reasons for holding back. The most important one was that he wasn't sure Apple could be saved at all. This was a company whose share of the computer market had now sunk below *3 percent*.

Jobs was also the CEO of Pixar, which was having a stunning run of success. *Toy Story* had won awards, critical acclaim, and the love of millions; Pixar's IPO had proven Jobs's early investment to be insanely shrewd. He wasn't sure if it was wise, sustainable, or even legal to serve as CEO of two public companies simultaneously. He worried that the employees of both companies would see him as not fully committed.

And no, leaving Pixar was not in the cards. "I've never quit anything I've started," Jobs said.

Trimming

Apple had spread itself far too thin. It had too many divisions, too many products, too much chaos. Rubinstein and Tevanian's first act was to lop away the broken parts.

First to go: the Performas, that archipelago of low-end Macs. "It was crappy product," Rubinstein says. "I mean, the failure rate was enormous. So I'm like: I'm just gonna flush 'em."

Apple had been attempting to break into the corporate world—again—with the Apple Network Server, a tall beige Unix box packed with slots for hard drives, which would dish out files to corporate networks. It went away, too. So did the Power Mac 9700, a six-slot tower that was underway but behind schedule, and a group in Paris that was working on an address book and other apps.

Then came the printers. Canon made Apple's printers, and Rubinstein thought they were terrible.

Apple Masters

It didn't escape Gil Amelio's attention that the Mac shone brightest in artistic and scientific fields. So in April 1997, with the company seemingly on its deathbed, he had an idea: What if he could tap into the Mac fandom of famous people from the arts and sciences?

The AppleMasters program was born. These masters would lend their endorsements to the Mac and serve as a celebrity advisory council. In return, Apple would give them Macs, host special dinners twice a year, and offer the world's best tech support.

At the inaugural event, 25 masters joined Amelio and the board for a private dinner at San Jose's Silicon Valley Capital Club: Nobel prize–winning scientists, astronauts, environmentalists. Boxing legend Muhammad Ali. Bestselling author Michael Crichton. Broadway tap dancer Gregory Hines. And, of course, Hollywood types: Jennifer Jason Leigh, Richard Dreyfuss, and *Jurassic Park* producer Kathleen Kennedy.

The AppleMasters program would eventually include novelists like Douglas Adams and Tom Clancy, actors including Harrison Ford and James Woods, and directors like Sydney Pollack and Terry Gilliam. Apple incorporated many of the "masters" in company events and ads, including its extended "Think Different" campaign.

Eventually, in 2002, after accumulating 75 AppleMasters, the company quietly discontinued the program. By that point, Apple was generating star power all by itself.

He discovered, furthermore, that Apple was sitting on a billion dollars' worth of unsold printers in a Sacramento warehouse. "I mean, it looked like *Raiders of the Lost Ark*—that warehouse scene."

He sold the entire business to Epson and HP.

Rubinstein also asked Jobs for an expert product-marketing guy. Jobs knew just the man: Phil Schiller, who'd been at Apple from 1987 to 1993, working mostly on PowerBook product marketing.

Phil Schiller
Born: 1960, Boston, MA
Schooling: Boston College
Before Apple: MGH, Howard Hughes Medical Institute, Nolan, Norton
Between Apple: FirePower, Macromedia
Apple: 1987–1993, 1997–present

As a teen, Schiller worked at the Discovery Dolphin and Sea Lion Pavilion at the New England Aquarium, and once volunteered for his childhood idol Jacques-Yves Cousteau.

Tevanian, for his part, shut down the Advanced Technology Group. In its barely viable condition, Apple could not afford a dedicated research arm that rarely created shipping products. "I hated to do it, but I had to do it," says Tevanian.

Jobs also advised Amelio to kill the Newton. He'd always hated the idea of using a stylus with a computer.

Amelio instead spun it off into its own subsidiary—Newton, Inc. It lasted one year.

Woolard

Ed Woolard, the 63-year-old former CEO of DuPont, had joined Apple's board in June 1996. Amelio described Woolard as a southern gentleman who vowed to take an active role in saving Apple—but he couldn't have anticipated *how* active.

Woolard had never been impressed with Amelio's performance. At the January 1997 shareholders' meeting, where Amelio had confronted yet another hostile audience, he'd been even less impressed. "He couldn't answer the questions, didn't know what he was talking about, and didn't inspire any confidence," he said.

Woolard had called Jobs to sound him out about the Amelio situation. Jobs didn't hold back, calling Amelio "the worst CEO I've ever seen. If you needed a license to be a CEO, he wouldn't get one."

In the spring, Woolard called Fred Anderson, as he usually did before every board meeting, just to get a check on Apple's financial pulse.

"How's morale in the company, Fred?" Woolard asked.

"It sucks, Ed."

"Are we gonna make the plan this year?"

"Probably not."

"Are you gonna stay if we don't make a CEO change?"

"I don't know, Ed. Probably not."

On the Fourth of July weekend 1997, the board convened a special weekend executive session; Amelio was not invited. Woolard offered the directors a chilling assessment.

"If we stay with Gil as CEO, I think there's only a 10% chance we will avoid bankruptcy. If we fire him and convince Steve to come take over, we have a 60% chance of surviving."

The board voted to end Gil Amelio's time at Apple—and to ask Steve Jobs to become CEO.

From his hotel room in London, where he and his wife attended the Wimbledon tennis tournaments every year, Woolard called Jobs to break the news. Amelio was about to be fired, and the board wanted Jobs to replace him.

But Jobs declined. He would continue as an unpaid adviser, and he'd accept a seat on the board. But he didn't want to be chairman, and definitely not CEO.

"We'd just taken Pixar public, and I was happy being CEO there," Jobs explained. "I was enjoying spending more time with my family. I was torn. I knew Apple was a mess, so I wondered: Do I want to give up this nice lifestyle that I have?"

Jobs finally agreed to take a more active role for 90 days, long enough to find a permanent CEO.

Woolard's next call was to Gil Amelio, who was spending the Independence Day weekend at his Lake Tahoe house with his family. "We think you need to step down," he said.

Amelio felt blindsided. Hadn't he just rescued Apple with the NeXT deal?

"I told the board it was going to take three years to get this company back on its feet again," Amelio said. "I'm not even halfway through!"

Woolard wouldn't budge. "We need somebody who's going to drive the sales, and we know sales and marketing isn't your primary strength."

Amelio was growing angrier. "Who knows about this?"

"Well, of course, all the board members know," Woolard said. "And Steve Jobs."

Now Amelio was livid. Why would Jobs be involved? He wasn't even on the board!

But the next day, Amelio gave a more conciliatory interview to Jim Carlton of the *Wall Street Journal*. "I have to confess: when it comes to consumer sales, that is not my background," he said. "I'm very proud of what I did in that company. I was the emergency room physician who rescued the patient. Now it is time for the general practitioner to come in and exercise long-term care."

On Monday morning, July 7, Gil Amelio took the stage at Apple's auditorium to say his farewells after only 18 months on the job. Steve Jobs spoke next.

"What's wrong with this place?" he said. "It's the products! The products suck! There's no sex in them anymore!"

It was as simple a statement of Jobs's business philosophy as anyone could wish for. Saving Apple wasn't about convertible bonds or restructuring; it was about the products. It would always be the products. Make them great, and Apple could come back.

Era's End

The board asked Fred Anderson to serve as Apple's acting CEO while a search firm looked for a permanent CEO. Whoever it was would become Apple's third CEO in four years. "I have to admit that our biggest weakness at Apple is managing succession," Mike Markkula says. "Lord knows we've tried, but this is a volatile business."

During the Spindler and Amelio years, Apple made every possible attempt to turn itself around: new products, new services, new alliances, new strategies, new executives, offers to merge or sell. None of it worked. "We did a lot of things that are, as we sit here, literally unexplainable. We did a lot of nutty things," says Greg Joswiak, now Apple's head of worldwide marketing. In the great story of Apple, "I literally tell people: What happened before '97 didn't count."

Greg Joswiak
Born: 1963, Detroit, MI
Schooling: University of Michigan (BS)
Apple: 1986–present

His college buddies joked that his nickname, Joz, was a perfect blend of "Jobs" and "Woz."

Today, some Apple fans call those years the Dark Years—or, because they were the years without Steve Jobs, the Interregnum. Robert Brunner and Ike Nassi like to say that they worked at Apple "between Jobs."

The common wisdom is that during that era, corporate executives ran the show, not passionate artists—largely true. Still, the company was not exactly asleep during those years. During Jobs's absence, Apple created the PowerPC alliance and delivered well-regarded Macs containing RISC chips, shipped the first mass-market digital camera, developed the first consumer speech-recognition software, moved to CD-ROMs across the line, and redefined the laptop computer with the PowerBooks.

Apple lost a heartbreaking number of talented engineers, designers, and managers during those years—not only because they worried about Apple's longevity but because internet startups were sprouting up and offering huge salaries. But for Jobs, the astonishing news was how many people *stayed*. "When I returned to Apple," he said, "a third of the people there really were A to A-plus people—the kind you'd do anything to hire. Despite Apple's troubles, they'd stayed, which was the miracle."

"The founder imparts a culture to the place," says Michael Tchao. "Apple is populated by people who are there to change the world. And I think that never left Apple, even when Steve did."

As for Amelio: For all the mockery of his mismatch with Apple's subversive culture, he was, in the end, the man who recognized what it would mean to bring Steve Jobs back to Apple.

PART 3

Steve 2.0

28. Turnaround

Knowing what you know now, you might have expected Apple's employees to be overjoyed when Steve Jobs returned to Apple on July 6, 1997. Who better to lead Apple than its founding genius, the charismatic visionary?

But not everyone cheered.

"Half the company thought: 'Wow, this is great—someone who actually knows Apple and believes and can fix the products!'" says former marketing communications head Allen Olivo. "And the other half was like, 'No way—the guy's been gone for so long. He already got booted once, and NeXT failed!'"

Plenty of people, in fact, left Apple *because* Jobs came back.

Nor was the press convinced that all would be well. *Wired*'s June 1997 cover story was "101 Ways to Save Apple." Its sage advice included "Admit it. You're out of the hardware game," "Invest heavily in Newton technology," and "Abandon the Mach operating system you just acquired." (A guest tip from Mac originator Jef Raskin: "Throw out the old and clumsy desktop, along with its operating-system-and-applications paradigm, and go for true task-centered design.")

Software makers were skeptical, too. "The idea that they're going to go back to the past to hit a big home run is delusional," said developer Dave Winer.

Morale was at zero. Talented people were leaving in droves. The board was old and weak. The products were shoddy; the Mac model explosion was baffling. The marketing was splintered and ineffectual. Bozo managers were at war with each other. The clones ate away at what few sales Apple could have racked up. The distribution channel was a mess, with broken, dusty Macs on back shelves, ignored by uninformed, apathetic salespeople.

The company had no CEO, no strategy, and, Jobs felt strongly, no soul.

He tackled all of it at once. He threw himself fully, relentlessly, exhaustingly, into his nameless and unpaid role. "It was pretty bleak those first six months," he said. "I was running on vapor."

Stock Options

First, Jobs turned his attention to the brain drain. Apple's attrition rate was 33 percent, meaning that it was losing a third of its workforce each year.

Apple's top employees all had stock options as part of their compensation. Stock options represent your right, after a few years of working, to buy stock at whatever its price was the day you got them—$30 a share, for example. If the stock goes up to $40, then you make an instant $10 profit on every share when you sell. Stock options generally inspire employees to work hard and stick around.

But if the company's stock price goes *down*—if it's now, say, $20 a share—then your options are worthless. That's what had happened at Apple. The options were "underwater."

During Jobs's very first week back at Apple, therefore, he asked the board to reprice the employee options much lower, so that they were actually worth something.

To his tremendous annoyance, the board wanted a couple months to study the issue.

"Are you nuts?!?" he said. "If you don't want to do this, I'm not coming back on Monday. Because I've got thousands of key decisions to make that are far more difficult than this."

The board agreed to reset the options at $13.25, the price the day Amelio was removed.

The Board

But Jobs was still furious that he'd had to *convince* anybody about repricing the options. He didn't see how he could succeed if the board was going to be a drag on every decision he made. So now he made another demand: "I don't have time to wet-nurse the board. So I need all of you to resign—or else *I'm* going to resign."

Jobs specified that Ed Woolard could remain on the board—"he was a prince, one of the most supportive and wise people I've ever met"—as well as Hughes International president Gareth Chang, who'd joined only a few months earlier.

The departing members were Katherine Hudson, Bernie Goldstein, Delano Lewis—and Mike Markkula.

Markkula had been Apple Computer, Inc.'s cofounder, its chairman, its longest-running director. He'd been the first investor. He'd hired and fired every CEO. He'd always been a paternal figure to Jobs. Jobs drove to Markkula's mansion to speak to him in person.

The two cofounders took a walk. Jobs explained that he wanted to start fresh. "He was worried that I might take it poorly, and he was relieved when I didn't," Markkula says.

From the beginning, Markkula had intended to stay at Apple for only a few years, while the tiny startup got on its feet. But one crisis had always led to another. "It always seems like there's something awful that needs to be attended to. And how can I leave when that's going on?"

Next thing he knew, 20 years had gone by. It was finally time to step away.

To replace the outgoing directors, Jobs's first invitation went to his friend, Oracle CEO Larry

Ellison. Next, he invited another friend and fellow long-walk taker: former Apple marketing head Bill Campbell, now CEO of Intuit.

Ed Woolard recommended Jerry York, who'd been the CFO at Chrysler and IBM and brought financial expertise. With Jobs, Apple now had a six-man board.

The Structure

Almost as soon as he returned to Apple, Jobs created a radically integrated administrative structure that endures to this day.

In Apple's new arrangement, the seven divisions were hardware engineering (Rubinstein), software engineering (Tevanian), sales (Mitch Mandich, joining from NeXT), marketing (Guerrino De Luca), operations (George Scalise), finance (Anderson), and legal (Nancy Heinen). Their leaders were known as the executive team, the ET. (Jony Ive would soon join the ET as well.)

Jobs was still spending two days a week at Pixar. For his three and a half days at Apple, he set up regular weekly meetings: Tuesday afternoon, marketing. Wednesday, hardware with Rubinstein.

The most important meetings were Mondays, from 9 a.m. to 1 p.m. The ET convened in Infinite Loop 1 to discuss every product in detail. Jobs went around the table, getting updates and asking questions—no PowerPoint allowed.

"I hate the way people use slide presentations instead of thinking," Jobs said. "People who know what they're talking about don't need PowerPoint."

The Monday ET system "blew up the competing strategies of the products with all different road maps," says Schiller. "It created a culture of collaboration, a culture of shared purpose, a singular focus on what we're doing."

This structure also created what Apple people describe as a debate culture. "We're all going to fight as loud as we want, as much as we want, about whatever it is we think. Say it in this room. And then when we leave the room, we all own the decision together," Schiller says. "You can't come back six months later and go, 'Well, I never thought that was a good idea.' No, you were in the room. You didn't say that!"

There was plenty of fighting, often with Jobs, and it got plenty loud. "And sometimes we switched positions, and would argue the opposite way. That's how we brainstormed stuff out," says Rubinstein. "I mean, it was almost comedic. He was always screaming about something or other. Once, we were in this huge fight. I'm standing in Target down in Cupertino, pushing my cart around buying toilet paper or whatever the hell it was, and Steve and I are on the phone, yelling at each other."

Jobs often told a story from his childhood, still cited at Apple today, about a widowed 80-year-old neighbor who had a rock tumbler: "a motor and a coffee can and a little band between them," according to Jobs. "We got some regular old ugly rocks, and we put them in the can with a little bit of liquid and grit powder, and he turned this motor on." When they opened the can the next day, the rocks were smooth, shiny, and beautiful. "That's my metaphor for a team working really hard on something they're passionate about," Jobs would say. By "bumping up against each other, having arguments, having fights sometimes, making some noise and working together, they polish each other and they polish the ideas."

Not every executive thrived on the debates the way Schiller and Rubinstein did. But in the end, the system gave the company a critical advantage: It weeded out situations where people knew a project wouldn't be great but kept their mouths shut for fear of retribution.

The Quadrant

In the strange, liminal period from February to July 1997, when Amelio was CEO but Jobs was in the background, Jon Rubinstein and Avie Tevanian had begun scaling back Apple's archipelago of initiatives. The printers and Performas, the OpenDocs and ATGs—all gone.

But even when Jobs took charge in July, Apple was still trying to keep 15 product lines alive: The 1400, 2400, 3400, 4400, 5400, 5500, 6500, 7300, 7600, 8600, 9600, the Twentieth Anniversary Mac, eMate, Newton, and the Pippin.

Their model numbers signified nothing at all. "I mean, I couldn't figure out the damn product line after a few weeks," Jobs said. "I kept saying, 'Well, what is *this* model? How does this fit?' And I started talking to customers, and they couldn't figure it out, either."

Finally, in a strategy session with his top lieutenants, Jobs interrupted a presentation of new product plans. "Stop! This is crazy!" he said.

He jumped up, grabbed a dry-erase marker, and impulsively drew two intersecting lines, forming a quadrant. Jobs wrote "Consumer" and "Pro" above the two columns, and "Desktop" and "Portable" beside the two rows.

"Here's what we need."

A reproduction of Jobs's famous product quadrant on the whiteboard.

Apple, he said, should make only *four products*: two laptops and two desktops. No separate model lineups for Sears, CompUSA, and ComputerLand. No cameras, scanners, printers, monitors, servers, hard drives, CD players, game consoles, or handhelds. No more redundant bureaucracies, no more duplicative fiefdoms. No 15 product lines—just four Macs.

The room was silent. Surely this was going much too far.

The "quadrants" idea would mean shutting down 70 percent of the company's projects. It would mean flushing away millions of dollars of investment and thousands of person-years of effort. It would mean throwing away promising, sometimes truly thrilling initiatives.

But Jobs was emphatic. "If we only had four, we could put the A-team on every single one of them," he said. "And if we only had four, we could turn them all every nine months instead of every 18 months. And if we only had four, we could be working on the next generation or two of each one."

The cancellations upset many of Apple's people deeply. "There was huge turmoil, because you were killing products that people were working on," says Eddy Cue, now senior VP of services. "It's like: 'We're gonna go from all these different products for everybody to, like, *two*? Are you guys crazy?'"

Some people quit. Some refused to play along.

"Some people struggled with meeting Steve's energy level or with the very high expectations that he set," says Deirdre O'Brien, now Apple's head of retail and HR, who worked in operations at the time. "And a lot of people left, either because they had opportunities at other growing companies around the Valley, or because they just felt like it was impossible to do what Steve was pushing us all to accomplish. You needed to have the fire in the belly for the very ambitious mission. If you didn't have it, it was time to leave."

Some, on the other hand, were thrilled. "To me, it just felt like a fresh start," says Myra Haggerty, who'd joined Apple in 1993 to write printer drivers, and is, today, Apple's VP of sensor technologies. "Somebody coming in, making super-clear decisions, giving us direction, caring about the products—I just felt unleashed as an engineer."

Killing the Newton

Of all the victims of the Quadrant overhaul, the Newton upset Apple's faithful fans the most. By 1997, it had nearly become the technological marvel Apple had always intended. The handwriting worked, the features were solid, and a cult of superfans had arisen. With the MessagePad 2000, the Newton was finally breaking even.

For months, Jobs was torn. He shut down the Newton, Inc. subsidiary and brought the project back to Apple while he pondered—but soon thereafter, he killed it altogether.

The Newton's fans swiftly took to the press and the internet to express their outrage. About 75 people gathered at Infinite Loop to stage a rally. It happened to be the first day of work for Apple's new worldwide operations head, Tim Cook.

"I had to cross a picket line to get in the building—they're out with signs, and yelling," he remembers. "At IBM and Compaq, where I had been working, I had been involved in helping with thousands of product introductions and withdrawals—and, I have to say, very few people cared about the withdrawals. . . . I had never seen this passion that close up."

Jobs watched the protesters from a window above. "They have every right to be upset," he told Phil Schiller. "Get them coffee and doughnuts, and send it down to them. Tell them we love them and we're sorry."

A common theory holds that Jobs killed the Newton because it was Sculley's pet project—the man who'd stabbed Jobs in the back in 1985. But Jobs insisted that the Newton was a distraction for talented engineers. "My gut was that there was some really good technology, but it was fucked up by mismanagement," he said later. "By shutting it down, I freed up some good engineers who could work on new mobile devices. And eventually we got it right when we moved on to iPhones and the iPad."

The Newton was the most ambitious project of the Sculley era—and, at $500 million, its most expensive—and the company would study the lessons of its failure for years: too early with the technology, too early with the announcement, too late with the product.

But even after the Newton's demise, a global community of superfans continued to use, repair, upgrade, discuss, and love their Newtons. To this day, they cherish the creativity, charm, and ingenuity of Apple's only pen-based computer.

In the words of Newton fan Victor Rehorst: "Newton never dies. It just gets new batteries."

Think Different

When Jobs arrived at Apple in 1997, the company was running 12 different ad campaigns. They weren't coordinated; in fact, their messages often conflicted. An agency review was already underway, involving 23 competing ad agencies submitting their proposals for a new campaign. The best BBDO—Apple's agency at the time—could come up with was "We're back."

Jobs didn't have the time to sit through days of presentations. In July 1997, he shut down the broader review process and solicited ideas from only three companies. He favored an idea from Arnold Communications: "What computer would God use if he only had seven days?"

But he also wanted to invite Lee Clow, director of the firm then known as TBWA/Chiat/Day. Clow, his old pal from the early Apple years, had brought the "1984" ad to life.

Jobs wanted a campaign that would pay tribute to creativity, independence, rebelliousness—the spirit of the old Apple and the new one. He wanted to honor the creative people and educators who'd stuck by the Mac even through its dark times. He wanted something that would inspire not just the public, but, critically, Apple's own employees as well.

Back in L.A., creative director Rob Siltanen asked four of his teams to prepare some campaign ideas. They tacked up their ideas on wallboards: photos, pencil sketches, taglines. "But there was one campaign that jumped out at me, and it jumped out in a big way," he says.

It was an idea for a poster-and-billboard campaign, featuring black-and-white photos of revolutionary people and events: Einstein, Thomas Edison, Gandhi. One was the famous Vietnam War protest photo of flowers in a gun barrel. Above each photo was the striped Apple logo—the only color on the image—and the words "Think Different."

Art director Craig Tanimoto had created it as sort of a rebuttal to IBM's "Think IBM" laptop ads.

"There was a purity about that I will never forget," Jobs said. "I cried in my office as he was showing me the idea."

He worried that viewers would think that Apple was comparing itself to Gandhi and Edison, and would find that a preposterous overreach. But overall, he loved it. Chiat/Day was back. Siltanen put all hands on deck to develop the campaign.

(As Jobs often observed, "Think Different" was not a grammatical error. Apple was telling you not *how* to think, but rather what to think *about*. It followed in the line of phrases like "Think big," "Think thin," or the skiers' bumper sticker "Think snow." Saying "Think differently" would make no more sense than "Think bigly" or "Think snowly.")

The first prototype TV ad ran two minutes—much too long. It set a slideshow of famous inspiring people to the propulsive beat of Seal's "Crazy," whose chorus went: "We're never going to survive unless we get a little crazy." But Siltanen's team found it impossible to cut the song down to a 60-second TV ad. There simply was no elegant way to chop it up or shorten it.

Frustrated, Siltanen toyed with using narration instead. He'd always been moved by the monologues in the Robin Williams movie *Dead Poets Society*—for example, "Despite what anyone might tell you, words and ideas can change the world." So he started writing in his journal: "To the crazy ones. Here's to the misfits. The rebels. The troublemakers." And, his favorite part, which he envisioned for the closing: "The people who are crazy enough to believe they can change the world are the ones who actually do."

Siltanen recorded his own voice as a placeholder. But when Jobs saw the prototype ad, he went ballistic. "I thought you were going to write something like 'Dead Poets Society'! This is crap!" he said. "It's advertising-agency shit!"

Siltanen, furious and disappointed, told Clow to find someone else to finish the ad; he was done with Jobs. Ken Segall, newly hired at the agency, took it over.

Then, to the agency's amazement, Jobs called to say he'd changed his mind. He wanted to proceed with the "crazy ones" script.

At this point, the agency had only 17 days to create the TV ad. They had the idea and Jobs's blessing. Now came the hard part: securing the rights to use the famous people's images. Most had never before allowed their images to appear in ads.

Jobs plied his own connections. He called the families of John F. Kennedy and Jim Henson himself, and flew to New York to discuss the John Lennon clip with his widow, Yoko Ono. Almost all of his heroes, or their estates, agreed to participate. (Every participant received money and Apple products to donate to their favorite causes.)

Robin Williams was the agency's first choice for the narration, but he maintained a long-standing policy of not doing ads. The agency hired a parade of L.A. talent to try their hands at the narration: Richard Dreyfuss, Peter Gallagher, Sally Kellerman, and even Phyllis Diller.

Lee Clow thought that Jobs himself should record it, which he did, reluctantly—in three takes, in the Apple auditorium—as a backup.

On September 28, 1997, the "Think Different" ad debuted on ABC's *The Wonderful World of Disney*, which happened to be airing the network premiere of *Toy Story*—from Pixar, of course. Until the last moment, Jobs was torn between the Richard Dreyfuss version and the one he narrated himself. In the end, he went with Dreyfuss's. "If we go with mine, it'll become about me," Jobs said. "And this can't be about me. It's about the company."

This is the full version (the gray type, added by Ken Segall, appeared only in the print ads):

> *Here's to the crazy ones.*
> *The misfits. The rebels. The troublemakers.*
> *The round pegs in the square holes. The ones who see things differently.*
> *They're not fond of rules. And they have no respect for the status quo.*
>
> *You can praise them, disagree with them, quote them, disbelieve them,*
>   *glorify or vilify them.*
> *About the only thing you can't do is ignore them.*
> *Because they change things.*
> *They invent. They imagine. They heal.*
> *They explore. They create. They inspire.*
> *They push the human race forward.*
>
> *Maybe they have to be crazy.*
> *How else can you stare at an empty canvas and see a work of art?*
> *Or sit in silence and hear a song that's never been written?*
> *Or gaze at a red planet and see a laboratory on wheels?*

> *We make tools for these kinds of people.*
> *While some see them as the crazy ones, we see genius.*
> *Because the people who are crazy enough to think they can change the world . . .*
> *are the ones who do.*

Jobs himself contributed the line "They push the human race forward."

In 60 seconds, backed by a gentle piano-and-strings theme, the ad presented clips of Albert Einstein, Bob Dylan, Martin Luther King Jr., Richard Branson, John Lennon (with Yoko Ono), Buckminster Fuller, Thomas Edison, Muhammad Ali, Ted Turner, Maria Callas, Mahatma Gandhi, Amelia Earhart, Alfred Hitchcock, Martha Graham, Jim Henson with Kermit the Frog, Frank Lloyd Wright, and Pablo Picasso. Many of them were Jobs's own heroes.

(As for the young girl who opens her eyes in the closing shot: Weirdly enough, Chiat/Day lifted that clip from the middle of an obscure Baegu-language music video called "Sweet Lullaby," by the French rock group Deep Forest. Her name is Shaan Sahota, and she's the director's niece. Today, she's a writer and doctor in London.)

The ad said nothing about computers. It didn't even *show* computers (not that Apple had any new computers to show). Furthermore, most people didn't recognize many of the featured figures. Who would recognize, for example, the faces of Buckminster Fuller, Frank Lloyd Wright,

Martin Luther King Jr., opera star Maria Callas, Muppets creator Jim Henson, Pablo Picasso, and Shaan Sahota were among the 18 "crazy ones" in the ad.

Turnaround • 307

or Martha Graham? But Chiat/Day considered their obscurity a feature, not a bug. It prompted people to talk about the ad, to replay it, to research it—"Who was that guy?"

What the ad *did* say is that Apple did have a soul—and it had been there all along. All the fumbling during the Dark Years didn't count. All the creative people who'd stuck with the Mac knew what they were doing. All the employees who kept the faith should be proud.

The ad connected Apple's rebellious early years with the company Steve Jobs intended it to be again. Viewers didn't interpret the ad to mean "Apple thinks that it belongs in the pantheon of these legends." They understood that it meant "We're inspired by these rebels and geniuses. We aspire to be like them."

The ad was another historic success for Apple and Chiat/Day. It won one advertising award

The Secrecy Protocol

Steve Jobs had always had a natural distaste for the press, perhaps as a result of the *Time* "Computer of the Year" episode; especially during his second term at Apple, he wielded *secrecy* like a weapon. There were to be no leaks. Prototypes were protected in locked rooms with badge-only access. Employees were categorically forbidden to discuss their work—even with coworkers outside their teams, even with spouses.

Secrecy worked for Apple on every level. First, the mysterious silence about what Apple was working on was itself a marketing tactic, fueling speculation and buzz. That way, at the product's launch, the surprise produced waves of media coverage and consumer excitement.

Second, if the timeline slipped or the product had to be canceled, there was no negative PR fallout. Nobody even knew the thing had existed.

Third: No cannibalization. No risk that people would hold off buying the PowerBook G3 because they knew that the PowerBook G4 was about to come out.

There was a morale factor, too. Anyone working on a secret project couldn't help feeling that they had exclusive access to something legendary and important.

Finally, of course, the secrecy prevented competitors from catching up. There would be no repeat of the Newton situation.

Jobs and his PR chief, Katie Cotton, strictly limited interviews and reviews to a few select publications, usually the *New York Times*, *Wall Street Journal*, *Time*, *Newsweek*, *BusinessWeek*, and *Fortune*. They handpicked the journalists that got access to each new product, or who'd get an Apple interview.

The special treatment of the chosen few didn't guarantee positive coverage. But it ensured that those lucky reporters listened carefully to whatever story Apple told.

after another, and an Emmy. It was endlessly parodied and imitated (CBS made a "Think Funny" ad; ABC made a "Think Dharma" ad to promote its sitcom).

Best of all, as Jobs had hoped, the ad gave everyone at Apple a new sense of pride and hope. "Think Different" photo posters hung all over campus.

Apple wound up spending $100 million on the campaign, on back covers of magazines (never on the inside). Billboards (only classy locations). Bus shelters (only those with illuminated panels). Walls of buildings in major cities. Over the years, dozens more famous creative people would appear in the print ads, which ran in various forms for five years.

Macworld Boston

For a man who'd been so reluctant to get officially involved with Apple, Jobs's attention to fixing the company was all-consuming. Fred Anderson ran the administrative side of Apple. Jobs, from a small office in the City Center building, threw himself into everything else: product strategy, hardware, software, industrial design, marketing.

But if there was any doubt at all that Steve Jobs was in charge, it evaporated at the Boston Macworld Expo in August 1997, his first public appearance since Amelio's departure.

He walked onto the stage to a one-minute standing ovation. "Thank you very much for that warm welcome," he said. "Now, I guess I'm gonna have to give a speech."

His talk was 32 minutes long. No new products, no demos, no teleprompter, no note cards—just Steve Jobs, speaking with candor, clarity, and humor.

"After four weeks, here's what I've found: Apple is executing wonderfully—on many of the wrong things." Apple's employees, he said, "can't wait to fall into line behind a good strategy. There just hasn't been one."

To get there, he said, "this change needs to start at the top." The crowd cheered as he announced the new board of directors. "I am joining the board as well," Jobs said. Bigger cheers.

The highlight was Jobs's juxtaposition of Apple's brutal press coverage with some of its realities. "It's crazy. Absolutely crazy. The Mac OS is still the best thing in the world. There are over 20 million active users. There are thousands of developers, and there's a billion-and-a-half-dollar software industry built around this thing."

He pointed out that Mac OS 8 had sold 1.2 million copies in its first two weeks. He noted that 80 percent of computers in creative fields were Macs; 60 percent of teachers used Macs; and 64 percent of all websites were created on Macs.

At this point, the crowd would have followed him to hell and back. That's when a rare miscalculation almost cratered the goodwill.

Gil Amelio had spent his last six months trying to reach détente with Microsoft. In principle, Microsoft needed Apple (Mac fans bought a *lot* of Microsoft software), and Apple needed Microsoft (without Microsoft Office, Macs would be unviable). But even after the "look and feel" lawsuits, the two companies were still hung up on lingering patent and copyright disputes.

Immediately upon returning to Apple, Jobs had called Bill Gates directly.

"I'm going to turn this thing around, but I need help," he said. "If we kept up our lawsuits, a few years from now we could win a billion-dollar patent suit. You know it, and I know it. But Apple's not going to survive that long if we're at war; I know that. So let's figure out how to settle this right away."

As late as the Macworld dress rehearsal, Jobs had still been on the phone with Gates hammering out the deal. It boiled down to four points:

- Microsoft would commit to making Office for the Mac for at least five years—at least one Mac version for every new Windows version.
- Apple would make Internet Explorer the default web browser on Macs.
- Both companies would cross-license all of the disputed patents for the next five years. No more lawsuits, no more bickering. (Microsoft reportedly paid Apple $100 million as part of this agreement.)

Gates/Big Brother addresses the Macworld Expo crowd.

- Microsoft would buy $150 million in nonvoting Apple stock, so that it would have an interest in Apple's success.

After a decade of hostilities, Gates and Jobs had finally recognized that their companies would make better collaborators than enemies.

The Microsoft deal was supposed to be the grand finale of Jobs's Macworld talk. "Apple lives in an ecosystem, and it needs help from other partners," he began. "I'd like to announce one of our first partnerships today—a very, very meaningful one—and that is one with Microsoft."

A wave of confusion washed over the crowd. There was applause, laughter, and booing, all at once. Microsoft? The enemy? The Borg?

When Jobs got to the part about making Internet Explorer the Mac's default browser, the boos grew louder. "The user can, of course, change their default should they choose to," Jobs reminded the audience, as though addressing a child.

Finally, Jobs invited Bill Gates himself to appear, via satellite. His face appeared on the big screen, dominating the room, his glasses gigantic, before rows and rows of captive spectators—like Big Brother in Apple's own "1984" commercial. Nobody in the audience could miss the parallel. It was practically a remake.

Gates, of course, had no idea how he was being projected in Boston. He went on gamely. "We're pleased to be supporting Apple," he said. "We look forward to the feedback from all of you as we move forward doing more Macintosh software."

In the end, the audience applauded. Still, Jobs chided them. "We have to let go of this notion that for Apple to win, Microsoft has to lose, okay?"

Jobs would later describe the setup as his "stupidest staging event ever," but it didn't matter. The keynote had been a master class in transparency and persuasion. It gave hope to Apple's fans, employees, and investors. Apple's stock shot up 33 percent in a day.

Store-Within-a-Store

In 1997, Apple's slow death was visible in two places: in print, and in computer stores.

If you walked into a CompUSA and asked to buy a computer, the odds that a sales rep would suggest a Mac were approximately zero. And if you asked a *question* about a Mac, they wouldn't know the answer. The Macs in CompUSA's 170 stores were broken and sad-looking.

Jobs wasn't the only one to notice. During the Amelio era, Apple had hired the McKinsey consultancy to offer its advice on saving the company. The experts observed that Apple's share of computer chain retail sales had dropped from 4.7 percent in 1996 to 1.8 percent a year later—thanks primarily to a lack of "in-store advocacy." In short, nobody at the stores gave a damn.

Apple sales head Mitch Mandich, who'd come from NeXT, came up with a solution: the store-within-a-store at CompUSA.

The idea was that each store would dedicate 15 percent of its floor space to a dedicated Apple area. Apple would hire its own sales staff, trained in Cupertino; they'd be paid by the hour, instead of on commission like all other CompUSA reps. The Apple staff would maintain the area, making sure the latest Macs, software, and accessories were neatly on display and in working order.

Mandich hired Eight Inc., a production company that had designed the Apple presence at several Macworld Expos, to design the mini-stores. There were software kiosks, neatly arrayed rows of working PowerBooks, and "Think Different" banners hanging from the ceiling, so you could spot the Apple store from the moment you entered CompUSA.

In under a year, the Mac's percentage of CompUSA sales rose from 3 to 14.

The Best People

Jobs may have been a whirlwind of decision-making, but the company still needed a real CEO. He, Ed Woolard, Mike Markkula, and acting CEO Fred Anderson made up the search committee. They hired headhunter John Thompson to scour the universe of executives.

But the candidates with the most potential—executives from Kodak, IBM, Sun, and so on—believed that Apple couldn't be saved, and declined. The ones who *were* tentatively interested were clearly second-tier.

"I don't think we were getting a world-class CEO pipeline of talent," Anderson says. (He himself wasn't a contender. "I didn't think I had the skill set to turn around the product and that sort of thing. I was a finance and operations guy.") Thompson kept looking.

Meanwhile, Jobs was unhappy with Apple's workforce—what was left of it. About a third, he felt, were bozos. "Not only were they not doing the right things, but they were instructing everybody else to do the wrong things, too," he said. "It was time for them to leave."

In March 1997, Apple announced yet another round of 4,000 layoffs. The company that had once employed 18,000 people was down to 6,658. It was Apple's lowest head count in a decade—and the smallest workforce Apple would ever see again.

Clone Death

Licensing the Mac OS had triggered raging controversy from day one, but nobody had to ask which side Jobs was on. "It was the dumbest thing in the world to let companies making crappier hardware use our operating system and cut into our sales," he said.

"And it wasn't expanding the market," adds Fred Anderson, "which it was designed to do."

> ### The OS 8 Maneuver
>
> The original Mac OS licensing agreement required the clone companies to pay Apple $50 per machine. But Anderson calculated that Apple could have been making ten times as much selling those same computers itself. And so, in the waning days of his time at Apple, Amelio executed an ingenious legal ploy.
>
> The contract gave the cloners permission to include, on their machines, System 7.5 and all of its updates (7.6, 7.7, and so on). When Mac OS 8 was ready—understood to be the Copland OS—Apple would require higher pricing.
>
> But when Copland was canceled, Apple's lawyers seized the day. Yes, the agreement said that the cloners would have to renegotiate for Mac OS 8, but it didn't say what Mac OS 8 *was*.
>
> And so, in June 1997, as Apple was preparing to ship System 7.7 (code-named Tempo), the company simply renumbered it. What was supposed to be System 7.7 was now called Mac OS 8.
>
> The licensees were, of course, furious—but a deal was a deal. They had no alternative but to agree to a higher price: $150 to $350 per machine, depending on its speed.
>
> Little did they know that much worse news was about to come.

In 1997, Apple sold 1.8 million Macs; the clone companies sold 600,000. Often, the clones were both faster and cheaper than "real" Macs; because the licensees were small, they could begin using the latest chips before they were available in the quantities Apple needed.

Jobs told Anderson to wind the whole thing down.

Motorola, grumbling, backed out, taking a huge loss. Apple gave Umax another year, since its low-end machines didn't compete with Apple's.

But Power Computing was a $400 million business, breaking ground for a $28 million Texas headquarters, filing for its own IPO. Its CEO threatened to sue Apple for breach of contract.

In the end, Apple made a deal with Power. For $110 million in cash and stock, Apple bought Power's customer database, some excellent engineers—and the company's brisk disappearance from the earth.

iCEO

Within only four months, Jobs had made swift, decisive, controversial changes to Apple's product line, board of directors, corporate structure, ad campaign, retail presence, cloning program, and Microsoft relationship. It was a stunning amount of change in a shockingly small time. It also took a terrible toll on him.

"It was rough, really rough—the worst time in my life. I had a young family. I had Pixar. I would go to work at 7 a.m. and I'd get back at 9 at night, and the kids would be in bed. And I couldn't speak, I literally couldn't, I was so exhausted," he told biographer Walter Isaacson. "It got close to killing me. I was driving up to Pixar and down to Apple in a black Porsche convertible, and I started to get kidney stones. I would rush to the hospital and the hospital would give me a shot of Demerol in the butt and eventually I would pass it."

By summer's end 1997, it was clear that the CEO search was going nowhere. For months, the board had been encouraging Jobs to accept the title himself.

So on September 16, he agreed to extend his 90-day commitment. He still refused payment and would not sign a contract. But he did tell the board he'd relieve Fred Anderson as interim CEO. (He would joke that his business card should say iCEO. But it was a very, very inside joke, given that Apple hadn't yet introduced a product with a lowercase *i* prefix.)

Inside and outside Apple, his admirers rejoiced that at least *somebody* was making decisions and making changes. But nobody could say whether it would save Apple. For Apple's 1997 fiscal year (which ends in September), Apple had broken its own record for losses yet again: $1 billion. That number, of course, included the $427 million the company had spent to buy NeXT, severance payments for its massive restructuring, and the millions paid for Power Computing.

Even so, the faithful hadn't wanted to hear about more losses. In two years, Apple had now lost almost $2 billion.

Dell CEO Michael Dell poured salt into the wound when, at the annual Gartner IT Symposium in October, a moderator asked him what he'd do if he were Apple's CEO. "I'd shut it down and give the money back to shareholders," he said.

Jobs, furious, never forgot that dig. In 2006, when Apple's total stock value passed Dell's, he sent an email to all Apple employees. "It turned out that Michael Dell wasn't perfect at predicting the future," he wrote dryly.

Jobs announced the first signs of life onstage at the San Francisco Macworld Expo in January 1998, where Amelio had gone down in flames a year earlier. He was nearly unrecognizable in his new mustache and full beard, streaked with gray. This time, his "one more thing" moment at the end of the keynote did not involve a product. Instead, he took the wraps off an Apple creation many thought they'd never see: a profit.

"Every group at Apple has been burning the midnight oil over the last six months," he said. "Apple Computer last quarter made a $45 million profit."

The Credit

Who saved Apple?

In one year, Steve Jobs turned Apple from a dying enterprise to a profitable, promising one. Without his exhausting year of swift action, there would be no Apple today.

But there was also John Sculley's $3 million investment in ARM, which appreciated 27,000 percent. He couldn't have known it at the time, but without the $800 million ARM payoff, there would have been no money to buy NeXT in 1997.

There was also Fred Anderson, whose swift financial maneuvers in the summer of 1996 pulled Apple back from bankruptcy.

There were the Mac fans who continued to root for the company they loved—and joyously mock the tasteless mediocrity of Windows—even when Apple was headed for disaster.

There were the developers who stayed the course, believing in Macs—and coding for them—even when the sales figures looked dismal.

There were the employees of Apple who maintained the faith. "Those of us who got to 1997 kept the place alive—barely—and kept really good people there," says QuickTime engineer Peter Hoddie.

And there was, yes, Gil Amelio, who had been warned of bringing Jobs into the tent, yet still chose to buy NeXT for the sake of Apple's future.

Luck, strategy, faith, and people all played a role. Had they not unspooled exactly the way they did, there would have been no Apple left for Steve Jobs to save.

29. Ive

Robert Brunner ran Apple's Industrial Design Group (IDg) for eight years. During that time, the studio's work won more design awards than all other tech companies combined. It established the IDg's tradition of secrecy, of setting up shop away from the main Apple hub. "The creative process really is about having space, and the ability to iterate and experiment and try things that may fail," Brunner says. "And if you've got someone in your face all the time, it's not happening."

He and his team produced Apple's most eye-catching products during those years, including the Macintosh LC, Color Classic, Newton, StyleWriter II inkjet printer, PowerBook, and the Twentieth Anniversary Macintosh.

But never mind all that. Brunner's epitaph, he likes to say, will be: "Here lies the guy who hired Jony Ive."

Jony

Jonathan Ive was born in London in 1967. Young Jony (pronounced "Johnny") loved to accompany his father, a silversmith, woodworker, and design educator, on visits to studios and design schools. He absorbed his father's belief in design as a profession and a way to change people's lives. By the time he was a teenager, Ive already knew what he wanted to do with his life: "Making sculpture on an industrial scale," he says.

He studied industrial design at Newcastle Polytechnic. He had his first hit product while he was still in college and interning for a design firm: a ballpoint pen with grooved rubber grip strips on either side of the barrel that became a bestseller in Japan.

In 1989, he won a travel bursary (money to cover educational travel), which he used to visit design firms in Silicon Valley. He was thrilled to see their designs, and they were thrilled to see his. Job offers poured in from frogdesign and Brunner's Lunar Design, among others.

"Here comes this kid," Brunner says. "Spiky hair, and really sweet. And he started showing his portfolio." ("I was unspeakably shy," Ive says.)

Brunner was blown away to discover that, in designing a futuristic home cordless phone, Ive had gone to the trouble of modeling all the electronic components *inside it*, just to show that the design would work.

In May 2012, Princess Anne made Ive a Knight Commander of the Order of the British Empire. Technically, he's Sir Jony Ive.

At 23, Ive started up his own firm, Tangerine. It took off almost immediately; Ive designed power tools for Bosch, sinks for Ideal Standard, electronics for Hitachi. Brunner, now starting up Apple's design studio, tried to hire him again; once again, it wasn't the right time.

Brunner never again wanted to go through another frantic, PowerBook-style design crunch. So in late 1991, he staged an "investigation" called Juggernaut. The idea was to explore new concepts in mobile electronics—laptops, handhelds, cameras—*before* they became Apple products on a deadline. He invited designers both inside and outside Apple to contribute Juggernaut designs.

Brunner loved what Ive came up with. "They didn't rely on anything we had done or seen before," he says. "They had an emotional maturity that's rare for someone Jonathan's age." In September 1992, Brunner tried to hire Ive a third time—and this time, Ive bit. He joined Apple at 27 years old.

His first job was the MessagePad 120, the second-generation Newton. The original Newton's lid had been intended to flip up and around to the back. But if you inserted a PC card (a modem, for example), it protruded too far for the lid to close. The clumsy solution: You had to remove the lid from its hinges. Ive's solution: a double-hinge mechanism that cleared the PC card. Now you could flip the cover over to the back, as though on a stenographer's pad.

Ive soon became VP of design, and when Brunner decided to leave Apple in 1997, he chose Ive as his successor. He could already see the talent—and the work ethic. "I mean, the amount that he would put into something—literally to the point of jeopardizing his health," he says. "He had this craftsperson's mentality—a higher calling just to make something right."

Ive, however, wasn't happy at Apple. These were the years of CEO shuffling and desperate corporate flailings. "It was a really creative-adverse environment. It was a very unpleasant place to be," he says. "Being part of a company devolving into irrelevance and becoming so dysfunctional—I still have PTSD from those five years."

During those years, design was not paramount; it was, Ive says, "a service function." Gil Amelio met Jony Ive only once during his year and a half as CEO.

By 1997, Ive was so disheartened that he was on the verge of quitting.

The Collaboration

September 1997 marked the occasions of two funerals: Princess Diana and Mother Teresa. But in his design lab in Valley Green 2, Ive was nervous; Steve Jobs was coming to visit. Ive had heard rumors that Jobs intended to bring back Hartmut Esslinger. Years later, Ive would learn that Jobs had come to the studio to *fire* him.

Jobs was unimpressed by what he found in the studio: foam models of prospective new laptops, all-in-ones, and handhelds. But he liked Ive instantly. "I could tell after that first meeting that Amelio had wasted his talent," Jobs said.

Ive spent two hours showing Jobs around the design studio. "And he went from somebody

Ive and Jobs (pictured in 2005) spent hours together every week.

thinking, 'I'm going to fire you,' " Ive says, "to saying, 'Apple is shipping a huge amount of shit, and the products are awful, but I also see the work that you've done and the way that you think.' "

Jobs began spending hours in the design studio; eventually, he visited almost every day. Jobs and Ive began having lunch together. Their families became friends. They took the long walks so characteristic of Jobs when his brain was in high gear.

"I clicked with Steve in a way that I had never before done with someone and never have since," says Ive. "He understands what we do at our core better than anyone," Jobs said. "If I had a spiritual partner at Apple, it's Jony."

At Apple before Jobs's return, the life of a new product began with electrical engineers (EEs), who decided what the machine would be: what logic board, screen, processor, and so on. Next, mechanical engineers (MEs) ensured that the components would work together. Finally, as a low-priority last step, industrial designers (IDs) were asked to wrap a skin around the result.

Jobs radically revamped the process. No longer would the engineering team hand the design studio a box of components, with the expectation that they'd design an enclosure to hold them.

Now, instead of reporting to hardware head Jon Rubinstein, Ive reported directly to Jobs. "He understands that Apple is a product company," Jobs said. "That's why he works directly for me. He has more operational power than anyone else at Apple except me."

Over the years, this revised product-development pipeline became formalized as an internal playbook called the Apple New Product Process. Now, all three teams—EE, ME, ID—sit together from the very birth of a new product, without any consideration of cost or manufacturing. Distinct phases follow—concept, prototyping, engineering, production, and shipping—each with its own milestones. Any one of the three teams can stop the train at any time. "And maybe there'll be dead ends, and we'll start again, until we get to something that we all are going to make. But they're doing it collectively as one team," Phil Schiller says.

Project C1

In 1997, the hottest concept in computing was the *network computer*, or NC. Thanks to the internet, the thinking went, individual computers no longer needed storage, disks, and software. The *internet* could be the disk. The *internet* could run the software. All you needed on your end was a terminal: screen, keyboard, and mouse. Your computer could be much smaller, less complex, and less expensive.

Better yet, there would be no more separate Mac and Windows versions of software, no more wars for developer attention. Software could all just be on the internet, written in Sun's Java programming language, which ran on any kind of computer.

The industry's chief proponent of NCs was Jobs's best friend, Larry Ellison; Jobs was sold on the idea. And so, on the day of Jobs's very first meeting with Ive, the two sat down in a conference room to talk about creating an NC.

Ive was floored by Jobs's passion for design. "What was amazing for me was to hear Steve articulate what I felt," he says. "And not only could he articulate it—he could say, 'And I'm the boss, so go and do it!' And so for me, this was amazing. This was like *turbo* redemption."

As the world now knows, what Ive and his team came up with did not look like any computer that had come before; it barely looked like a computer at all.

Under Brunner, the design studio had been experimenting with translucent plastic; it first appeared on the eMate "laptop." The idea was an offshoot of a standard step in the design sequence, where designers ordered clear plastic parts so that the engineers could evaluate airflow and other aspects of the internals. Those transparent prototypes were always coveted by engineers and designers.

But Ive took the notion of see-through to a radical new scale for project C1 (named because it was the first consumer desktop): The *entire* computer was translucent. Every part of it: the case, the keyboard, the keys, the mouse/keyboard connectors, even the power cord. You could even watch the mouse ball spin as you moved the see-through mouse, because Ive had made the ball half dark, half light.

Apple named the color Bondi blue for the turquoise waters of Bondi Beach near Sydney, Australia.

And then there was the shape. After inspecting dozens of foam models in Ive's studio over the weeks, Jobs had fallen in love with one in particular. It was part egg, part pyramid, all curves—and a royal pain for the engineers and manufacturing leads, who preferred a clean, simple box.

"They came up with 38 reasons they couldn't do it," Jobs said. "And I said, 'No, no, we're doing this.'"

"I would argue the way that things *ground*"—how they interact with the surface they sit on—"tells you an awful lot about what they are," Ive says. "The combination of the [C1's] handle and the form made the product look like it's just arrived, or it's just going. And to me, that made it feel vital and alive."

And the color! This computer was *blue*. No personal computer had ever been blue!

Hardware

The C1 was designed at breakneck speed; since the demise of the Performa line, Apple had had no low-cost Macs to sell, nothing even under $2,000. To meet the deadline, the IDg replaced its usual design process—sending blueprints to tooling companies—with early 3D CAD (computer-aided design) programs like Alias Wavefront.

Jon Rubinstein, however, had concerns about the network computer concept. In 1997, dial-up internet speeds weren't anywhere near fast enough to run online software. Furthermore, he and Jobs had tried eliminating the hard drive once before, at NeXT (the base model came with only a recordable optical disc). It hadn't gone well.

"Steve, you already made this mistake once at NeXT! And we had to fix it!" Rubinstein told Jobs.

The staff finally "ganged up" on Jobs and convinced him to add a hard drive to the C1. Now it wouldn't be an NC, but a full-blown computer. Ive's original C1 design had a horizontal tunnel at the bottom, where you could slide and hide the keyboard when you weren't using it. Eliminating that hole gave Rubinstein's team room to add the hard drive and a CD-ROM drive.

The hole, however, was not the only thing Rubinstein eliminated. This Mac would not have a SCSI port (for external hard drives), a serial port (for printers), or ADB connectors (Apple's standard keyboard/mouse jacks). Rubinstein wanted to replace all of those with a new jack nobody had heard of: USB. The C1 computer would have three of them: one on the keyboard, two hidden behind a tidy door.

USB (Universal Serial Bus), invented in 1995 by a coalition of Compaq, IBM, Intel, and others, was intended to replace a motley universe of computer connectors. Into this one universal jack, you'd be able to connect keyboards, mice, disk drives, cameras, printers, music players, network adapters, and so on. It was 30 times faster than the existing serial ports, and hot-pluggable: You didn't have to restart the computer when attaching or detaching gadgets.

USB had another advantage, too. The number one tech-support call topic, both at Apple and Microsoft, was peripheral gadgets that didn't work—because people hadn't installed the *drivers*, the software bits that taught the computer to recognize the gadget. The beauty of USB, though, was that it could auto-recognize what you'd plugged in, auto-download the proper driver from the internet, and offer to auto-install it. That would be the end of "My thing doesn't work" calls. (It was also the dawn of the downloadable software updates that Apple's products offer to this day.)

Rubinstein knew that Mac fans would have to buy adapters for their existing drives, printers, and scanners. But USB was already starting to appear on Windows machines. As long as Apple was banking everything on this new machine, shouldn't it have the latest technology—and set the stage for the future?

But there was one more element Rubinstein, Ive, and Jobs wanted to kill: the floppy drive.

Yes, floppies were slow and ancient and had tiny capacity. But every desktop Mac ever made had had a floppy drive! Without one, how were you supposed to back up files or transfer them?

"You had external floppy drives. You had Zip drives. You had cloud storage. You had external hard drives. I mean, it was gonna be an inconvenience," Rubinstein says, but ultimately, "it was not gonna be an issue."

Losing the floppy drive, SCSI, and ADB represented the birth of a new philosophy at Apple: a willingness, even an eagerness, to kill its darlings—old technologies, even ones that Apple had developed and promoted—in the name of advancing the state of the art. "Skate to where the puck's going, not where it's been," Jobs would say, quoting hockey star Wayne Gretzky.

Naming C1

As C1's release drew nearer, Apple's marketing executives needed a real name for their revolutionary, cool-looking new computer. "We need a really great name for this," Jobs told them. He kept saying that he wanted something iconic, memorable, and simple. "Like the Sony Walkman—that's a great name. We need a name like that."

The very next meeting included Apple's marketing communications team, PR team, and the ad agency.

"We got the name!" Jobs announced as he strode in. "It's MacMan!"

For the agency, says Chiat/Day creative director Ken Segall, "it was an awkward moment. Steve was all-in on MacMan, but we did not respond well to it." There were two problems: "MacMan" had a gender bias, and also seemed too derivative of Sony's Walkman name.

But Jobs held firm. "Sony is a great consumer electronics company," he said. "If we can get a little rub-off by calling it MacMan, that's fine with me."

By meeting's end, Jobs gave the agency two weeks to come back with its own suggestions. "If you can't do better," he said, "we're going with MacMan."

Segall took on the task of finding name candidates. The only rules were that they had to contain the word "Mac" and imply "internet." A few days later, the agency presented, on poster boards, the five best names they'd come up with: MiniMac, iMac, eMac, Macster, and EveryMac.

Jobs hated them all.

Segall kept "iMac" in the mix for another round of names. Jobs claimed to hate these, too, but he did have *iMac* silk-screened onto a C1 model and asked around for people's opinions.

Next thing anyone knew, the computer had been named. It would be iMac.

The Bet

Project C1 was radically, dangerously, stunningly different from anything any company had done before. The design was risky. The concept was risky. Eliminating every existing jack and the floppy drive was *really* risky.

And manufacturing the thing was nearly impossible. "The product could not be built," says mechanical engineer David Hoenig, who'd been charged with leading the effort. He didn't just mean that the iMac couldn't be mass-produced. "You couldn't build one in the *lab*," he says. Among other problems, the handle wasn't anchored to anything.

Jobs, furious at all the pushback, blew his stack. "Ninety-five fucking hundred jobs are depending on you, and you've failed!" he snapped at one manufacturing meeting. He announced that he would be sending the iMac design files to Acorn, his favorite external design consultancy. "And if they tell us they can do what you say you can't do, you guys are all outta here."

In the end, even Acorn declared the iMac to be unmanufacturable as designed. Ive agreed to make structural changes, including adding hidden mounts to support the handle.

When Jobs had returned to Apple, he instituted a periodic off-site meeting at a resort in Santa Cruz called the Top 100. To each one, he invited the 100 people from Apple he felt were most important to the company at that moment: the executive staff, the heads of product lines, the heads of Apple's international offices, and so on. Here, Apple leaders unveiled new products, offered updates, and set Apple's directions for the coming year. (Apple continues the Top 100 tradition today, now at the Carmel Valley Ranch, 90 minutes south of Cupertino.)

"We can keep selling 4 percent of the market in beige computers," he said at a Top 100 retreat in early 1998. "But how do we get beyond, to the people who aren't using beige computers?"

> ## The Handle
>
> The first iMac weighed 35 pounds. It was not intended to be portable. And yet it had a handle: a prominent, reinforced crossbar on the back, translucent white against the blue of the case.
>
> Ive got pushback from all sides. "It'll cost a fortune to engineer!" "It's going to drive up the price!" "What's the purpose of a handle on this thing?"
>
> "Well, it's not to carry the product!" Ive replied.
>
> In 1997, Apple's research showed that people were still fundamentally nervous around technology. But when you see a handle, "you know that you're given permission to move it. You understand something about its physicality."
>
> Steve Jobs got it immediately. "That is the best reason for this handle," he said.
>
> Soon after the iMac shipped, Ive visited a computer store to watch people crowding around the iMac on display. "I remember watching somebody absentmindedly, without thought, holding the handle while talking to somebody else. And there was just almost a physical affection."
>
> That moment, to him, justified the entire exercise – and confirmed the similarity of taste that he would have with Steve Jobs for the next 13 years.

A drape covered an iMac prototype. "So I'm going to take an approach like what Swatch is doing with watches, and making this a fashion statement." He pulled away the cloth. Apart from the people who'd worked on it, it was the first time anyone had seen the iMac.

When the Top 100 attendees learned that there would be no floppy drive or traditional Mac jacks, "it was a very spirited debate," says Allen Olivo, then Apple's director of marketing communications. "Steve had to work a little bit to sell the thing in."

After the presentation, Olivo walked with Jobs back to the hotel.

"Well, Steve, that's different," Olivo said. "Do you think it's going to work?"

"I don't know. We'll find out in a few years," Jobs told him.

30. iMac

On May 6, 1998, Jobs took to the Flint auditorium stage. By no coincidence at all, this was the precise spot where he'd stood 14 years earlier to pull the original Mac out of its carrying case. Apple's invitation called it a special event, and that was putting it mildly—every customer, employee, developer, shareholder, and board member was watching to see if bringing Jobs back to Apple had been a good idea.

Jobs began by acknowledging some of Apple's original leaders in the crowd: Mike Markkula, Mike Scott, Steve Wozniak, plus Andy Hertzfeld and many of the original Mac team.

Then he recapped for the audience Apple's progress in the ten months since he'd become iCEO. Apple's annual turnover rate had dropped from 33 percent to 15, below the Silicon Valley average. Apple had just reported its second consecutive quarterly profit: $55 million. And Apple's market cap (the total value of all its shares) had shot from $1.8 billion to $3.8 billion.

The Quadrant

Next, Jobs introduced the public, for the first time, to the 2 x 2 grid of Macs that would become Apple's focus: consumer and pro, laptop and desktop. "This is it. We've got to make four great products," he said.

One quadrant, he said, was already filled in: the professional desktop. It was the Power Mac G3, which had come out in November 1997. It was the fastest Mac ever made *and* the fastest selling in Apple history: 500,000 in six months.

Power Macintosh G3

Sold: November 1997–January 1999
Price: $2,000 (desktop), $3,000 (tower)
Processor: PowerPC 750 (G3) at 233 MHz (up to 333 MHz)
Memory: 32 MB, expandable to 192 (desktop) or 384 (tower)
Equipment: 4 GB (desktop) or 6 GB (tower) hard drive, 24x CD-ROM, 1.44 MB floppy drive

The motherboard was identical in the horizontal G3, tower G3, and PowerBook G3 models. A small "personality card" held the differentiating circuitry.

The Megahertz Myth

In the late nineties, Intel enthusiastically promoted the idea that megahertz (MHz)—how many million times a second a chip can perform tasks—was a simple measure of a computer's speed.

The truth was more complicated. You can't compare megahertz among different chip families, because many other design variables affect a chip's speed. A 500 MHz Intel Pentium chip was substantially *slower* than a 500 MHz Power Mac.

So Apple asked Chiat/Day to create some ads that would communicate that *very* nerdy footnote to a general audience. The result: two TV ads that became legendary for their snark and humor. One features an actual snail inching its way across the screen with a Pentium chip on its back, as a slow, old-timey waltz plays. "The chip inside every new Power Macintosh G3 is up to twice as fast," the narrator said.

The other was a parody of Intel's own ads, which featured dancers in "bunny suits"—clean-room suits worn by technicians in chip-fabrication facilities—dancing to disco music. Even if viewers didn't recognize what those outfits were, the ads seemed to suggest that Pentium had pizzazz.

Apple's response ad began with a bunny-suited Intel technician being blasted by a firefighter with a fire extinguisher, as "Disco Inferno" plays ("Burn, baby, burn!"). The technician looks down sadly at the wisps of smoke coming off of their appendages. "Apple Computer would like to apologize for toasting the Pentium II processor in public," says the narrator. "But the fact remains: The chip inside every Power Macintosh G3 is up to twice as fast." The technician tries to exit the frame with a dance move, but stumbles.

Mac fans loved the ads, but Intel was furious. Apple was basing its claim on speed tests that *Byte* magazine had conducted in its labs. Intel complained to the editors about its testing methods.

Byte stood by its findings.

Burn, baby, burn.

Technically, Apple had already filled in the lower-right square of the product grid, too, in November 1997, with the PowerBook G3. But now, after only five months, Jobs announced that Apple was replacing it with a different laptop—but keeping the name: PowerBook G3. It was the dawn of Apple's practice to maintain one name for successive Macs through the years.

The new machine was the fastest laptop ever made—by anyone. "It eats Pentium notebooks for lunch," Jobs said.

PowerBook G3 ("Wallstreet")
Sold: November 1998–January 2001
Price: $2,300 to $3,500
Processor: PowerPC G3 (233 to 292 MHz)
Memory: 16 MB, expandable to 192 GB
Equipment: 2 GB hard drive (up to 8 GB); 12-, 13-, or 14-inch screen; hot-swappable bay on each side, for any combination of batteries, floppy drive, Zip drive, CD, or DVD

The batteries had light-up charge-level indicators that worked in or out of the machine.

What the audience saw was one of three hand-built iMacs; LG's factory in Korea had not yet finished setting up its production line.

But now, he announced that he was going to fill in the third quadrant: the consumer desktop.

"Today, I'm incredibly pleased to introduce iMac, our consumer product." A slide appeared, bearing a big lowercase *i* next to five words: *internet, individual, instruct, inform, inspire.* It was the closest he ever came to explaining the name iMac.

"This is what computers look like today," he said, projecting a photo of a hideous-looking, wire-infested beige PC. "This—is iMac." He whisked away a black cloth.

The iMac G3 boasted a gorgeous 15-inch color screen, a 233 MHz PowerPC 750 ("G3") processor, 32 MB of memory, and a 4 GB hard drive. Stereo speakers faced forward, flanking the CD-ROM drive—and there were *two* headphone jacks, so two people (students in a classroom, for example) could listen to a CD together. There was even an infrared lens, for those rare occasions when you wanted to beam files to and from PowerBooks.

The Slot-Load Crisis

One day before the iMac's unveiling, Jon Rubinstein was meeting with a key supplier when Jobs's assistant broke into the room. "Steve needs to talk to you right now. *Right now!*"

The first iMac test units had just arrived from the factory. Jobs was on the Flint stage, rehearsing, and he noticed something about the CD-ROM drive: It was a tray-loader. When you pressed the blue button on the front of the iMac, a black CD *tray* slid out.

"What the fuck is this?!" Jobs exclaimed, enraged. He had expected a *slot*-loading CD, like the ones in high-end cars, where the disc slides smoothly in without a tray, naked and unsupported.

But slot-loading drives were twice the height of the tray-loading units and would have taken up too much space inside. Furthermore, in the march of progress, slot-loading drives were always months behind the tray ones—in getting thinner, in getting faster, and, eventually, in gaining the ability to burn CDs.

Rubinstein was sure Jobs had *known* that the first iMac would have a tray, but did his best to calm Jobs down.

"I'm only going to go ahead with the launch if you promise we're going to go to slot mode as soon as possible," Jobs said tearfully. Rubinstein promised. And sure enough: The third-generation iMacs (1999) came with slot-loading drives.

That decision, though, would cost Apple dearly in 2001. When CD burning became a thing, "we got caught with our pants down," Rubinstein says—because, of course, the first available CD burners were tray-loaders. Apple introduced iTunes, the software that let you make song playlists and burn them to homemade CDs—but "we couldn't release iTunes on the iMac, which was the product that really needed to have that. We had to release it on the [Power Mac] tower first."

During their many years of collaboration, Jobs and Ive occasionally chose designs that favored cool form over practical function. The slot-loading iMac was one of them.

Most important, the iMac had a 33.6 Kbps dial-up modem *built in*—a rarity in 1998, and one of the ways the iMac still bore the hallmarks of its origins as a network computer.

Onstage, a camera followed Jobs around the iMac. "The back of this thing looks better than the *front* of the other guys'," Jobs continued. "It looks like it's from another planet. A good planet. A planet with better designers!"

A slideshow revealed close-ups of design details. "Look at this mouse. It's the most wonderful mouse you've ever used," he said.

(It was not. In fact, the "hockey puck" mouse became notorious as "quite simply the worst mouse ever made," as ZDNET put it. It was too small to support your palm, too round to know by touch when it had rotated out of alignment. Accessory makers leapt into the void with more traditionally shaped mice, better-shaped shells that snapped over the puck, and ADB-to-USB adapters that accommodated Apple's older mice.)

When the iMac hit the stores three months later, the reviews were mixed. Critics didn't know what to make of its radical shape, comparing it to a gumdrop, the Volkswagen New Beetle, or the AMC Gremlin.

"I was able to set it up in just five minutes, and I was cruising the Internet about 10 minutes after that," wrote the *Wall Street Journal*'s Walt Mossberg.

"Most of the consumers Apple is trying to appeal to live in a world where floppy disks are important," said the *New York Times*'s critic.

And the *Boston Globe* asserted that "the iMac might have done Apple a world of good if it had appeared a few years ago, when far fewer homes had PCs in them. Today, the iMac will only sell to some of the true believers. [It's] clean, elegant, floppy-free—and doomed."

The iMac Era

The iMac had gone from concept to availability in an astonishing 13 months. By any measure, the result was a revolution. Next to ordinary Windows PCs—ugly, modular, beige, strangled by cables—the iMac looked like a blue bullet from the future.

Apple received 150,000 orders before it even shipped. In the first year, Apple sold nearly two million iMacs; in the first three years, five million of them. It was the bestselling computer in America. It was the fastest-selling computer in history. It was on the "Best of the Year" lists of *Time, Wired, Fortune, Newsweek, USA Today, Consumer Reports, BusinessWeek,* and *Popular Science*.

Apple's research showed that two-thirds of buyers got onto the internet the very day they turned on their iMacs. A third of the sales went to people who'd never bought a computer before; 12 percent were people switching to the Mac from Windows. Apple's long-suffering market-share number leapt from 2.4 percent to 5 percent. Apple's stock hit $40, its highest value in three years.

Apple spent $100 million to market the iMac. Chiat/Day's billboards in big cities featured a photo of the iMac bearing only the words "Sorry, no beige." (What would Jerry Manock think?)

In one TV ad, seven-year-old Johann Thomas and his border collie race against Adam Taggart, a Stanford MBA student, in unboxing a new computer and getting online fastest. Johann finishes in eight minutes; Taggart, fumbling with an HP Pavilion 8250 PC, is still plugging in cables as the commercial ends.

Within months, translucent blue plastic products flooded the market: cordless phones, game controllers, blenders, toasters, clock radios, boom boxes, staplers, tape dispensers, electric toothbrushes, and hair dryers. Even IBM and Dell began offering laptops with colorful shells.

But as Jobs told *Newsweek*'s Steven Levy, "The iMac isn't about candy-colored computers. The iMac is about making a computer that is really quiet, that doesn't need a fan, that wakes up in fifteen seconds, that has the best sound system in a consumer computer, a superfine display. It's about a complete computer that expresses it on the outside as well. And [competitors] just see the outside. They say, 'We'll slap some color on this piece of junk computer, and we'll have one, too.' And they miss the point."

Descendants

The very day after the iMac shipped, Jobs directed his teams to begin work on its successor. They announced the second-generation iMacs only four months later, at the January 1999 Macworld Expo.

Apple's consumer surveys had revealed a surprising/unsurprising fact: When asked which aspect of a computer was most important to their buying decisions, consumers said that it was the *color*. So these January 1999 models came in a choice of colors. "Collect all five," Jobs joked.

For Jony Ive, the second round was a chance to address the things that had bugged him in the original. Now, for example, he had the chance to tidy up the layout of the internal components.

A week after announcing the five color iMacs, Apple announced its earnings for the holiday season. They had tripled from the previous year.

Ive and Rubinstein pose with the second-generation iMacs: Strawberry, Blueberry, Grape, Lime, and Tangerine.

iMovie

Only nine *more* months later, in October 1999, Jobs returned to the Flint auditorium to reveal a *third* generation of iMacs. Their smaller cases offered a clearer view to the internals; the metal shielding that had hugged the inner surface was now gone. So was the internal fan. (Apparently, Jobs hated fans.)

The new iMacs also had, at long last, the slot-loading CD drive. Jobs, onstage, was so excited about it that he inserted and removed a CD in his demo *twice*. "*I* think it's really cool," he said.

But the real grabber was the price. It was Apple's first $1,000 Mac in nine years.

There was a "one more thing," too: a $1,300 model called the iMac DV. It added a video-output jack (for monitors or projectors); a DVD drive (instead of just CDs) that could play movie DVDs; and a new, high-speed data jack called FireWire.

Apple had been developing FireWire with Sony, Panasonic, Philips, and six other companies for 13 years. The partner companies had been including a FireWire jack on every digital camcorder since 1995. You could hook up your camcorder to the iMac DV, import footage, edit it, and play

Because iMovie 1.0 was free and drop-dead easy to use, it ignited an explosion in video-editing popularity.

> ### Glenn's Medal
>
> Jobs hired Glenn Reid, a veteran of Adobe and NeXT, to write iMovie. Reid had zero experience with digital video, but Jobs somehow sensed that he'd be perfect for this app.
>
> Reid soon discovered, however, that QuickTime, the crown jewel of Apple's digital video empire, did not recognize the DV (digital video) format spoken by DV camcorders. To import camcorder footage, QuickTime had to bring in audio and video separately, in analog format, and then marry them together.
>
> Reid's second-in-command, Joe Holt, solved the problem by writing a new, modernized replacement for QuickTime. Now the iMovie prototype could import digital audio and video in real time, without dropping a frame; you could even watch the video coming in, full-screen. No video app, including Adobe's, had ever managed that feat.
>
> Then one day, Reid got called into Jobs's office, where he found Phil Schiller and other executives waiting. "So I understand you're not using QuickTime," Jobs said. "And that's a big fucking problem."
>
> "Well, that's true," Reid said. But when he explained why, Jobs's demeanor changed.
>
> "Is that true?" Jobs asked the QuickTime manager. "That you can't import video and audio at the same time?"
>
> "Yeah, that's basically true," was the reply.
>
> Suddenly, Reid was no longer the target of Jobs's annoyance.
>
> "You know what I think? I think *this* guy deserves a fucking medal!" he said, pointing at Reid. Jobs berated the QuickTime manager for another five minutes before ending the meeting.
>
> A few weeks later, Reid was summoned to the human resources director's office—never a good sign. But as he entered the room, the director whispered to him, "Don't worry."
>
> To honor Reid's subversive ingenuity, Jobs had orchestrated a little medal ceremony. He presented Reid with an actual, physical medal that he'd had made, which hangs on the wall of Reid's office to this day.
>
> The words engraved across its face say: "Fucking Medal."

back a professional-looking home movie. Never again would friends have to sit through interminable showings of unedited vacation footage.

For Jobs, the idea of using the Mac as a video-editing station carried overwhelming marketing logic. In 1998, the tech industry was still infatuated with the concept of network computers: cheap computers that relied on the internet instead of a hard drive. The iMac, of course, had narrowly missed becoming one itself. But now that every Mac contained a hard drive, Jobs felt compelled to reinforce the importance of having storage built into your computer. And you *definitely* needed a hard drive if you wanted to edit videos.

But to perform that editing miracle, you'd need software. Apple's old friend Adobe was already selling a program, only for Windows, called Premiere—and declined to make a Mac version of Premiere. Jobs was furious. Apple had *launched* Adobe! "I put Adobe on the map, and they screwed me," he said.

Jobs never again wanted to feel beholden to another company to reach his objectives; Apple would have to make its own video app. Besides: This way, Jobs could ensure that its app was clean, beautiful, and drop-dead simple.

Glenn Reid wrote iMovie with two colleagues in nine months. It was the simplest video-editing software imaginable—a "crap-removal tool," as Reid called it. The app was perfect for removing all the footage that some dad shot accidentally by forgetting to hit Stop, or a parent yelling at their kid.

Jobs demonstrated iMovie's simplicity onstage by creating a complete, edited scene in a matter of minutes: kids soapily washing a dog on the lawn, with a title ("Ruffy's Bath") and sweet guitar music underneath. The demo made it clear: For the first time in human history, *everyone* would be able to make movies. The audience gave him what was likely the longest ovation ever given for a guy working at a computer.

The Millionth iMac

As the day approached that the actual one millionth iMac would roll off the line, Jobs had what he thought was a great marketing stunt: He'd put a golden ticket inside the box, just like Willy Wonka in *Charlie and the Chocolate Factory*. Whoever found the ticket would get the iMac for free, fly to Cupertino for a tour of the campus, and get to meet Jobs in top hat and tails, à la Wonka himself.

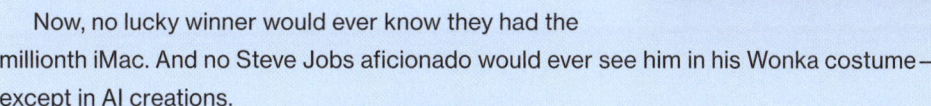

This was going to be a headline-generating moment, and Jobs was all in. (His enthusiasm for this random act of delight was reminiscent of the "Mr. Macintosh" episode of 1982.) Apple's designers created the actual golden ticket.

Unfortunately, according to California law, every sweepstakes must have a "No purchase required" stipulation. And without the option to hide the golden ticket in an actual iMac box, the whole idea fell apart. Apple's lawyers shut the idea down.

Now, no lucky winner would ever know they had the millionth iMac. And no Steve Jobs aficionado would ever see him in his Wonka costume—except in AI creations.

"Imagine the Steven Spielbergs of the world, being able to use this technology when they're kids," Jobs said. "It's the convergence of five technologies: iMac DV, FireWire, QuickTime, the amazing iMovie software, and digital cameras. And four of these five are from Apple."

In July 2000, there would be a fourth round of iMacs—now in Indigo, Ruby, Sage, and Snow. Each would come with a sleek, transparent, oblong optical mouse; the hockey-puck mouse was dead. "We have many faults, but we do listen," Jobs said.

The Indigo model sold for $800: the lowest-priced Mac in history.

In these first two years of the iMac era, Apple induced two self-fulfilling prophecies. First, the iMac turned USB into a universal, ubiquitous standard. Second, despite all the grumbling, the public soon accepted Apple's removal of the floppy drive. It *was* kind of ancient and small, wasn't it? Soon, the floppy began disappearing from Windows PCs, too.

Apple's directors, delighted with the iMac's success, begged Jobs to lose that "interim CEO" business. At one point, they offered him a million shares of stock, and six million options, if he'd agree to become Apple's permanent CEO.

Jobs continued to refuse. "This is not about money," he said. "I have more money than I've ever wanted in my life."

31. Keynotes

Today, every tech executive announces products the same way: on a stage, live-streamed, usually stiff and scripted. But in the back of their minds, they're all trying to be Steve Jobs.

Jobs never used a script or a teleprompter. He always seemed to be speaking off-the-cuff—but in fact, each presentation had been dissected, polished, and perfected over many months. "I've never seen anyone else who could transfix an audience, whether it was in front of one person or thousands, tens of thousands of people," remembers Jon Rubinstein. Jobs built each slideshow himself, using graphics he requested from the engineering and design teams, in a slideshow program called Keynote that NeXT engineers had originally written (then called Concurrence) just for him, and rehearsed in an Infinite Loop 3 replica of the auditorium.

Upon arrival at the actual venue, Jobs dress-rehearsed his entire presentation twice a day, start to finish. Even so, Jobs had crippling stage fright. He could rarely sleep the night before a keynote. His team installed a porta-potty just offstage for his exclusive use, which he used before and after each of his segments onstage.

Sculley, Jobs, Bud Tribble, and two product team members prepare for a presentation in 1984.

Virtually nobody outside his team knew what he was going through. Onstage, he was a natural, persuasive speaker—and, often, hilarious. At the WWDC 2002 keynote, for example, Jobs's job was to persuade software companies to write their programs for the new Mac OS X operating system. "Anyone else would've just said, 'Okay, on this date, we're getting out of Mac OS 9 and moving to Mac OS X. You gotta get on board,'" Phil Schiller says. But Jobs had a much more memorable idea.

The sound of an organ boomed in the hall: Bach's famous, gloomy Toccata and Fugue in D Minor. On the screen, a towering stained glass church window appeared. As thick white fog rolled across the stage floor, an actual casket rose from a trapdoor that Apple had built into the stage. "We literally had a person under the stage running a dry-ice machine," Schiller remembers.

Jobs strode over to the casket, opened its lid, reached inside, and propped up its occupant so that the entire audience could see: It was a gigantic, four-foot-tall boxed copy of Mac OS 9.

Jobs began to read solemnly from a sheet of paper. "Mac OS 9 was a friend to us all," he began. Yes, it was a eulogy. For an *operating system*.

"He worked tirelessly on our behalf, never refusing a command, except occasionally when he forgot who he was and needed to be restarted," Jobs went on. "He's now in that great bit bucket in the sky, no doubt looking down upon us with that same smile he displayed every time he booted. Mac OS 9 is survived by his next generation, Mac OS X—and thousands of applications, most of them legitimate."

No developer could miss the message: Mac OS 9 was dead to Apple.

Jobs closed the casket, laid a single rose on top, and gave a little wave as the casket sank back down into the stage floor, to applause and laughter. Not a soul could have missed the point.

iBook

To Jobs aficionados, the July 1999 Macworld Expo was Peak Keynote.

"This is going to be a great Macworld," he began, stalking the stage in his customary black turtleneck, Levi 501 jeans, and New Balance sneakers. "We've got some great new products—some really great new products. Some *insanely* great new products. Some really, totally, *wildly insanely* great new products. We have got . . ."

The audience lost its mind—because it was not Steve Jobs at all, but an impostor. It was Noah Wyle, who had just *played* Steve Jobs in the 1999 TV movie *Pirates of Silicon Valley*.

The real Jobs barged onto the stage in mock indignation. "That's not me at all! You're blowing it! You're supposed to come over here, open a water, get the slide clicker, *then* you can put your hands together. This 'insanely great' thing—we stopped using that a hundred years ago!"

The crowd roared.

The real Jobs then offered an update on the company's health. Apple had just cleared $200 million, its seventh consecutive quarterly profit. It now had $3 billion in cash. And since the iMac's stunning debut, 4,000 *new* apps had come to the Mac.

Finally, he filled in the last remaining square of the 2 x 2 product grid that he'd announced a year earlier: the consumer laptop square.

It was an iMac to go: a two-toned laptop, almost round; there wasn't a single straight edge on this thing. A spring-loaded handle along the hinge snapped discreetly out of the way when you released it. Most of the lid, and all of the interior, was white polycarbonate (bulletproof) plastic; the bottom and sides were rubberized blue or orange, for grip. Later, Apple would offer Indigo, Graphite, and Key Lime options.

iBook

Sold: July 1999–September 2000
Price: $1,600 or $1,800
Processor: PowerPC G3 (300 or 366 MHz)
Memory: 32 or 64 MB
Equipment: 3.2 or 6 GB hard drive; 12-inch screen; tray-loading CD-ROM; optional wireless card

The iBook's round power adapter offered grooves for wrapping the cable—a first in the industry.

One Giant Leap

As the iBook unveiling approached, Apple's demo team invited Jobs and his marketing execs to their lab in Infinite Loop 3. "Okay, Steve," one of them said. "Look out the window!"

An unseen staffer on the roof lowered a four-foot doll of Flik, the blue ant from *A Bug's Life*, on a bungee cord. The guts of an iBook were duct-taped to the doll; in its USB jack was a g-force sensor. When the staffer threw the doll off the roof (shown here in an AI reimagining), a PowerBook in the lab displayed a real-time graph of the g-forces as Flik bounced on the cord. The point was to show, as dramatically as possible, how AirPort could transmit data in real time.

"This will be great for the keynote!" Jobs said. "Except we're not gonna use a doll." He looked at Schiller. "Phil, you're gonna do it. Trust me—this will go in the demo hall of fame. Just do this!" Schiller reluctantly agreed, on one condition. He told Jobs he wouldn't sign any release of liability.

At New York's Javits Center, the production team built a 25-foot platform at the side of the stage. An air bladder of the type that stuntpeople use would break Schiller's fall. The weekend before the show, a stunt coordinator, who'd worked with Mel Gibson and Arnold Schwarzenegger, taught Schiller to fall safely and dramatically, kicking his legs for effect.

On the day of the event, a record crowd chanted for the show to begin. "Steve! Steve! Steve!" But word of the stunt had spread—and at that moment, the Javits lawyer burst backstage, took one look at the platform, and shouted, "*No!* No fucking way!"

The Apple team pushed back. "We have to do it! This is our big introduction!"

"Then you're going to sign a release form," the guy said. "The Javits is not going to be liable for this." Schiller reiterated that he would not sign a waiver. "Well, then, I guess you don't have a show," said the Javits guy. "Do not let this show start," he told the stagehands.

The product team sent for Apple chief counsel Nancy Heinen. Fine, she said—*Apple* would sign a waiver. She pulled out her business card and wrote one sentence: "Apple releases Javits Center." Reluctantly, the Javits guy let the show begin.

When the time came, Schiller stood on the platform, clutching the iBook, its motion sensor transmitting to the big screen—and leapt, kicking his legs all the way down. The crowd loved it. "Thank you," Jobs said to Schiller, clapping him on the back. "You know, nothing is beyond the call of duty at Apple."

> **Leaving Macworld**
>
> The Macworld Expo conferences alternated on the East and West Coasts, six months apart, meaning that Apple had to have something new to announce twice a year. The annual January show in San Francisco, in particular, made life miserable for Apple staff because it meant working straight through the holidays. So many Apple engineers spent so much time away from their families that their bosses implemented a program they sardonically called the Divorce Avoidance Program, which involved either emergency time off or $10,000 checks if a project made its deadline.
>
> After the January 2009 Macworld, Jobs finally pulled Apple out of Macworld. From now on, Apple would stage its own product-launch keynotes, on its own timeline.

It also bore a solid-colored Apple logo. On that day, Apple retired the multicolored, striped logo.

"So that is iBook," Jobs said finally. "But there is one more thing." And then he pulled off one of the greatest, funniest, most joyous keynote stunts of his entire career.

The Stunt

Jobs started surfing the web at the podium. "I'm gonna go to CNN Interactive here and see what's on CNN. . . . And maybe I'll go to Disney here. . . ."

The audience was confused. *This* was his grand finale? *This* was his "one more thing"—browsing web pages?

Jobs as magician.

And then Jobs picked up the laptop. "Why don't you come on over here?" he said to the cameraman over his shoulder, strolling casually across the stage. A realization ripped through the crowd like voltage. Laughter and joy broke out as the magic trick hit them: *There were no cables!*

"Oh, you noticed something!" he said.

The iBook was online wirelessly.

Just as the ovation hit its peak, Jobs grabbed a magician's hoop and passed it over the iBook. Apple had just introduced the world to Wi-Fi.

The Making of Wi-Fi

Hard though it may be to imagine, there was no public internet until 1993—and there was no Wi-Fi for the masses until the iBook.

In the corporate world, computers connected to the internet over ethernet cables. Professional installers wired each computer to a router in a utility closet through holes in the walls and floors.

But Apple's biggest markets were homes and schools. You couldn't wire schools because the walls were full of asbestos. You couldn't wire homes because who's going to want to open up their walls?

The iBook engineering team considered every networking system available: infrared, Home-PNA (networking over the building's existing phone wiring), PowerLine (over the existing *electrical* wiring). All had serious limitations.

Jony Ive's flying saucer design played off the AirPort's sci-fi premise.

Meanwhile, a group of engineers from the recently disbanded Advanced Technology Group had been looking into a futuristic option: networking over radio waves.

In 1986, NCR (National Cash Register Co.) wanted to use a radio signal to connect its department store cash registers to the back office—without drilling holes in walls and marble floors. By 1996, engineers from NCR and other companies had formed a subgroup of the IEEE (Institute of Electrical and Electronics Engineers) in hopes of hammering out a standard. If every manufacturer adopted the same technology, then every computer could speak the same wireless language. Someday, you might even be able to get onto the internet wirelessly in airports and hotels!

By early 1998, the IEEE was still working on the wireless standard, but its member companies couldn't wait. They began selling wireless networking gear, even without a settled standard—with

The Guy in the Aisle Seat

In the fall of 1997, Jobs planned to unveil the first Macs with PowerPC 750 (G3) chips. To prove that the Power Mac G3 was the fastest desktop ever sold, he conceived of a race against the fastest Windows PC on the market. The computers would run identical Photoshop automation scripts, applying the identical 30 image-processing filters to the identical photo.

Less than a week before the event, Jobs asked Phil Schiller, newly installed as VP of product marketing, and PowerBook line manager Greg Joswiak to show him the demo. It was their first time working with Jobs. They proudly showed him the speed test: Sure enough, the Mac scorched the PC.

Unfortunately, what you get when you apply every Photoshop filter to a photo is, as Joz calls it, digital mud. "You run every single filter, it destroys the image!"

Jobs stared at the two young marketers. "I'm sure you're good at a lot of parts of your job. This isn't one of them," he said coldly. "It looks like *shit*! I want it to be beautiful! I want to see something great by Monday."

Joz and Schiller had four days to come up with a better demo. But they were not artists or Photoshop masterminds; they were marketing guys about to lose their jobs.

And then Joz remembered the guy on the plane.

Three months earlier, Joz had been stuck in a middle seat on a flight back from Boston. He'd started chatting with the guy on the aisle, whose name was Bert Monroy. "And when I asked him what he did, he basically explained that he was one of the world's foremost experts on Photoshop!"

Joz called Monroy and offered him a deal: If he'd spend the weekend at Apple fixing the Photoshop demo, Joz would give him a high-end PowerBook.

Monroy agreed, and wound up flipping the script. Instead of starting with a photo that Photoshop's filters systematically degraded, the Photoshop script would systematically *build* an image from a blank canvas: a gorgeous, detailed rendition of the Mac OS 8 poster.

Jobs loved it. But, ever the teacher, he wanted to impart a lesson. "Look, I want to tell you where I was coming from the other day," he said. "I once worked with a guy named Mike Markkula. And Mike knew a thing or two about marketing. He said that the old saying goes, 'Don't judge a book by its cover,' but everyone does. And so we're going to spend a lot of time not just on the book, but on the cover. And that's where I was coming from."

At a media event on November 11, 1997, Schiller presented the Photoshop race onstage. The Power Mac G3 finished building Monroy's OS 8 poster 24 seconds before the PC.

Years later, Schiller would become Apple's worldwide head of product marketing (and an Apple Fellow); years after that, Joswiak would succeed him.

"But without that call, without me being in the middle seat," Joz says, "I'm probably not here today."

nosebleed prices: around $2,000 for the transmitter, plus $700 for the receiver card on each computer.

Apple decided that wireless was the way to go—and that the AT&T spin-off Lucent had the best technology.

Jon Rubinstein and Tim Cook flew to Utrecht, the Netherlands, to negotiate a deal. Cook, who had steeped himself on every aspect of Lucent's costs for the wireless cards, made a brazen offer: Apple would give Lucent the deal to make the wireless cards—if Apple could have them for $50 apiece. That was *93 percent less* than Lucent's price for its own WaveLAN product ($700).

Also, Apple would need them in 11 months.

It was a jaw-dropping demonstration of armed negotiation. "It was basically a proctology exam," says Rubinstein. Lucent agreed to the terms.

The heart of Apple's new system was a $300 base station, wittily called AirPort, which connected to the internet with a dial-up modem inside—or over ethernet, if you were one of those rare broadband customers. Antennas inside the iBook lid picked up the base station's signal, and passed it to a receiver card ($100) that you could install under the iBook's keyboard.

Pursuing this project was an enormous risk for Lucent, since nobody knew how many iBooks Apple would sell. It was a risk for Apple, too, because it meant buying antennas for every iBook—and every subsequent Mac model—whether customers used them or not.

The iBook teemed with charming details. The power cord wrapped, yo-yo-like, around its own chrome disk charger. A two-stage hinge spring held the laptop open or shut like a wedding-ring box.

Of course, Apple had made sure that there were still reasons to buy the more expensive PowerBook. The iBook had no card slots, microphone, video-output jack, or floppy drive. The iBook was bigger and heavier (6.6 pounds) than a PowerBook G3, too. And not all the critics loved the radical shape. Some compared it to a toilet seat, a purse, or a makeup compact. "The only thing missing from the new Apple iBook is the Barbie logo," snarked *PC Magazine*.

But once again, the buying public had a different opinion. In one month, Apple sold 300,000 iBooks, which were a hit with schools. In the first three days, Apple sold nearly half a million of the $100 AirPort cards—so many that it had to pause taking orders. By the 1999 holidays, the iBook was the bestselling consumer laptop in the U.S.

Two months after the iBook keynote, the IEEE finally published the specs for its new wireless standard. Immediately thereafter, a group of manufacturers hired a branding company to come up with something catchier than the official name, IEEE 802.11b Direct Sequence.

Today, as everyone knows, the name they chose was Wi-Fi.

32. Millennium

At the January 2000 Macworld Expo, Jobs opened his keynote with a recap of Apple's historic feat of corporate turnaround. "We've reinvented almost everything about the way we run our business, about the way we market our products, about the way we distribute our products, about the way we engineer our products." He'd fired up the cult of Apple, become the subject of business school case studies, driven the stock from $14 a share to $102, and made Apple the place where everyone wanted to work.

In the holiday quarter, he said, Apple sold 1.35 million Macs—the most in Apple's history. "One Mac every six seconds," he noted.

He mentioned that Apple had invested $12.5 million in Akamai, whose technology reduced streaming video lag. It had gone public, making Apple's stake worth over $1 billion. "So far," Jobs said, "we have made more profits off internet streaming than any other company in the world!"

Next came a demo of a suite of internet tools Apple was calling iTools: iReview, a database of reviews *of websites*; KidSafe, a list of 50,000 sites deemed safe for children; iCards, beautiful emailed greeting cards; free email addresses ending in @mac.com; HomePage, for designing your own websites; and iDisk, a virtual 20 MB "hard drive" on your desktop that *actually* resided on the internet, for ease of backups and file sharing. (The only features that survive today are the free email address and the iDisk—now iCloud Drive.)

The Tech Bubble

In 1989, working on a NeXT computer, British computer scientist Tim Berners-Lee invented something he called the World Wide Web: a potentially infinite array of linked documents on the internet. He was working at CERN, the European Organization for Nuclear Research, and designed the web for global CERN scientists to access one another's research.

It was such a good idea that on April 30, 1993, CERN released the web to the general global public. A gold rush resulted. Investors poured money into anything with ".com" in their names—whether or not they had profit or even a business plan. Between 1995 and 2000, the tech-heavy Nasdaq index rose from 1,000 to over 5,000. Everyone was going to be rich!

Hello, Dave

As the nineties came to a close, humanity prepared for the Y2K apocalypse. They stocked up on food, water, and guns, bought generators, and withdrew cash.

In the early years of mainframe software, memory had been so expensive that programmers used only two digits to represent the *year* in date calculations. Nobody was thinking ahead to the year 2000, when we'd need four digits to distinguish 1900 from 2000.

On New Year's Eve 1999, the computers that ran the world's banks, utilities, power plants, factories, and security systems would roll over to year "00." Worst case: Credit cards would stop working, power grids would fail, cellphones would stop working, and chaos would prevail.

But for Mac fans, the "Y2K bug" episode was only another reason to feel smug. The Mac could handle dates from 30,081 BC to AD 29,940, which, for most people, pretty much covered it.

Steve Jobs, still new at the iCEO job, called TBWA/Chiat/Day's creative director Ken Segall. "Maybe we should do an ad about this," he said.

Segall proposed a parody of HAL, the creepily affectless computer in *2001: A Space Odyssey*. "It struck me that because HAL lived in the year 2001, he could actually look *back* at what happened in 2000," Segall says. "HAL could literally be the voice of experience."

The entire ad was a slow zoom in to HAL's famous red "eye," as we hear HAL speaking to astronaut Dave: "Hello, Dave. You're looking well today. Dave, do you remember the year 2000, when computers began to misbehave? . . . Only Macintosh was designed to function perfectly, saving billions of monetary units. You like your Macintosh better than me, don't you, Dave? Dave?"

After securing permission from director Stanley Kubrick, who'd created the HAL character, Segall's next challenge was reproducing the voice of HAL. The original actor, Douglas Rain, declined to participate; he was "an angry man," Segall says. "He resented that we were intruding on him."

A casting call turned up a spectacular audition tape from Tom Kane, the voice of C-3PO in the Star Wars video games. He nailed HAL's voice, right down to the trace of Canadian accent.

The Chiat/Day team built a flawless replica of the HAL computer's faceplate, which involved renting a $100,000 specialty lens to represent the red "eye."

Apple aired the ad on the Super Bowl during the first ad break after the kickoff. The next day, Stanley Kubrick called Jobs directly, congratulating Apple and its agency for getting HAL just right.

> ### OptionsGate
>
> From his return to Apple until his death, Steve Jobs accepted a salary of only $1 a year—but the board compensated him in other ways. When he dropped the "interim" title from the CEO job, they gave him a Gulfstream V jet and $20 million in stock options.
>
> When the tech bubble burst in 2000, Jobs's options became worthless. So in August 2001, he asked the board for another 7.5 million options to replace the first grant.
>
> But in the time it took to cancel the original grant, Apple's stock price went *up*. So in December 2001, chief counsel Nancy Heinen recorded the second gift as though it had been granted to Jobs in *August*, thereby making him whole.
>
> Backdating stock options isn't illegal; a board can put any date it likes on an options grant. But failing to disclose a grant *is*.
>
> In 2006, the Securities and Exchange Commission (SEC) had begun investigating the tech industry's practice of backdating options. When Apple's board did its own investigation, it found and reported some options irregularities—but noted that Jobs hadn't benefited from them, didn't even know what backdating was, and had actually returned one of the questionable grants. It was a big, complicated, high-profile mess.
>
> Jobs was never charged, but Heinen and CFO Fred Anderson paid a price for the scandal; both settled with the SEC without admitting wrongdoing.
>
> The following year, the iPhone came out, and OptionsGate became old, old news.

It all ended in March 2000, when investors realized that the emperor had no clothes. Tech companies' stocks crashed; the sillier ones, like Pets.com, Boo.com, and eToys, went bankrupt. In two years, the Nasdaq plunged by nearly 80 percent, wiping out $5 trillion in market value.

Apple was hit hard, too. Fewer people working in fields like graphics and advertising meant fewer people buying computers. Its stock fell from $53 to $14 in three months, and Apple reported a $45 million loss for its quarter ending in September.

Dell laid off 8 percent of its workforce. HP, 7 percent. Compaq, 15 percent. But Steve Jobs laid off nobody.

"We're going to innovate our way out of this," he said. "We're going to invest in ourselves."

His thinking was that all of this was just one big business cycle. Market slumps were perfect times to invest in new projects; once the economy bounced back, Apple would be sitting pretty. Jobs *hired* engineers to write new apps, to accelerate Mac OS X, to open new Apple Stores, and especially to work on laptops. Nobody ever accused Jobs of traditional business thinking.

Mac OS X

Jobs had originally announced that adapting the NeXT OS for the Mac might take a year or two. In the end, it took five years, for all kinds of reasons:

- **Technical difficulty.** Wrestling NeXTSTEP into something that would run on the Mac was a staggeringly complex challenge. It wasn't replacing a car engine; it was a brain transplant.
- **Politics.** When Apple bought NeXT, hundreds of Apple engineers had already spent years working on Copland. They had written their own modern kernel, called NuKernel, and did not intend to go quietly. "People had their own agendas," says Avie Tevanian, who led OS X development. "There were definitely people who thought we were still gonna use NuKernel. I had to fight that battle."

Classic, Carbon, and Cocoa

From the Rhapsody experience, Apple had learned how important it was to give software companies an easy way to adapt their apps to the modern new OS. For Mac OS X, Apple offered them a choice of three paths. They could:

Do nothing (Classic). Thanks to an emulator, existing apps would run fine, in their own simulated Mac OS 9–style bubble. They would not enjoy the pleasures of a rock-solid modern OS—its new look, crash protection, and so on—and if one app misbehaved, the whole bubble would crash. But at least you'd no longer have to restart your Mac.

Update the existing programs (Carbon). If programmers were willing to put *some* effort into getting with the OS X program, they could rewrite about 10 percent of their apps to gain most of the OS's advantages, using a set of tools called Carbon. Jobs assured developers that Carbonizing their apps would take only a couple of months. The resulting software would have the crash protection, the good looks, the cool-looking graphics. Apple itself used this option for AppleWorks, iMovie 2, iTunes, FileMaker, and the Finder.

Write new programs from scratch (Cocoa). As Mac OS X became a bigger hit, Apple assumed that developers would create new programs exclusively for it, using the NeXT development tools. These would be known as Cocoa apps. They looked like Carbonized programs, but felt smoother and more solid, and came with a bonus suite of features, like a special mini-window for finding and using your fonts, and sharper text on Apple's Retina (very high-resolution) screens. Apple rewrote many of its own programs as true Cocoa apps: iMovie 3, iDVD, Safari, iChat AV, iPhoto, TextEdit, Stickies, Mail, and so on.

Today, Cocoa apps are the only ones still running on Macs.

Over and over, engineers complained to Tevanian about the existing features that he was retiring, or new ways of doing things in the new OS. "The Mac faithful aren't used to that," they'd say. Tevanian always had the same answer: "We can keep making the two percent market share happy, or we can go after the other 98 percent. Which sounds smarter to you?"

Jobs often settled the battles unceremoniously. "This is the plan. If you don't support it, you have to leave." Sometimes, they did.

- **The Great Rhapsody About-Face.** Apple originally told software companies that they'd have to rewrite their apps from scratch, using a new programming language, to run on the new OS, code-named Rhapsody. Not a single developer committed to doing so. Why would they spend years writing programs for an untested OS almost nobody would be using?

  So in May 1998, Jobs renamed Rhapsody—the modern new OS would be called Mac OS X (pronounced "OS ten")—and offered a way for developers to *adapt* their existing apps instead of rewriting them anew.

  Software companies were much happier with this new approach. It looked as though the Mac, for the first time in 16 years, would finally have software as exceptional as the hardware.

The First Demo

The Mach kernel was the titanium heart of the new OS. But the user interface—the parts humans would see on the screen—was another challenge.

As was his fashion, Jobs was personally involved in every design decision. Every Tuesday afternoon, he met with the designers to hash out ideas—and he had lots of them. "At one point, Steve wanted to do all of our error messages as *haikus*," says Mac OS X design lead Cordell Ratzlaff. "We would all think, 'What is he smoking?' "

The OSes of the day, like Windows 95, NeXTSTEP, and even Mac OS 8, had gray color palettes; their buttons were sharp-cornered rectangles. Ratzlaff went the opposite way. "Where it's dark and gloomy, let's make it light and colorful. Where things are really rectangular, let's make them more organic and rounded and softer."

By July 2000, Jobs was ready to show off the fledgling Mac OS X at the Macworld Expo. The new interface design, called Aqua, looked clean and modern. Menus were semitransparent. Buttons and scroll bars looked like they were made of colorful, reflective, 3D water droplets. "One of the design goals," Jobs said, "was when you saw it, you wanted to lick it."

The Dock was a strip of icons hugging the edge of the screen for quick launching of your favorite apps, documents, and web pages. A new, multicolumn Finder view, carried over from NeXTSTEP,

The early version of Mac OS X had faintly striped window trimmings, which echoed the visual texture of the iMac.

let you view the contents of a deeply nested folder without leaving your starting point. You could view pictures, watch videos, play music files, and preview documents right on the desktop, without having to open their corresponding programs.

For long-suffering Mac fans, a minor part of the demo made the biggest impression. It was the Bomb app, a custom program designed to crash after five seconds. Jobs started a QuickTime movie playing (the trailer for *Mission: Impossible*)—and when the Bomb app crashed, the video kept right on playing. "When an app crashes in Mac OS X, it does not bring the system down, like we all are used to," Jobs said.

Memory protection had come to the Mac at last.

Jobs concluded the keynote by reminding the audience that Apple was the last company on earth that made "the whole widget." Only Apple, with its control of hardware *and* software, could have produced iMovie and FireWire, for desktop movie editing. Or AirPort wireless networking. Or the iTools internet features. "We don't have to get ten companies in a room to agree on everything to innovate," he said.

But before closing the keynote, there was, per tradition, "one more thing."

"After two and a half years," he said, "I'm pleased to announce today . . . that I'm gonna drop the 'interim' title."

The audience was on its feet, hollering, full-throated with joy. They wouldn't let Jobs continue. Only Jobs could have pulled off the greatest turnaround in business history—and now, he seemed to be saying, he wasn't going anywhere. "You guys are making me feel funny now," he said, trying to keep his composure. "I accept your thanks on behalf of everybody at Apple."

The Cube

Jobs's 2 x 2 grid of Macs was brilliant in its simplicity. So it was a surprise when, at the July 2000 Macworld Expo in New York, he announced that he was expanding it to accommodate a fifth Mac: the G4 Cube.

Reviewers described it as "gorgeous"; "sculpture"; "a work of art." *Macworld* called it "an ice sculpture with a block of platinum trapped inside." *Wired* went with "a toaster born from an immaculate conception between Philip K. Dick and Ludwig Mies van der Rohe." Jobs himself said it was like "a brain in a beaker."

And then there was this, from the *New York Times* architecture critic in full grad student mode: "Perhaps this is not, after all, a machine, but a box of emptiness, a chunk of force-field that has been captured from the event-horizon of a black hole and returned to earth, where its power to warp time, space, and gravity has been harnessed to serve consumer needs."

The design was peak Jony Ive. DVDs popped vertically from the top slot like Eggos. The power button was a touch-sensitive, self-illuminating *spot*, activated by skin. To indicate when the machine was asleep, a tiny LED light pulsated 12 times a minute, mimicking human breathing. Creating the effect was, Ive says, "so ridiculously hard." The design studio in Valley Green 2 had no darkened room, so designer Duncan Kerr spent hours "with a black sheet over my head, tuning the algorithm just to try to get the slope of the curve up and the slope of the curve down."

The Cube contained the miniaturized guts of a Power Mac G4.

To add memory, you flipped the machine upside down, where a pop-out handle lifted the guts out of the acrylic shell. Impressively, the Cube had been engineered not to require a fan. (Jobs hated fans.) It was a perfect complement to Apple's flat-panel Cinema Display, which carried power, USB, *and* video over a single cable. (Jobs hated cables.)

Ive relied on a few hacks to reach the Cube's diminutive size. First, the power supply was not built into the machine itself; it was an enormous metal-clad brick on the cord. Second, there was no built-in speaker; you got a pair of spherical, clear-acrylic Harman Kardon external speakers. Third, the ports were on the bottom (USB, FireWire, ethernet, modem, video out, cable), which meant that you had to turn the computer onto its side to plug or unplug.

The Cube may have been the most beautiful personal computer ever made—it's in the permanent collection of the Museum of Modern Art in New York City—but it was, in Cook's words, "a spectacular failure commercially." In the fourth quarter of 2000, while Apple sold 308,000 iMacs, it sold only 29,000 Cubes. Apple dropped the Cube's price by $500 until all of them were gone.

The Cube had ignored the wisdom of the Quadrant: It didn't have an obvious audience. Consumers were much more attracted to the iMac, which cost $1,000 less; pros naturally went for the Power Mac, which had slots and *still* cost $200 less.

Still, "I would argue that the Cube was successful in a lot of ways," says Jon Rubinstein. "We learned how to do clear plastics. We learned how to do cooling with no fans. So: commercial failure, successful product."

Some of the lessons of miniaturization paid off almost immediately, in fact, with the "sunflower" iMac G4 (2002), which concealed all of its computery guts inside the dome-shaped base. Then, in 2005, Apple introduced the Mac mini: a *really* tiny square box (2 inches tall, 6.5 inches square) that, once again, contained all the components of a real Mac. It was a BYODKM machine: Bring Your Own Display, Keyboard, and Mouse.

This weird, off-kilter little machine was a side hustle for Apple. But at $500—the least expensive Mac in history—it became a cult hit. People grabbed it as a second Mac, a home-theater hub, a cheap software development machine, a server for websites. Over the years, Apple redesigned it a few times; once CDs went away, it got even smaller. It shrank to 5 inches square, a solid block of aluminum—with rounded corners, of course.

*The Mac mini made a perfect second Mac—
or a tempting first Mac for Windows users.*

33. Retail

In February 2001, at the Tokyo Macworld Expo, Jobs unveiled a fifth generation of iMacs. Now they had *patterns* on the case: white puffy dots against a blue background (Blue Dalmatian) or a psychedelic floral pattern (Flower Power).

The truth was, the iMac wave had crested. Sales growth was cooling off. "I mean, you know you're stretching it when you're putting psychedelic flowers on the outside of a computer, and that's your Macworld announcement," says Ron Johnson.

Johnson, as home products director of Target, had helped it shift to "cheap chic" in the 1990s by striking up collaborations with hot designers: Michael Graves, Philippe Starck, Todd Oldham, Isaac Mizrahi. From 1980 to 1999, the company's sales grew tenfold, and it earned a pop-cultural nickname: Tarjé.

Apple, meanwhile, hadn't fixed the "in-store advocacy" problem. The poorly paid, poorly trained salespeople at most big-box stores still knew nothing about Macs, leaving them broken and dirty on the shelves.

In 1998, Apple's new operations chief, Tim Cook, had had enough. He ended Apple's contracts with Best Buy, Circuit City, Computer City, Sears, Montgomery Ward, and OfficeMax.

The CompUSA stores-within-stores weren't the solution, either. Once the initial excitement over the program died down, they sank into neglect. Local Mac fans did what they could, sneaking into CompUSA stores to "volunteer," spending each Saturday answering questions, and fixing icons that customers had renamed "@&#^(@#*&." But even so, Macs still weren't getting the in-store respect they deserved.

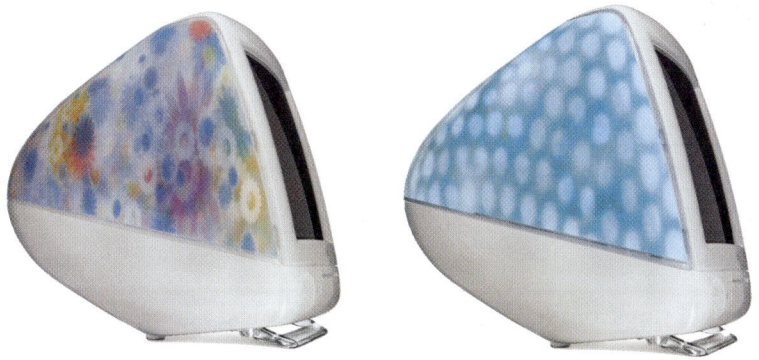

The 2001 iMacs were less transparent, more patterned.

Apple had opened its online store in December 1997 (built with WebObjects, of course). But only 40 percent of Americans had internet access, and for many of them, shopping online was still a frightening prospect.

Steve Jobs decided that Apple would have to open its own stores.

The idea struck most people as insane. Retail was a massively risky, complex enterprise. It involved mastery of real estate, rents, architecture, zoning, staffing, pricing, policies—Apple had no experience in any of it. Furthermore, the idea had already been tried. Gateway opened its first of 300 stores in 1997. They were a money-losing disaster; by 2003, they had all closed. HP and IBM had experimented with stores, too. None of them took off.

But Jobs had few alternatives. "Apple was not gonna succeed without its own stores," says Johnson. "They had run that experiment for 25 years! If the stores didn't work, Apple wouldn't work."

And so, in 1999, Jobs invited Gap CEO Mickey Drexler to join the Apple board. In 15 years, he had transformed the Gap from a sleepy regional $400 million business to a $14 billion powerhouse. Like Jobs, Drexler was a control freak and a stickler for attractive presentation.

Now Apple needed an executive to create the stores. When they met just before Thanksgiving 1999, Ron Johnson and Jobs hit it off. "All of a sudden there's a torn pair of jeans and turtleneck, and he's off and running about why he needed great stores," Johnson says.

Ron Johnson
Born: 1958, Minneapolis, MN
Schooling: Stanford, Harvard Business School
Before Apple: Target
Apple: 2000–2011
After Apple: JCPenney, Enjoy Technology

Johnson spends equal time in the U.S. and France, where he maintains a second home.

Johnson wasn't concerned about repeating Gateway's fate. Those stores were just showrooms; they didn't actually *sell* anything. Customers were expected to order their PCs online and then return to the stores to pick them up. What Jobs and Johnson spent the next two years creating was totally different.

- **Each store would stock and sell the product.** Gateway had a good reason for using the showroom model: Managing inventory across hundreds of stores was a *nightmare*. How do you forecast, ship, track, and stock eight different colors of iMac—in every store?

But Jobs was insistent that the Apple stores should carry inventory. You should be able to buy your Mac and then immediately get help setting it up.

- **Each store would do repairs.** Trained technicians would fix your machine right in the store, on the spot or overnight. This idea, too, was controversial. Wouldn't a visible repair center imply that the product itself might give you problems? But Johnson wanted to convey that the company was standing by to help.

- **It would be hands-on.** In 1999, fewer than half of Americans owned computers; even fewer were on the internet. Apple Stores, therefore, were designed to be internet cafés. People could *try* computers and find out what the web was all about, with staffers on hand to guide them. Many people tried the internet—or even computers—for the first time in Apple Stores.

    Of course, leaving unattended tables full of online Macs to the public exposed them to idiots. They'd open porn sites, mess up the desktop, create dumb documents. So Apple developed software that automatically wiped every Mac clean every night and rebuilt it with a fresh install of the OS.

- **There would be geniuses.** Johnson loved great hotels, and especially the friendly bartenders in the hotels' bars. "So what if we built our service [department] in a bar, and we dispensed advice instead of alcohol?"

    Inspired by the parade of geniuses in Apple's "Think Different" ads, Johnson came up with a great name: Genius Bar. That name would also make it easy to attract the best technical staff in each city. What would-be employee didn't want to be known as a genius?

    "But what if our genius can't answer the question?" asked a team member at an Apple off-site. Johnson's solution was to create a sort of Batphone hotline, a bright red phone that connected geniuses directly to Apple's own engineers in Cupertino. (The red phones no longer exist, but in-store geniuses can still call Cupertino engineers.)

- **The store would be huge.** At the time, Apple had only four products to sell: its four Mac models. "The entire Apple product lineup at that time would fit on a conference-room table," Johnson says. But if the store was small, it would make *Apple* seem small. Instead, he decided that Apple stores would resemble the Gap stores: clean, open, airy floor plans, white walls, wooden tables—and at least 6,000 square feet of space.

    The front of the store would *sell* Macs. The back would be dedicated to customers *after* they'd bought their Macs: special areas for learning about photos, music, and movies; tables of camcorders, MP3 players, and Palm organizers to try out; a theater with a projection screen for free classes; kids' tables for trying games; and, of course, the Genius Bar.

The Apple Store on Regent Street, London, was the first Apple Store in Europe.

- **It would be gorgeous.** Apple hired top architectural firms like Bohlin Cywinski Jackson and, later, Foster + Partners, to create stunning store designs. Some would have two levels, connected by glass staircases or cylindrical elevators; glass bridges connected some sections. (Jobs was obsessed with glass staircases. His name even appears as the lead inventor on two patents for them.)

  Apple spared no expense. The glass came from a German specialty fabricator. The stainless-steel staircases were bead-blasted near Tokyo. The blue-gray marble flooring of second-generation stores came from a single quarry in Italy. Jobs required the marble supplier to lay out the entire floor's worth of slabs for each store, to make sure that there were no visible shifts in the blue-gray hue of adjacent tiles, and to align the veins of the marble. Only then were the tiles numbered on the back and shipped to the U.S.

- **There would be teaching.** Everyone who bought a Mac was invited to a "New to Macs" class on Saturday mornings. If you needed more customized training, you could sign up for Apple One-on-One: $100 a year for *unlimited* private lessons. At its peak, the program had 1.4 million members.

  "The Palo Alto store alone had 2,000 members," Johnson says. "We taught the digital arts to more people in Palo Alto than Stanford University taught freshmen the liberal arts!"

- **It would hire interesting staff.** Johnson wanted *interesting* people, with their own ideas, hobbies, and outside lives. They should have the confidence to go toe-to-toe with Jobs, in case he ran into them. And they'd be so social and competent and charismatic, "the Ritz-Carlton would want to hire them to work in the hotel." Every store-manager candidate flew to Cupertino to meet with Johnson. Genius Bar staff spent an entire month in Cupertino for training.

- **Hourly pay.** At most retail stores, salespeople typically earn minimum wage, plus a commission on each product they sell. But commissions, Johnson felt, tend to make salespeople pressure customers. Furthermore, "it was never a belief that we would sell someone the first time they walk in the store," he says. "They were gonna come in four or five times." A commission system wouldn't work, since you might not encounter the same salesperson each time.

- **No registers.** Jobs hated the idea of a checkout stand; at one point, he even proposed prohibiting the use of cash altogether. In the end, the store had cash drawers, but they were hidden in the handsome wooden tables.

Apple was about to change a *lot* of retail variables all simultaneously, all for the first time. Jobs proposed setting up a few temporary stores and running them as a test.

Johnson refused. He didn't think he'd be able to attract top talent if they knew that Apple wasn't fully committed. He insisted on opening 25 stores in the first year, one every two weeks, and not stopping.

But at Drexler's suggestion, Johnson rented a warehouse on Bubb Road, about three miles from the Apple campus, to create a prototype of the store. "You have to feel a store," Drexler told him. "You've got to feel the space."

In the main space, 50 by 120 feet, a team of 15 began to hash out a design. "Okay, what kind of floor do we want? Metal? Wood? And what kind of wood? We'd get all these samples of wood. What color should the walls be? White. What kind of fixtures? What's the layout? It was really like an art project," Johnson says. Jobs and Drexler came over every Tuesday to inspect the progress. Jobs was, as usual, deeply involved in every decision. In one meeting, he spent half an hour choosing a shade of gray for the bathroom signs.

Once word leaked that Apple was working on retail, the reaction was swift and decisive. *BusinessWeek* ran an article called "Sorry, Steve: Here's Why Apple Stores Won't Work." It included a scathing critique by David Goldstein, president of a research firm called Channel Marketing: "I give them two years before they're turning out the lights on a very painful and expensive mistake."

Skepticism reigned within Apple, too. "This is crazy," board member Art Levinson said. "I'm not sure I can support something like this."

Board member Ed Woolard agreed. "Gateway has tried this and failed." Woolard had already grown weary of the Apple roller coaster, and this was the last straw. After four years on Apple's board, he resigned.

But Apple's fans loved the stores from the start. When Apple opened its first two stores in May 2001—in Tysons Corner, Virginia, and Glendale, California—over 1,000 people camped out overnight to be among the first admitted inside.

News helicopters flew overhead. Jobs and Johnson said a few words, and then threw open the doors. "Let's high-five everybody when they come in," Johnson told the staff. "Let's make it a moment!" The employees spontaneously applauded the customers as they entered, and handed out T-shirts.

Apple opened 25 stores in the first year, each averaging 20,000 visitors a week. (The Gateway stores had averaged 700 visitors a week.) The first New York City store, built in a former SoHo post office, got a thousand shoppers an *hour*. Within three years, the Apple Store chain had become the fastest chain in history to attain $1 billion a year in sales. Design patents and architectural awards poured in.

Over the years, the squiggly white tabletops gave way to handsome rectangular Parsons tables; the cash drawers gave way to iPod Touches and then iPhones, so that any employee could check you out wirelessly. The same person who'd helped you find what you needed could also check you out.

Today, sales staff don't spend time in Cupertino; they get a week of local training, and then a week of shadowing a current employee to learn the ropes. And these days, the 6,000 square feet that once seemed excessive isn't nearly enough. The days when Apple had only four products are long gone, and some of Apple's products, like the Vision Pro goggles and Apple Intelligence, require longer demos. Crowding is a growing problem. No longer can you just walk up to the Genius Bar without an appointment. Increasingly, people wind up ordering what they need online and then just picking it up from a dedicated area.

But by the end of 2025, Apple was operating over 530 stores in 27 countries. A typical store sells about $45 million of Apple goods a year; the flagship stores generate over $100 million a year.

Apple had finally licked the in-store advocacy problem.

34. iPod

Jobs was right about iMovie. Suddenly, video editing was something anyone could do—and every one of them would need a Mac to store it, edit it, or upload it. But what about all the other electronic gadgets that were flooding the market? Digital cameras, music players, and organizers all had the same problem. To be useful, they all needed a central computer.

And so, at his January 2001 Macworld keynote talk, Jobs announced a new direction for the Mac. It would become the digital hub for cameras, music players, and camcorders.

"For a few reasons. One, because it can run complex applications," he said. "Two, because the PC has got a big screen on it. That doesn't just mean you can see more information; it means we can make much better user interfaces. Most of these digital devices have pretty brain-dead UIs." Finally, the computer could get online, to make sharing your edited masterpieces easy.

iMovie was the first phase of the digital hub. iMovie, he said, "could make a digital device called the camcorder worth ten times as much. It's ten times more valuable to you. So what are we gonna focus on next?"

Burn

Jobs's gift for seeing the future didn't fail him often. But he did miss the CD-burning thing.

At the end of the nineties, two technologies radically transformed the recording industry: computers with CD-burning drives, and the internet.

A music CD is nothing more than a disc containing music data. Young music lovers began *ripping* their music CDs: copying the song files, compressing them into a compact file format—MP3s—and storing them on their hard drives. The conversion to the space-saving MP3 format degraded the audio quality slightly, but most people were happy with the tradeoff.

They could also upload the MP3 song files to websites like Napster, LimeWire, and Kazaa, which became global repositories of free pop music—*stolen* music, the record companies said. At its peak, Napster had 80 million members enjoying the thrill of free music.

Or they could burn (record) their MP3 songs onto blank CDs, creating their own playlists made of songs from different albums, for use in the car, stereo, or portable Discman player. "The music compiled the way *I* want, not the way some record company wants," Jobs said. "And people

> ### SuperDrive
>
> At the January 2001 Macworld Expo, Jobs also revealed that he'd been working in secret with Pioneer to develop a new drive that could read and record CDs—*and* read and record *DVDs*. Nobody else had this do-everything technology, which Apple called the SuperDrive.
>
> And what could you burn onto those blank discs? Well, files and folders you wanted to back up or share, for example. But what about your own home movies?
>
> Most people thought of DVDs as something you put into home-*theater* DVD players, to watch movies on your TV. But *creating* video DVDs was not something ordinary people could do. That process—designing interactive menu buttons, processing all video files—required specialized "authoring" software that cost $5,000 or more. Only movie studios did it.
>
> But Jobs loved nothing more than smashing down the barriers to creative expression. So on the very day he announced the SuperDrive, he announced a new, free, single-window, drop-dead-simple app called iDVD. (It was based on a DVD-burning app from a German company called Astarte, which Apple bought in April 2000.) You dragged your QuickTime and iMovie creations into place, where they became buttons. Now, with one click, you could burn the result onto a blank DVD, which would play on any of the nation's ten million home DVD players.
>
> Those kids today, with their instant uploads to YouTube! They'll never appreciate what life was like in the days of delicate 5-inch, silver-reflective CDs and DVDs. But in its day, iDVD was miraculous.

are doing this like crazy. They're making all these compilations. Drive-to-work music, workout music, this music, that music, hot-date music." In 2000 alone, music fans had bought 320 million blank CDs to hold their custom, ripped music.

By the end of 2000, 40 percent of new PCs came with CD-RW (rewritable) drives—but none of them were Macs.

In 2001, Apple fixed that lapse. Over the course of the year, Apple equipped every Mac model with a CD-RW drive: the Power Mac G4 in January, iMacs in February, iBooks in May, and PowerBooks in October. Now Macs would have the hardware to burn music CDs. All they needed was the software.

iTunes

Plenty of Windows programs could organize your music and burn playlists onto CDs: RealJukebox, Windows Media Player, MusicMatch, and so on. They were complex and homely.

Apple had no time to write a whole new music app from scratch. Jobs asked Sina Tamaddon, who was in charge of Apple applications, to seek an existing program to buy and adapt. In mid-2000, he found SoundJam, a Mac program that was designed to load music onto the Diamond Rio, the first commercially successful MP3 player. (It had ten buttons and held six songs.)

SoundJam's creators, Bill Kincaid, Dave Heller, and Jeff Robbin, were all former Apple engineers, and knew how to write elegant Mac software. Tamaddon bought their company and brought all three back to Apple.

Jobs set them up in Infinite Loop 1, directly across a skybridge from his own office, so that he could pop in often. Over the next four months, Jobs worked with them to carve their app into something much simpler. The revised app, now called iTunes, gained only one new feature: burning music CDs. The rest of the work was ripping *out* features that Jobs thought added complexity.

In the resulting app, with one click, you could sort your music collection by song, track length, artist, or album name. A single Search box found text in any part of a song's description: its name, band, or album. A screensaver mode filled your screen with psychedelic visuals that danced to the music.

TBWA/Chiat/Day created a TV ad in which a twentysomething young man takes a seat in an otherwise empty theater. On the stage is a huge cluster of famous musicians, a who's who of nineties pop: George Clinton, Lil' Kim, Iggy Pop, Ziggy Marley, and so on. The young man asks each musician for a song he loves ("How about something for, you know—the ladies?" he asks Barry White), and each musician responds submissively ("Now you're speaking my language," White replies).

iTunes 1.0 adopted the faux brushed-metal look of iMovie.

As music swells, the words "Rip. Mix. Burn." appear.

The ad struck some as cringey, with the world's most admired musicians awkwardly agreeing to serve the young man. But nobody hated it more than the record companies. They were already seeing their industry's life flash before its eyes. It appeared to them that Apple was promoting *stealing* music, and profiting from it, at their expense.

Jobs disagreed. "Ripping, of course, means you have a physical CD," he'd say. You bought it. You paid for it. You can do with it what you like.

The record companies and their trade group, the RIAA (Recording Industry Association of America), ran ad campaigns of their own. They attempted to add copy protection to CDs. They sued individuals who'd been caught sharing music files. They pressured Congress to regulate file sharing. And they continued to lose money.

Consumer Electronics

By 2000, Jobs's quadrant of Mac models was full. Apple was off life support. But Mac sales still represented only 5 percent of the market.

Jobs had long thought about branching out. If Apple lent its gift for elegance to electronics *beyond* computers, it could sidestep the whole zero-sum PC game.

He and Jon Rubinstein looked into digital cameras, organizers, camcorders, and cellphones. Video cameras weren't the answer; they weren't enough of a mass-market phenomenon. Cameras probably weren't it, either; how would Apple improve on the models already coming from Canon, Nikon, and Sony? "And besides, we thought cameras were gonna end up in phones anyway," Rubinstein says. "Same thing with organizers."

What about music players?

Everybody loves music. The Sony Walkman portable cassette players had proved that; over its 20-year run, people had bought 185 million of them. And in this field, there was obviously a role for Apple to play: The existing digital players were *terrible*. They were ugly and complex, studded with buttons, saddled with baffling software.

Rubinstein put together a team to investigate the technologies they might need to build a music player. The ingredients weren't hard to imagine: a battery, a screen, a processor, audio circuits, and storage.

Storage was the showstopper. The players of the day held your song files in one of two ways: in flash memory, with a capacity of only a handful of songs, or on hard drives, which held more songs, but were huge and heavy and prone to skips in the music when they got bumped.

For a moment, Rubinstein was encouraged: IBM had developed the Microdrive, an incredibly tiny spinning hard drive only 1 inch in diameter, intended for mobile devices. It could hold 340 MB of data (about 68 songs)—a big improvement over the memory-based units.

But when Rubinstein told the Microdrive team the price he'd be willing to pay for hundreds of thousands of the drives, "they kinda laughed at me," he says. IBM's answer was no.

Toshiba

In February 2001, Jobs, Rubinstein, Tim Cook, and Apple procurement chief Jeff Williams flew to Japan for the annual Macworld Tokyo conference, where Jobs was the keynote speaker.

Whenever he was in Japan, Jobs spent a few days visiting Apple's suppliers, even companies that weren't then working with Apple; the idea was to see what they were working on, to track upcoming new technologies. So on this trip, Jobs gave Phil Schiller an assignment: Meet with Sony about collaborating on this music player. Sony, with its expertise in miniaturized electronics, would make the hardware; Apple, with its successful QuickTime media technology, would do the software.

But Sony's head of consumer electronics shut down that idea. Sony, he said, was working on its own music players.

Meanwhile, Rubinstein and Williams were visiting Toshiba. "At the end of the meeting," Rubinstein says, "they go: 'Hey, we've got this thing. We don't really know what to do with it.'"

Toshiba had developed a hard drive the size of an Oreo, only 1.8 inches across and one-fifth of an inch thick. It was slightly wider than IBM's drive, but the capacity was insane: 5 *GB*. Since a typical song file was 5 MB, this drive could hold *a thousand* of them.

Rubinstein and Williams knew *exactly* what to do with it. They said nothing to Toshiba about a music player. But right there, on the spot, they offered to buy as many of those drives as Toshiba could manufacture. Toshiba even agreed to an exclusive—no other company would have this drive.

On the same trip, they found a great black-and-white screen that could show multiple lines of text—a big advance over the single-line screens on the cellphones of the time.

Rubinstein reported back to Jobs. "I can do this now. I need ten million dollars."

Jobs wrote the check, Fred Anderson confirmed that it wouldn't bounce, and the music-player project was underway. Its code name: P68 Dulcimer.

Fadell

Through the grapevine, Rubinstein heard about a guy who seemed to be ideal for leading the P68 project: a brash, energetic go-getter named Tony Fadell.

Fadell had learned the thrill (and exhaustion) of small-team, skunkworks engineering during four years at General Magic. At Philips, he worked on handheld computers, learning to build a product in a corporation with 300,000 employees. After a six-week stint at RealNetworks ("I could see the writing on the wall: I was going to hate that job"), he'd launched a startup to build a home-stereo MP3 player, only to see his funding vanish in the 2000 tech-bubble collapse.

And that's when he got a call from Rubinstein.

Fadell assumed that he'd be working on an organizer, maybe a Newton-type device. Only after he signed a contractor's nondisclosure agreement did he learn that Jobs wanted to build a pocket music player.

Tony Fadell
Born: 1969, Detroit, MI
Schooling: University of Michigan
Before Apple: General Magic, Philips
Apple: 2001–2010
After Apple: Nest, Build Collective, lead designer of Ledger Stax, author of *Build*

Fadell's first job was selling eggs door-to-door from his wagon in third grade.

He began work on a rough technical design. He bought screens and batteries, mocked up a tiny circuit board, and calculated how much space it all might take. As his six-week contract expired at the end of March 2001, he was supposed to show Jobs the fruits of his efforts.

Fadell figured that a physical mockup would be more illustrative to Jobs than drawings or 3D renderings. He cut up and stacked some foam board, printed graphics for the front and back, and, to give it some heft, embedded some fishing weights from his grandfather's tackle box.

Rubinstein had coached him on the technique of presenting to Jobs: You offer three options, but save the one you *really* like for last.

On the day of the big meeting, Fadell first showed Jobs a mock-up that was designed to use a removable memory card and AA batteries. The second one had built-in memory and a rechargeable battery. Finally, Fadell lifted up a wooden bowl to reveal his favorite: a rechargeable model containing the tiny Toshiba hard drive. Sure enough, that was the model Jobs loved.

Fadell's foam-board mockup.

He gave Fadell and Rubinstein the green light—and a deadline. He wanted the music player ready for the 2001 holiday buying season, now *eight months away*.

The P68 team suddenly found itself in one of the most intense time crunches in electronics history.

The Race

Fadell quickly joined Apple full-time. To him, getting the music player out by the holidays was important for three reasons.

First, because Jobs wanted it. Second, because Fadell still had PTSD from his time at Philips,

> ### The Scroll Wheel
>
> Fadell's music-player design had four navigation buttons, plus a round control pad below them. Pressing a Down button to scroll through a list of songs might have worked on a player that held only ten songs. But a thousand?
>
> "Steve, you are not going to sit there and press a button 800 times and go, 'I want to get to *Yes* there at the end of my list,'" Schiller told Jobs. "This doesn't work!"
>
> Only days later, Schiller was visiting the Bang & Olufsen store at the Stanford mall. As he studied the BeoCom 6000 cordless phone, he discovered an ideal manifestation of this idea: a spinning ring surrounding an OK button, with a few other buttons around its perimeter. The longer you spun the dial, the faster you moved through lists of options. It was exactly what the P68 needed.
>
> He bought the phone and showed it to Jobs, who enthusiastically agreed.
>
> To make sure they weren't stepping on any patent toes, Schiller and Rubinstein flew to Europe to meet with Bang & Olufsen. Since the two companies admired each other, it was the work of only minutes to draw up a one-page contract. B&O would give Apple freedom to use the wheel idea for its music player, provided B&O could keep using it on phones. No money changed hands—only some technical expertise. ("The funny thing is," Rubinstein adds, "that I licensed that technology *back* to B&O. They didn't know we'd stolen it from them!")
>
> Six years later, ironically, Apple considered using the same dial on the iPhone—and decided that it would never work on a phone.

iPod • 363

where most of the products he'd worked on were canceled just at the finish line. "99 out of 100 projects would get canceled before they ever shipped," he says. "Philips was the place where any good idea went to die." Fadell wanted to get *this* product so far along that nobody could change their mind.

Finally: the Sony threat. "Sony was number one in every audio category for devices—number one," Fadell says. If Sony's upcoming player beat Apple's to market, it would be lights out for P68. Apple *had* to beat Sony.

The only way to make the deadline was to hire some outside help: music-player circuitry from a local company called PortalPlayer, and middleware (for managing lists of songs, albums, and artists) from Pixo, which had been founded by former Apple engineer Paul Mercer.

But the interface—how you interacted with the hardware and software—was all Apple, much of it Jobs. "Steve sat there with the human-interface designers, picking the font, picking the lines, picking the speed of how it would scroll. He was all into every detail of that," remembers Schiller.

Fadell and product manager Stan Ng crisscrossed Asia to find the screens, batteries, and other components they'd need. Of course, they never told their suppliers what these components would be *for*.

In the end, the ingredients formed a global stew: hard drive from Japan, memory from Korea, audio chip from Scotland, FireWire connector from Texas Instruments. In Taiwan, a factory called Inventec, who'd built the Newton, would do the manufacturing.

Jeff Robbin raced to make iTunes iPod-ready. How would the Mac speak to the player? How would it sync your music files onto it? "Making it into a product that my mom could handle—that was the goal," he says.

The effort was reminiscent of the push to finish the Mac: a small team of gifted engineers, on impossible deadlines, driven by Jobs's demands for excellence, in hopes of changing the world.

In August, a tiny circuit board was ready to test. The team plugged in a pair of Walkman headphones and listened to the first song ever played on an iPod: "Groovejet (If This Ain't Love)," a driving dance track.

Everyone—Jobs, marketing, Jony Ive's team, and Chiat/Day—worked for weeks on coming up with a real name for P68. But time was running out; packaging and marketing materials had to go to press. With

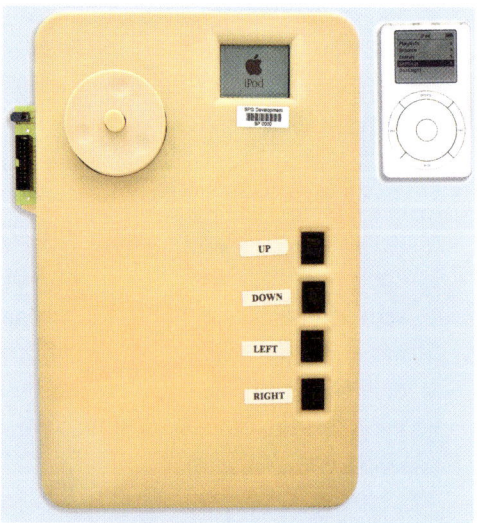

This early prototype test mockup would not have fit in your pocket. (Finished iPod shown for scale.)

little fanfare and even less time, they just kept their working title "iPod" as the actual product name. "I admit—at the time, I thought it was a wacky name. I kept thinking of pod people, you know, in *Aliens*," Schiller says. "You know, I trusted Steve; if he's right, he's right. But yeah, I remember the first time thinking: 'That's a stretch!'"

Design

"If you think of nearly every consumer product of that size, it's instantly forgettable. It's another plastic box," Ive says. The iPod would definitely not be that. It was the size of a deck of cards, 2.4 x 4 inches, with rounded-off back corner edges to feel good in your hand. Its back shell would be made of highly polished stainless steel—not plated, not painted. It was scratchable, but gorgeous. Each iPod would be hand-polished in Japan with a buffing wheel.

It would have no on/off switch; when not in use, it would simply go to sleep. Touching any button would instantly wake it.

The front of the iPod would be a gleaming double-shot plastic: a clear layer molded onto a white layer—*very* white. At a time when *nobody* was making white consumer gadgets, Ive decided that the earbuds would be white, too.

Ive remembers the feedback—"You can't do white headphones! They're going to look ridiculous!"—but he held firm. "I saw the headphones as being like little white flags, because the product may be in your pocket," Ive says. "Let's not apologize. Let's be forthright and clear and honest that you've got 'em!"

Steve Jobs backed him up.

The Drive

The iPod would move in and out of your pocket. It would get bumped and dropped. The ruggedness of the Toshiba hard drive became a key question, which Apple answered by setting up robotic arms to drop-test hundreds of drives and prototype iPods. They easily survived 30-inch drops, which seemed enough.

To prevent skips in the music while you were running, PortalPlayer's engineers designed the iPod to scoop up 20 minutes' worth of music at a time from the hard drive into memory, and then feed it to the playback circuitry from there.

At first, the plan was for the iPod to sync music with iTunes on the Mac in both directions. You could rip songs onto your Mac, and they'd wind up on the iPod; but you could also plug a friend's iPod into your Mac and slurp in *their* music. But in the end, that feature felt too pirate-friendly.

The iPod would sync only one way—from Mac to player. And as a final acknowledgment of the iPod's purpose—carrying your *own* music around—Apple shipped every iPod with a sticker on the screen that said "Don't steal music."

The Reveal

Apple did indeed create the iPod, top to bottom, in time for the holidays. On October 23, 2001, Apple invited journalists to a special event at its own auditorium. (The invitation said "Hint: It's not a Mac.") It was only a few weeks after the 9/11 attacks, and the world was still on edge—a tricky time to make a major product announcement.

Jobs explained how the iPod broke ground in many ways at once: the tiny hard drive, the scroll wheel, and its tiny piezoelectric speaker that *clicked* through the list items as you scrolled.

And there was FireWire. All previous players connected to the computer with a USB cable; copying 1,000 songs from the computer over USB would take five hours. FireWire, on the other hand, could sync a 1,000-song collection in ten minutes—*while* charging the iPod. Copying one whole CD took ten seconds.

The interface design was clean and idiotproof: You scrolled through lists of lists and clicked to drill down, artist > album > song. Once a song was playing, the wheel adjusted the volume. The battery life was 10 to 13 hours. A bright backlight let you read the screen in low light.

To applause, Jobs announced: "A thousand songs in your pocket." To no applause, he mentioned the price: $400.

That day, the iPod opened up a market where Apple wasn't the eternal also-ran. "We got to be number one at something," Schiller says. "We weren't the underdog, we weren't the better-product-but-not-selling-as-much. It was: here's a new thing."

For Jobs, the iPod's impact was existential. "If there was ever a product that catalyzed what's Apple's reason for being, it's this. It combines Apple's incredible technology base with Apple's legendary ease of use with Apple's awesome design. Those three things come together in this," he said. "So if anybody was ever wondering why is Apple on the earth, I would hold up this as a good example."

The scroll wheel was surrounded by four buttons, all designed to be operated with one thumb.

iPod for Windows

The world remembers the iPod as an instantaneous success—a gorgeous, disruptive, industry-shifting triumph.

In fact, its sales were . . . okay.

It took a year to sell 350,000 of them. The central problem was that the market for the iPod was a subset of a subset of a subset. You could only use an iPod if you had a computer—and it was a Mac—and it had a FireWire jack.

Making the iPod work with Windows would broaden the potential market from 7.5 million to 500 million. But to Jobs, that was a hard no. He was a firm believer in the halo effect: If the iPod worked only with Macs, more people would buy Macs.

As it turns out, he was right. The iPod did boost Mac sales somewhat.

But Rubinstein and Schiller were convinced Apple was leaving 90 percent of the money on the table. Meeting after meeting, they argued for bringing the iPod to Windows, but Jobs would not budge. Halo effect! Halo effect!

Finally, Schiller, Rubinstein, Joswiak, and Ng ganged up on Jobs. One day, they brought to him a spreadsheet based on market-research data. It compared the two universes: one where the iPod was just for Macs, one where it was also available to Windows. The model did indeed show that Apple would lose some Mac sales. But iPod sales to PC owners would hugely outweigh the losses.

Jobs, weary of the battle, got up from the table. "Screw you guys. I don't care what you do—it's on you. I'm outta this discussion," he said. And he left the room.

In principle, Rubinstein and Schiller had won the argument. But bringing the iPod to Windows PCs wasn't quite that simple, because Windows didn't have iTunes.

Rubinstein and Schiller engineered a quick solution: They'd buy an existing music-loading program. They cut a deal with the makers of MusicMatch Jukebox, a popular app for Windows. It wasn't an especially lovely program—in fact, it was one of the apps Jobs had mocked to illustrate how beautiful iTunes was by comparison—but it would do for now.

iTunes Store

Early in the life of the iPod, Jobs dreamed of being able to buy new songs right in the iTunes app. But the rights to virtually every commercial song on the planet belonged to five major record companies—AOL Time Warner, BMG, EMI, Sony Music, and Vivendi/Universal.

Eddy Cue had come to Apple in 1989 right out of Duke to become John Sculley's personal tech-support guy. Soon, he was on Apple's internet team; from there, at Jobs's request, he took charge of building Apple's first online store. (In typical fashion, Jobs told Cue he wanted the store up and running within three months—a sheer impossibility. Cue had it running in five.)

Now Jobs asked Cue to open talks with the record companies about selling their music through iTunes. Nobody had ever sold music over the internet, but Cue hoped that the appeal would seem obvious: If music fans had a delightful, frictionless, one-click way of buying songs, they'd be much less inclined to pirate music.

But that wasn't how the record labels saw it. "They told us to go pound sand," Cue says.

Eddy Cue
Born: 1964, Miami, FL
Schooling: Duke
Apple: 1989–present

Cue is an avid car collector, and serves on Ferrari's board of directors.

As it turned out, the labels were planning to open their *own* online music stores, called Press-Play (Sony and Universal) and MusicNet (Warner, BMG, and EMI).

From the day they were conceived, these stores were designed to fail. Each store offered only a subset of its respective companies' songs, and none of them had anything from Madonna, Led Zeppelin, the Rolling Stones, Pearl Jam, and the Beatles. The prices were different for every song. These sites charged $10 or $15 monthly subscriptions, and the offerings were hobbled by restrictions. You could burn ten songs a month to a CD—but no more than two songs from the same artist. You could download 50 songs a month, but they self-destructed after 30 days. If your subscription expired, your music files expired, too, turning into worthless icon carcasses on your hard drive. And you couldn't put your music onto *any* music player.

What Jobs envisioned was radically different: every song from every catalog in one place. Every song a dollar. Own it forever. Burn it, iPod it, copy it, whatever. No copy protection.

In the fall of 2002, Jobs sent his assistant James Higa and Eddy Cue to visit Roger Ames, CEO of the record company he thought would be easiest to convince: Warner.

Negotiations took months. The record company wanted restrictions; Jobs wanted none.

In the end, they hashed out a compromise: Apple would design a copy-protection scheme, called FairPlay, but it would be radically less restrictive than what had come before. You'd buy your song for $1, and you'd own it forever. You could play it on up to three computers (later changed to five). You could burn a song to a CD as many times as you liked, but no more than ten burns of a single playlist. And, of course, you could download your music to your iPod.

What finally convinced Warner was Jobs's cleverest, most persuasive argument of all: All of this would be only a small, constrained trial. It would be only in the U.S., only for a year, only for a subset of the Warner catalog—and above all, only for Macs, 5 percent of the market. What harm could come from so inconsequential a test?

If Warner was unhappy with the results, they'd just shut it down. Warner was in.

Once Warner was on board, Universal, BMG, EMI, and, finally, Sony also signed on. (Under the Warner deal, Apple would pay Warner 67 cents per song sold. The other labels held out for 70 cents; eventually, Apple agreed. Eddy Cue had the not-so-unpleasant task of telling Warner that they'd now be getting *more* per song than they'd negotiated, just to make things fair.)

Then there was the matter of the musicians. Many of them, having witnessed the horror of free music sharing, were skittish about any kind of online distribution. Jobs got involved, calling, meeting, and giving demos to Bono, the Eagles, Sheryl Crow, Led Zeppelin, and Madonna, or their agents, his charisma on high. One by one, they agreed to sell their music digitally for the first time. It took the Rolling Stones until 2005, and the great white whale, the Beatles, gave in in 2010. But Jobs eventually got virtually everybody.

When he wasn't chasing down music rights, Jobs was sitting with Cue, Robbin, and designers to create the store itself. He pored over every detail, as he had with every aspect of every iPod.

Finally, on April 28, 2003, the iTunes Store went live. It was clearly a leap ahead of what had come before. You could easily search the entire universe of songs, click to listen to a 30-second sample, click Buy, and watch as the song appeared in your iTunes library, complete with album art and track details.

It wasn't just easier than PressPlay and MusicNet; it was also easier than file-sharing sites. With those, Jobs pointed out, it might take you an hour of hunting to find a playable copy of an illegally posted song. "It's got the last four seconds cut off the end, and it's got a glitch in the middle," he said. "You'll spend an hour at that rate—to get four songs that cost under four bucks from Apple . . . you are working for under minimum wage!"

The iTunes Store sold a million songs in the first six days, and that was just to Mac owners. After six months of further negotiations, Jobs convinced the record companies to expand the experiment to include the other 95 percent of the market: Windows users.

Not all musicians were delighted. They had put a lot of thought into the sequence and structure of songs on their albums. The songs weren't meant to be cherry-picked and rearranged according to the whims of people who didn't appreciate the greater creative goal. It was a deeply disruptive

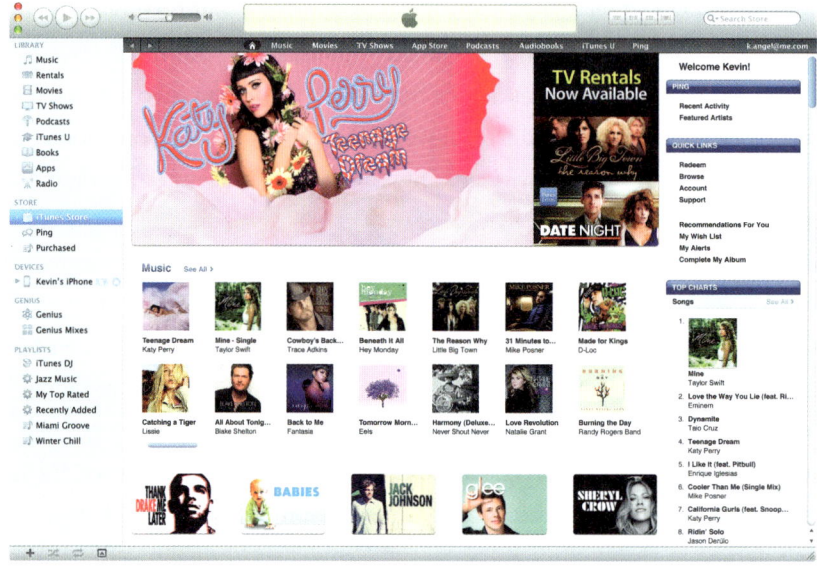

The iTunes Store eventually grew to include audiobooks, podcasts, TV shows, and movies.

development. "Fewer and fewer music lovers are willing to buy the music industry's shopworn business model: $17.99 for a recorded CD that contains only a couple of tracks they like," noted *Time*.

It didn't matter. In the first year, Apple sold 70 million songs. On February 23, 2006, Jobs called 16-year-old Alex Ostrovsky to let him know that his purchase of Coldplay's "Speed of Sound" was Apple's *one billionth* downloaded song. Apple gave the kid ten iPods, an iMac, a $10,000 gift certificate for more songs, and a Juilliard music school scholarship in his name.

Apple now had 85 percent of the paid-music market. The iPod's ubiquity led directly to the creation of podcasts. They were still a fledgling entertainment format in June 2005—but once Apple added podcasts to iTunes, a million people downloaded episodes in two days. (Microsoft, furious that podcasts were named after the iPod, started a campaign to call these recorded audio shows "netcasts." It went nowhere.)

In October 2005, Jobs pulled off another rights-acquisition miracle. The iTunes Store started selling network TV episodes, for playback on video iPods, for $2 each—no ads.

Once again, Apple started selling something that had never before been available to buy à la carte. Who decided the value of a TV episode? Steve Jobs did: $2.

Silhouettes

Chiat/Day's first TV ad for the iPod depicted a middle-aged white guy bopping awkwardly alone in his apartment. Online, people called it the "iClod" ad.

But the second campaign, which debuted in October 2003, was a world-changer. Chiat/Day art director Susan Alinsangan came up with a campaign called "Silhouettes." This time, the dancers were visible only as energetic black shadows against bright backgrounds of lime green, yellow, fuchsia, bright blue, and pink. The only trace of white was the iPod and its trademark white earbuds. "Instead of asking you to buy this device, Apple was asking you to buy the emotion," says Chiat/Day's Ken Segall.

They were *good* dancers, dancing to the coolest new rock, techno, and hip-hop songs of the day, often chosen with Jobs's input: U2, the Black Eyed Peas, Daft Punk, Nelly Furtado, Coldplay, and others. In 2005, Eminem and Wynton Marsalis appeared in variations where their faces were distinct even in silhouette.

Animators hand-painted the white cords into the footage.

Every song featured in an iPod ad shot up the charts. In 2006, Jobs persuaded his hero Bob Dylan to allow Apple to sell a $200 digital "box set" of every Dylan song ever recorded—and to appear in an iPod ad. Dylan's new album hit number one on the *Billboard* chart—his first in 30 years.

In October 2003, Apple finally opened the iTunes music store to the Windows universe. Jeff Robbin's team had undertaken an unthinkable task: making a version of iTunes for Windows. As Jobs unveiled the app at the Macworld Expo, the joke on the big screen behind him wasn't lost on anyone. It said, "Hell freezes over."

Apple had just opened the doors to its music store to hundreds of millions of more people, and the results were mind-blowing. During the fourth quarter of 2003, Apple sold 733,000 iPods, a 235 percent increase. Apple's revenues jumped by 36 percent, to $2 billion for the quarter. "Happy holidays" took on a whole new meaning.

Generations

The iPod team had driven itself to exhaustion getting the iPod ready for Christmas 2001, out of fear that competitors would eat its lunch. But no serious competitors appeared by Christmas 2001, or even Christmas 2002. By the end of 2003, there were some rivals, but they were janky-looking and clumsily engineered. Sony's much-feared MP3 player was so hobbled by corporate paranoia that it could not play MP3 files, which could be considered a drawback in an MP3 player. "We thought we would have a year lead on Sony and everyone else. We didn't figure it was going to be *five* years," says Rubinstein.

What rivals didn't see coming was Apple's aggressive plan to cannibalize its own bestsellers. The iPod team intended to introduce a new model for the holiday season *every single year*. Millions of people would buy iPods repeatedly so they'd always have the latest. They'd pass last year's models to kids and siblings, who themselves became part of the iPod economy.

"'September, September, September.' That was the mantra," says Fadell. "It was just this steady drumbeat, unrelenting." The thinking was: "If we leapfrog ourselves, then no one else can leapfrog us."

The iPod engineering team had grown from 35 people to hundreds. The spreadsheets showed sales in the millions. One day in 2004, the massiveness of the effect hit Jobs as he walked the streets of New York City. "On every block, there were one or two people wearing white headphones," he said. "And I thought: *Oh, my God, it's really starting to happen*."

Year by year, the iPod morphed. There were bigger models, smaller ones; color screens, no screens. But all of them were a year or two ahead of competitors.

The scroll-wheel concept evolved on the second-, third-, and fifth-generation iPods.

- **Touch-wheel iPod (July 2002).** The second-generation iPod worked with Mac *or* Windows, cost $100 less than the original, and came in a 10 GB size. It also had a *stationary* scroll wheel that didn't actually rotate. Instead, the touch of your finger "turned" it—an improvement over the original, mechanical dial, whose wheel could get dust or sand in the seams and become glitchy.

- **Dock-connector iPod (April 2003).** The third-generation iPod was thinner and more rounded-off. The four navigation buttons around the wheel became four dime-sized, touch-sensitive, light-up buttons below the screen. The base model's price dropped by 25 percent—it was now $300, with a 20 GB drive inside.

   Most important, this model no longer had a FireWire jack. Instead, it introduced the dock connector: an inch-wide, 30-pin connector on the bottom. To that, you could attach either a FireWire cable or a USB cable. There wasn't much of a speed difference anymore, now that USB 2.0 had come out—and now using an iPod with Windows was a sure thing. Not every PC had FireWire, but they all had USB.

   Apple began licensing the specs of that 30-pin connector to accessory makers. Suddenly, you could buy iPod docks, speakers, and microphones that attached to that jack. There were iPod strollers, treadmills, belts, and boom boxes. Hotels offered clock radios with the 30-pin connector. The Sharper Image offered a $700 massage chair with an iPod socket in the armrest. 80 percent of all cars offered the connector in their new models. By 2007, more than 3,000 iPod accessories were available.

The iPod mini was the world's smallest music player—at the time.

Then there were the cases, priced from $10 for plastic to $200 for Coach python skin. One analyst calculated that iPod owners were spending on accessories one-third of what they'd spent on the iPod itself. The iPod economy grew to a billion dollars' worth of accessories a year—and Apple got a piece of it all through "Made for iPod" licensing fees.

- **iPod mini (January 2004).** In the iPod's fourth year, Jony Ive became intrigued by the qualities of brushed aluminum. "Unlike with stainless steel, you could blast it and then anodize it—which is a form of dyeing—and then you could do color in an unusual way," he says. For the first time, an iPod had fully rounded edges, like a gently flattened tube, in a choice of five colors. The machine was 3.6 inches tall and 2 inches wide, about half the volume of the original iPod. It was so small and so beautiful that it came with a clip so you could wear it on a belt. In fact, you could buy an armband holder for $30: Apple's first wearable device. But on an iPod this small, there wasn't room for the standard scroll wheel. Ive's team removed the four buttons around the wheel; instead, the four compass points *of* the wheel were clicky.

    On paper, the math made no sense: At a time when a regular iPod cost $300 for 20 GB, the Mini cost $250 for only 4 GB. "People were like, 'They've lost their mind again!'" says Greg Joswiak. "Let's prove them wrong," Jobs said.

    He did. The iPod mini was a monster hit, selling about 300,000 a quarter. In a year, it doubled the iPod's market share from 31 percent to 65. (It helped that Carly Fiorina, CEO of HP, wanted to sell iPod minis through its own channels. Apple wound up selling 30 million iPods through that deal alone.)

    It was the bestselling product in Apple history.

- **iPod photo (October 2004)** brought color to the iPod's screen, so that you could see your photos, although at the size of a Wheat Thin. The $500 iPod photo had a 40 GB capacity, although "8,000 songs in your pocket" never became a slogan.

- **iPod shuffle (January 2005).** This iPod was a thin, plastic, one-ounce slab—almost a bookmark. To get the price down to $100, Apple eliminated the hard drive, the wheel, the 30-pin connector, *and the screen*. You couldn't see what song was playing, or choose which would play next.

    Even Apple's biggest fans had to chuckle at the cheekiness of Jobs's assertion that losing the ability to choose songs was a *good* thing. Apple advertised the shuffle with the tagline "Give chance a chance."

    The iPod shuffle was the first to contain flash memory, which retains its contents even when turned off. It held only about 240 songs. A removable plastic cap concealed a USB jack for plugging directly into a Mac or PC. A new Autofill button in iTunes chose a random 120 songs from your collection to load onto the iPod each time you clicked it.

Apple made very little money on the iPod shuffle itself; it was designed to provide the cheapest possible on-ramp to the iTunes music store. "The whole thing was to get users of the music store to try and think, 'Oh, I love this so much, I'm gonna buy up. I'm gonna upgrade my iPod,'" Fadell says.

- **iPod nano (September 2005).** The iPod mini, from 2004, was a bestseller of a size that Apple had never seen before, flying out the door by the millions. So naturally, at the very peak of its popularity, Steve Jobs canceled it.

  He wanted to make room for something he thought was even better: the iPod nano. This "kill your darlings" act—replacing a year-old megahit with an unknown quantity—would have given a more conventional CEO screaming red hives. But Jobs would say, "We can't fall so in love with our success that we're afraid to take a risk and make a new success."

  "If it were up to me, I would've been a little more conservative," says Rubinstein. "But he wanted to go for it, always—so we shut down the mini."

  At a special event in the Apple auditorium in October 2004, the camera zoomed in on Jobs's jeans. "Now, *this* pocket's been the one that your iPod's going in," he said, tapping the pocket. Then he pointed to that weird little inner, upper coin pocket. "You ever wonder what *this* pocket's for? Well, now we know!"

  And from it, he pulled out a shiny, tiny iPod nano.

 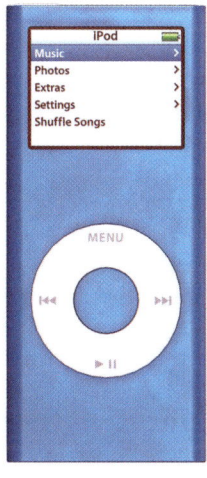 

The iPod nano was the bestselling product in Apple's history.

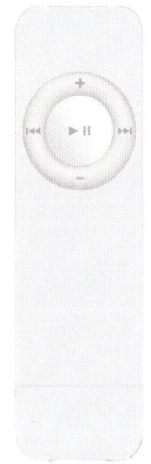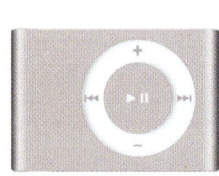

The first-, second-, third-, and fourth-generation iPod Shuffles.

Microsoft Fights Back

Microsoft could not sit still and watch Apple waltz away with the digital-music market. In 2004, it developed a software standard called PlaysForSure, which it licensed to other music-player companies. Dell, Samsung, and Creative made players; Yahoo, Rhapsody, Napster, and MTV made the online stores.

It bombed. It was balky and complex; sometimes, your music wouldn't play, an unfortunate feature of a system called PlaysForSure.

Microsoft tried again in 2006 by creating its *own* music player: the Zune. It was an obvious knockoff of the iPod of the day—$250, 30 GB drive. But it was noticeably thicker, taller, and heavier, and it didn't have a scroll wheel. It was available in black, white, or *brown*.

There was a Zune music store, but it didn't have TV shows, movies, audiobooks, or even podcasts, and the music catalog was much smaller: 2 million vs. 3.5 million songs on iTunes. The kicker: The Zune could not play songs in Microsoft's own PlaysForSure format.

The Zune's flagship feature was wireless song sharing: You could beam a song to another Zune-owning friend within 30 feet. That person could play the song three times within three days before it turned into a pumpkin. None of it mattered: Microsoft pulled the plug on the whole thing in 2011.

Witnessing the iPod become the dominant platform must have been a sobering experience for Microsoft. For once, *Microsoft* learned what it was like to have a very tiny market share.

It had a high-resolution color screen, and it was gorgeous, but it stored *fewer songs than the model it replaced*. Like the iPod shuffle, it had 2 or 4 GB of flash memory (good for 500 or 1,000 songs) instead of a hard drive; Apple had signed a deal to buy Samsung's entire production of flash memory until 2008.

"Flash was a completely bonkers idea, especially if you've been marketing a product on a thousand songs in your pocket," says Ive.

So firmly did Jobs believe in the iPod nano that he bought enough parts to make 14 million of them for the first holiday season. "We had not sold that many iPods *period* through that time," Joswiak says. "And now we're putting a plan in place in one quarter to sell 14 million?!"

Jobs was right. iPods had sold briskly before, but nothing like the nano. It was a towering, colossal, breathless hit that made the iPod mini look like a warm-up act. It easily sold all of those 14 million Nanos—and by 2009, Apple had sold *100 million* of them.

"That was a huge change for Apple," Ive says. "The iPod began to change people's sense of the relevance of the company."

There would be seven generations of the iPod nano. The third-gen model, known internally as Nano Fatty, was a squat little guy with a bigger screen for better video viewing. It was the only model with disappointing sales. The fifth generation introduced a tiny video camera, tiny speaker, tiny microphone, which all disappeared the following year.

- **iPod, fifth generation (October 2005).** Now five years old, the mainstream iPod was beautiful, thin, and capacious. Its hard drive was 30 or 60 GB, and it came in shiny white or black. But the biggest news was that its color screen could play *video*—like the TV shows from the iTunes Store.

    In June, Apple offered a black version with a bright red click wheel, known as the U2 Special Edition. It had the band members' signatures engraved on the back, and came with a link to a 33-minute video with U2 interviews and music.

- **iPod shuffle, second generation (September 2006).** The first iPod shuffle, a plastic USB stick for $100, had sold well even without a screen. Its replacement was the size of a big postage stamp. It had a metal spring clip on the back, for clipping to your clothes—and cost even less: $80.

- **iPod shuffle, third generation (March 2009).** In 2005, *Saturday Night Live* parodied the parade of ever-smaller iPod models. Fred Armisen, playing Jobs in a black turtleneck, unveiled the iPod micro, the size of a soda pull-tab; the iPod pequeño, the size of a pinky fingernail; and, finally, the iPod invisa, which he claimed to be holding between his thumb and forefinger. "It holds *eight million songs! Every photograph ever taken! Pong!*"

But Apple continued to do precisely what the parody had foretold. The third-gen iPod shuffle was little more than a tie clip, only 1.8 inches long. There was no room for controls; you were supposed to click the buttons on the earbuds cord to play and skip tracks and adjust the volume. If you held down the earbud button, you'd hear the name of the song or playlist you were listening to.

- **iPod shuffle, fourth generation (August 2010).** The tie-clip iPod might have been a little *too* minimalist, because its successor ($80) brought back the playback buttons. It resembled the second-generation, clip-on model—except now it was the size of a *small* postage stamp.

The iPod Effect

Anything that changed society as radically as the iPod was bound to leave controversy in its wake. Parents fretted that kids weren't paying attention. Lawsuits asserted that the FairPlay copy protection locked music buyers into Apple's ecosystem. (Apple won.) The rapid pace of new iPod introductions prompted cries of planned obsolescence.

Most of all, the record company executives seethed. Before their eyes, the damn iPod was destroying the joyride of the CD era. But what could they do? The iTunes Store was their best alternative. They were getting two-thirds of all the income from the biggest music store in the world—without paying a penny for rent, utilities, employees, or inventory.

But those objections were just gnats to the iPod tsunami. This thing was *cool*, and everybody wanted one. Banks gave away iPods as premiums for opening new accounts. Birthday, Christmas, and bar mitzvah presents were iPods. Crime reports ticked up as people stole iPods.

The white earbuds were everywhere: planes, trains, and buses; sidewalks, gyms, beaches. "What's on your iPod?" became a conversation starter, a talk show topic, and a regular celebrity feature in magazines. President Bush and VP Cheney both had iPods—and revealed their contents. iPods showed up as props on TV (*The Office, Scrubs, One Tree Hill, Medium, Queer Eye for the Straight Guy*) and movies (*Legally Blonde 2, War of the Worlds, The Italian Job, Blade: Trinity, Agent Cody Banks, First Daughter*).

In all, there would be 18 generations of various iPods over 20 years, during which Apple's stock rose from $15 a share to $635. At one point, Apple was buying 45 percent of the entire world supply of flash memory just to keep up with iPod nano sales.

iPods and the iTunes Store came to represent nearly 60 percent of Apple's income; the Mac was now Apple's *lesser-known* product.

The iPod triumph changed Apple, too. If it hadn't been clear from the iMac's success, now it was crystal clear: Design wasn't an afterthought at Apple. It was the *starting* point of product development.

In all, Apple sold 450 million iPods. Jobs's commercial instincts, reality-distortion skills, and maniacal attention to detail had produced something nobody had seen coming: the most popular electronic device in the world.

After all that, it would be hard to imagine a device that could be a bigger hit, a greater triumph, a more influential cultural force. But one was coming.

The Color-Changing Mac

The eMac (2002) was an oddball, one-off Mac marketed to schools. It resembled an all-white iMac, minus the handle. It was the last CRT-screen computer Apple ever made—and came close to being the coolest.

During his 25 years at Apple, industrial designer Duncan Kerr created some of the company's most magical features, including the "breathing" sleep light, the first backlit keyboard, the Cube's touch-sensitive, illuminated power button, and the multitouch tablet that led to the iPhone.

In 2001, he experimented with lining the eMac's translucent white housing with colored LEDs. "We'd done all of the iMacs with the different shell colors," he says. "So one of the ideas was: Could we create essentially infinite color variations?"

Steve Jobs loved the idea and gave it the green light (and blue, orange, pink . . .).

The chameleon eMac project got fairly far along. But in the end, it never saw the light of day, partly because of cost and partly because the light was hard to see in a well-lit room.

For a few years, Apple also offered the eMac to consumers; but in 2006, Apple turned out the lights on the eMac for good.

35. Back to the Macs

The iPod dominated the Apple headlines in the early 2000s, but the brain trust of Jobs, Ive, Rubinstein, Tevanian, and Schiller didn't ignore the Mac. A steady stream of replacements for the four Macs in the Quadrant were progressively faster, thinner, and more beautiful.

- **Power Mac G3.** When Jobs returned to Apple in 1997, the sprawling archipelago of Mac models that he inherited included three beige Power Mac G3 models, all with similar guts: a minitower, a desktop, and an all-in-one model for the education market that earned the nickname "Molar."

  At the January 1999 Macworld Expo, Jobs revealed their successor. The influence of the iMac, which Jobs had unveiled eight months earlier, was obvious: This 30-pound machine was all blues and translucent plastics, with arched handles at the top.

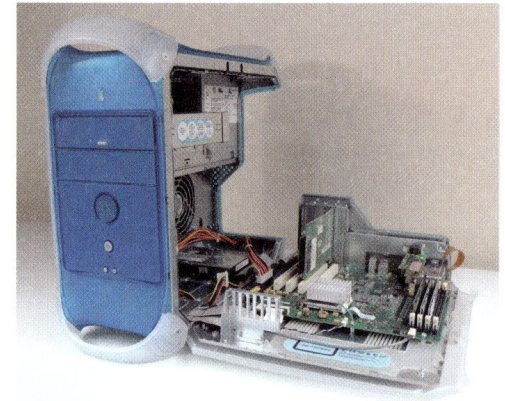

The Power Mac G3 (Blue and White) opened like a drawbridge for easy access to the guts.

- **PowerBook G4 (January 2001).** The black plastic PowerBooks had had a good run, but Jony Ive was in charge now. At the January 2001 Macworld Expo, Jobs spoke about the difference between Apple's laptops and Windows laptops. "We have the most powerful notebooks in the world," he said. "But *they* have the *sex*. We want both. So today, we're introducing a totally new PowerBook."

  The PowerBook G4 contained a PowerPC G4 chip, 15-inch screen, five-hour battery life, a slot-loading DVD drive, a translucent keyboard—and a titanium body. "Like the spy planes," Jobs noted. It was only an inch thick and weighed 5.3 pounds.

  Titanium, as it turned out, was beastly difficult to engineer and manufacture; over a dozen mechanical designers quit in frustration. The unpainted titanium of the lid and underside of the laptop looked great, but some "TiBook" fans were dismayed to see some discoloration and peeling of the paint on the painted portions, including the palm-rest frame and the hinges.

  But the idea of metal lived on. Apple would never again make a rounded plastic laptop.

The Flipped Apple

It seemed obvious that the Apple logo should sit in the center of the original PowerBook's lid. What *didn't* seem obvious was which *way* it should sit.

Option A: The Apple logo is upright when the laptop is closed in front of you. Pro: You see that the laptop is oriented correctly for opening. Con: Once the lid is open, the logo is upside down to everyone passing by.

Option B: The Apple logo is upside down when the laptop is in front of you. Pro: Once the lid is open, the logo is upright to everyone passing by. Con: The logo is upside down to *you*. In user testing, people kept trying to open the laptop from the wrong side.

On the original PowerBook, the designers went with option A. But one day in 2000, someone asked on Apple's internal messaging system: "Why is the Apple logo upside down on laptops when the lid is open?" Suddenly, the issue was reopened.

And that is why, when Apple introduced the PowerBook G4 (Titanium) in 2001, Jobs flipped the logo. From that day, it has appeared upside down to you, but upright to the rest of the world.

As for opening the laptop from the wrong edge: Jobs figured that you'd learn your way around in just a few attempts. Meanwhile, the rest of the world would see the apple the way God intended it to appear.

- **iBook (May 2001).** After two years, Apple retired the "toilet seat" iBook design, with its two-toned plastic. In its place was a simple, white slab, polycarbonate plastic outside, magnesium on the inside. It was 30 percent lighter (4.9 pounds) than the colorful iBooks, and occupied less than half the volume.

  As fond as Jobs was of the Quadrant of four Mac models, behind the scenes, there was always more to it than that. "Each quadrant, of course, had Good/Better/Best models," says Rubinstein. "It wasn't really four. It was more like 12. It was a complicated matrix of things to do, but much simpler for the customer." The new iBook was a case in point. It came in four different versions, each with a different disk drive. You could get one with a CD-ROM drive for $1,300, a DVD drive for $1,500, a CD-RW model for $1,600, or the "Combo drive" (DVD/CD-RW) for $1,800.

  By mid-2001, Jobs and Ive had redesigned all four of the classic Mac quadrant boxes. Translucent plastic and sleek, uncluttered design were the hallmarks across the entire line.

Closed or open, only the name of the second-generation iBook resembled the original.

Mac OS X

In March 2001, Apple finally released the product for which it had bought NeXT in the first place: Mac OS X. It had taken five years to reach usability.

The Aqua interface was beautiful, and its little animations were cool. But on the whole, it was "dog-slow, feature poor," as Ars Technica put it. More than one fan noted the irony in the first code name: Cheetah.

But version 10.1, "Puma," appeared six months later, much faster, more solid, and complete. Suddenly, OS X was growing into the solid, modern OS that Jobs and Tevanian had envisioned. "Version 1.1 of Apple's new operating system is a triumph," noted *BusinessWeek*. The *Detroit Free Press* wrote that it "runs circles around Windows XP."

At a time when Windows virus attacks made frequent headlines, Mac fans were also quick to point out another virtue of OS X: It was free from the grief of viruses, Trojan horses, and spyware. (Although much of the credit went to Mac OS X's secure design, critics noted that another factor might be the Mac's tiny market share. What enterprising bad guy would bother writing a virus with such a small payoff?)

Another six months later, there was 10.2, "Jaguar," which Apple sold as a DVD for $130, with a new Address Book app, a Search box in every Finder window, iChat text messaging, and a big set of accessibility features. Annual releases followed ever since:

- **10.3 Panther (June 2003, $130)** introduced a brushed-metal look to Finder windows; faxing; a new web browser called Safari; and iChat AV, Apple's first video-calling app. It was especially handy if you bought Apple's first video camera, the iSight ($150).

- **10.4 Tiger (May 2004, $130).** Another summer, another update—now with Spotlight (a single search box for the entire computer), Dashboard (a set of small, floating "widget" windows

iSight

It was always fun to see Apple turn its mighty eye on something that everybody knew was cruddy and produce its own better, cleaner, prettier version. In 2003, it was the webcam.

Webcams were ugly, plastic, low-quality attachments that gave computers vision, mostly for the video chats. Apple's was not ugly, plastic, or low-quality. "This thing is gorgeous," *Macworld* said. "It's *unnecessarily* good."

Most webcams produced jerky, low-res video over USB; the iSight delivered full TV smoothness (30 frames per second). Most webcams required multiple cables to the PC (power, audio, video); the iSight's single FireWire cable carried it all.

Privacy was built in, too: A green LED lit up whenever the camera was on. And with the twist of a ring at the front, you could close the internal lens cap—a three-leaf iris behind the glass. Just plugging in the camera opened iChat AV automatically.

Starting in 2006, Apple built cameras right into its laptops, iMacs, and monitors; nobody had any reason to buy an external one. Nowadays, iSights live on only on eBay.

for weather, stock prices, sticky notes, and so on); VoiceOver (reads what's on the screen aloud for people with vision impairment); and parental controls.

At Apple's 2004 WWDC, where conversations often turned to Microsoft's tendency to copy Mac OS features in new Windows versions, Apple banners said "Redmond, start your photocopiers" and "This should keep Redmond busy."

 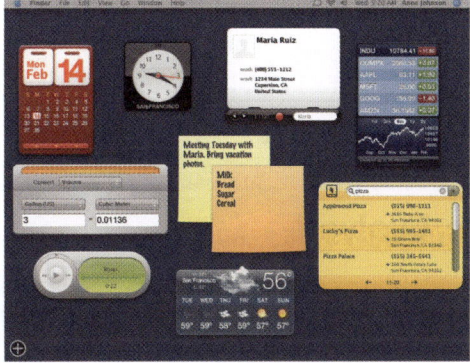

Mac OS X 10.4 introduced group video chats (left) and Dashboard widgets— the great-grandchild of desk accessories.

Photo Booth kept children up to the age of 90 entertained for hours.

- **10.5 Leopard (June 2005, $130).** This update included, by Apple's count, 300 new features. The big ones were Time Machine (a continuous backup system); Photo Booth (a selfie app with hilarious built-in visual effects); Quick Look (preview what's in a desktop icon without actually having to open it); and Screen Sharing (lets gurus assist newbies over the internet).

  On Macs with Intel processors, a new feature called Boot Camp let you start up your Mac in—get this—Windows. Somewhere, every Apple CEO who'd campaigned to make Macs run Windows wept softly.

- **10.6 Snow Leopard (June 2008, $30).** When Apple software head Bertrand Serlet (Avie Tevanian's successor) introduced Snow Leopard at WWDC 2008, his big-screen slide said: "0 New Features." Apple had instead spent the entire year working on speed, stability, and underlying technologies for developers' pleasure.

iMac, Evolved

The 1998 iMac put the new Apple on the map and taught millions of people the pleasures of simplicity and good design. But at its heart, it was a beautiful shell wrapped around a bulky, aging technology: the cathode-ray tube, like the one inside an old-timey TV set.

I'm a Mac. I'm a PC.

For decades, Mac fans felt like underdogs. They *knew* they had a better computer. Why was its market share so small?

Apple's "Switch" ad campaign of 2002 featured people talking about how much happier they were after switching to Macs. It ran for a year. (One of them featured a slightly stoned-seeming Ellen Feiss talking about losing a class paper when her PC crashed. She became an instant internet celebrity.)

But nothing worked like the "Get a Mac" ad campaign, which ran from 2006 to 2009. It became a popcorn-munching form of comeuppance. Most began the same way:

"Hello, I'm a Mac." Actor Justin Long was dressed casually, long hair, chill and loose, with a certain Steve Jobs–ness.

"And I'm a PC." Comedian John Hodgman was pudgy, conservative, in a suit and glasses, with vague Bill Gates vibes.

In each 30-second skit, the PC came across as an insecure corporate doofus, jealous of the Mac's coolness and simplicity.

In one, PC sneezes. "I have that virus that's going around. In fact, you better stay back—this one's a doozy!"

"That's okay—I'll be fine," says Mac.

"No, no—do not be a hero. Last year, there were 114,000 known viruses for PCs."

"PCs. Not Macs," Mac says.

"I think I gotta crash," PC says, falling backward like a tree.

The 66 ads, created by the Media Arts Lab division of TBWA/Chiat/Day, were a huge hit with Mac fans—and, apparently, with undecided computer buyers; after a year, Mac sales were up 39 percent. The campaign also spawned a thousand parodies.

Microsoft was not so thrilled. It counterattacked with its own "I'm a PC" ads, in which very cool people with very cool jobs (swimming with sharks, NASA) say that *they* are PCs.

It spawned no parodies.

It was no secret that flat screens were the future. For a decade, they'd been getting better, bigger, thinner, and cheaper. They were brighter and crisper than CRTs; they didn't flicker or emit radiation like CRTs; and they used only a fraction as much power.

By 2000, Jobs, Ive, and Rubinstein felt that the world was ready for a radical redesign of the iMac. Creating it would take two years, but they would bring flat screens to the masses.

The obvious way to create a flat-screen, all-in-one computer would be to mount the circuitry and drives behind the screen. This, in fact, was exactly what Gateway and IBM had done on their all-in-one computers. It was Ive's first instinct, too.

But Jobs didn't like the foam models Ive produced. What was the point of having a flat screen if you glommed a fat box onto the back of it? The screen wasn't flat anymore; the computer looked like it had a reverse beer gut. Worse, that design required mounting the circuit board vertically, with the jacks coming out the side of the screen. Did anyone really want cables sprouting out of the *screen*?

The guts/screen design would also mean mounting the CD/DVD drive on its side, which, as the Twentieth Anniversary Mac had painfully demonstrated, meant slower disc speeds.

"This is the official death of the CRT," Jobs said of the iMac G4.

Jobs invited Ive to his home to contemplate other approaches. As they sat in his garden, Ive was taken by the bright, tall sunflowers that Laurene Jobs had planted. He flipped open his sketchbook and began to draw.

"What if the screen were separated from the base like a sunflower?"

What Ive came up with was a white, domed base, 6 inches tall and 10.5 inches in diameter, that looked something like a bike helmet on your desk; it concealed all of the computer's guts. From it, a gleaming, chrome-plated strut emerged, topped by a very thin, very bright 15-inch flat-panel screen. That stainless-steel arm was a fantastically complex, rotating, bending piece of engineering. It concealed 40 cables and cost Apple $120 apiece—but Jobs and Ive adored it.

With only a fingertip, you could change the screen's angle, tug it down to keyboard level for closer scrutiny, or even swing it completely away when you need a look behind it. The support arm was usually hidden behind the screen, making the display seem to float in midair.

The flat-screen iMac brought the G4 chip to the iMac line for the first time, but, at $1,300 to

$1,800, it was far more expensive than the original iMac. Nor was everyone in love with its design; it reminded some people of a sunflower or a desk lamp, but others found the white domed base disproportionately big, like a volleyball half-submerged on your desk.

iMac G5

The "sunflower" iMac sold fairly well. But after only two years, technology had marched on, components had gotten smaller, and Apple felt comfortable with exactly the design it had mocked in 2002: putting the guts behind the screen.

This third-generation iMac design (August 2004) was called the iMac G5. Squeezing the electronics and drives behind the monitor *did* make the flat-panel screen less flat—the new iMac was two inches thick—but it was still the thinnest desktop computer on earth. Its snow-white plastic body, containing a 17- or 20-inch screen ($1,300 to $1,900), floated off the desk on a thin L-shaped aluminum foot; the double-shot white acrylic bore distinct iPod overtones. Actually, it looked for all the world like a monitor that had lost its computer. "It's around here somewhere!" Phil Schiller joked at the Paris Apple Expo, pretending to search for the CPU.

Jony Ive hadn't wanted to mar the white front with speaker holes, so the speakers were on the bottom edge, where you couldn't see them, bouncing sound waves off the desk. A hole in the vertical part of the aluminum foot let you pass cables from front to back. If you used Apple's Bluetooth keyboard and mouse, and got yourself an AirPort card, you'd have yourself the world's first *one-cable* computer (the power cord).

You could tilt the screen, but there was no height adjustment and no side-to-side adjustment

The fundamental form of the modern iMac began in 2004 (acrylic), evolved to aluminum (2007), and almost vanished (2023).

except turning the entire machine. There were gripes about the connectors on the back, too, all stacked up in a single column.

But reviewers admired the thing's purity of design. "No buttons, trays, or protrusions," *InfoWorld*'s reviewer wrote. "This machine is front-office material when viewed from any angle."

There were small updates—thinner body, built-in camera—and major ones, like replacing the white acrylic with aluminum in 2007.

Then, in 2023, the iMac's body got *ridiculously* thin and came in a choice of colors.

But none of them messed with the fundamental arrangement of the white acrylic ancestor: aluminum foot, guts behind the screen.

iPhoto

The dot-com crash of 2000 affected every computer maker, but it was an especially stressful time at Apple. "Steve realized that if we don't take this into our own hands, we may not be here," says Xander Soren, the first iTunes project manager. "He bet big on software because he had to. If Adobe wasn't gonna be there with Photoshop and their creative apps, we needed to have iPhoto. We needed to have iMovie in case Avid leaves. So, it was all kind of a hedge."

iMovie, iTunes, and iDVD were all hits, and all bore a family resemblance for ease of learning.

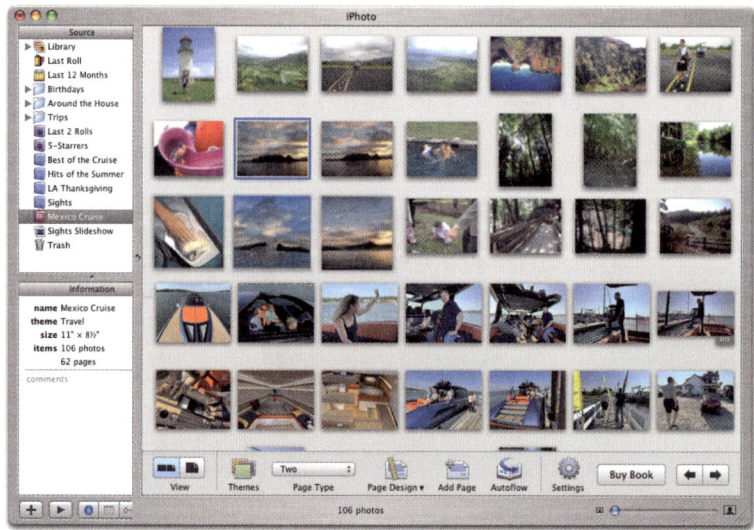

"Thus," Wired's reviewer wrote of iPhoto, "an archive of otherwise forgotten pictures and anonymous folders is transformed into a living, breathing gallery of pictures."

Jobs had expanded the iMovie team, across the skybridge from his office, into a full-on apps group, dedicated to bringing his digital-hub strategy to life.

But managing your music collection, editing your camcorder footage, and burning DVDs were only the beginning. Digital cameras were rapidly becoming a thing. In 2002, tech-savvy consumers and photographers snapped up over 18 million of them.

Once you'd taken your photos, you encountered what Jobs called a "chain of pain." To get your photos off the camera, you'd have to connect a USB cable or eject its memory card and slide it into a card reader (sold separately). Then you'd drag the pictures into a folder on your hard drive. Naming and organizing them was left to you. If you wanted to *do* something with the photos, like sharing, posting, or printing, you were again on your own.

So Jobs charged iMovie author Glenn Reid with the creation of iPhoto: a digital shoebox, he called it, for organizing, printing, and sharing your photos.

iPhoto shortened the chain of pain. Just connecting the camera's USB cable auto-opened the app and started downloading your photos. A couple of clicks let you make a slideshow, a print, or a web page gallery. You could order, by mail, printed posters, cards, prints, calendars, or—the pièce de résistance—linen-bound, professionally printed, hardcover photo books.

Reviewers dinged iPhoto for its 1.0 sparseness: color-correction features, an Email button, and CD burning would have to wait for later versions. But, as *Wired* declared, "The software is the last piece in the puzzle in Apple's digital hub strategy."

It wasn't.

GarageBand

iTunes could organize your music files, but it didn't *create* music—and the Mac was supposed to be a tool for creative people.

Like many employees, Xander Soren had been hired at Apple without knowing what the job would be. "They'll say: 'Just trust us. You'll work on something cool.'" He wound up serving as product manager for iTunes 2.

In 2001, he pitched another project to the apps group: a tool that could help even nonmusicians create professional new songs.

He proposed a simple, user-friendly music app. It would be the first one ever to combine music elements from three sources: a microphone, an electronic keyboard, and *loops* (prerecorded snippets, designed to sound good together, that could repeat seamlessly for as long as you liked). You could drag and drop drum, bass, guitar, and piano loops, and boom: an instant backing track for your new song.

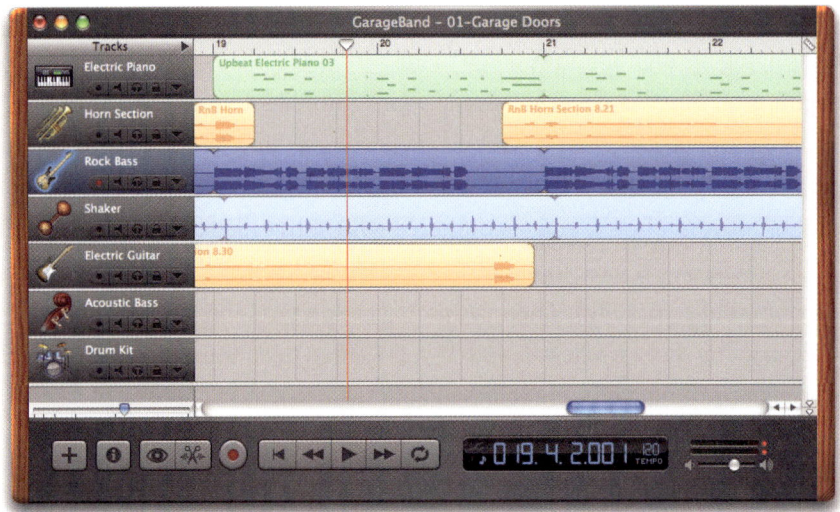

GarageBand combined live recordings with canned loops.

Jobs had a name for the new app immediately: GarageBand.

Apple bought a German company called Emagic, whose Logic software was already popular with pro musicians. Apple would maintain both apps, GarageBand and Logic—consumer and pro, just as it did with iMovie and Final Cut Pro, or iPhoto and (for a while) Aperture.

Only nine months after the acquisition of Emagic, GarageBand became the final app in the suite Apple was now calling iLife (iTunes, iMovie, iPhoto, iDVD, and GarageBand). Jobs demonstrated it with obvious delight onstage at the 2004 Macworld Expo, with pop star John Mayer on hand to create a new song in real time.

Within months, recognizable GarageBand loops began popping up in actual pop songs, like Rihanna's "Umbrella" and Usher's "Love in This Club." But GarageBand's target audience was songwriters who'd never had the tools to produce polished recordings. "Someday," Jobs would say, "some kid is gonna see that little guitar icon in their Dock, and click on it, and teach themselves about music. And they're gonna become the pop star of their generation."

In 2015, a 13-year-old girl in L.A. did exactly that. She found GarageBand on her Mac, started fooling around with it, and, with her brother, posted a couple of songs on SoundCloud.

Her name was Billie Eilish.

Intel

By 2005, all of Apple's hardware and software had been Jobsified: gorgeous, easy to use, beloved. But even Steve Jobs could not budge the needle on the market-share problem. Apple sold 3 million computers a year; Wintel sold 200 million.

For Motorola, which made the PowerPC G4 chips, and IBM, which made the G5, Apple's business constituted a minuscule fraction of their revenue. They had little incentive to invest in newer PowerPC chips, and hadn't been releasing chips on the schedules they had promised.

Jobs had spent a year and a half attempting to light a fire under its two chip suppliers. But during that time, he'd also been taking his customary long walks through the Stanford campus with Paul Otellini, Intel's president, discussing the nuclear option: Macs with Intel inside.

Windows on Mac

From the day Apple announced in 2006 that all new Mac models would come with Intel inside, geeks and bloggers salivated. "Macs and PCs now use exactly the same memory, hard drives, monitors, *and processors*," they thought. "It shouldn't be very hard to make a Mac run Windows!"

That idea would appeal to potential switchers, who were tempted by the Mac's sleek looks, yet worried about leaving Windows behind entirely. Or the people who loved Apple's iLife apps, but had jobs that relied on some piece of Windows-only corporateware. Even true-blue Mac fans occasionally looked longingly at certain Windows-only games.

In 2005, Apple finally opened the door to that long-fabled situation—Windows on a Mac—just enough to make it possible. Boot Camp was a free program that could restart a Mac into Windows. At that point, it would be a full-blown PC—with full speed and full app compatibility—with no trace of the Mac on the screen. You'd have to restart the Mac *again* to return to the world of OS X.

You could also buy, for about $80, a "virtualizer" program like VMware or Parallels (shown here), which ran Windows *in a window*, without leaving Mac OS X. Thousands ran Mac and Windows programs side by side, even copying and pasting documents between them.

It had taken only 13 years, but the dream of the Star Trek project had finally come true.

Back to the Macs • 391

Of course, switching to Intel would mean yet *another* brain transplant for the Mac. Apple had pulled off similar transitions twice already—once when it moved from Motorola to PowerPC (1994), and again when it switched to Mac OS X (2001). But each transition entailed pain and inconvenience for Mac fans and developers.

Fortunately, Apple had had a huge head start. Since NeXTSTEP had already been adapted to run on Intel, Mac OS X on Intel wouldn't be such an overwhelming project. In fact, as Jobs took the stage at WWDC 2005, he was ready to spill some beans.

"I have something to tell you today. Mac OS X has been leading a secret double life for the past five years," he said. On the screen behind him, an aerial view of Infinite Loop appeared. "In that building right *there*, we've had teams doing the just-in-case scenario": developing a version of Mac OS X for Intel processors. "Today, for the first time, I can confirm the rumors that every release of Mac OS X has been compiled for both PowerPC and Intel. This has been going on for the last five years—just in case."

There was not much applause at this announcement—first, because Jobs himself had spent years casting the Wintel alliance as the evil empire. All those ads mocking Intel! The snail, the bunny suits! Second, the audience of developers really didn't know what this meant for them. They'd all written their apps for PowerPC processors—*now* what? Would they have to rewrite their apps yet again?

Yes and no. Apple built into Mac OS X Tiger an emulator called Rosetta, a reference to the Rosetta stone, an artifact discovered in 1799 that let archaeologists translate Egyptian hieroglyphs. When you opened a Mac app that had been written for an older processor, Rosetta sprang into action, invisibly and automatically, to translate the code instructions into Intel's chip language. Most existing programs ran fine that way, at about 75 percent as fast as they did on PowerPC Macs.

Overall, the transition to Intel went remarkably smoothly. The first two Intel machines, an iMac and a MacBook Pro (Apple retired the name PowerBook), looked identical to their predecessors. You wouldn't even know they contained Intel—Jobs refused to put the blue "Intel Inside" sticker on his machines, as Intel generally required—except that these Macs were much faster and used less power.

Diagnosis

Under Steve Jobs, Apple's string of successful design, technology, and marketing achievements seemed almost miraculous. But inside Apple, there was a darker story.

During 1997, the Year of the Turnaround, Steve Jobs exhausted himself. It was a year of putting out brush fires to save Apple, but it was also the year he got kidney stones: excruciatingly painful, jagged calcium oxalate crystals that form in the kidney and get lodged in the ureter. Dehydration, stress, poor diet, irregular eating, and reduced physical activity are all contributors, and Jobs endured all of it.

Five years later, when Jobs's urologist recommended a follow-up scan of his kidneys, something shadowy appeared on his pancreas.

Commencement

In February 2005, Steve Jobs turned 50. Stanford University invited him to give the commencement speech to its 2005 graduating class in June. Jobs never gave this kind of talk—standing at a podium, reading from a script, no slides, no products—but he accepted.

He wrote the speech himself, with feedback from Tim Cook and Jobs's wife, Laurene. She and the three Jobs kids—Erin, Eve, and Reed—were in the audience.

"I never graduated from college. Truth be told, this is the closest I've ever gotten to a college graduation," he began. "Today, I want to tell you three stories from my life."

The first story was about "connecting the dots": how discovering the calligraphy class at Reed College led him to champion fine typography on the Mac. The second was about being evicted from Apple in 1985, "the best thing that could have ever happened to me," he said. "It freed me to enter one of the most creative periods of my life."

The final story was about his cancer diagnosis. He was healthy now, he said, but the episode gave his final advice particular authority: "Your time is limited, so don't waste it living someone else's life. Don't be trapped by dogma, which is living with the results of other people's thinking."

He finished by describing the final issue of the *Whole Earth Catalog*, the counterculture magazine he'd loved as a teenager. On the back cover was a photograph of a country road, with the words: "Stay Hungry. Stay Foolish."

"And now, as you graduate to begin anew, I wish that for you," Jobs said. "Stay hungry. Stay foolish."

As Apple approached its 50th anniversary, the talk has been viewed 75 million times on YouTube.

It was the only commencement speech Jobs ever gave.

Pancreatic cancer is one of the deadliest and most aggressive cancers; it's usually not diagnosed until after it has already spread. At that point, not even 2 percent of patients live more than five years.

Jobs's cancer was, however, a rare, slow-growing type—pNET (pancreatic neuroendocrine tumor)—that's easy to remove with surgery. Survival is measured in years, not months.

But in 2003, Jobs declined the operation. "I really didn't want them to open up my body," he told biographer Walter Isaacson. Over the pleadings of his doctors, he spent nine months trying to heal himself with carrot and fruit juices, acupuncture, and herbal remedies.

By July 2004, the cancer had grown. Now, at last, Jobs agreed to the recommended surgery: a modified Whipple procedure, which involved removing part of the pancreas.

The day after the operation, Jobs revealed his situation to the company, and mentioned that his recovery was likely to take months. "While I'm out, I've asked Tim Cook to be responsible for Apple's day to day operations, so we shouldn't miss a beat," he wrote. "I look forward to seeing you in September. PS. I'm sending this from my hospital bed using my 17-inch PowerBook and an Airport Express."

His recovery was slow, difficult, and painful. The cancer had spread to his liver. He began chemotherapy.

Tim Cook was a rigorous, demanding, and principled interim CEO. In personality, he was the anti-Jobs: calm, quiet, unflappable, with little interest in the spotlight.

Change was coming to the rest of the ET, too. Fred Anderson stepped down in 2004.

Avie Tevanian and Jon Rubinstein left in the fall of 2005. Tevanian had been working with Jobs for 16 years. Now, with a wife and young kids, he felt ready to move on from launching another product every nine months (and taking urgent calls from Jobs on vacation).

Rubinstein, too, was tired. When it was becoming clear that Jobs would have to step down, Rubinstein proposed that he and Cook could run the company jointly. When Jobs declined that idea, Rubinstein planned an 18-month winding-down plan, and finally left in April 2006.

Jobs's executive team now consisted of Cook, newly promoted to chief operating officer, Jony Ive (design), Phil Schiller (marketing), Scott Forstall (software), Bob Mansfield (hardware), Eddy Cue (services), and Peter Oppenheimer (chief financial officer).

When Jobs returned to Apple that September, he dove into work with renewed urgency.

36. iPhone

The invention that turned Apple into a world-beating, billion-selling, society-changing colossus was not a laptop or a music player; it was the iPhone. It seemed to appear in 2007, fully formed, beautifully conceived, self-assured, and conceptually obvious.

But behind the scenes, the iPhone we know today was made possible by more than bold bets, fanatical attention to detail, brilliant design, and a vision for the future; there were also false starts, last-minute redesigns, and a few strokes of luck.

For starters, the product Apple set out to build first was not a phone. It was a tablet.

The Multitouch Demo

Interdisciplinary teams at Apple are always experimenting with fledgling technologies. "There's hundreds of little startups that are just poking around, doing stuff," says sensors VP Myra Haggerty. "Sometimes someone's like, 'Hey, come look at what we're working on!' Then you go into some random lab somewhere, and they're doing this really cool thing. 'What could we do with this?'"

Take, for example, Duncan Kerr's projector demo.

Phil Schiller, Tony Fadell, Jony Ive, Steve Jobs, Scott Forstall, and Eddy Cue.

In 1999, Kerr, a British designer with a polymath design background—engineering, technology, industrial design, interface prototyping—had joined Ive's ten-person industrial design studio.

In early 2003, he began holding Tuesday meetings with interface designers and input engineers to explore new ways of interacting with computers; after all, the old "point mouse, click button" routine was 25 years old. Kerr's team experimented with technologies like camera-driven systems, spatial audio, haptics (vibrating feedback), and 3D screens. "We'd invite research people in, or companies who had some curious technology. We did a lot of demos, tried stuff out," he says.

Kerr was especially intrigued by the idea of manipulating on-screen objects with fingers. But mocking up ideas on paper could take the team only so far. He, along with interface designers Bas Ording and Imran Chaudhri, wanted to build a real-world multitouch display to continue their explorations. Enter: the iGesture NumPad mouse/touchpad.

It was a flat, black trackpad, 6.25 x 5 inches, made by a Delaware company called FingerWorks. Wayne Westerman was a piano player and repetitive stress sufferer; with his professor John Elias, he'd invented a set of keyboards that required barely a feather's touch. Because they could detect and track multiple finger touches simultaneously, they could also interpret *gestures* that you drew on the surface, replacing mouse actions. For "Open," for example, you could twist your fingertips on the surface as though opening a jar.

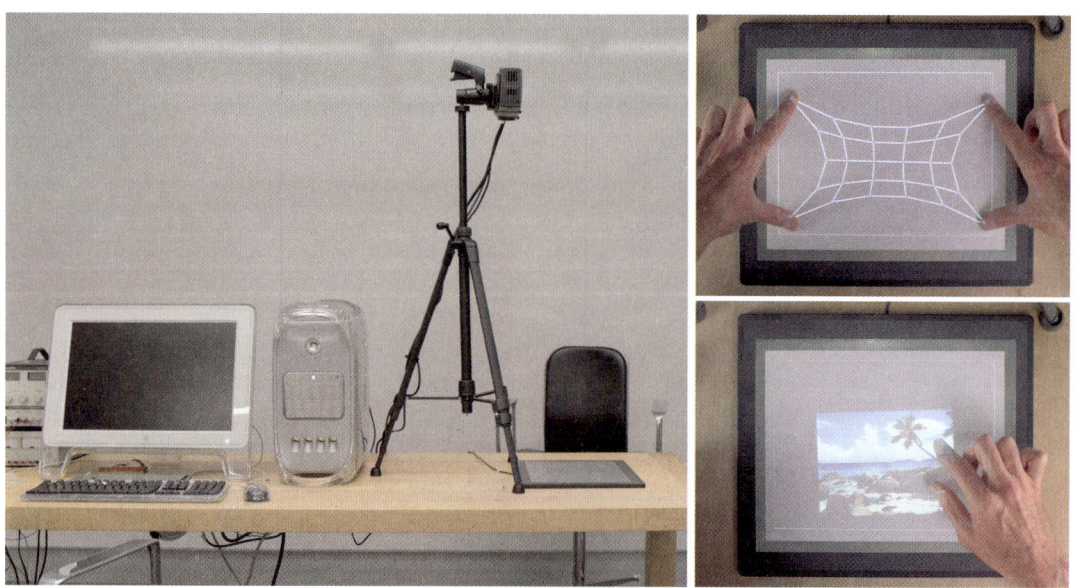

In 2018, to assist Jony Ive with a presentation about creativity and nurturing ideas that don't yet address a defined problem, Kerr's team precisely re-created their 2004 projector/multitouch touch pad demo.

In late 2003, Apple commissioned FingerWorks to build a bigger version of their multitouch pad: 12 x 9.5 inches, a better approximation of a computer screen's size. In the design studio of Infinite Loop 2, Kerr's team set up a test rig in the corner of the conference room. They mounted an LCD projector on a tripod, shining directly down onto the trackpad. They taped a sheet of white paper over it so that the projector's image—generated by a nearby Power Mac—would be bright and clear.

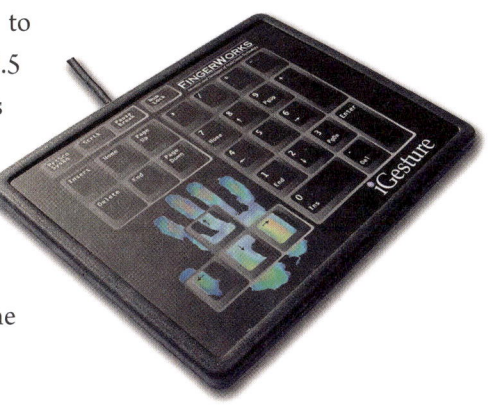

The FingerWorks iGesture was intended for people with repetitive stress injuries.

Then the fun began: developing ways to interact with the on-screen elements. You could slide a finger to move an icon in the projected image. You could spread two fingers apart to enlarge a map or a photo. Using both hands, you could tap, move, and stretch objects.

It was magical.

In November 2003, Kerr's team showed the demo to Ive, who showed it to Jobs. *Everyone* who saw the multitouch demo loved it, swore that it was the future. Of what, they weren't yet sure.

The Tablet

In late 2005, Jobs attended the 50th birthday party of a Microsoft engineer, the husband of a friend of his wife, Laurene. Over dinner, the guy lectured Jobs on how Microsoft had solved the future of computing by inventing a tablet with a stylus: portable, powerful, untethered.

"But he was doing the device all wrong," Jobs said later. "This dinner was like the tenth time he talked to me about it, and I was so sick of it that I came home and said, 'Fuck this. Let's show him what a tablet can really be.'"

Jobs came in to the Monday-morning ET meeting riled. "We need to show the world how to create a *real* tablet," he said, and by that, he meant no stylus. "God gave us ten styluses," he often said, waggling his fingers.

The FingerWorks demo suddenly seemed as though it would be incredibly useful. Apple bought FingerWorks outright, bringing Westerman, Elias, and their patents to Cupertino.

Using iBook laptop components, Ive's team built prototype multitouch tablets that ran Mac OS X, but they weren't compelling. In 2005, a page-sized touchscreen required a fast processor, which required a fat battery; the prototypes were disappointingly heavy and thick. Worse, the Mac's OS wasn't a good fit for operation by finger touch.

The First Phone

Plenty of people take credit for planting the idea of a phone in Steve Jobs's head, but that idea didn't actually need planting. By 2005, cellphones could already play music. They were crude and limited, but the writing was on the wall: Nobody wanted to carry two different devices. The iPod's days were numbered.

But Apple had zero experience with phones—no engineers, no designers, no contacts in the cellular industry. At the board's recommendation, therefore, Jobs partnered with a veteran phone maker: Motorola.

It was a logical choice. Moto's processors had powered Macs for years, and its thin, shiny RAZR flip phone was a huge bestseller. The plan was to add iPod software to a phone that Motorola had already designed. It would be the first phone that could play music from the iTunes Store, the main source of music purchases for 80 percent of the music-downloading public.

When rumors swirled that Apple was working on a phone, Apple's stock spiked, photoshopped concept images circulated online, and Apple fans got foamy at the mouth. Would it look cool, like an iPod? Would it have a click wheel? Would it hold thousands of songs? Would you be able to download songs directly to the phone? Would you be able to use your songs as ringtones, so that the phone bursts out in "Da Ya Think I'm Sexy?" when your partner calls?

Apple's first phone, built with Motorola.

The answer to all of those questions was no. The Motorola ROKR E1, as it was called, was a homely, cluttered, plastic phone. No matter how much room was left on its memory card, it could hold only 100 songs. Without FireWire or USB 2, transferring songs took forever. It was internet-connected, but couldn't download songs.

"The frustrating part was, people kept calling it the Apple phone or the iTunes phone," says Greg Joswiak. "It was like, 'Trust me: We had nothing to do with this. We created iTunes; they created the phone.'"

When Jobs unveiled the ROKR at a September 2005 special event, his disgust was evident. "It's a pretty cool phone," he managed, to no applause.

And the ROKR wasn't alone. *All* cellphones were awful.

"We just hated them; they were so awful to use," Jobs said. The market, he noted, was huge—a billion phones a year—four times the number of PCs sold. "It was a great challenge: Let's make a great phone that we fall in love with," Jobs went on. "We've got the technology. We've got the miniaturization from the iPod. We've got the sophisticated operating system from Mac."

Of course, an Apple music-playing phone would shoot the iPod business in the head. But better Apple, he reasoned, than a rival.

P1 and P2

The shortest route to a music phone would be adding phone features to the iPod. Jobs asked Tony Fadell, who'd been running the iPod business, to mock something up.

Fadell's little team came up with various approaches. One would be a full-screen video iPod—"like a tiny little iPad before the iPads," Fadell says. It could have a virtual click wheel: It could show up on the screen when you needed to scroll through lists, and then disappear for video playback.

Another prototype was a standard iPod with cellular guts.

In both cases, the problem was the wheel. It was fantastic if you were scrolling through a list of phone numbers—but entering text was a nightmare. You'd have to scroll to one letter at a time. Even with an assist from word predictions, it was an exercise in insanity. "We tried for weeks and weeks and weeks to try to make that happen, but it never worked," Fadell says.

And then someone remembered the multitouch experiment that the design studio folks had assembled a couple of years earlier. Maybe a shrunken-down version of that technology could be the interface for a *phone*. Maybe the phone would be *a full multitouch screen*—no tiny thumb keyboard! *Nothing* on the front surface but screen!

"Put the tablet project on hold," Jobs said. "Let's build a phone."

By this point, the multitouch group, including Kerr, Ording, and Chaudhri, no longer needed their projector setup. They had developed stand-alone hardware: 12-inch multitouch iBook screens, still tethered to Power Macs. To represent a phone screen, they limited the "live" area of the screen image to a phone-sized rectangle.

Ording had been an interface designer at Apple since 1998, specializing in cool animations. In the animation program Macromedia Director, he mocked up a demo of a Contacts app containing 200 names. You could flick your finger on the screen to scroll the list; tap someone's name to open their "card"; and then tap a phone number to open a fake dialing screen.

The best part, though, was *inertial scrolling*. Flick your finger on a web page, for example, and it would keep scrolling as though on its own momentum. You could keep flicking to scroll faster; if you stopped flicking, it slowed to a stop, as though obeying physics.

When you reached the end of the page, "instead of just stopping abruptly, it would actually have sort of a bouncy feel to it," Ording says. "So it's not only more fun to use, it's also functional." The little bounce *told* you that you've reached the edge. It was an overwhelmingly compelling, joyous, natural-feeling bit of tech magic.

Now there were two teams working on phone concepts. Fadell led the iPod+phone project, code-named P1. Jobs assigned the all-screen phone, P2, to senior software designer Scott Forstall, who'd joined NeXT in 1992 right out of Stanford, had followed Jobs to Apple, and had led the design of Mac OS X's Aqua interface and Safari web browser.

Scott Forstall
Born: 1969, Hawaii
Schooling: Stanford
Before Apple: Microsoft, NeXT
Apple: 1997–2012
After Apple: consulting, investing, mentoring, philanthropy, consulting, producing, investing, photography

Forstall has become a producer of documentaries and Broadway shows, winning an Emmy for *How to Survive a Pandemic* and a Tony for *Fun Home*.

Jobs encouraged both teams to charge forward at full tilt for six months. The competition between the two projects became either friendly or toxic, depending on who's telling the story.

Finally, it was time to choose one of the two projects as the winner; Apple would put all its chips on the most promising one.

After watching the latest demos from both groups, Jobs found P2, the multitouch, all-glass phone, vastly more challenging and complex—but much, much cooler. "We all know this is the one we want to do, so let's make it work," he said.

Fadell would oversee hardware; Forstall would develop the software. There was one phone now, and its code name was Purple.

From the beginning, however, the all-screen phone idea had a towering, potentially insurmountable problem: *How would you type on this thing?*

The giant of smartphones at the time was the BlackBerry, a business phone that was half screen, half tiny plastic keyboard. "Crackberry" fans adored that fast, accurate keyboard. How could a phone with *no* keyboard hope to compete? Each on-screen key was one-third the width of your fingertip. Surely entering text would be an exercise in frustration and typos.

But Jobs insisted that a virtual, on-screen keyboard could be a revolutionary advantage. The keyboard could appear whenever you needed to type, and then disappear the rest of the time. The keys could change depending on what you were doing: numbers for a calculator, symbols for passwords. And you could change languages or writing systems on demand. No BlackBerry could do any of that stuff.

OS

What OS should run on the new phone? Early on, Fadell advocated adapting the Pixo software that already ran the iPods. Rubinstein favored Linux, a Unix-like OS that already came in a low-power mobile version. But Forstall and Avie Tevanian championed Mac OS X itself as the OS for the new phone. It would require massive shrinking; the Mac version was bloated overkill for a phone. But OS X was solid, polished, and mature, crash- and virus-resistant.

"Nobody had ever thought about putting operating systems as sophisticated as OS X inside a phone," Jobs said later. "We had a big debate inside the company whether we could do that or not. And that was one where I had to adjudicate it and just say, 'We're going to do it. Let's try.'"

Secrecy

Jobs had always believed that secrecy was critical in making a product successful. But the secrecy on Purple was next-level.

Jobs told Fadell and Forstall that they could poach anyone in the entire company for their teams—but could not hire anyone from outside. And when they did interview someone, they couldn't say for what. "You're going to have to work insanely hard hours for years—like seven-days-a-week kind of thing," Forstall told candidates. "But if you join this team, you're not just going to brag to your kids that you were on this original team to build this product whose name I can't tell you—you're going to brag to your *grandkids* one day."

Jobs set up the software team in an off-campus building, soon called the Purple Dorm, equipped with multiple badge checkpoints and security cameras.

Ive's studio in Infinite Loop 2 was already a heavily protected sanctum, with badge access, cameras, and frosted exterior windows. But now the glass doors were frosted, too. Black cloths covered every prototype, and curtains hung around any areas where the user interface was being tested. To represent the screen image, the hardware designers were given a 3.5-inch piece of paper with a fake printout of icons on it.

In hopes of reducing the chance that engineers realized that they were working on the same project, Apple used several different code names. The number of people who knew about the whole Purple project, hardware and software design, you could count on one hand.

Apple told its suppliers, like Samsung (processor), Marvell (the Wi-Fi chip), and CSR (Bluetooth chip), that their components would be going into a new *iPod*; Apple even supplied fake schematics and design documents to support the lie. Apple reps sometimes made appointments at those companies under the names of *other* companies so that nobody would see a bunch of security badges lying on the receptionist's desk that said "Apple."

Most of the people who worked on the iPhone at Apple never saw the product they'd worked on until the moment Jobs held it up onstage at the Macworld Expo.

For Forstall and Fadell, the all-screen phone concept triggered a thousand questions. Nobody had ever done a multitouch screen. How would you select text with your big fat finger? How small could app icons be while still being tappable? How would the screen distinguish between intentional and accidental touches?

Ording and Chaudhri spent months of experimentation, trying to make all of this feel natural, physical, and precise. To simulate what you'd see as you tapped various controls, they built up layers in Photoshop, or mocked up animations in Macromedia Director.

Eventually, Display Engineering built for the Purple team a rig they called the Wallaby: an iPhone-sized, multitouch screen tethered by a thick cable to a blue-and-white Power Mac G3, which did the actual computing. (As software engineer Ken Kocienda notes, that means that the first iPhones ran on PowerPC chips!)

On the Wallaby, they could experiment with the swiping, tapping, and pinching gestures that would make the iPhone famous. They could see how things would *feel*.

Forstall was intensely focused on making the phone feel responsive: no delay between your finger dragging and the image on the screen following it. The video used to create this illusion ran at 60 frames a second—twice as smooth as TV video.

Two generations of the Wallaby. The earlier model (left) had only a dimple where the Home button would be.

Two Weeks

Forstall's team created mockups and demos to show Jobs, five or six options for each one. Jobs would say, for example, "I like number four, and let's then add this piece of number two." But at one of these design review meetings in 2004, Jobs announced: "This isn't good enough."

"Yeah, we're working on it," Forstall replied.

"No, this just isn't *nearly* good enough. I don't know if you guys are *capable* of making it good enough."

Jobs threw down an ultimatum. He would cancel all of their meetings for two weeks. "I'm coming back here Monday two weeks from now, and you better have something a lot better."

Forstall booked hotel rooms across the street for the entire design team. "They worked like 168 hours a week for two weeks," he says. "It was insane, the amount of caffeine that was imbibed there. Everyone was there the whole time, working insanely hard hours, brainstorming, trying stuff—and we had *so* many breakthroughs in those two weeks."

In those two weeks, the team had come up with, Forstall says, 70 percent of what makes an iPhone an iPhone: the Home screen grid of apps, the way a list (for example, Contacts) slides to the left when you tap to open a Details screen, the Back button that returns you to the list, inertial scrolling of song and contacts lists, and much more.

Jobs was uncharacteristically silent as he watched the demos. He didn't jump in along the way, as he usually did. After watching the presentation twice, he finally smiled. "We've got it. This is it," he said.

"It was great work," says human-interface VP Greg Christie. "And our reward for doing a great job on that demonstration was to kill ourselves over the next two and a half years."

The Keyboard Solution

Text entry, however, remained a showstopper. At one design-review meeting in the fall of 2005, Forstall found himself unable even to type his name; the keys were simply too small to tap with precision. Apple was betting the company on this phone, and the phone was betting everything on the virtual keyboard.

Phone software engineering director Henri Lamiraux once again stopped the train. "Starting from now, you're all keyboard engineers," he told the entire 15-person software team. They would focus *exclusively* on the keyboard until it worked.

A month later, they gathered in a conference room for a keyboard derby. Each designer presented his on-screen-keyboard variations.

They had played with a long list of approaches. In one, you tapped dots and dashes, like Morse code. Another let you generate entire words at once by pressing clusters of keys simultaneously. One proposal featured huge buttons, impossible to miss even with fat fingers, with multiple letters on each one; you'd tap repeatedly to cycle through its letters, as though on a 2003 flip phone.

Forstall's favorite design displayed all the letters on 11 keys. The phone compared your typing

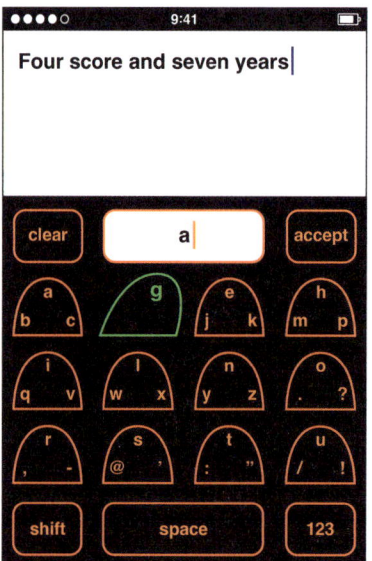 

Two of Kocienda's keyboard designs. Left: To enter the uppermost letter on a key, you'd just tap; to enter one of the lower letters, you'd slide to that side. Right: In the derby-winning design, the software figured out the word you're thinking of.

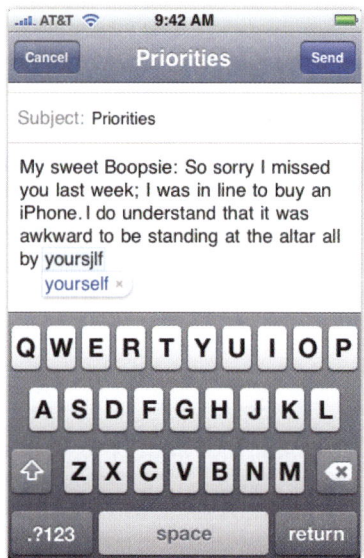 

Left: Autocorrect took care of most typos. Right: The landing areas of the keys changed size (invisibly to you). There is no word "weej" or "weel," so the K key's target area grew to permit sloppier aiming.

Kim's Lock-In

"Intense" doesn't even begin to describe the development of the iPhone. Everyone knew that Jobs was betting the company on this phone. "So you put all these supersmart people with huge egos into very tight, confined quarters, with that kind of pressure, and crazy stuff starts to happen," says Andy Grignon, who led the phone's radio engineering.

One Saturday night shortly before Christmas, everyone was still at the office. Kim Vorrath, Forstall's chief of staff, had been in a heated hallway discussion with an engineer. Flustered, she ran back to her office and slammed the door so hard that its handle mechanism broke. She was now trapped inside.

Somebody called security—but nobody responded. It was, after all, a Saturday night during the holidays.

Word reached Forstall, who headed over to Vorrath's office. "There's a whole crowd around outside, and Kim's kind of smiling and laughing inside now; now it's become absurd," he says.

The team jiggled, turned, pushed—anything to get the door unstuck. One engineer offered to climb into her office through the ceiling; then, of course, there would be *two* people locked inside.

Finally, someone produced an aluminum baseball bat. Forstall realized that he could use it as a lever. He put part of the bat into the door handle, pulled hard, and burst the entire handle apart. At this point, he could reach inside and pull out the entire lock apparatus. The team cheered, grinning. Vorrath was free.

In many tellings of the story, Forstall used the bat to bash the door down, rather than simply using it to pry out the lock. "No, I did not smash it like that, because smashing wouldn't have done a lot," he says.

So why do so many people remember it that way?

"It's probably one of those Mandela effect things," he says.

---

with an invisible word list, and calculated the word you meant. Forstall promptly made its author, Ken Kocienda, the new head of keyboard design.

Unfortunately, when you tried to type words *not* on the word list, like unusual names, Kocienda's idea broke down. And if you got distracted partway through a word, you'd have to backspace and start over, since the phone wasn't yet showing the word you were going for.

The most approachable design, he concluded, would be the standard QWERTY layout. Yes, the keys were too small—but maybe computational logic could make them "bigger." He developed algorithms that invisibly *changed the sizes* of the keys as you typed, enlarging the "landing area" of each key based on probability. If you typed *tim*, the phone enlarged the strike zone of the E key, greatly diminishing your chances of mistyping that last letter. The more likely the next letter, the larger the invisible target area.

There would be autocorrect, too—if you typed *imsane*, the phone would realize you meant *insane* and replace it automatically—and autocomplete: If the phone recognized a word you were typing, it would propose finishing it for you. There was a good deal of art to the science of these features; every word was weighted according to its English-language frequency, so that autocomplete would assume you meant *ball* and not *balk*.

Kocienda also created a list of words that the phone would not propose as corrections: hate speech. "We resolved that we would never provide software assistance for attempts to slur or demean," Kocienda says.

Design

Jony Ive and his team on the first floor of Infinite Loop 2 built hundreds of designs. Jobs and Ive spent hours picking apart prototypes, sketching on whiteboards, or critiquing ideas.

Ive dreamed of a sleek, silver slab with a seamless black face: an infinity pool in your hand. Jobs spontaneously decided that the screen would measure 320 x 480 pixels, exactly twice the size of the color iPod's screen. There would be no visible antenna, mouthpiece, earpiece, buttons, or keys.

At one point, Duncan Kerr was asked to develop a notification light on the front. "It didn't make it to the final thing, thank goodness," he says.

The leading design had a rounded-edged back, made of aluminum to feel cool and strong in your

Ive's team generated hundreds of prototypes. Some were flattened aluminum tubes, like iPods of the time; some had rounded top/bottom or side edges.

Slide to Unlock

In 2006, Steve Jobs's cellphone was a Palm Treo, which had a stylus-driven touchscreen and a thumb keyboard. Jobs found that it was prone to butt-dialing.

"I'd be somewhere, I'd pull out my phone, it's ringing—caller ID says 'Steve,'" says Forstall. "I answer, I hear him talking to someone else in the background—he butt-dialed me a fair bit."

The Purple team had to ensure that its all-touchscreen phone wouldn't butt-dial in your pocket or purse—or butt-open an app, butt-enter random taps, or even butt-delete emails. There had to be some big, deliberate gesture to confirm that you really did mean to wake the phone.

Forstall's team tried all kinds of gestures: a two-finger slide, a two-finger twist (as though turning a door handle), and pressing two dots simultaneously. But Jobs wanted a one-finger gesture, and it had to be a horizontal one, so you wouldn't accidentally trigger it when you slid your hand into your pocket.

The solution was "slide to unlock": a finger-height "track" whose text, graphic, and animated words communicated what you were supposed to do.

"Slide to unlock" was intuitive, joyous, and effective. It also became the heart of Apple's decade-long patent lawsuits with Samsung and others. After various wins, losses, appeals, and overturns, Apple won a $120 million victory against Samsung, which became moot when, in 2018, the companies quietly settled their case out of court.

---

palm. Unfortunately, radio waves don't travel through metal; Ive had to compromise by cladding the last top inch of the back in black plastic. Behind it were the cellular, Bluetooth, and Wi-Fi antennas.

On the top edge was a Sleep/Wake switch and a headphone jack. On the left edge: a pair of volume keys and a silencer switch—a gift to anyone whose phone had ever rung at a bad time. It was a tiny, fingernail-operated flipper switch whose hinge was parallel to the phone's length, so that you couldn't accidentally flip it while sliding the phone out of your pocket.

Apple's debate culture was in full force on the topic of physical buttons. Ive felt that *any* button on the face of the phone would interrupt the infinity pool. Software VP Bertrand Serlet favored *four* buttons across the bottom, as on the PalmPilot.

Forstall held out for *one* button, centered beneath the screen. No matter how lost you'd become within some function, the Home button would take you back to the Home screen, the phone's version of the Finder on the desktop. The Home button ensured that the iPhone would be the easiest-to-use, least-frustrating phone on the market.

The Network

Apple didn't own a network of cell towers, but Jobs detested the thought of having to cede control to a cell carrier. He recoiled at the thought of Apple's brilliant, innovative ideas being squashed by a bunch of corporate suits who dictated the design of the phones on their networks. There was brief discussion of becoming an MVNO—a mobile virtual network operator, where Apple would buy airtime from an existing carrier, so that it could offer cell service on its own terms. But in the end, partnering with a brand-name carrier would offer better profit margins and chains of phone stores in which to sell the Apple phone.

Jobs asked Eddy Cue to start setting up meetings with carriers.

Five companies dominated the U.S.: Verizon, AT&T, Cingular, Sprint, and T-Mobile. All of them ran on one of two network flavors:

- **GSM** was the dominant standard outside the U.S., the obvious choice for a global phone. The American GSM networks were Cingular, AT&T, and T-Mobile.

How the iPhone Got Glass

The day after the iPhone unveiling at the Macworld Expo, Jobs showed his team an upsetting twist: Somehow, in the preceding days of pulling it out of his pocket, its hard-coated plastic screen had gotten visibly scratched.

"We can't do plastic! We need to do glass!" he told them urgently.

The team was aghast; the phone would be shipping in six months. All the testing, antenna tuning, and component ordering had been completed, based on a plastic screen. They'd also *tried* glass, six months earlier; in the one-meter drop tests, it had shattered every time.

"Within three to four years, technology may evolve, and we can do that," operations chief Jeff Williams told him.

"No, you don't understand," Jobs said. "When it ships in June, *it's going to be glass.*"

Jobs was relentless. He contacted Wendell Weeks, CEO of Corning, one of the world's largest producers of industrial and specialty glass, and explained that he needed glass that was incredibly thin, strong, and scratch-resistant.

Weeks remembered that Corning had developed just such a glass—50 years earlier. The secret was science: When aluminosilicate glass takes a bath in molten salt (potassium nitrate) at 752 degrees Fahrenheit, potassium ions from the salt replace the much smaller sodium ions in the glass. The bigger ions cram into the glass structure, tightening it like a drum skin and making it exceptionally tough. Corning had had a cool name for this miracle glass, too: Chemcor.

- **CDMA** offered better call quality, and could handle more calls simultaneously. Verizon and Sprint used this network type.

The problem with GSM phones was their requirement for a SIM (subscriber identity module) card, a tiny memory card that stored account information. Jobs and Ive hated SIM cards. A SIM card required an openable tray, an unsightly interruption in the phone's gleaming edges. Its compartment took up valuable space inside. And it was a clunky component that Apple didn't design or control.

Jobs wanted CDMA; he wanted Verizon. Unfortunately, Verizon imposed thick binders full of requirements for the phones on its network. Each phone tier—basic, flip phone, smartphone—required a prescribed list of features, implemented in specified ways. Verizon required that its logo must appear on the phone, and that certain Verizon apps must come preloaded. When Cue told Verizon's execs that those were nonstarters, "they just threw us out like a dog," he says.

Apple had pioneered delightful new phone features. Your text messages, for example, would appear as speech balloons in screenplay format, so you could review your entire history of exchanges

---

"We need all of that glass you can make," Jobs told Weeks.

Trouble was, Corning wasn't making *any*. Chemcor had been too expensive to find many uses besides racing-car windshields; by the early 1990s, Corning had shelved it. Weeks explained that it would be impossible to crank up a new factory, resurrect the Chemcor process, and deliver millions of screens in time for the iPhone's release.

Jobs was unrelenting. "Get your mind around it. You can do it," he said.

Weeks put his top scientists and engineers on the project. In a matter of weeks, he converted a Corning LCD-screen factory in Kentucky to making Chemcor glass. Meanwhile, product-design engineer Steve Zadesky had to find all the components needed for a switch from plastic to glass touchscreens: one company to cut the glass, another to make an electrically conductive layer, and so on. Fadell says it was like landing 200 jets on an aircraft carrier within minutes—"and all the jets were running out of fuel."

Apple pulled it off. The iPhone shipped in June with scratchproof glass, now called Gorilla Glass.

In subsequent years, Corning (with a $200 million research contribution from Apple) made steady improvements to Gorilla Glass, making it increasingly scratchproof.

The Steve Jobs reality-distortion field has had some epic successes. But convincing one of the world's largest glass manufacturers to convert a factory, resurrect a 1962 chemical process, and replace the iPhone's screen in a matter of weeks would be hard to top.

with someone. And there was Visual Voicemail, Jobs's pride and joy: For the first time in history, you'd be able to see a *list* of your voicemail messages in an inbox, like email, so that you could listen to them in any order. But these features required some accommodation by the cell carrier, and Verizon wasn't willing to make them.

Cingular's executives, however, were thrilled to work with Apple. They agreed to accommodate Visual Voicemail, to give Apple control over the design, pricing, and marketing of its phone, and to allow changes in the sign-up procedure. Cingular also agreed to offer an unheard-of, flat-fee, unlimited-data subscription—$20 on top of the voice plan—instead of charging by the megabyte.

Cingular was also willing to collaborate blindly. Its engineers would not see either the hardware or the software of the new phone; they'd see only individual pieces of the whole on a need-to-know basis. In fact, Apple designed a *fake* user interface to ensure that Cingular engineers remained in the dark. Cingular's CEO, Stan Sigman, did not see the phone himself until the day Jobs unveiled it to the public.

Jobs even talked Cingular into giving Apple about 10 percent of iPhone customers' cellular payments—something no carrier had ever done.

All Cingular asked in return for this enormous leap of faith was a four-year exclusive.

As a happy bonus, GSM circuitry was more compact than CDMA; by choosing Cingular, Apple's phone could be thinner than it would have been on Verizon. Jobs and Ive regretted the SIM card, but they immediately began dreaming of an electronic, SIM-less GSM technology. (Apple brought the eSIM to market in 2018.)

The phone would go on sale in June 2007. But as a cellular device, it would require FCC approval, which meant making public filings. Jobs didn't want the phone's existence to be revealed by some paperwork leak—so in a rare breach of his "don't announce till it's shipping" mantra, Jobs planned to unveil the phone at the January 2007 Macworld Expo, six months before it was ready for sale.

But in September 2006, with the release four months away, one last radio frequency test revealed a showstopping design flaw: The antennas at the top of the phone emitted too much radiation. A recently completed study had suggested that cellphones could increase brain cancer risk (a theory that 63 subsequent studies have since largely debunked). The engineers spent every day of those last four months moving the antennas, and the black plastic portion of the case, to the *bottom* of the phone, no longer right next to your brain.

At long last, the iPhone's design seemed complete.

The Show

January 9, 2007. San Francisco's Moscone Center. At this early stage, only 100 hand-built iPhone prototypes existed, buggy and incomplete. Behind the scenes, Apple had done a good deal of jerry-rigging. Each demo phone was rigged with a tiny video-output board so that the audience could watch on the big screen. AT&T (which now owned Cingular) had built a temporary cell tower just for the show. And because the iPhone's fledgling circuitry sometimes dropped the cell signal and then reconnected, the engineers hard-coded its signal-strength indicator to show four bars at all times.

Jobs had rehearsed the presentation for five days, but it hadn't been glitch-free even once. The prototypes continued to drop calls, go offline, freeze, or restart.

The engineers had worked up a "golden path": a set of demo procedures for Jobs to follow in a tested sequence. It was a tightrope through the minefield of early, eminently crashable software. For the finale, Jobs intended to play music, show how it auto-paused when a call came in, and then go to Maps. "It was absolutely terrifying," says hardware manager Nitin Ganatra. "There's always that chance that he's going to hit some bug that nobody has ever seen before, on stage during the keynote, and it's going to be in one of the things that I'm responsible for."

Finally, it was time.

The 4,000 Expo attendees were stoked. Among them: the entire iPhone crew, along with Andy Hertzfeld, Bill Atkinson, Steve Wozniak, and the 1984 Macintosh team. Jobs looked thin, a result of the cancer surgery and subsequent medications.

"Today, we're introducing three revolutionary products," he said. "The first one is a widescreen

The first iPhone was surprisingly tiny.

> ### The Name "iPhone"
>
> iMac, iMovie, iTunes, iPod... Of course Apple would name its phone iPhone. The problem was that it didn't own the trademark; Cisco Systems, the networking-gear maker, did.
>
> Worse, only weeks before Apple's launch, Cisco introduced its own iPhones. These were voice-over-internet phones—they carried sound over the internet instead of the phone network, giving you free calls forever.
>
> When Apple called its new phone iPhone, Cisco sued for trademark infringement.
>
> In the end, the companies settled: Both could use the name. For Cisco, it was no big deal; its own iPhone was a minor product, and it wasn't worth a protracted fight with Apple's mighty legal team.

iPod with touch controls. The second is a revolutionary mobile phone. And the third is a breakthrough Internet communications device." With each phrase, a corresponding icon appeared on the big screen. With each phrase, the audience cheered.

"An iPod. A phone. And an internet communicator," Jobs repeated mischievously. "An iPod... A phone... Are you getting it? These are not three separate devices!"

A wave of hysteria broke over the crowd as they got the joke. "This is *one device*! And we are calling it iPhone."

Jobs made phone calls, sent texts, composed email, zoomed in to photos. He played music and video on the biggest screen ever put on a cellphone. The Safari browser did not show crude, stripped-down "mobile" web pages, like other phones; it showed *entire* websites, exactly as you would see on a computer screen. Email was fully formatted, complete with graphics and attachments.

Using Google Maps, Jobs looked up the closest Starbucks; then, with a tap, he called it. When the employee answered, he said, "I'd like to order 4,000 lattes to go, please?" Yes, the first public call ever made on the iPhone was a prank call.

The audience lost it.

"Just kidding—wrong number!" Jobs said, and then he hung up. ("My first impression was that he was just being humorous," Starbucks employee Hannah Zhang would say later. "He sounded like a gentleman.")

The technology was jaw-dropping. An ambient-light sensor brightened the display in sunlight and dimmed it in darker places. A proximity sensor detected when the phone was next to your face, shutting off the screen to save battery power and avoid cheekbone taps. A tilt sensor detected when you'd rotated the phone—to watch a video, for example—so that the screen image was always upright.

In the end, nothing crashed. Nothing glitched. "We had three backup phones," says Forstall. "He didn't go to any of them."

The demo was so dazzling that few noticed how much the first iPhone *didn't* have: A front camera. Zoom, flash, or autofocus on the back camera. Video recording. Copy and paste. GPS. MMS (sending photos as text messages). Voice dialing. The base model, with 4 GB of storage, cost $500 with a two-year AT&T contract—twice as costly as its rivals from BlackBerry and Palm.

But Jobs was so confident in the iPhone's prospects that he made one other announcement that day: The company would be changing its name. It wasn't Apple Computer anymore.

From now on, it was just Apple.

Aftermath

After the introduction, the Apple team fell into hugs, champagne, and tears. Some went out drinking; some went out for a meal. There was a lot of sleep to catch up on.

The iPhone news lit up the internet. Shares of Palm, BlackBerry, and Motorola began sinking even before Jobs's presentation was finished; Apple's rose 8 percent. Everyone had an opinion,

At the Fifth Avenue store in New York, the line was 1,000 people long.

The first iPhone still had room for more apps on the bottom row.

including Microsoft CEO Steve Ballmer. "There's no chance that the iPhone is going to get any significant market share," he told *USA Today*.

On June 29, 2007, after six months of breathless anticipation, the day of launch finally arrived. The iPhone had been the subject of 11,000 print articles, and a Google search turned up 69 million mentions. Starting a week before the release, Apple fans camped out in front of Apple and AT&T stores, hoping to be the first to get their hands on what some called the "Jesus phone."

Jobs, Ive, and Forstall went to an Apple Store to watch the crowds. The next night, they threw a party at San Francisco's Palace of Fine Arts for everyone who'd worked on the iPhone—and their significant others. Laurene came with Jobs.

Tech reviewers focused on the pure magic of what the iPhone's creators had achieved: the multitouch thing, the direct manipulation of objects on the screen, the finger-only navigation. Making the entire phone a screen unleashed a world of possibilities. It was a huge canvas for photos, videos, maps, and web pages. It could change its look for each app.

Critics did note the high price, of course, and pointed out that there was no memory-card slot, no way to swap the battery, and no way to add apps.

It took 74 days for Apple to sell a million iPhones; in the first year, it sold six million of them.

That was good, but nowhere near iPod good. Part of the problem was AT&T's spotty coverage, made worse by all the unlimited data all those iPhones were using. People complained of dropped calls, delayed messages, and miserable download speeds. "There's just no parallel for the demand," said AT&T's harried chief technology officer.

The iPhone's 16 apps included internet features (web, email, stocks, weather), cellphone features (messaging, calling, camera), organizer features (calendar, address book, calculator, notes), and, of course, iPod features.

But it lacked one thing that would have made it a universe-denting grand slam—and Steve Jobs himself had kept it out.

37. Apps

From the day Apple introduced the iPhone, software developers salivated over its potential. A real computer, online, in your pocket—it could be capable of so much more than 16 functions! It should have games, messaging, social media, streaming music, document editing, banking and investing apps. Why didn't Apple let people write new apps for the iPhone?

Jobs resisted the idea with every cell in his body. His mantra: Open systems = loss of control.

If he opened the meticulously crafted iPhone to random faceless developers, it could become a junkyard of ugliness, porn, and malware. "Someone could write an app that brings down the entire AT&T network," he would say.

But Jobs wasn't opposed to expanding the universe of apps—if Apple wrote *all of them*.

"I want you to make a list of every app any customer would ever want to use," he told Forstall. "And then the two of us will prioritize that list. And then I'm going to write you a blank check, and you are going to build the largest development team in the history of the world, to build as many apps as you can as quickly as possible."

Forstall, dubious, began composing a list. But on the side, he instructed his engineers to build the security foundations of an app store into the iPhone's software—"against Steve's knowledge and wishes," Forstall says.

Officially, Apple gave developers one bit of freedom: They could write "web apps," which were essentially iPhone-shaped *web pages*. They worked fine for accessing internet data, but they couldn't tap into the phone's camera, microphone, or sensors, and wouldn't work offline.

It wasn't enough. Two weeks after the iPhone's release, someone figured out how to "jailbreak" the iPhone: to hack it so that they could install custom apps.

Jobs burst into Forstall's office. "You have to shut this down!"

But Forstall didn't see the harm of developers spending their efforts making the iPhone better. "If they add something malicious, we'll ship an update tomorrow to protect against that. But if all they're doing is adding apps that are useful, there's no reason to break that."

Jobs, troubled, reluctantly agreed.

Week by week, more cool apps arrived, available only to jailbroken phones. One day in October, Jobs read an article about some of the coolest ones.

"You know what?" he said. "We should build an app store."

Forstall, delighted, revealed his secret plan. He had followed in the footsteps of Burrell Smith (the Mac's memory-expansion circuit) and Bob Belleville (the Sony floppy-drive deal): He'd disobeyed Jobs and wound up saving the project.

In March 2008, Jobs announced the App Store: a single, central catalog of every available app. Most were free; for paid ones, Apple would take 30 percent of the payment. And the kicker: To ensure their quality, Apple would hand-test *every single app* before listing it, and an automated testing app would search its code for improper activity and bugs. No app would bring down the AT&T network on Jobs's watch.

500 apps were available on the day the App Store opened—and within three days, iPhone owners had downloaded them *ten million times*. Within a month, they were spending $1 million a day on apps.

After one year, the store offered 75,000 apps, downloaded 1.8 billion times. The iPhone had become a game machine, remote control, musical instrument, chat terminal, star finder, newspaper, task manager, ebook reader, ticket kiosk, language translator, birdsong identifier, and on and on. It would go on to decimate sales of maps, consumer cameras, calculators, alarm clocks, camcorders, watches, dashboard GPS units, tape recorders, flashlights, scanners, answering machines, and radios—and, of course, iPods. Nokia, Palm, BlackBerry, and Radio Shack never recovered.

Today, the app store lists almost two million different apps, which iPhone owners download 38 billion times a year. 95 percent of them are free to download; either they show ads, or you pay to unlock more features. Apple has paid developers $320 billion over the years.

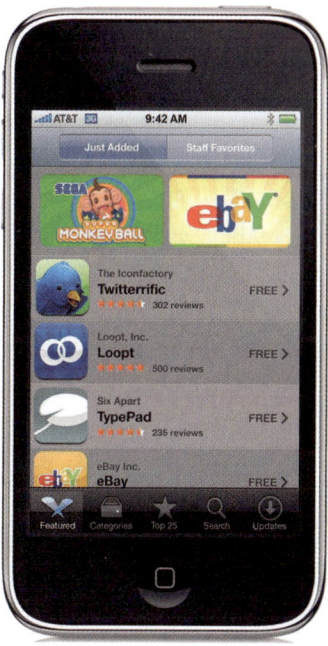

iPhone owners download 38 billion apps a year.

iPhone 3G

The first iPhone was sluggish; the camera was weak; the screen was only 3.5 inches. Worst of all, AT&T's ancient EDGE network was dog slow. It took a whole minute to pull up the *New York Times*'s home page; two minutes for Amazon.com.

Jobs wanted to release a new iPhone model every fall, just as the iPods had, and for the same reasons. Work began immediately on a second-generation phone: the iPhone 3G, named for AT&T's

Android

You can't have a success as big as the iPhone without catching the eyes of competitors—and Google was definitely paying attention.

In 2005, Google had bought a phone operating system created by two former Apple (and General Magic) engineers, Andy Rubin and Chris White. "Android" was the company's name, the OS's name, and Andy's nickname.

Google originally planned a phone with a tiny plastic keyboard. But once the iPhone went stratospheric, the Android team started over with a look-alike: multitouch screen, apps on a black home screen, swipe gestures, Visual Voicemail, and so on.

Jobs went ballistic. "I will spend my last dying breath if I need to, and I will spend every penny of Apple's $40 billion in the bank, to right this wrong," he seethed. "It's a stolen product. I'm willing to go to thermonuclear war on this." Google CEO Eric Schmidt soon left Apple's board.

Apple went on to sue Samsung and HTC for patent and design infringement. After seven years of litigation—over 50 lawsuits, countersuits, and appeals in ten countries—the companies finally settled out of court. None of it stopped Android.

Google did the Microsoft thing: It makes the software, other companies make the hardware. In fact, Google gives away Android, but charges for the inclusion of Google apps and services. (Google would eventually start making its own phones, too.)

Jobs had seen this movie before. Google's open platform resulted in exactly the kind of fragmentation and jankiness he'd feared—but rapidly claimed market share. Today, in the U.S., iPhones dominate, with 56 percent market share. But globally, where cheaper phones are in greater demand, Android holds 71 percent of the market.

---

much-faster third-generation network, which was going live in major cities. It loaded websites two to three times faster than before.

For the first time, you could now be on a phone call and on the internet simultaneously. And the phone had GPS. Google Maps still didn't offer turn-by-turn directions, but at least it could now show your current location as a blue dot. And the new phone could connect wirelessly to ActiveSync, Microsoft's system of synchronizing data with your corporate Windows computer. Maybe an Apple product would finally make inroads into the business world.

The iPhone 3G's up-front price was lower, too: $200 instead of $500. (On the other hand, the service price went up $15 a month, but few consumers did the math.) Meanwhile, Apple and AT&T had renegotiated. AT&T would no longer get the iPhones at a discount, but Apple would no longer collect a percentage of its cell-plan fees.

MacBook Air

"We love words that end in *est*," says Apple marketing head Greg Joswiak. "If it should be small, we want the smallest. Should it be thin? We want the thinnest. If it should be fast, we want the fastest."

At the January 2008 Macworld Expo, Jobs held up a standard interoffice envelope and pulled out of it the thinnest laptop in the world: the MacBook Air. It was little more than a thin

aluminum wedge, an airfoil. As the audience gasped, he pointed out that its *thickest* point, near the hinge, was still thinner than the *thinnest* point of its Sony rival. "This thing looks as if it's descended from a spatula," wrote the *New York Times* tech critic. "When it's on a table, you might mistake this laptop for a placemat."

Sacrifices had been made, of course. There were no Ethernet, audio input, or FireWire jacks, and only one USB port. There was no modem, no CD drive. And the Intel Core 2 Duo processor made the Air the slowest Mac available.

But compared to its Windows rivals, the Air was generously appointed: illuminated, full-size keyboard, built-in video camera, and a MagSafe magnetic power cord, which detached harmlessly if someone tripped on it.

The most significant advance, however, was only a footnote at the time: For $1,000, you could get a MacBook Air with a solid-state drive—nothing more than a big memory chip—instead of a spinning one. With no moving parts, an SSD is extremely rugged, better for battery life, and faster, especially in starting up the Mac and opening programs.

At the time, solid-state drives were much too expensive for the mainstream. But in time, they became available as a storage option in all Apple laptops. Today, you couldn't get a MacBook with a spinning hard drive if you wanted one.

Above all, there was now the App Store. It did for the iPhone what the Windows version of iTunes had done for the iPod: shot it into the stratosphere.

Apple sold its first million iPhone 3Gs in *three days*. Apple was now the second biggest-selling phone maker in the world. The iPhone 3G goosed Apple's revenues by 53 percent in 2008—to $37.4 billion a year. After three years, Apple had sold more than 90 million iPhones, and it was collecting more than half of the entire cellphone industry's profits.

The iPhone Era

The invention of the iPhone may not have rivaled the wheel or fire. But it did become a culture-rocking, existence-altering, universe-denting phenomenon.

It changed how we travel, socialize, and shop. It became our phone, TV, radio, camera, camcorder, newspaper, computer, game machine, and music player. It introduced the term "screen time," and raised concerns about effects on mental health, privacy, relationships, attention span, and driving. Sociologists blamed touchscreen phones for a new loneliness epidemic, for a decline in young people having sex, and even for a slowing birth rate.

The iPhone made possible business behemoths like Uber and Lyft, DoorDash and Grubhub, Venmo and Cash App, Instagram and TikTok, Instacart and Airbnb. By one analysis, the iPhone Effect generated $1.3 trillion in business in 2024.

The camera *alone* changed everything. When Jobs presented the iPhone in 2007, he barely even mentioned its small, low-resolution camera. But within months, the iPhone had become the most used camera on the planet. What Jobs had introduced wasn't "an iPod, a phone, an internet communicator" after all; today, making phone calls isn't even one of the top five uses of an iPhone. A better description would have been "a computer, an internet communicator, and a *camera*."

Everyone involved with creating the iPhone remembers knowing that they'd made something magical. But none of them, not even Jobs, knew that they'd unleashed so powerful a cultural force.

38. Sequels

Apple's iPhone division quickly fell into the iPod's release cycle: a new model every year in time for the holidays, accompanied by a new version of the software (iOS).

By 2026, Apple had released 50 iPhone models. The biggest technological leaps took place in the early years, when there were many improvements to be claimed in batteries, cameras, screen quality, and features. Video capture and copy/paste arrived with the iPhone 3GS (2009). A fingerprint sensor became part of the Home button on the iPhone 5S (2013), removing the need for a passcode every time you woke the phone.

But six models in particular marked milestones in the evolution of the world's most popular phone.

iPhone 4: Selfie Camera, Flash

The iPhone 4 (2010) introduced the iPhone's first major redesign. Jony Ive replaced the rounded back with shiny, parallel slabs of Gorilla Glass on both the front and the back, joined by a band of stainless steel—"the thinnest smartphone on the planet," Jobs pointed out.

You could no longer tell by touch which way the iPhone 4 was facing in your pocket.

The iPhone 4 also introduced the Retina screen, with so high a resolution (326 pixels per inch) that the human eye can't distinguish individual pixels at reading distance. The new LED on the back served as a camera flash, video light, and flashlight.

Finally, this model introduced a front-facing camera, making video calls possible. By 2013, "selfie" was the *Oxford English Dictionary*'s word of the year.

AT&T's four-year iPhone exclusive had finally ended, bringing the iPhone to Verizon and the other carriers at last. Jobs could finally get rid of the SIM card slot.

Software was coming along, too. There was iMovie (video editing on a phone!), multitasking (music while doing email!), and the option to organize your apps in folders.

For all of these reasons, the iPhone 4 became the bestselling model yet. Apple sold 1.7 million in the first three days of its release. After a year, iPhone 4's represented half of all the 250 million iPhones Apple had ever sold.

GizmodoGate and AntennaGate

The iPhone 4 was the subject of the two most controversial episodes in iPhone history.

First, there was the *Gizmodo* affair. In March 2010, after celebrating his 27th birthday at the Gourmet Haus Staudt beer hall in Redwood City, Apple engineer Gray Powell accidentally left his very secret iPhone 4 prototype on a barstool. A college kid found it and sold it to the gadget blog *Gizmodo* for $5,000.

Gizmodo took it apart and posted photos, six weeks before its official launch. Jobs called *Gizmodo* editor Brian Lam personally. "I really want my phone back," he said. Lam emailed back that he wouldn't return the phone unless Apple confirmed that it was a genuine prototype.

Accordingly, Apple's lawyers formally claimed the phone, and Lam returned it. But a week later, California police arrived with a warrant at *Gizmodo* writer Jason Chen's house, broke down his door, and seized his laptops, iPhone, and iPad—a PR disaster all around.

(Onstage at WWDC 2010, Jobs let the audience know that he was about to show them the iPhone 4. "Stop me if you've already seen this," he cracked.)

The second controversy was AntennaGate.

A few iPhone owners discovered that if they held their phones in what became known as the "death grip," with slightly sweaty palm flesh covering the gap between segments of the stainless-steel band, they'd drop calls. Hysteria reigned; class-action lawyers sprang into action.

Jobs was upset and, at first, defensive. "Just avoid holding it in that way," he wrote to a tech blogger.

iPhone 4S: Siri

In 2003, a nonprofit in Menlo Park called SRI International started a project to create an AI assistant, funded by DARPA, the U.S. military's research arm—your tax dollars at work.

In 2007, three SRI researchers left to commercialize the technology. They called it Siri, short for Sigrid (old Norse, "beautiful victory"), and released it as an iPhone app in February 2010. Two months later, Apple bought the company.

In October 2011, Scott Forstall took the stage to demonstrate how Siri, now beefed up and integrated into iOS, could respond to normal speech. It didn't matter if you said, "What's the weather going to be like in Tucson this weekend?" or "Will I need an umbrella in Tucson?"; Siri spoke her reply either way. (The original American voice had been recorded by Susan Bennett, a former backup singer for Roy Orbison and Burt Bacharach; the original British male voice was provided

In a press release, Apple said that the problem was not the actual antenna design; it was the on-screen signal-strength indicator. "We were stunned to find that the formula we use to calculate how many bars of signal strength to display is totally wrong," it said. "Their drop in bars is because their high bars were never real in the first place."

A software patch fixed the bars display, but didn't solve the death-grip problem. Nail polish, tape, or a case all succeeded in keeping your skin off the band gap, but those didn't seem like perfect solutions. Finally, Jobs decided to address the problem head-on at a press event.

With evident exasperation, he explained that *many* phones exhibited the signal drop when held in the death grip, including BlackBerry, Windows Mobile, and Android phones; he played videos to prove it. He noted that the problem seemed to affect only about 0.5 percent of iPhone 4 owners.

But in any case, he said, Apple would offer a free bumper (a case that covers only the phone's edges) to any iPhone 4 owner who wanted one, and would give a full refund to anyone who returned their iPhone 4.

AntennaGate evaporated rapidly thereafter, and the iPhone 4 became the fastest-selling product in Apple history.

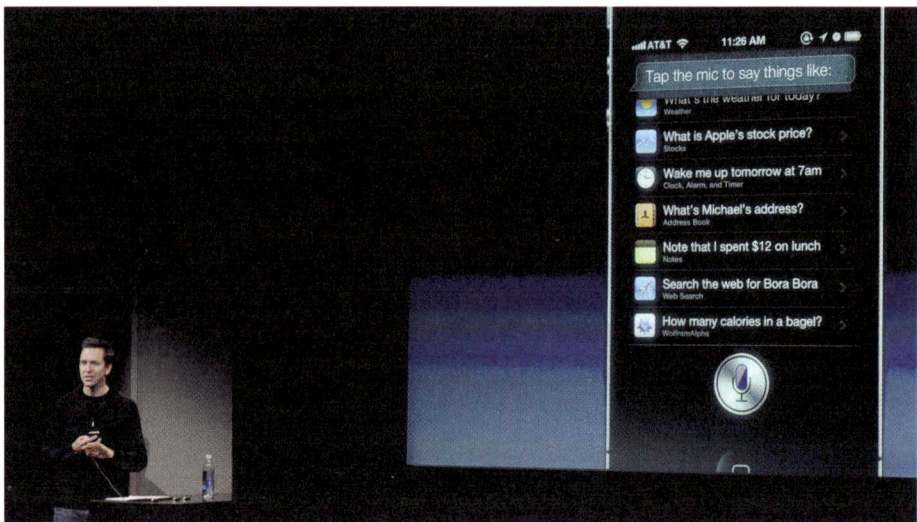

Forstall introduced the world to Siri at a special press event in Apple's auditorium.

by TV and radio host Jon Briggs, who'd been the narrator for the BBC quiz show *The Weakest Link*. Neither knew at the time that their voices would be used for a voice assistant.)

Right out of the gate, Siri could set alarms ("Wake me at 7:35"), set timers, look up numbers ("What's Robin's work number?"), make calls, send texts ("Tell Cindy I'm running late"), check or add things to your calendar, get directions ("How do I get to the airport?"), look up businesses, create location- or time-based reminders ("Remind me to water the plants when I get home"),

Siri was one of Apple's earliest AI experiments.

consult the Find My Friends app ("Is my dad home?"), search the web ("Google Benjamin Franklin"), or just ask questions ("What's a 17 percent tip on $62 for three people?").

Soon, iPhone owners were delighted to find that she'd been equipped with sass. ("What is the meaning of life?" you might ask. "I can't answer that now, but give me some time to write a very long play in which nothing happens," she might reply. Or, "All evidence to date suggests it's chocolate.")

The early Siri had trouble understanding accents, and bombed out without a strong internet connection. A subset of iPhone fans never did feel comfortable using Siri; there is, after all, something odd about conversing with a machine. But for most of those who stuck with it, Siri became an instinctive tool for getting things done, even if it was just for setting alarms and placing calls.

iPhone 5: Lightning

2012's iPhone 5 didn't offer any new *features*, but it did make news. Its screen grew from 3.5 to 4 inches diagonal, and it was 18 percent thinner. But the real headline was the death of the classic 30-pin charging connector. In its place, Apple had designed a new connector it called Lightning, what Phil Schiller called "a modern connector for the next decade."

Lightning was far smaller than the inch-wide 30-pin connector, meaning that phones and tablets could be thinner. It conducted data faster. And it was reversible: You no longer had to make sure it was right-side up.

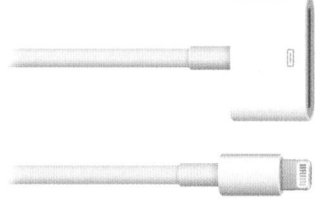

The 30-pin and the Lightning connector.

But hundreds of millions of iPods, iPads, and iPhones, cars and alarm clocks, boom boxes and charging stands were now incompatible without an adapter.

Once again, Apple had chosen newer and better over backward compatibility. Over the years, Apple had dumped a long line of connectors that it had once hailed, including data jacks (ADB, FireWire, 30-pin), monitor connectors (ADC, Mini-VGA, Mini-DVI, Mini DisplayPort), and power adapters (MagSafe, MagSafe 2).

Each time, people grumbled, but somehow muddled through.

iPhone 6: The Plus Size

The 2014 iPhone introduced another redesign: aluminum body, screen glass that wraps around the curved edges. But its real significance was its size—*sizes*, actually. Apple now offered, for the first time, a 5.5-inch model called the 6 Plus. Samsung had been crowing about its jumbo phones; Apple had responded.

USB-C: The Jesus Jack

USB was the most successful jack in the history of electronics—or at least the most common. Over 20 years, USB wound up in everything from computers to toothbrushes.

Unfortunately, nobody liked USB. Half the time, you tried to plug it in upside down. And it had three different possible shapes at the end, so you always grabbed the wrong cable.

In 2012, a consortium of electronics makers began working to develop a supercable: a tiny connector that could carry power, video, audio, and data—simultaneously. It could replace a laptop's power cord, USB, video output, and headphone jacks. It could connect to monitors, drives, phones, keyboards, and mice.

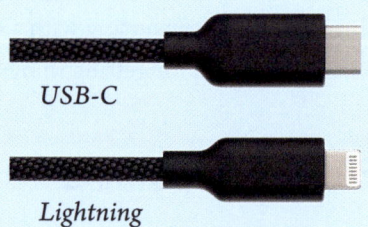

USB-C

Lightning

The connector was symmetrical, so there was no upside down. It was identical end for end, too, so it didn't matter which end you grabbed first. It charged your gadgets and transferred data faster. And the brand didn't matter. My Samsung USB-C cable could charge your MacBook and his Surface tablet. One cable to rule them all.

USB-C arrived in 2014. Apple began adding it to its laptops in 2015 (as both a charging port and a data jack), and to the iPad in 2018.

USB-C was so successful that in 2022, the European Union ruled that *all* cellphones, tablets, cameras, headphones, e-readers, keyboards, mice, GPS systems, and earbuds sold in the EU would require USB-C chargers. The era of boxes full of obsolete, incompatible chargers was ending.

That ruling was the nudge Apple needed to make its last charging-jack switch: It replaced the iPhone's Lightning jack with USB-C in 2023. The era of proprietary Apple chargers was over—and a new age of convenience and interoperability had begun.

All of this would have been a hard no from Steve Jobs, who'd died three years earlier. He'd preferred to offer only a single model—nobody had fond memories of the Performa era—and he liked his phones small. Tim Cook was putting his own mark on the lineup.

iPhone 7: No Headphone Jack

In the fall of 2016, Phil Schiller took the stage to unveil the redesigned, very thin iPhone 7. Now the phone was water- and dust-resistant. Now it had stereo speakers. On the Plus model, there was a second camera lens, so that, for the first time on an iPhone, you could zoom in. (There'd

By comparing the images from its two lenses, the iPhone 7 Plus could identify the closer image (the subject)—and blur the background.

been "software zoom," but that was just blowing up the image.) Apple had even contrived a way for these two cameras, working together, to simulate the soft-focus backgrounds of professional cameras—Portrait mode.

But then Schiller dropped the bomb: The headphone jack was gone. "Now, some people have asked why we would remove the analog headphone jack," he said. "It really comes down to one word: courage. The courage to move on, do something new that betters all of us."

In truth, Apple's engineers had been wanting to ditch the jack for years. It was an ancient, analog component that limited the phone's thinness. And *inside* the phone, the receptacle ate up space that would be better used for a bigger battery, more camera lenses, and so on.

Schiller reassured fans that the iPhone 7 would come with a pair of white Lightning earbuds, plus an adapter for regular headphones; but iPhone fans howled. Now they couldn't listen to music *while* charging the phone, because there was only one Lightning jack.

Unless, of course, they paid $160 for Apple's new, wireless AirPods.

Samsung and Google ran ads, mocking Apple for removing the jack. Bloggers had a field day. "Taking the headphone jack off phones is user-hostile and stupid," went one editorial. Cynics theorized that Apple eliminated the headphone jack to *force* people to buy AirPods or Beats. "The biggest winner from removing the headphone jack is Apple," wrote one.

In time, of course, the panic subsided. Within a year, Samsung and Google had removed *their* phones' headphone jacks, too.

AirPods

Steve Jobs always hated wires. Give him wireless networking (AirPort), wireless keyboards (Magic Keyboard), and wireless mice (Magic Mouse) any day.

But the idea for the AirPods—earbuds so wireless, they weren't even corded to each *other*—didn't come from him. In 2012, a skunkworks team of audio engineers at Apple had observed how much people despised their earbud cords: untangling them with every use, plugging and unplugging, getting them snagged.

Bluetooth headphones were already outselling wired ones, but the audio quality wasn't great, and the "pairing" procedure was a chronic nightmare. Maybe, the engineers thought, they could solve both problems.

After months of effort, they proudly presented the executive team with their prototype wireless earbuds. The cable to the iPhone was gone, but a wire still connected the two buds, dangling beneath your chin or behind your neck. A small box hung between them, containing the antenna and battery.

The executive team wasn't sold. It wasn't elegant.

Back in the lab, the engineers considered a more radical approach: Maybe the earbuds should be *completely detached*—no wires at all. All the audio gear would be self-contained in each bud—an approach accommodated by a low-power Bluetooth chip developed by a company called Passif Semiconductor, which Apple bought in 2013.

The industrial design team proposed that it might *just* be possible to cram, into each bud, audio drivers (speakers), accelerometers (to detect finger taps), optical sensors (to pause the music when you removed the AirPod), and a circuit board. The battery, antenna, and microphones would have to squeeze into a 1-inch stem.

But that idea presumed the existence of a battery that could *fit* into a tube thinner than a straw. Fortunately, there was one: a tiny, specialized, cylindrical battery called a pin coin cell, originally developed for laser sights on bow-and-arrow equipment.

Getting rid of the wires, however, presented two gigantic problems. Either of them could have killed the project. The first danger was obvious: The detached buds might fall out of your ears, especially when you were running.

Apple's original 2001 wired iPod earbuds were dime-sized disks that wedged into your ears. They weren't especially comfortable. So in 2009, Apple commissioned scientists at Stanford to make 3D images of hundreds of people's ears, using MRI scanners. The goal was to find a mathematically determined average shape that would fit the broadest number of ears. Using a CAD database, Apple engineers manipulated the earbud's surfaces to maximize their skin contact across the greatest range of ear shapes. The result, the EarPods, were still wired. But they were a far more comfortable, secure fit for most people.

For the wireless buds, hardware engineering VP Kate Bergeron and her team expanded their ear-shape database into the thousands. They built an SLR camera rig that could scan people's ears without the time and hassle of an MRI: Just brush the hair out of the way and snap. (Today, Apple uses iPhones to scan people's ears.)

Eventually, Apple engineers could be spotted all over campus wearing prototype AirPods, violently shaking their heads in an attempt to dislodge them. They rarely succeeded.

As for losing them when they *weren't* in your ears: Apple's wired earbuds already came with a plastic case that you could use to keep them tidy. You could tuck the buds into cutouts and wrap the cord around the edges.

Nobody actually bothered. But the concept gave the AirPods team an idea: If you had a case to hold the wireless buds when you weren't using them, you'd be far less likely to lose them. Nobody worried about losing contact lenses or hearing aids, after all; they just went into a case when not on your head.

As a bonus, the AirPods case could have a battery of its own. You'd charge the case, and the case would charge the buds between uses.

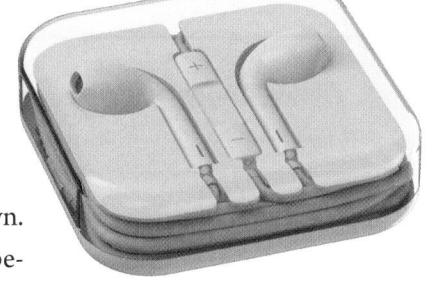

The wired EarPods case inspired the AirPods case.

(There may be no story more Apple than the effort the design studio put into the *sound* the case's lid made when it snapped shut. "That also had to be lovely," says industrial design VP Molly Anderson. "Magnet placement, magnet positioning, magnet force, the force of the spring, the choice of material, the finish of the material—all of that was carefully considered to achieve a delightful experience.")

The second problem was audio quality. The music transmitted from the phone in your pocket would have to reach receivers on each side of your head. In between was your body—basically, as the engineers liked to say, a bag of water. Radio waves don't travel through water.

"Keeping latency [audio lag] as low as possible and audio in sync if you're playing from your phone, which is in your back pocket, is a really hard technical problem to solve," says Bergeron.

The final product would only succeed, of course, if Apple could also solve the Bluetooth Hell problem. Pairing Bluetooth devices at the time—wirelessly mating them with your music player or phone—was a frustrating ritual, different on every product. It usually involved holding buttons down, entering codes, and swearing.

In the finished Apple version, you just held the case near the iPhone and tapped Connect. "The pairing went from being horrible to magical," Bergeron says.

The AirPods looked shockingly weird at first—but soon became a status symbol.

The AirPods looked like nothing that had come before them—and the reaction followed a now-familiar pattern. First, the memes: The case looked like dental floss, they said. The buds looked like Q-tips, or cigarette butts, or electric toothbrush heads. Parodies popped up. (Conan O'Brien's show made a spoof of the old iPod "Silhouette" ads. It ended with the tagline "Apple AirPods: Wireless. Expensive. Lost.")

iPod Touch

The iPod Touch (introduced in September 2007) was, in essence, a noncellular iPhone. It ran apps, did calls and texts, played movies and music, displayed ebooks, took pictures and videos, got online—but didn't contain a cellular antenna and didn't require a service plan. You used it over Wi-Fi, for free. At a time when the two-year cost of an iPhone was about $2,000 with service, the iPod Touch cost $200, paid only once.

Without cellular, you didn't get GPS navigation. But millions of teens and preteens used Touches as a sort of My First Cellphone, saving their parents thousands a year. Between school, home, and friends' houses, they were always on Wi-Fi, so why not?

By the time Apple retired the Touch in 2019, savvy fans had bought 100 million of them. Parents, kids, schools—everybody won.

Except the cell carriers, of course.

Then, acceptance: The radical design shifted to a mark of coolness. It really *was* a blessing not to have to mess with cords. The magnetic lid click turned the case into a global fidget toy.

AirPod sales, slow at first, took off with the second-generation models, which let you speak to Siri hands-free ("Hey, Siri—next track"), and with the Pro models. They gained noise cancellation and, in 2025, heart-rate monitoring.

By that point, AirPods alone were a $7 billion business; in 2020, AirPods' peak, Apple sold 114 million pairs.

iPhone X: FaceID

The Home button had been critical to the success of the original iPhone. One button to take you home! You couldn't get lost! People loved it. When Apple added a tiny fingerprint sensor to it in 2013 (Touch ID), the Home button became an even more beloved element.

And so, in 2017, Apple got rid of it.

The problem was screen space. As long as the Home button hogged the front panel, the iPhone couldn't fulfill its true destiny as a giant screen, edge to edge.

In prototype after prototype, Apple's sensor team sought other places to put the finger sensor. They tried it on the back, for example. "That would have been the easy way," says iPhone marketing VP Kaiann Drance. "But ergonomically, we just thought that was really odd. It was awkward."

Some phone manufacturers were embedding the sensor *behind* the screen glass, but the accuracy was only mediocre.

Face recognition seemed like a dead end, too. Camera-based systems were easily fooled, didn't work in all lighting, had to be retrained as your look changed, and didn't work on all faces.

But Apple had one more option in its back pocket. In November 2013, it had bought, for $360 million, an Israeli 3D motion-sensor company called PrimeSense. (The most famous use of its technology was Microsoft's Xbox Kinect: a horizontal bar, set in front of your TV, that watched your whole body to control the movement of on-screen characters.) Face recognition wasn't even on Apple's radar; at the time, the company was more interested in PrimeSense's technology and its people.

So when all potential Touch ID replacements seemed to be bombing out, one of the former PrimeSense engineers piped up: "Hey, we could solve this with depth!"

Instead of using a camera to detect the *image* of your face, he was proposing using the PrimeSense technology to map its *shape*. That way, it would work in any lighting, even in pitch dark, and couldn't be fooled by a photo.

Sensors VP Myra Haggerty and her team began building prototypes. By 2015, they'd made enough progress that they were ready to present their prototypes to the executive team, including Jony Ive and Dan Riccio, head of hardware engineering, at an off-site meeting.

The executive team loved the premise. Face ID would let the screen fill the entire front of the phone. It would take fewer steps to unlock the phone; your phone would just seem to *know* you.

Haggerty guessed that the Face ID could be ready for the 2018 iPhone, but the ET wanted it for the *2017* model, the tenth-anniversary iPhone—in two years instead of three. "We're going for this," Riccio said.

The sensing had to be rock-solid. It had to be secure and private. It had to be shrunken from Kinect size—10 inches wide—to a tiny dot. And it had to be perfected in under two years. "It's a common story at Apple: Do the impossible," says Haggerty.

Myra Haggerty
Born: Syosset, NY
Schooling: SUNY Stony Brook
Apple: 1993–present

Haggerty worked her way through college shucking clams and oysters.

Engineers from all over the company joined the effort. To train their neural network to recognize faces, they needed a machine-learning lab in a hurry; Apple became the number one buyer of second-hand NVIDIA graphics cards on eBay. Apple's facilities team rushed to install air-conditioning systems that could handle the heat of all those machines.

If Face ID was going to become the new log-in process for the iPhone, it would have to work on every face, every time: ethnicity, skin tone, gender, hairstyle, beards, glasses—everything. That meant testing on every conceivable face.

Haggerty's team held regular "test fests" in the auditorium, where employees on their lunch hours stepped up to a line of phones, trying to fool the system. On Makeup Mondays, employees were asked to show up in various makeup and costume styles.

The sensors team received permission, for the first time ever, to take iPhone prototypes out of the country so they could test humans in different global regions. Back in California, they persuaded volunteers to grow a beard for six months, then shave it off, so that the Face ID team could capture data from all stages of the process.

They attended a motorcycle rally with a mobile data-capture rig, hoping to find a lot of guys with creative facial hair. (They did.) They took their measuring equipment to a twins conference.

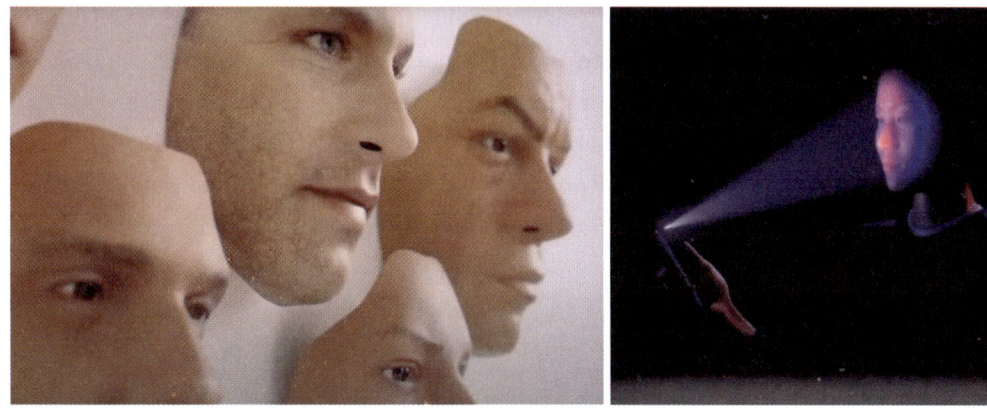

The Face ID's lab, hung with Hollywood-made fake faces, wasn't creepy at all (left). The iPhone measures the reflections of infrared light off of your facial contours.

They took it to ski resorts, to see if the infrared sensors would be fooled by light reflecting off of snow. (Apple had no shortage of employee volunteers for the ski-resort trips.)

But what about masks? Could someone get into your phone by taking a mold of your face while you were unconscious and building a latex mask of your face? To rule out that possibility, Apple commissioned Hollywood effects artists to create realistic human masks. "They were making dinosaurs one week, and then they're making masks for us the next week," Haggerty says.

In the lab, they dressed up the masks with hats or sunglasses. They bought an artificial sun—100,000 lux, four times as bright as full daylight—to expose the masks to different kinds of light. Little by little, the AI learned.

In the beginning, not everyone was even sure that face recognition would be a good idea. Debate consumed the engineering, management, and executive teams. Removing the Home button might trigger a customer outcry. "Pressing that physical Home button might've been one of the most used interactions with any product in the world," says interface-design VP Alan Dye.

How would you go back to the Home screen without a Home button? By swiping up the screen from the bottom. How would you trigger Siri with no Home button to hold down? By holding in the power button on the side.

In the end, Face ID made its deadline. On September 12, 2017, Phil Schiller unveiled the iPhone X (pronounced "ten") and Face ID. It works, he said, by shining 30,000 dots of infrared light onto your face; an infrared camera reads the distortion of their spacing and shape to find its contours. If the infrared camera confirms that you're you—if the mathematical model of your facial contours matches what it captured when you trained it—then the phone unlocks. (That model of your face, he said, is never transmitted, not even to Apple.)

Face ID can't be forced on you when you're sleeping, because you have to be *looking* at the phone. It can't be forced on you by a police officer, because you can disable it with a quick, secret button press. If you grow fat, skinny, old, or hairy, it will still work, because it gradually updates its model of your face as you use it. (If you have plastic surgery, you have to retrain it.)

Apple said that Face ID could be fooled only once in a million tries, versus once in 50,000 times for the fingerprint sensor. Bloggers and YouTubers pounded on Face ID looking for weaknesses; they found, as Apple had explained, that twins and young kids could sometimes trick it. For them, the traditional passcode was still available.

The full screen and advanced Face ID technology drove the iPhone X's price to an eye-popping new high—$1,000. But that didn't stop fans from buying 16 million iPhone X's in three months.

After the early years' initial leaps in technology, Apple found fewer game-changing new features to add; the iPhones improved each year primarily in their camera technologies, speed, and color choices. The most dramatic design experiment was 2025's shockingly thin iPhone Air ($1,000), which managed to fit virtually all of the electronics into the camera bump at the top; the rest was battery. The Air imposed some battery-life compromise and had only one lens—but was a gleaming hint at the shape of future iPhones, including a folding model.

The iPhone 6, X, 16 Pro, 17 Pro, and Air: a compressed timeline of design.

39. iPad

Steve Jobs's pancreas surgery in 2006 was not the end of his health battles. Reporters commented on how gaunt he appeared at the iPhone 3G launch in September 2008. He had lost more than 40 pounds. The internet whipped itself into a frenzy of speculation about his health.

But Apple PR chief Katie Cotton told them that "a common bug" was responsible for Jobs's weight loss. "If there's more, we'd tell you," she said.

The questions were more than idle curiosity. Under U.S. securities law, companies are required to disclose information that could affect investors' decisions, and Jobs was the CEO and driving force of Apple. (The SEC opened an investigation. No charges were filed.)

Then, in January 2009, for the first time since his return to Apple in 1997, Jobs skipped the Macworld Expo. In an open letter, he explained that he owed his weight loss to "a hormone imbalance." A week later, he announced that he'd be taking another six-month medical leave. Once again, Tim Cook would lead Apple until his return.

In truth, the cancer had spread. In March, Jobs received a liver transplant in Memphis, Tennessee. By June, he was back at work, and in September, he was onstage once again, introducing the fall's new iPod models. The standing ovation when he appeared seemed to go on forever.

"I'm *very* happy to be here with you all," he said with a wry smile.

Prototypes

In fiction, tablet computers had been kicking around for decades. They're in *2001: A Space Odyssey* (1968), for example, and *Star Trek: The Next Generation* (1987). In 1972, Alan Kay wrote a paper about a slate computer called the Dynabook (half screen, half keyboard). John Sculley's 1987 Knowledge Navigator video was about a tablet, too.

In 2000, Microsoft tried to will tablets into existence by creating a tablet+stylus version of Windows. It wound up on a few thick, heavy, button-studded models.

Publicly, Jobs told reporters that he had no interest in making a tablet. "People want keyboards," he'd say. (He even hated the *word* "tablet." When Apple Marketing proposed the name iTablet, Jobs's reaction, Dan Riccio remembers, was "visceral.")

But internally, the topic came up often. Jobs hadn't forgotten the multitouch tablet project that he'd set aside in late 2004 to create the iPhone. And in late 2008, he assembled his team to discuss a new product, code-named K48. For the first time anyone could remember, he argued that this one wouldn't require much work. "Just think of it as a big iPod Touch," he'd say.

He and Ive envisioned a magical slate, all screen, with enough margin to accommodate your gripping thumbs. It would be an incredible viewer for newspapers, magazines, ebooks, videos, web browsing, email, maps, games, and photos.

But the prevailing wisdom, even among many members of Jobs's team, was that tablets had been done, and they'd bombed. Tablets were, as the trope went, "made for people without laps"—people who never sat down, like UPS delivery people and warehouse workers.

Jobs, however, was convinced that this could be Apple's next big hit. Jony Ive worked up 20 prototype designs of various sizes. The new product couldn't *actually* be a big iPod Touch, because the aspect ratio (proportions) of its tall, skinny screen weren't a good fit for newspapers, magazines, and TV viewing. Ive and Jobs settled on 4:3 proportions, which suggested a sheet of paper or a book—and when rotated 90 degrees, a TV screen.

This time, recovering from his liver transplant, Jobs didn't have the health or the energy to oversee every detail, as he had on the iPhone. When he was too sick to come in to work, the team visited him at home to show him their progress.

"And as we started showing him slides and designs, the energy just came," Forstall says. "He would stand up and be like, 'Oh, what about this?' You talk about people who are extroverts and introverts—extroverts get their energy when they're in a crowd, and introverts get their energy from being alone? Steve was like a designovert. He got his energy when he was designing."

From a hardware standpoint, the iPad was easy. It used the same chips, the same memory architecture, and the same OS kernel as the phone.

The software, though, required some work. The iPad could run most iPhone apps, but they appeared at phone size: a little rectangle centered in the tablet's much bigger screen. App creators could adapt their iPhone apps into iPad versions without much fuss, but in the meantime, a "2X" button on the iPad screen could blow up iPhone apps to fill the iPad frame.

Keynote

As 2010 dawned, rumors had been swirling for weeks. "Last time there was this much excitement about a tablet, it had some commandments written on it," noted the *Wall Street Journal*.

On January 27, Jobs held a special press event in San Francisco. Among the audience members: Laurene, Reed, his biological sister Mona Simpson, and his surgeons.

"It's so much more intimate than a laptop, and it's so much more capable than a smartphone," Jobs said.

He began by noting that Apple had become "the number one mobile devices company in the world," bigger than Sony, Samsung, or Nokia.

A new slide revealed an iPhone on the left, a laptop on the right, and a question mark in between. "All of us use laptops and smartphones now," he said. "And the question has arisen lately: Is there room for a third category of device in the middle?"

Everyone in the room already knew the answer.

The iPad he unveiled was half an inch thick: aluminum back, 9.7-inch glass screen with a Home button beneath, no camera. The price was $500 for a Wi-Fi version, $630 (plus $30 a month) for a cellular version that could get online anywhere.

Jobs sat in a leather Le Corbusier easy chair and demonstrated web browsing, videos, email, calendar, address book, music, maps, photos.

He revealed iBooks, the ebook app and store that Apple had been quietly developing for months. When he swiped an electronic page corner, the animated page edge curled to follow his finger, as a paper page would.

As Jobs concluded the presentation, he summarized the iPad: "Our most advanced technology in a magical and revolutionary device at an unbelievable price."

No Apple introduction had ever been as polarizing. Techies spat on it. "Laughably absurd," said one online. "How can they expect anyone to get serious computer work done without a mouse?" Then there were the jokes about the name iPad, and its evocation of a feminine-hygiene product. ("Yes, the iPad is small, lightweight, and slim. But can you swim with it?" asked the *Los Angeles Times*.)

iBooksGate

As 2010 approached, Amazon ruled the ebook market. It had snagged that position, in part, by selling bestsellers at a loss—a flat $10 per book. The five major book publishers panicked, fearing that the practice devalued their books and deprived them of revenue.

For Apple's new iBooks store, Jobs proposed a different arrangement, known as the agency model: The *publishers* would determine their books' prices; Apple would take a simple 30 percent commission. It was quite a switch from the prevailing wholesale model, where a store like Amazon bought ebooks at a discounted price and then sold them for any price they liked.

The problem with Apple's deal, though, was a clause in the contract: Once a publisher had set the price for a book in the iBooks store, they could not offer any *other* store lower prices.

To the U.S. Department of Justice, that clause would "protect Apple from having to compete on price at all, while still maintaining Apple's 30 percent margin." The DOJ said that Apple and the publishers were guilty of price fixing—competitors conspiring to raise prices.

In July 2013, a federal judge ruled that Apple had violated antitrust laws. The five publishers settled, paying consumers a total of $166 million. Apple settled a separate class-action suit by reimbursing consumers $400 million. Apple denied wrongdoing to the end, arguing that its plan would have fostered competition by breaking Amazon's near-monopoly.

In some ways, however, everybody won. Despite the lawsuit, Apple's agency model soon took over the entire industry—with one difference. Today, publishers can negotiate book prices independently with each ebook store. That gives publishers a hedge against Amazon's aggressive control over ebook pricing: *They* can decide what Amazon will pay for each book title.

Of course, the name didn't really matter. As tech columnist Andy Ihnatko pointed out, "Apple could call it a 'mangled baby duck,' and people would still buy it."

And buy it they did. Sales didn't spend a year slowly ramping up, as with the iPod and iPhone; on April 3, 2010, the day iPads went on sale, 300,000 people snapped them up (along with 250,000 ebooks). Apple sold almost 20 million iPads in the first year, 50 million in the second—faster than the iPod, the iPhone, or any consumer-electronics product in *history*. "It's on track to becoming the fourth-largest consumer electronics category by the end of next year, right below TVs, smartphones, and laptops. And unlike those devices, the iPad is *a CE device category of one*," noted the *Wall Street Journal*.

"It's very easy, when something new launches, that naysayers come out of the woodwork and beat it down and say why it's not going to work," Schiller says. "No one, *no one* predicted, in their highest estimates, that it would have been as successful as it has been."

Launching the iPad in 2010 gave it the built-in momentum of the App Store. An early criticism of the iPad, notes then–iPad product marketing VP Michael Tchao, was: "What's it for?" But thanks to the App Store, people answered that question themselves. "It's for everything," he says. "And so this thing that people didn't know what to do with became a huge instant success."

For most people, manipulating digital materials by touching them was a new, deeply satisfying experience. The iPad became a perfect, goof-proof computer for technophobes, older folks, and younger ones. Caretakers called it a revolutionary communication and socialization tool for children with autism. It cut down on paperwork and records-checking time in hospitals. Musicians used iPads as digital music stands.

It was an efficient business tool, too; iPads wound up in half of the Fortune 100 companies within 90 days. The iPad became a digital checkout register at stores and restaurants. Magazines and newspapers raced their apps to the App Store.

Airline pilots saw a sudden drop in *back injuries* as they adopted the iPad to replace the 25-pound stack of manuals, navigation charts, and taxi charts that the FAA requires them to carry. ("The single biggest source of pilot injuries: carrying those packs," said American's operations VP.)

Clearly, the iPad was an ideal consumption device—for books, music, video, photos, web, email, and so on. But as Apple fell into its usual pattern of constantly upgraded iPad models, it worked steadily to make it a better *creation* tool. There would be docks, keyboards, screen covers with keyboards, and

Under Tim Cook, the iPads splintered into multiple sizes and prices, from the Mini to the Air and Pro.

FlashGate

The iPad seemed to be the world's greatest web-browsing machine, with one confounding omission: It couldn't play Flash videos.

Flash, a browser plug-in, permitted the playback of animations and videos on the web, including sports clips, news shorts, games, and all the TV shows on Hulu. Flash was standard on computers. But in an April 2010 open letter called "Thoughts on Flash," Steve Jobs explained why the iPhone, iPod, and iPad would never use Flash.

Flash, he said, was buggy, crashy, full of security holes, and a big battery drain. Many of its features didn't work unless you held your cursor still over a button—and touchscreen devices like Apple's didn't *have* cursors. Most of all, Jobs never wanted Apple to be beholden to a third party for an important software component. "We cannot be at the mercy of a third party deciding if and when they will make our enhancements available to our developers," he wrote.

He noted that YouTube, Vimeo, Netflix, Facebook, NPR, *Time*, the *New York Times*, the *Wall Street Journal*, and all the TV networks' sites had already switched to open standards like HTML5, CSS3, and JavaScript.

Adobe's executives were livid. They denied the accusations, ceased all work on Flash for iOS, and stepped *up* work on Flash for Android, Palm, BlackBerry, and Windows Phone 7.

Jobs's letter caused a firestorm. Hundreds of thousands of websites used Flash—and now, if they wanted to be accessible by iPhones and iPads, they'd have to be reworked.

In the end, though, Jobs was right. For all the reasons he had identified, Flash rapidly began disappearing from the web. By December 2020, Adobe killed off Flash for good.

---

the Apple Pencil (a stylus!). There would be iPad versions of Apple's creative apps, like GarageBand, iMovie, Logic Pro, and Final Cut Pro. In 2025, the iPad even got a menu bar and overlapping windows.

By the end of 2025, Apple had sold an estimated 750 million iPads. For Tchao, who had worked on the Newton, the iPad's success closed a personal story arc. "I used to say that my career goal was to work on a product that was even more successful than Newton," he says, laughing. "And I believe I have done that."

Convergence

The iPhone and iPad became such nutty, global, white-hot phenomena that many Mac owners soon owned iPhones and iPads, too. Now Apple not only made the whole widget; for millions of people, it made *three* widgets.

In 2008, Steve Jobs had begun thinking about ways to harness that opportunity. What if the

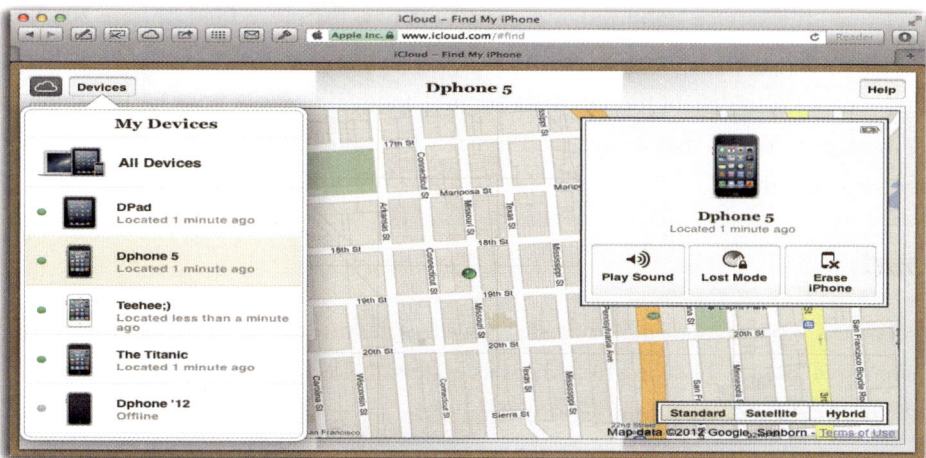

Find My iPhone located your phone across the country, across town, or in your couch cushions.

changes you made on one device—adding a calendar appointment, changing a phone number, adding a web bookmark—magically replicated across all your other Apple gadgets (and even Windows machines) via the internet?

In the summer of 2008, Apple unveiled the first iteration of this syncing idea: a $100-a-year service called MobileMe. It worked by storing the master copy of your information in the cloud. Whenever your machines were online, they connected to the mother ship and updated themselves.

Make a change on your Mac, and it appeared on your iPhone and even your PC. Change an appointment in iCal on the kitchen Mac, and it wirelessly sprouted onto your traveling spouse's iPhone four states away. Add a new friend to the address book on your Windows machine, and the same entry appeared on your Mac and iPhone. Your web bookmarks were the same everywhere. On Macs, MobileMe also synced your passwords and preference settings.

That was the idea, anyway. MobileMe's debut in July 2008 was one of the most ham-handed launches in Apple's history. There were bugs, glitches, and error messages. Some subscribers had no email at all for a day or two. A few people lost all the email they'd ever accumulated. Apoplectic customers flooded the Apple support site with angry messages, calling it "ImmobileMe" and "MobileMess."

Jobs was livid. In a tense meeting in Apple's auditorium, he berated the MobileMe team. "You've tarnished Apple's reputation," he said. "You should hate each other for having let each other down!" He replaced MobileMe's project leader with Eddy Cue, who was already managing Apple's internet services.

Apple took the unusual step of apologizing, saying that the launch "was a lot rockier than we had hoped." It offered a free one-month extension to every subscription.

Over time, the service stabilized and grew. One new feature, added in June 2009, became an instant classic: Find My iPhone, a website that showed you your lost iPhone's location on a map. You could make your distant phone play a sound (handy if it was in your house somewhere), display a message ("You've found my lost iPhone! $25 reward!"), change its password remotely, or even erase it completely.

iCloud

In the iPad era, Apple replaced MobileMe with a free, far more ambitious service called iCloud. At its unveiling in 2011, Jobs didn't hide from its history. "Now, you might ask, why should I believe them? They're the ones that brought me MobileMe!"

As before, iCloud synced your email, address book, calendar, and bookmarks among your i-devices, Macs, PCs, and the web. But it also synced your iBooks, app purchases, song purchases, and photos. iCloud quickly became an essential perk of being an Apple customer. Today, millions take it for granted that the calendar, contacts, and bookmarks are identical on their Macs and iPhones—without even knowing what iCloud is.

Merging to Macs

As a fraction of Apple's total business, the Mac had shrunk to a distant third place, after the iPhone and iPad. (Today, it's below 8 percent, by revenue.) It was a strategic move, therefore, to begin

Upside-Down Scrolling

For Apple's Mac OS team, the marching orders for Lion seemed to have been "Raid the iPhone and iPad for useful features." No wonder, then, that the first time you tried to scroll a window using your laptop's trackpad, you felt gaslit.

Sliding your fingers *upward* made the page scroll *down*—backward from the behavior before Lion. Sliding your fingers *down* had always moved the contents of the page *up*. Why would Apple throw such a monkey wrench into your life?

On the iPad, of course, the screen acts like a sheet of paper on the desk. When you push upward, the page moves upward. Apple thought that the Mac should work like that, too—but with the trackpad as the stand-in for the iPad's glass.

Veterans of the old system had two choices: Spend a couple of days getting used to the new arrangement, or visit Settings and turn off a checkbox called "Scroll direction: natural."

Or, as some old-timers preferred to call it, "Scroll direction: unnatural."

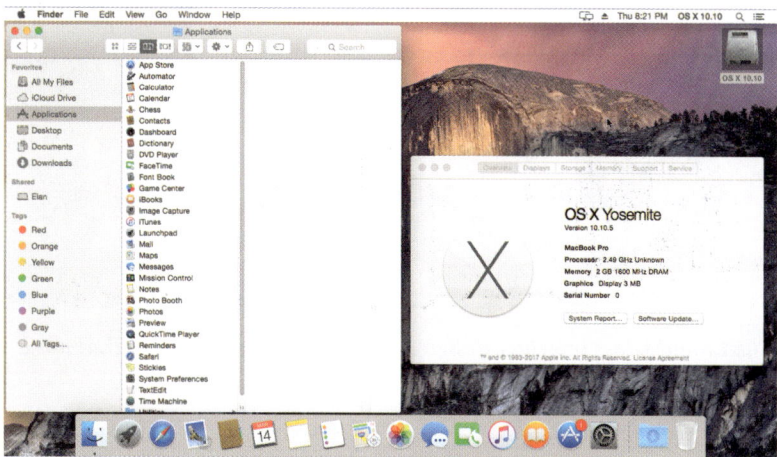

Yosemite eliminated the brushed-metal look for the Dock and windows.

bringing features, techniques, and design elements from iOS to the Mac. Making the phone, tablet, and computer look and work alike meant that the learning curve was shorter, no matter which machine you picked up first, and gave Apple's product line a consistent look.

For example, Mac OS X 10.7 Lion (July 2011) brought an App Store, push notifications, autocorrect when typing, emojis, FaceTime video chat, and AirDrop (sending files wirelessly among Macs, iPhones, and iPads, without any setup) to the Mac. In 2012, Mac OS X 10.8 Mountain Lion inherited Reminders, Notes, Notification Center, and the Messages app, so that you could text people's phones from your Mac, with the pleasures of a full keyboard. Maps and iBooks came over to the Mac in 10.9 Mavericks (October 2013).

In Yosemite (October 2014), Apple redesigned the Mac's interface to look like iOS's—no more brushed metal or 3D gelatin buttons—and launched the era of Macs, iPhones, and iPads working together as a system. A feature called Handoff let you begin a task on the Mac—open a web page, compose an email message, look up something in Maps, start working in a Pages, Numbers, or Keynote document—and complete it on the iPhone. Or vice versa. Instant Hotspot allowed the iPhone to serve as an internet antenna for the Mac, and the Mac could now be a speakerphone for iPhone calls. A new app called Photos, modeled on the iPhone app, replaced the beloved iPhoto and Aperture apps.

OS 26

In 2025, Apple's designers introduced a single, unified interface design for everything that had a screen. The screens of the Mac, iPhone, iPad, Watch, Apple TV, and Vision Pro all now look alike.

The Liquid Glass interface design unifies every Apple product that has a screen (left). Buttons appear to refract light (right).

The design was called Liquid Glass, and the premise was that all interface elements—buttons, widgets, notification bubbles, pop-up menus—were now largely transparent, as though printed on real, refractive glass.

The iPad, meanwhile, took a radical leap toward Macdom: In its fall 2025 update, it gained a menu bar (with keyboard shortcuts), overlapping windows, Exposé, Close/Expand/Minimize window buttons, the Preview graphics app, and an arrow cursor—all inherited from the Mac. After decades of refusing to build a touchscreen Mac, Apple now had a product that came surprisingly close.

As the cherry on top, Apple implemented an idea that had been kicking around internally for years: unifying the OS *version names* across all of its platforms.

As 2025 dawned, Apple had still been numbering its operating systems sequentially. The latest macOS (as Apple began spelling it in 2016) was version 15; the latest iPhone software was 18; for the Watch, it was OS11. Without a spreadsheet, it was impossible to know if your software version was the latest.

Multiple overlapping windows made the iPad even more of a Mac than before.

But in the fall of 2025, Apple renamed all of its operating systems "26," a reference to the upcoming year: macOS 26, iOS 26, watchOS 26, and so on.

With every passing year, Apple's product line developed more of a family resemblance.

Death of iPhoto

When the idea for iCloud began to take shape at Apple, one idea seemed obvious: photo syncing. You should have a single photo library, stored "in the cloud" (remote data centers), whose changes auto-propagate to all of your Apple devices. If you deleted a photo from your iPad, or snapped a photo with your iPhone, or edited a photo on your Mac... that change would show up in seconds on all of your other machines.

At the time, however, Apple offered two different photo apps. There was iPhoto, free, for basic work; and there was Aperture, a $200 app for pros. Both had been designed to run on hard drives; their underlying architecture was utterly wrong for online storage.

"You reach a point of diminishing returns when you try to reengineer this hulking old database code that was never designed with cloud use in mind," says Joe Schorr, Aperture's product manager at the time.

The photo team realized that if they wanted to deliver the magic of photo-collection syncing, they'd have to write a whole new app. It would be clean and beautiful, and it would run identically on Macs, iPhones, and iPads—but it would mean killing off iPhoto and Aperture.

"Let me tell you, end-of-lifing a product is very, very painful," says Schorr, who had to end-of-life three Apple apps within a year. "You just start to feel like, 'I'm coming in every day to kill things.' At one point, I jokingly ordered a bloodied butcher's apron from Amazon and wore it in the office."

Sure enough: The cancellation of iPhoto and Aperture in 2015 produced torrents of furious emails and calls, and online editorializing along the lines of "Apple's really made a mistake here."

Apple did what it could to soften the blow. The new Photos app made it simple to import your existing iPhoto collection, and many of Aperture's best features eventually found their way to Photos. Schorr's team even collaborated with Adobe, whose Lightroom app was Aperture's archrival, to create a software tool that imported Aperture libraries into Lightroom.

In the choice between embracing newer, better technologies and propping up aging, less capable ones, Apple chooses the newer ones every time. "Would you rather that Apple didn't innovate as much because we worked to support five years' backward compatibility? It just doesn't make sense," Schorr says.

In other words, end-of-lifing will never die.

40. China

When Steve Jobs came back to Apple in 1997, he found smoldering messes in every corner of Apple's operations. Two months' worth of computers were sitting in warehouses, unsold and rapidly losing value. He needed a crackerjack operations officer—badly.

Tim Cook was born in Mobile, Alabama, in 1960, the second of three sons; his dad was a shipyard worker, and his mom was a pharmacy worker. He majored in industrial engineering at Auburn University, then got his Duke MBA at night while working for IBM by day; over 12 years, he worked his way up to IBM's director of fulfillment (order processing). Already, his tireless work ethic became apparent: He routinely worked at the plant between Christmas and New Year's to make sure IBM didn't miss any orders.

In 1994, he joined the reseller division of Intelligent Electronics, a computer wholesaler. During his three years as chief operating officer, he became a ninja in matters of supply chain, distribution logistics, and inventory management. Compaq snapped him up in 1997 as the head of corporate materials, a job that boiled down to getting the right materials to the right places, at the right times, at the right prices.

He'd been at Compaq for only six months when Apple recruiters began calling.

In 1998, accepting a job at Apple was by no means a no-brainer. "The company had been losing sales for years and was commonly considered to be on the verge of extinction," Cook says. "The people who knew me best advised me to stay at Compaq."

But his meeting with Jobs has all the hallmarks of direct reality-distortion-field exposure. "We started to talk, and I swear, five minutes into the conversation, I'm thinking: I want to do this. And it was a very bizarre thing, because I literally would have placed the odds on that near zero, probably at zero," he says. "My intuition already knew that joining Apple was a once-in-a-lifetime opportunity to work for the creative genius, and to be on the executive team that could resurrect a great American company." At age 37, Cook became Apple's head of operations.

In some ways, Cook was the anti-Jobs: polite, reserved, soft-spoken, unemotional, but capable of becoming steely, relentless, and demanding.

And intensely private. In 2014, he discussed his sexuality for the first time. "Being gay has given me a deeper understanding of what it means to be in the minority," he wrote in a *BusinessWeek*

Cook and Jobs had wildly different personalities—but a similar drive.

essay. "It's also given me the skin of a rhinoceros, which comes in handy when you're the CEO of Apple."

Otherwise, he says, "I hate talking about me. You know, it's not something I do well or do a lot." But he allows that he's "an engineer, an uncle, a nature lover, a fitness nut, a son of the South, a sports fanatic," who packs energy bars so that he can work through his lunch hours.

In matters of drive, work ethic, and insistence on perfection, Cook and Jobs were perfectly matched. Cook awoke at 3:45 a.m., did an hour of email, hit the gym, and got to work at 6. "When you love what you do, you don't really think of it as work," he says.

In meetings, he asked question after question, or dove deeper and deeper into the spreadsheets, until he fully understood the issue—or exposed that his subordinate didn't.

"That man has a fast mind. And a grasp. And a memory—honestly, it's borderline photographic," says Joe O'Sullivan, the acting head of operations who showed Cook around.

Cook had little patience for political bickering among his staff. "I despise politics. There is no room for it in a company," he'd say. He surrounded himself with cool customers like himself. He put Deirdre O'Brien, who'd come to Apple straight out of business school, in charge of forecasting. And he invited Jeff Williams, whom he'd known at IBM—like Cook, an even-keeled southerner with a Duke MBA—to assist with parts procurement. Williams flew to Cupertino for interviews out of courtesy, but remembers thinking that "there is zero chance I'm joining that three-ring circus."

Williams was initially unimpressed by Jobs, who said things like "We're gonna be the Sony of the PC industry" and "We're gonna bring fashion to computers. We are going after the consumer."

Williams, of course, knew from his time at IBM that profit margins on computers are slim, and on *consumer* computers even thinner. "I thought: 'Guy, you do *not* know what you're talking about.'"

But soon, he, too, fell under the spell. "It was just this amazing, contagious enthusiasm. It felt to me like what I would imagine a Silicon Valley startup would be, maybe on steroids," he says. He took the job. "It was exciting to be the underdog."

Cook, who once called inventory "fundamentally evil," calculated that unsold inventory loses 1 or 2 percent of its value every week. He closed half of Apple's 19 warehouses, consolidated Apple's 100 supplier companies down to 24, and got Apple fully out of manufacturing its own computers. Within a year and a half, he'd whittled Apple's inventory from 30 days down to two. Apple had become a just-in-time computer company, building Macs as orders came in, carrying almost zero inventory.

Cook also began shipping Macs out of Asia on planes instead of ships. He didn't go quite as far as Jobs once proposed ("We should have a fleet of Apple planes that are just flying around the world full of iMacs")—but customers got Macs they'd ordered within two days.

Cook's operational efficiency, supply-chain mastery, and business execution were the perfect complement to Jobs's design thinking, product imagination, and market intuition. In 2004, Jobs added the Mac division to Cook's portfolio; in 2005, Cook became Apple's chief operating officer.

Foxconn

Until Apple's near-death experience in 1996, Apple did its own manufacturing, mostly in its Fremont, Colorado, Singapore, and Ireland factories. But when Apple contracted Sony to manufacture the PowerBook 100 in 1991, the company was amazed at the speed, quality, and low price of the result. The Asian-manufacturing seed had been planted.

Apple gave the iMac manufacturing contract to LG in Korea. But in 1999, when LG couldn't expand fast enough to keep up with demand, Apple got a call from Uncle Terry.

Taiwanese entrepreneur Terry Gou founded Hon Hai Plastics in 1974, when he was 23. He ran it out of a shed, with ten elderly employees, and he delivered the finished parts on his bike. He visited 30 U.S. states to drum up business, and gradually built Foxconn (the company's international name) into a computer-parts powerhouse.

Gou set up shop in Shenzhen, China, a "special economic zone" established by the Chinese government in the late seventies with lower taxes, looser regulations, and funding for building factories and worker dorms. By the end of the nineties, factories lined both sides of the road for 100 miles. Foxconn—"Fox" for Gou's favorite animal, "conn" for connectors—was building parts for HP, IBM, and Compaq.

In that 1999 phone call, Gou made Cook an astonishing offer: Foxconn would manufacture iMacs for far less than LG was charging. Just to accommodate Apple, Foxconn would build and equip new factories in California and the Czech Republic.

Apple soon found out the meaning of the term "China speed." Setting up the tooling for a new computer had taken LG twelve weeks; it took Foxconn 25 days.

In 2003, Foxconn told Apple that it would build a new factory just to make the iPod nano out of anodized aluminum, which has a hard, colored oxide layer that doesn't chip or peel like paint. Jon Rubinstein flew to China to inspect the new factory. But when Gou took him to the site, Rubinstein was horrified. "It was, like, this field filled with garbage! I had a heart attack."

But only six months later, Rubinstein returned to find a five-story structure in place. Only the second floor was finished, but inside it, the assembly lines were already running. "In the U.S.," Rubinstein says, "you couldn't even get the permits approved in that time."

Foxconn's low cost, speed, and reliability were freakishly good. Within a few years, Apple closed its manufacturing operations in Mexico, Wales, Singapore, South Korea, and Taiwan; even Foxconn's Czech and California plants shut down. All of it moved to Foxconn in China.

By the mid-2000s, the Foxconn campus included movie theaters, grocery stores, hospitals, restaurants, and dozens of dorms; 500,000 workers lived there at a time. Raw materials, fabrication equipment (stamping, molding, forging), assembly, and testing were all in one place, making problems easy to address and orders quick to fulfill. Steve Jobs had to retire his long-held dreams of making Apple products in robotic U.S. factories; they couldn't possibly compete.

Foxconn workers assemble iPhones in 2021.

The Foxconn arrangement allowed Cook to engineer ingenious competitive strategies. For example, Apple bought its own automation equipment and installed it in Foxconn's factories, for Apple's exclusive use. If things weren't going well, Apple could reclaim the machine and truck it to a different factory—or threaten to.

Cook also took pride in locking up an entire industry's worth of materials or equipment, preventing competitors from even getting started. When Jobs predicted huge sales for the iPod nano, which would store songs on memory chips, Apple prepaid $1.25 billion for all the flash memory Samsung and Intel could make for years. And in 2008, Jony Ive designed the MacBook Pro to be milled from a solid block of aluminum (a technique previously used only in aerospace, high-end car engines, and medical devices), which created a thinner, stronger, sleeker laptop. Apple bought 10,000 CNC (computer numerical control) milling machines—all that equipment maker FANUC could make for three years.

Apple sends hundreds of engineering and operations people to China to supervise each production line. They dictate every aspect of every tool that machines every part: how fast its bit moves through the material, at what angle, and with how much lubricant. "Their focus on quality is like nothing I've ever seen," says a U.S. supplier in Shenzhen. "If there's issues, they'll shut a line down in a heartbeat. They don't screw around."

Jony Ive spent long stretches at the factories, supervising fabrication. When Foxconn locked down its campus during the SARS epidemic in 2002, he lived in Foxconn's dormitories for six weeks—and spent seven days a week overseeing the production of the new aluminum Power Mac tower.

Crisis

Hundreds of thousands of young Chinese workers came to Shenzhen from poorer rural regions, hoping to make some quick money. Many of them had to be taught such basic city skills as crossing a road and using a bathroom. For Foxconn, the result was enormous worker churn; in some factories, the entire workforce turned over four times a year.

The work was grueling by American standards: 12-hour shifts, standing, monotonous work, no chatting allowed, overseen by almost militaristic managers. Many of the workers were teenagers, far from home for the first time, lonely and depressed.

Most of this was invisible to western consumers until 2010, when a series of worker suicides at Foxconn put Apple into one of the worst public relations crises in its history. By the end of the year, 18 workers had jumped from their dormitory rooftops; four died.

The news ripped through the global media. Investigations, including Apple's own, shone light on systemic problems: illegal amounts of overtime, underage workers, unsafe handling of hazardous waste, and industrial accidents like aluminum-dust explosions. Westerners were learning for the first time how their products from Apple, Dell, HP, IBM, Samsung, Microsoft, Sony, Toshiba, and others were made.

Jobs sent Cook to China with a team of experts to investigate. Publicly, Jobs insisted that "Foxconn is not a sweatshop," and pointed out to a customer that Foxconn's suicide rate was "well below the China average."

Foxconn hung yellow mesh nets around the buildings to catch potential jumpers—a move that critics called tone-deaf. But Apple's consultants stressed that suicide is opportunistic: Less opportunity means fewer deaths. "If the nets save one person's life, I don't care," Jobs said. The nets stayed up.

Cook promised to dive deeper. "What we will not do—and never have done—is stand still or turn a blind eye to problems in our supply chain," he wrote to the entire company.

When it came to labor conditions, Apple vowed to become the industry's most transparent company. It began publishing monthly reports about its suppliers' compliance with its 60-hour workweek policy and annual reports on supplier responsibility. It became the first tech company to join the Fair Labor Association (FLA), a nonprofit monitoring group, and invited it to inspect its suppliers' factories. It appointed a new head of supplier responsibility, whose team grew to over 100 people.

The audits continued to find some violations of Apple's policies and Chinese law, and pay, hours, and working conditions are still nothing like what an American worker would expect. But the FLA says that the reforms and Apple's involvement in worker well-being have had a huge effect.

Today, 1,500 suppliers in 50 countries make parts for Apple's products. But 90 percent of Apple manufacturing still takes place in China, where Foxconn's assembly lines produce over 3,000 iPhones a day.

The factory-conditions crisis was the first time Apple's reliance on Chinese factories turned out to be an Achilles' heel. But it would not be the last.

41. Apple Park

On June 7, 2011, Steve Jobs made his last public appearance. The audience this time was not a theater full of fans and journalists; it was the Cupertino city council. Jobs was seeking permission for a new Apple headquarters.

The company now employed 12,000 people in Cupertino alone; Infinite Loop had long since maxed out. "So we're renting buildings. Not very good buildings, either," Jobs said, his voice weak.

He narrated a slideshow of the new building. It would be a massive, glass-and-metal ring, almost one mile around—bigger than the Pentagon. "It's a little like a spaceship landed," he said. Most parking would be underground so that 80 percent of the property could be greenery.

It was up to the mayor and three council members to permit Apple to build. "We're the largest taxpayer in Cupertino," Jobs pointedly reminded them. If he didn't get the town's approval, "then we have to go somewhere like Mountain View."

He got the approval.

Origins

Jobs's health was failing; he spent much of his final two years immersed in the Apple campus project. He wanted the new campus to feel like the Stanford University quad, where low-slung buildings, connected by open-air passages, surrounded a central green. He wanted an open design: open to the outdoors, with an open floor plan that would foster spontaneous encounters among employees.

The superstar architect Norman Foster's firm, Foster + Partners, would design the building's structural, environmental, and logistical aspects. Jony Ive was, in effect, both the client and the designer, responsible for the minimalist design philosophy, the interior, and tens of thousands of details.

After considering building outlines like an amoeba and a three-leaf clover, Jobs, Ive, and Foster settled on a perfect circle. "This is not the cheapest way to build something," Jobs acknowledged. "Every pane of glass in the main building will be curved." But, he said, "we have a shot at building the best office building in the world."

Ive chose seamless wooden wall panels made of quarter-cut blond maple, harvested when the sugar content was lowest. No visible hardware would attach the custom-milled aluminum door

An aerial shot of Apple Park shows the tunnels heading to the underground parking (lower right).

handles to the glass doors. In the gleaming, all-white parking garage, the pipes and electrical conduits would be hidden in the beams. There was a "big fight" over the design of the button panels in the elevator.

The building would be powered entirely by renewable energy, much of it from the 800,000 square feet of solar panels on the roof.

In 2010, Apple got a lucky break: A 98-acre plot of land, only a mile from Infinite Loop, came up for sale. In a fine twist, it was HP's former computer-design center.

Jobs hired Stanford's chief arborist, Dave Muffly, to bring in native, drought-tolerant plants, including 9,000 trees. The pines came from two abandoned Christmas tree farms in the Mojave Desert. ("Who knew there were Christmas tree farms in the Mojave?" says Muffly.) At one end of

The roof of the Steve Jobs Theater is supported solely by the glass walls.

The corridor wall is all glass on one side, office entrances on the right.

the park inside the ring, he planted 800 apple, apricot, pear, plum, and cherry trees, to evoke the fruit orchards that stood in the area when Jobs was a kid.

To ensure that there would always be attractive ripples on the surface of the 160-foot pond, landscape architect Laurie Olin developed a hidden concentric-wave mechanism. "Apple being Apple, they decided to build a full-size mockup of a portion of the pool to study different kinds of pebbles and wave effects," he says.

For earthquake protection, Jobs wanted to use a system he'd seen in Japan, rarely used in the U.S., called base isolation. Apple Park sits on 692 enormous, stainless-steel saucers so that, in an earthquake, the building can slide safely as much as four feet in any direction, like an ice cube on a plate.

The building's dominant material was, of course, glass. The inner and outer walls—all four stories' worth—were made of 800 colossal, unbroken, curved glass panels, each 45 feet wide. The German glass foundry had to build a new, gargantuan oven—the largest in the world—to handle panels that size, and Apple had to charter a 747 to fly the finished pieces to California.

The campus also included a 100,000-square-foot fitness center, two 300,000-square-foot R&D buildings, a visitors' center across the street, and an underground auditorium, the Steve Jobs Theater. The roof of its lobby is a 165-foot-diameter carbon-fiber disk, weighing 81 tons, manufactured in Dubai. Curved glass wall panes are the only support. The floor is made of oak from the

Pandemic

Only two years after Apple's 12,000 employees settled into Apple Park, they were sent home again. "We'd built this beautiful campus designed to have everybody really close together in one space," says GarageBand product manager Xander Soren. "Turns out, that's one of the worst environments when there's a pandemic."

In general, the COVID pandemic was very good to the business of tech companies. Huge swaths of the population, now working from home, had to buy computers, phones, tablets—and services. In 2020, Apple tallied its highest revenue ever; its stock doubled.

Of course, disruptions in the global supply chain affected hardware makers like Apple, and the company experienced periodic product shortages. "We learned to be very scrappy," says Deirdre O'Brien, then Apple's chief forecaster.

At a time when people were generally unenthusiastic about gathering in theaters, Apple's product keynotes and press events had to change, too. Phil Schiller's team began producing video versions of those events.

The video productions permitted faster pacing and participation by more presenters. "We can widen the aperture to a more diverse set of employees, younger employees, especially the engineers and people who work directly on these things," Schiller says. The video keynotes were so successful that even after the pandemic ended, Apple never did return to live keynote presentations.

The pandemic also prompted one of the coolest and least known projects Apple ever undertook. Software VP Bud Tribble worked directly with his counterpart at Google to create a contact-tracing technology for iPhone *and* Android phones. A Bluetooth beacon broadcast a code to any phones within about 15 feet. If someone, a few days later, reported testing positive for COVID-19, everybody they'd exposed would get notified. It was all optional and completely private; nobody's name or data was ever collected or transmitted. You, and only you, would simply know that you'd been close to somebody who was infected.

The app didn't catch on—it required that state health agencies adopt it—but it was a historic example of archenemies laying down their swords for the greater good.

As at most companies, Apple employees worked from home for the early part of the pandemic, interacting solely through Webex or Zoom calls. In 2022, as the pandemic ebbed, Cook asked workers to come to the office one day a week, then two, then three. There were complaints in some quarters; why should they come in at all, if they'd been doing such a great job during the pandemic?

In the end, the three-days policy prevailed—and continues to this day.

Czech Republic; the leather on each of the theater's 921 seats comes from an Italian company called Poltrona Frau, which makes the seat leather in Ferrari sports cars. The reported cost per chair: $14,000.

In 2019, Apple held a grand-opening party that featured a performance by Lady Gaga. Ive designed, for that event, Apple Stage. Its six colorful aluminum arches form a rainbow of colors that match the original striped Apple logo. It's meant to be, Ive says, "a positive and joyful expression of some of our inclusion values." The rainbow became a permanent campus fixture.

The 2.8-million-square-foot Apple Park opened at the end of 2017, spotless and sweeping, but it had its critics. Architectural reviewers said that it was too isolated from its surroundings, that it reinforced car culture, that it was more of a permanent monument than a flexible workplace.

Employees reported bumping into glass walls or getting lost; instead of a central lobby, there were nine identical arc segments, each with its own, identical four-story, sun-filled atrium entrance. No artwork or other landmarks distinguish one segment from another.

As had happened throughout the history of Apple's headquarters buildings, by the time Apple Park was finished, the company had already outgrown it. The company wound up building more office parks, with far less fanfare, in the nearby city.

Steve Jobs didn't live to see Apple Park. "Every time I come here, it makes me think of the past, as well—and just the sadness. I just wish he could have seen it," Ive says.

42. Loss

In January 2011, Jobs's medical team found new tumors. Once again, he announced a medical leave. Once again, Tim Cook would cover his absence.

But this time, Jobs didn't say when he'd be back.

In March, to the delight of the developers at WWDC, Jobs took the stage to unveil iCloud. He was loose, upbeat, and funny, but frighteningly gaunt.

By July, the cancer had spread to his bones. He had trouble eating and sleeping, and he was in pain.

In August, he asked Cook to visit him at home.

"There has never been a professional transition at the CEO level in Apple," Jobs told him. "The last guy is always fired, and then somebody new comes in. I want there to be a professional CEO transition. I have decided, and I am recommending to the board, that you be the CEO, and I'm going to be the chairman."

Cook had known that he would succeed Jobs—but not *yet*.

"Are you sure?" Cook asked.

"Yes."

"Are you *sure*?"

"Yes. Don't ask me anymore," Jobs said.

Cook asked how Jobs wanted to be involved as chairman. Jobs cited what happened at Disney when Walt Disney died in 1966. "People looked around, and they kept asking what Walt would have done," Jobs said. "I never want you to ask what I would have done. Just do what's right."

On August 24, Jobs attended his last Apple board meeting. From his wheelchair, he read a prepared statement to Apple's directors.

"I have always said if there ever came a day when I could no longer meet my duties and expectations as Apple's CEO, I would be the first to let you know. Unfortunately, that day has come. I hereby resign as CEO of Apple," he said.

He concluded with this: "I have made some of the best friends of my life at Apple, and I thank you for all the many years of being able to work alongside you."

There were brief remarks of support and admiration from the board. At lunch, Schiller showed him some mockups of future products. Scott Forstall showed him an early version of Siri. ("Are you a man or a woman?" Jobs asked. "They did not assign me a gender," Siri responded.)

Jobs had intended to continue offering advice as chairman of the board, but his health failed rapidly. He answered very few of the emails of love and admiration that poured in, and declined appearances and awards. He welcomed visits only from his closest friends.

On October 4, even as Cook and his team were introducing the iPhone 4S and Siri in Apple's auditorium, Jobs's life was waning. They rushed to his home in time to say their farewells.

In the midafternoon, surrounded by Laurene and his family, he faded into unconsciousness.

Steve Jobs died on Wednesday, October 5, 2011, at age 56.

Hundreds of millions of people read the news on iPhones, iPads, and Macs.

There were essays, TV segments, and social media posts by the thousands. Flags at Microsoft, Disney, and Apple flew at half-staff. Famous people, including Bill Gates and President Obama, offered tributes. People left apples missing a bite at Apple Stores.

In some cities, crowds held up candle-flame apps on their iPhones and iPads as a vigil. "The way that I communicate and the way that I interact with the world is through things that Steve Jobs has created," said one fan.

Jony Ive was gutted. Losing his closest friend was "brutal, heartbreaking." For Cook, "It was absolutely the saddest day of my life."

The next day, 50 of Jobs's closest family members and colleagues attended a private burial ceremony at Alta Mesa Memorial Park, a century-old cemetery in Palo Alto.

Three days after that, hundreds of special and famous friends attended a memorial service at a

Apple fans hold a vigil in Georgia.

Kindnesses

Everyone who ever worked with Steve Jobs can tell you a story about his fury. And despite his vast wealth, Jobs was uninterested in philanthropy. But the stories of his acts of kindness get less play: how he often visited Bill Atkinson in the hospital after his car accident, or how he paid for a nanny when Mike Murray needed help with his kids, or how, unbidden, he gave Pixar's John Lasseter a bonus to replace his dangerously rickety Honda Civic.

Jobs had a particular soft spot for people with tough illnesses—and none were tougher than Scott Forstall's. In 2004, a horrific virus sent Forstall to the hospital. He was throwing up every 15 minutes. He lost 5 pounds, then 10, then 30. Doctors gave him every anti-nausea treatment they knew, including drugs for chemotherapy patients. Nothing helped. "I wanted to die. For weeks, I wanted to die," Forstall says.

Jobs was deeply affected. He called Forstall every day, sometimes multiple times. When Forstall couldn't talk, Jobs spoke with his wife.

"Sometimes he was very kind, and sometimes he was exhausting," Forstall says. "Because he'd be like, 'You know what he needs? He needs to do a fruit diet for a few days.' And suddenly, all these things would arrive"—the best juicer on the market, baskets of fresh fruit. "He was so, so caring."

Finally, Jobs called at ten o'clock one night with a new idea. "I have the best acupuncturist in the world, and I'm going to bring her to you tonight. She's going to fix you," he said.

Forstall considered acupuncture to be junk science, but he was out of options.

For his part, Jobs worried that Stanford Hospital might not appreciate his bringing in an unaffiliated medical practitioner. "If I get stopped, I'm just going to dedicate a wing," Jobs joked.

At midnight, Jobs snuck into Forstall's room with the acupuncturist. The treatments lasted until sunrise—and for the first time in more than two months, he wasn't nauseous. After a second session, he felt well enough to drive home. "I was 100 percent dying, and Steve brought this person to me and saved my life. So I will always, always owe him that."

Eventually, he says, the Stanford team identified his illness as a rare virus that the Mayo Clinic had spotted only seven times in a decade. Most of those patients died.

Stanford University church. Some, Jobs had held close: Andy Hertzfeld, Steve Wozniak, Larry Ellison, Pixar's John Lasseter. With others, he'd had a spikier relationship: Bill Gates, Google founder Larry Page, Michael Dell, Adobe's Chuck Geschke and John Warnock. Family members spoke or read poems. Bono sang; Jobs's former girlfriend Joan Baez sang. Yo-Yo Ma played the cello. ("Steve wanted me to play cello at his funeral," he said, "and I said I would prefer he speak at mine. As usual, he got his way.")

Tim Cook hosted the memorial for Apple employees at Infinite Loop.

On October 19, Apple hosted a third ceremony, this time on the Infinite Loop lawn for Apple employees. Apple Stores all over the world closed so that their employees could watch remotely.

At this memorial, Tim Cook and board members Bill Campbell and Al Gore spoke warmly of Jobs and his legacy. Norah Jones and Coldplay performed.

Jony Ive spoke of the months he and Jobs would spend perfecting some component that nobody would ever see. "It cost us all, didn't it?" he said. "But you know what? It cost him most. He cared the most. He worried the most deeply. He constantly questioned, 'Is this good enough? Is this right?'"

Jobs's image appeared only in photos, in the form of huge black-and-white banners hanging on the Infinite Loop walls, and his voice was heard only once: In the "Think Different" ad, which Cook played as a tribute. It was the rare version of the ad narrated by Jobs himself, the alternative to the Richard Dreyfuss version.

Jobs had recorded it in 1997, when he'd just returned to Apple, when the company's prospects hung by a thread, when the future was full of possibility, but *only* possibility.

"While some may see them as the crazy ones, we see genius," his voice concluded. "Because the people who are crazy enough to think they can change the world . . . are the ones who do."

PART 4
Tim

43. Services

The company Tim Cook inherited had 61,000 employees, and was taking in over $100 billion a year. Its total stock value was $350 billion—more than Microsoft and Google combined—and continued to climb. Only five days after Jobs's death, Apple passed ExxonMobil as the most valuable publicly traded company on earth.

Cook was universally admired for his work ethic, discipline, and operational brilliance. What he was not, however, was a product person.

In the history of hardware companies, Jobs was a one-off: He was both the CEO and the primary product guy. In many ways, Apple existed purely to execute his ideas.

Cook, on the other hand, had never dreamed up a new product. He rarely visited Jony Ive's design studio, and declined invitations to check out the progress of the software-design team.

No sooner had he taken the helm than the questioning began.

CNN Money: "Without the creative genius of Steve Jobs, is Tim Cook capable of producing another 'unicorn' product?"

New York Times: "He will have to compensate for the absence of Mr. Jobs—and his inventiveness, charisma and uncanny ability to predict the future of technology and anticipate the wishes of consumers."

The *Onion*: "Apple Unveils Panicked Man with No Ideas."

The doubts weren't unfounded; history had shown that the successors of charismatic leaders rarely fare well. Microsoft faltered when Bill Gates stepped down; profits fell at Starbucks after Howard Schultz left; GE blew it under Jack Welch's successor. Why would this be any different?

Tim Cook in 2012.

In the short term, plenty of projects were already in the pipeline. In 2012 alone, Cook and his executive team (Cue, Ive, software chief Craig Federighi, and hardware head Bob Mansfield) oversaw the releases of a staggering number of new products: the iPhone 5, the iPad Mini, the third- *and* fourth-generation full-size iPads, the fourth- *and* fifth-generation iPod Touches, the seventh-generation iPod nano, the third-generation Apple TV, and updated versions of the Mac Pro (and Mac Pro Server), MacBook Air, MacBook Pro, iMac, and the AirPort Express Wi-Fi base station. The software engineers had been busy, too, putting together iOS 6, OS X Mountain Lion, and a redesigned iTunes.

But in rapid succession, two events tested Cook's crisis-management skills. First, there was the Foxconn labor-conditions mess. Then there was Apple Maps.

Apple Maps

While developing the iPhone, Apple had struck a brilliant bargain with Google: Google Maps and YouTube (another Google property) would come preinstalled on the phone—the only two non-Apple apps on it. (To maintain the secrecy of the project, Apple incorporated Google's data, but wrote the apps itself.) Good for Apple, good for Google.

After a few years, however, that deal turned sour. Google had agreed to update the iOS version of Maps in lockstep with its Android version. But in 2009, Google added turn-by-turn navigation instructions ("At the next light, turn left") to the Android app, but not the iPhone app. Roads in the Android version gained the smooth look of vector lines (mathematically drawn curves); the iPhone version was still made of blocky pixels.

Jobs was furious. Google was violating its agreement, yet still collecting vast amounts of valuable data from *Apple's* customers: how they were using Maps, where they were going, which roads actually existed.

As the Google contract came up for renewal, tensions flared. Google threatened to withdraw its iOS app altogether. Jobs, who had *always* hated putting Apple's fate in the hands of others, went with the nuclear option: building a new maps app from scratch.

Apple bought mapping companies, licensed road, traffic, business, and satellite-photography databases, and even chartered Cessna planes to fly over major cities and landmarks to create Flyover, a 3D view of those sights—something Google had never had.

But Google had had ten years to fix and meld those disparate data sources; Apple was trying to create the entire app in one year. "We set about building an entire worldwide mapping solution in one release. I mean, it was clearly insane," says one engineer.

The resulting app worked flawlessly when the Apple team tested it in Silicon Valley. An eight-person quality-assurance team spot-checked it in other places. Maps was gorgeous, it had turn-by-turn directions, and the Flyover feature was a knockout.

So in October 2012, Apple Maps replaced Google Maps on the iPhone.

Customer reports flooded in immediately. Entire lakes, bridges, and tourist attractions seemed to have been moved, mislabeled, or erased. Satellite images stitched together scenes from different

Skeuomorphism

From the beginning, Apple's goal was to demystify computers for everyday nontechnical people. It made sense, then, that its original interface designs were filled with pictures of real-world objects. Folders. Pieces of paper. Trash cans. Disks. You knew what they did because you recognized them.

The academic term for this design philosophy is skeuomorphism (skew-oh-MORPH-ism): designing a digital element to mimic a real-world one.

Apple's designs reached peak skeuomorphism in 2012, with iOS 6 and OS X Mountain Lion. The Notes app looked like a yellow legal pad, the Calendar had fake torn-off page edges, Voice Memos depicted an old-timey TV microphone, and Contacts resembled a leather-bound Rolodex, complete with stitching and leather binding. Later apps adopted skeuomorphic *materials*: the iBooks app had "wooden" shelves, the Game Center app's background was "green felt" (like a Vegas craps table), and Mac OS X window backgrounds were "brushed metal."

By the 2010s, though, skeuomorphism experienced a backlash. Jony Ive, in particular, didn't think that the public still needed the hand-holding of digital mimicry. Skeuomorphs struck him as cluttered, dated, and condescending, especially when compared with his sleek, modern hardware designs.

Jobs and Forstall had championed skeuomorphism, even though they'd never even heard the term. The sole goal, Forstall says, was to produce something that's "approachable, friendly. You can use it without a manual. It's fun."

When Ive became Apple's lead designer for software, his first act was purging all of the faux materials in iOS 7. The new design was called flat, because it eliminated the 3D, shadow-casting look of its predecessors. "There was an incredible liberty in not having to reference the physical world so literally," he says.

Today, the faux-materials era at Apple is long over. But here and there, skeuomorphism lives on: in the inbox icons in Mail, the shutter-click sound of the iPhone, and anywhere that you see a floppy-disk icon that means "save."

seasons, weather conditions, and even years. A Jacksonville hospital was in the right place, but the wrong decade; it had become a Publix supermarket 11 years earlier.

The Brooklyn Bridge was melted into the river, the road to the Hoover Dam plunged straight down into a canyon, and Auckland's train station was in the middle of the sea.

Cook, deeply concerned, summoned a meeting of the executive team. Forstall called in from New York City, where he'd gone for a four-day weekend. Cook said that he intended to post an apology for the Maps fiasco, and wanted Forstall to sign it.

Forstall, taken aback, pointed out that Steve Jobs hadn't apologized for AntennaGate. Instead, he'd said that the phone itself was terrific, but that a small number of people had seen problems, and that Apple had a plan to make things right. Why not say something similar here?

But Cook had made up his mind. "We are extremely sorry for the frustration this has caused our customers, and we are doing everything we can to make Maps better," his open letter said. He even suggested, for the first time in Apple's history, that iPhone fans should use *other* companies' apps or websites in the meantime. He mentioned MapQuest, Waze—and Google Maps.

A few weeks later, Apple announced Forstall's departure. Jony Ive would take over software design, Eddy Cue would take on Siri and Maps, and Craig Federighi would lead Mac OS X and iOS.

The disparate databases that drove Apple Maps did not always mesh neatly.

Apple Pay

In Cook's first ten years at Apple, Apple's sales more than tripled, profits nearly quadrupled, and its stock market value septupled. But it wasn't because he oversaw the creation of another iPhone-sized monster hardware hit. A huge chunk of the surge was services.

A service, in Apple's case, is an online, software-only, subscription product: music, TV shows, workout videos, and so on. You, the customer, get one-click sign-up, smooth integration with the gadgets you already own, and, often, better pricing, design, and features than rival services. What Apple gets is an enormous profit margin, as high as 75 percent. It has no raw materials to buy, nothing to manufacture, nothing to ship. It's just software. Apple's board and shareholders enjoy a consistent and predictable cash flow, month after month.

It all began in 2014, when Apple saw another calcified, crummy consumer category desperately in need of disrupting: paying for things.

Today, anyone under 40 knows that they can pay for things with their iPhones (or even Apple Watches). Double-press the button on the side, hold next to the cashier's terminal, and enjoy the tiny *ding!* that signals a successful payment.

Apple University

In 1983, Ann Bowers, Apple's first human resources director, started up an internal training program called Apple University. Its purpose was to teach Apple's values to current and future executives. It chugged along under various HR leaders until 1997—until Steve Jobs returned to Apple and, in his frenzy of bringing focus, shut it down.

But then, in 2008, four years after his cancer diagnosis, Jobs restarted the program. He wanted two things from Apple University: first, to spread his values about product design, and second, to develop a strong, well-trained leadership team rooted in Apple's core values, so that Apple would never again flail without him.

Jobs recruited Yale business school dean Joel Podolny to run the new program. Podolny hired a faculty of 17 business professors from Yale, Harvard, Berkeley, Stanford, and MIT, and set up classrooms in the City Center building. A typical classroom incorporated 15 office chairs around a U-shaped table, facing a big screen.

Every new Apple employee takes one key course: What Makes Apple Apple, now taught by Chris Espinosa and others. According to current human resources director Deirdre O'Brien, "It's about Apple's values. How do we show up? How do we all contribute to Apple's exceptional culture? How are we pursuing excellence every day?"

Other classes are optional. Some cover critical moments in Apple history: the decisions to make a Windows version of iTunes, for example, or to consolidate all iPhone manufacturing at Foxconn. One class, originally taught by Podolny himself with former Apple interface head Greg Christie, explores what makes an innovation revolutionary. Another covers the challenges of incorporating the employees of acquired companies into Apple's culture. Today, product-line managers and others teach the classes in conference rooms all over Apple—no longer in a single building.

Joel Podolny ran the university for 12 years. The pandemic, however, showed him some of the limitations of remote learning—basically, every student is just sitting watching a video—and he founded a startup dedicated to more interactive online classes.

Evidently, even the dean of Apple University learned something at Apple University.

The first advantage: no credit card to carry, demagnetize, or lose. The second one: far better security. Your thumb on the Home button confirmed your identity (today, Face ID does the trick).

No possibility of data breaches, either. The store doesn't see, receive, or store your name or card number. Instead, the phone transmits a one-time code that only your bank can translate. Even Apple doesn't see what you bought.

As usual in Apple Land, making things effortless for *you* had required vast amounts of effort up front. For almost two years, Apple worked with the major banks in total secrecy. It was a "codename frenzy," one bank rep said. Creating the new payment system involved hundreds of people at each bank, and 1,000 people at Visa. Most of them did not know that the client was Apple, and *none* of them knew the name of the service until Cook unveiled it onstage.

Even so, when Cook unveiled Apple Pay in September 2014, it seemed like magic that you couldn't have yet. Only about 3 percent of U.S. merchants had wireless terminals, only a few banks participated, and only the latest iPhone model had the necessary wireless NFC chip (near-field communication).

In time, of course, all iPhones would gain the NFC chip, all banks got with the program, and most stores would accept Tap to Pay. In 2016, only one in ten iPhone owners used it; by 2022, three-quarters of them did. Today, about 650 million people use it to pay for $6 trillion worth of things each year. Apple collects 0.15 percent from your bank with every tap.

Paying for stuff with your phone was faster and more secure than using your card.

Apple Pay was a classic Apple play. (1) Identify a broken system, loathed by consumers. (2) Wait until you can make it substantially better. (PayPal, Walmart, Target, and Google Wallet had all attempted pay-by-phone apps; all were clumsy, and all failed.) (3) Swoop in with a more polished solution. (4) Use your massive customer base to tempt partner organizations into favorable deals.

Apple Card

Five years later, Apple returned to another personal-finance realm filled with pain points: credit cards. The Apple Card, created in conjunction with Goldman Sachs, was intended to be more secure than regular cards—the 16-digit number did not appear on the card, so no sneaky restaurant waiter could copy it down—and more generous. It offered 1 to 3 percent cash back, the industry's lowest interest rate, and no late fees, foreign fees, returned-payment fees, or annual fees. The statement, in the app, listed plain-English vendor names instead of cryptic codes.

Goldman Sachs lost $1 billion on the arrangement in the first three years, and by 2025, was in talks to hand over the business to someone else. But consumers loved it: Within five years, 12 million people signed up.

Apple Music

When Jobs introduced the world to the iTunes music store in 2003, he made clear what he thought about the $15-a-month subscription music services. "One day, you stop paying your subscription fee, and your entire music library goes away," he said. "We think subscriptions are the wrong path."

By 2015, though, the world had changed. Spotify, Tidal, and similar services had demonstrated that people *liked* paying a flat monthly fee for unlimited music. That way, they didn't have to manage and sync downloaded music files. Streaming services were also light-years more joyous to use than their 2003 ancestors, with playlist suggestions based on your tastes and no rights-management headaches.

Tim Cook decided that Apple would get into the subscription music business.

Once more, Eddy Cue began making his rounds to the record companies. This time his proposal was an easier sell, since Spotify and its ilk had paved the way.

Jeff Robbin, the iTunes author who was now vice president of consumer apps, oversaw the creation of the streaming app itself. But when Apple bought Beats in May 2014 for $3 billion, it acquired not just its headphone business, but also its Beats Music streaming service. Now Apple had two potential starting points for Apple Music: the in-house version, with a simple design reminiscent of iTunes itself, and Beats' far busier, brightly colored, album-art-heavy design. Cue made the call: Apple would go with the Beats software.

The Beats deal had also brought record executive Jimmy Iovine and rapper/producer Dr. Dre to Apple, to help Apple build relationships with artists and nail down music licenses.

The resulting app, Apple Music, was much like Spotify—you could search for songs or bands, listen to ready-made playlists, and so on—but with an Apple twist: You could tell Siri things like "Play the top songs of 2005" or "Play some good running music" or "Play some Taylor Swift." (Apple Music was, at the time, the only service to have Taylor.)

There were even some free elements, like customized "radio stations" based on songs you like, and live, hosted internet radio stations. The app also included Connect, Apple's second attempt at a music-related social media service. (iTunes Ping, the first attempt in 2010, closed after two years; Connect lasted only one.)

Apple had clearly learned from its experience bringing the iPod to Windows. It made Apple Music not only for Windows, but even for *Android phones*. Somewhere, the ghost of Steve Jobs screamed.

The cluttered design of the app didn't win much love from critics. "Quick: What's the difference between New Music, Hot Albums, and Recent Releases?" asked Yahoo Tech's reviewer. "How is a Hot Track different from a Top Song, exactly?"

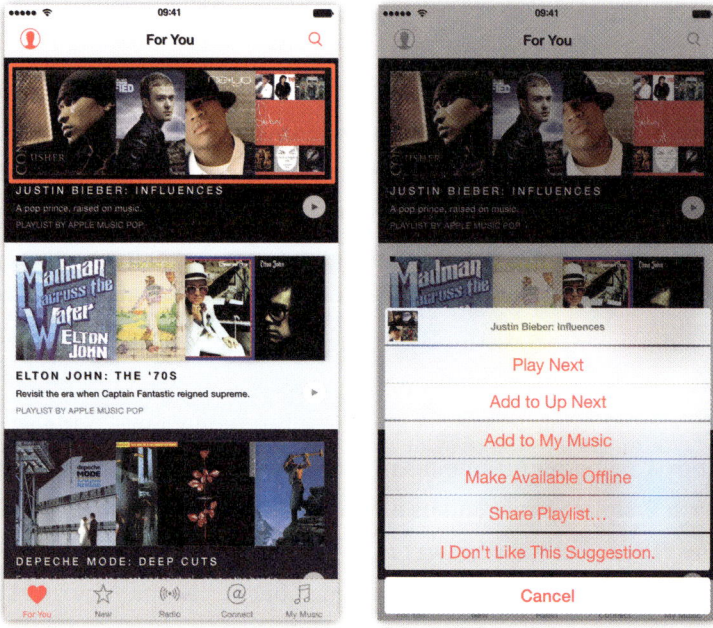

The original Apple Music app was well equipped with features.

But from a business perspective, it all worked. It took only six months for Apple Music to sign up ten million paying subscribers. After a year (and a redesign), 15 million; after four years, 60 million. In the summer of 2018, Apple Music out-subscribed Spotify.

Today, Apple Music is Apple's most successful subscription service by far, with over 100 million songs in its catalog and close to 100 million music fans paying $10 a month.

Apple News

From the dawn of the internet, Big Tech and Big Journalism have had a wary relationship. Craigslist, Google, and Facebook wiped out classified ads, a pillar of newspapers' income, and websites gutted print subscriptions.

Just after the iPad unveiling, Jobs spent two days in New York, visiting the publishers of the *New York Times*, the *Wall Street Journal*, and Time Inc. He hoped that they would create iPad app versions of their papers.

But the publishers didn't like the deal. They'd have to give Apple 30 percent of their revenue—and they'd get no subscriber details. Without the subscriber list, the publishers would have no way to develop relationships with their readers. "We need to be able to create online communities of those people, and we need the right to pitch them directly about renewing," Time Warner CEO Jeff Bewkes told Jobs.

Four years later, Apple returned to journalism again. This time, the goal was to create a universal news app on the Apple Music model: every newspaper and magazine in the world for a flat fee.

When the publishers once again hesitated, Apple proposed a test scenario: Its new app, Apple News, would offer a free set of sample articles from each publication—30 stories a day from the *New York Times*, for example. Readers could choose their areas of interest.

That was the premise of the Apple News app, which debuted in 2015. Four years later, a subscription version, Apple News+, was ready: selected stories from 300 newspapers, national and local (*Wall Street Journal*, *San Francisco Chronicle*...), and complete magazines (*People*, *New Yorker*, *Wired*...), for $10 a month—eventually, 400 publications for $13 a month.

It was a great deal for anyone who subscribed to two or more of them, even if it wasn't such a great deal for publishers: Apple split half of its revenue among all 300 of them. (In 2020, the *New York Times* pulled out of Apple News altogether.)

Apple Arcade

In early 2017, Cook reported to analysts that Apple's subscription services (a category that includes in-app purchases) had just pulled in $7.2 billion for the quarter. "Our goal is to double the size of the services business in the next four years," he said. Soon, those analysts found out what he meant; Music, Pay, and iCloud were only the beginning.

In 2019, the iPhone App Store offered thousands of games. All of the bestsellers, however, were free or $1 games that junked up the experience by displaying ads or upselling you on in-app purchases.

Arcade was an attempt to restore quality to the game catalog. It offered access to 200 iPhone games, without ads or in-app junkiness, for a flat $7 a month (for up to six people in a family).

Apple TV+

After spending years in the TV dealmaking trenches, one thing had become clear to Cue: Streaming was the next wave. He pitched a radical new business idea to Cook: Apple should launch its *own* Netflix-style streaming service. "We have a chance to build something," he said.

It would be a massively expensive and risky undertaking—billions of dollars just to get into the game—but Cue believed that Apple could pull it off. The company did, after all, have $257 billion in cash.

The Apple TV That Wasn't

Apple has dabbled in television set-top boxes three times. (That's not counting Jobs's idea for an *actual TV*; he died before he could implement it.) Two of them never saw the light of day.

The first attempt was the Apple Interactive Television Box, shut down by CEO Michael Spindler just before its release in 1993.

The second was a $300 box called iTV, which Jobs unveiled in 2006. (To placate the lawyers of ITV, a British network, Apple later renamed it Apple TV.) It was a silver-sided, 7-inch slab containing a hard drive, designed to play shows, movies, and music from the iTunes Store on your actual TV.

A 2010 version ($100) was 75 percent smaller and dedicated to streaming from sources like Netflix, HBO, cable channels, the iTunes store, and even your Mac, iPhone, or iPad. Jobs used to call the Apple TV product a hobby, but as it has become more capable over the years, fans buy a few million Apple TV boxes a year.

Then there was the most promising Apple TV product of all.

By the end of the 2010s, Jobs and Eddy Cue had achieved four astonishing feats of industry negotiation: with record labels, cell carriers, book publishers, and news publications. But there was one more industry left to conquer, the most hated in America: cable companies.

Cable boxes were hobbled by terrible interfaces and monthly rental charges. Apple's fantasy box would let you watch *live* TV in addition to streaming stuff. Its remote could pause or rewind live broadcasts. It could skip ads. It would recommend new shows based on your taste. It could record live shows for watching later. You could pull up stats on a sports team or player.

But this time, Cue wasn't trying to convince five record companies or four cellular companies to play ball. This time, he had to persuade five cable companies *and* three satellite companies *and* six entertainment conglomerates, like Time Warner and Viacom. Each entity had its own, incompatible technical standards and its own long-standing wars: Comcast despised Disney, Disney despised Comcast.

Despite years of effort, Apple just couldn't get the deals done. The cable and content companies themselves were among the biggest losers, as streaming services slowly began to eat their lunch. "There were just no standards and too many players," Cue says. "It's heartbreaking, honestly."

In 2017, Apple hired away the co-presidents of Sony Pictures Television, Zack Van Amburg and Jamie Erlicht. "If that's not enough to give Netflix and Amazon future nightmares," noted the *Hollywood Reporter*, "it's definitely going to wake them the hell up right this instant."

Cook and Cue decided that *everything* on Apple TV+ would be original, produced by a new division called Apple Studios. That way, Apple could avoid messy bidding wars for the rights to existing shows, and could maintain the rights to its own offerings forever.

Apple Studios intended to ramp up rapidly, but world events had other plans. The COVID pandemic halted TV production for almost two years; no sooner was it over than the Hollywood writers' strike shut things down again. "I was waiting for the locusts to arrive in L.A.," Cue says.

Apple TV+ finally launched in 2019, with only a handful of shows, but a correspondingly modest price: $5 a month. Anyone who bought an Apple product got a one-year free trial.

For the next few years, it was deals, deals, and deals—with stars, executives, and studios. (During the pandemic, Apple did wind up licensing some existing material, including *Fraggle Rock* and the *Peanuts* specials.) Apple asked show creators to avoid topics of politics and religion, and to avoid gratuitous sex and violence. Jon Stewart backed out of the third season of his Apple TV+ talk show when, he says, executives asked him to avoid certain hot-button topics.

In time, the operation found its stride and landed some hits, including *The Morning Show*, *Ted Lasso*, and *Severance*. A movie Apple bought at the Sundance Film Festival, *CODA* (short for "child of deaf adults"), became the first streaming-service movie ever to win the Best Picture Oscar. By the end of 2025, Apple had renamed the service, subtracting the "+." Apple TV was still losing hundreds of millions annually—but it had accrued 110 series, 45 movies, and more than 45 million customers paying $13 a month.

Apple Fitness

At the end of December 2020, Apple debuted Fitness+: a library of what became 4,000 prerecorded training videos for yoga, pilates, treadmill runs, cycling, meditation, and so on. In each, three trainers demonstrate the exercises at different difficulty levels. The tech twist is that the Fitness app connects wirelessly to your Apple Watch (required), so that your biometric stats appear superimposed on the video.

Apple One

When the cornucopia of subscriptions attained critical mass, Apple offered one more pair of velvet handcuffs: a single $20 payment for TV+, Music, and Arcade, or $38 for the whole services enchilada (TV, music, news, fitness, arcade). Each bundle cost less than half than signing up for multiple services individually.

Taken together, Apple's services—including advertising, AppleCare warranties, and iCloud storage subscriptions—now produce staggering numbers. They generate over $100 billion a year—only the iPhone division earns more—and grow 12 percent annually.

If Apple's services were a separate company, it would be number 40 in the Fortune 500, with greater revenue that Target, Johnson & Johnson, or Disney.

The clamor of concern about Tim Cook, Not a Product Person, began to subside.

HomePod

If Steve Jobs thought of Apple TV as "a hobby," what does that make the HomePod—a New Year's resolution?

It's a smart speaker, like a high-end Amazon Echo: rubber on the bottom, cloth mesh around the sides, touch-sensitive screen on the top.

But when it arrived in 2018 after some delays, critics found it more "speaker" than "smart." The sound quality was superb, thanks to seven tweeters and a gigantic woofer. But it could play music only from Apple Music—not Spotify, Pandora, and so on—and didn't work as a Bluetooth speaker.

An updated 2023 model lowered the price and the tweeter count. But the HomePod never snared more than about 6 percent of the smart-speaker market.

The HomePod Mini, a smaller, spherical 2020 follow-up in a choice of colors, is a happier story. It costs only $100 and sounds amazing for its size. It was, as *Forbes* put it, "as close as it gets to a no-brainer." The Mini became a far more popular gadget, selling about eight million a year—a perfect stocking stuffer, eye-catcher, and party-room filler.

44. Watch

For Jony Ive, Jobs's absence created both a personal and professional void. Jobs had called Ive his "spiritual partner." Ive called Jobs his "closest and most loyal friend."

Over Christmas 2011, Ive did some soul-searching, even contemplating a life after Apple. But he'd heard the voices chanting that Apple couldn't create new marvels without Steve Jobs, and he wanted to prove that they were wrong. He vowed to do one more major product. "I felt I couldn't leave unless there was another platform," he says.

To him, the trajectory of computers over time was obvious: They became steadily smaller, steadily more personal. Mainframe → microcomputers → personal computers → laptops → phones. The next step, naturally, would be something that you *wear*.

But wear *where*? The wrist seemed like a natural option. It's a spot that's easy to see and access. And wearing a small mechanism on your wrist was already culturally accepted.

Ive collected fine watches himself, but found the smartwatches of the day "abhorrent." To him, these watches—from Google, Samsung, Sony, LG—were chunky, awful things. Jobs himself had hated watches, and never wore one. Once, near the Valley Green 2 building, he'd taken off his watch and thrown it into the brush, and had never worn one since.

Still, Ive loved the idea of creating something so intimate that it's *on* you. He'd become increasingly concerned about the societal trend of burying our faces in our phones. He loved the notion of glancing at information without having to haul out your phone.

His teammates immersed themselves in watch history, watch lore, watch culture. They invited watchmakers, historians, and astronomers to give talks. They bought the most expensive watches in the world, taking care to have them shipped to a phony office to avoid arousing suspicion.

Ive, meanwhile, was stuck on the design. He knew that the watch would be a rectangle—"if you're basically working with lists, a round display's really inefficient"—but he struggled with the rest of it. None of his sketches and models seemed right. He asked his friend Marc Newson, a native Australian, to come by for some brainstorming. Newson was a highly regarded industrial designer who, as it happened, had decades of watch-design experience, and had founded a watch company called Ikepod. He came directly to Apple from the airport.

Apple's design studio, now on the first floor of Infinite Loop 1, was as secure as a launch silo. First, you entered a badge-accessed, camera-monitored, solid aluminum double door. Then you passed through a ten-foot stainless-steel corridor to a set of glass doors. Then you'd arrive at the receptionist, who confirmed your authorization to enter.

Once inside, a vast, bright, and airy expanse awaited: concrete floor, metal ceilings, huge windows (frosted, of course, to thwart spies outside). "It was like you had entered a spaceship," says Tony Fadell.

In the space that greeted you was a kitchen—the studio's version of a common room—and a set of long, Apple Store–style Parsons tables, where designers could present models to visiting executives. In an adjacent space, the company's small team of industrial designers sat at long tables topped by laptops or iMacs. Designers in 30 other disciplines populated nearby areas: sound designers, touch and vibration designers, and experts in color and typography. Software (human interface) designers occupied the other side of the room, often pinning up sketches for discussion.

Right in the center sat an all-glass cube, the only dedicated office in the studio: Jony Ive's.

Newson arrived at the studio at 2:30 in the afternoon, and sat down with Ive in the corner of the studio. Together, they began sketching.

The design studio also houses a complete fabrication facility, equipped with 3D printers, drills, lathes, and CNC milling machines, staffed by laser specialists, painters, and model makers. A designer can hand a sketch to a CAD (computer-aided design) artist, who will build a 3D model on the screen and then send it directly to the shop.

Alan Dye, vice president of design (center), discusses new typefaces in the design studio.

That's how Ive and Newson turned their drawings into a physical model. "We got a rapid prototype back," says Ive, "and it was just: 'Yep, that's it.'" It was, he says, essentially the Apple Watch we know today.

It would look like a miniature iPhone: a round-cornered rectangle, a black touchscreen flush with curved metal edges. You'd be able to swap in a different watchband just by sliding it in; nobody would need a tiny screwdriver for *this* watch.

Prototypes

While he was still in high school, Kevin Lynch started a Mac software company. He became a young OS engineer at General Magic, went to Macromedia (where he met Phil Schiller), then became chief technology officer at Adobe. In March 2013, Apple offered him a mysterious new job. "It's a new thing. It's hardware and software," was all hardware chief Bob Mansfield would tell him.

Kevin Lynch
Born: Chicago, IL
Schooling: University of Illinois Chicago
Before Apple: General Magic, Macromedia, Adobe
Apple: 2013–present

Lynch released his first software for the Mac in 1984, before *Inside Macintosh* was complete.

Lynch was in charge of the watch's software. The engineering, design, and marketing teams all brainstormed; everyone was expected to draw their ideas, no matter how poor they were as artists. Over and over, they'd ask: What is the purpose of this thing? Why does it exist?

The fact that you'd be *wearing a computer* posed a whole set of questions Apple had never had to answer before. What does it mean for it to be on your body? How should it try to get your attention, and how often? The screen was the size of a Wheat Thin—too small for multitouch gestures. How would you navigate it? The group was keenly aware that they were making the first Apple product that had no input from Steve Jobs, which only added to the pressure.

The multitouch gestures that worked so beautifully on the iPhone and the iPad—spreading two fingers to enlarge a photo or map, for example—wouldn't work on the watch. Your fingers would block your view of the tiny screen.

"I remember sitting around the art design table," says designer Duncan Kerr. "And I was like: 'We could have a digital crown! We've lived with crowns on watches for 100 years, so as an archetype, it's familiar. Put this thing on the side, and you turn it, and then you could get your zoom function and other things. Scroll and zoom with the crown, versus having your fingers on the display.' And Jony just lit up."

The digital crown—an off-center, knurled knob on the side, like the one you'd use to wind an old-time watch—soon became one of Ive's favorite aspects of the Watch.

The first concept was that you'd use the crown for navigating *everything*. You'd start with a bird's-eye view, seeing your entire watch's contents, tiny as pixels. Then you'd zoom closer and closer into an app, into its data.

The problem, Lynch says, was targeting. "Zooming out is easy, but zooming in—what are you zooming in on?" he says. "If you're looking at a list of meetings and you want to zoom into one, then you have to swipe on the screen to center it and *then* zoom. And now you've got this interaction going where you're panning and zooming. And it's too cumbersome."

A team of former iPod engineers had mocked up a set of crude prototypes, fitted with basic black wristbands. They were adapted from the sixth-generation iPod nano, the one that looked like a 1.5-inch square screen on a clip. They hacked each one to include a standard iPod 30-pin connector, which accommodated a homemade external accessory that offered a crown. But when Lynch decided that the team needed to start writing some actual software, the modified iPods wouldn't work. They didn't have the right processor.

Old iPhone models, however, did. So Lynch's team 3D-printed Velcro straps and phone holders. They'd strap the iPhone to their arms, mounted sideways, mostly hidden under their sleeves. It ran special software that lit up only a 1.5-inch square at the bottom of the screen, to simulate what you'd see on the watch. They walked around the office with this contraption, testing out early software builds.

The phone, of course, had no spinning crown. The engineers did have the external crown attachment they'd made for the iPod nano prototypes, but its 30-pin connector didn't mate with the iPhone's Lightning connector.

The final contraption, then, was a Velcro band on your wrist, supporting a phone holster, containing an iPhone, into which was plugged a 30-pin adapter cable, to accommodate the crown accessory. "It stuck out to about *here*," Lynch says, indicating a 3-inch extension from the watch. "It was like a Spider-Man web thing."

For the next year, the team wore these iPhone watches to develop the software. They redid the entire interface three times, working out what all the different input mechanisms should do: tapping the screen, turning or clicking the crown, pressing the side button.

Mansfield oversaw hardware teams all across Apple: Bluetooth, Wi-Fi, batteries, motion sensing, and so on. He had tried to retire from Apple three times; each time, Apple pulled him back in to help with urgent projects. (When he finally did retire, Lynch joked that his present would be a watch. And it was.)

In the summer of 2014, the first prototype watch arrived. The team had ten months to perfect the software.

AirTagGate

The AirTag, introduced in 2021, is a $30 tracking disk about the size of a half dollar. You can stick one in anything whose location you might want to know: luggage, car, bike, backpack, purse, pet collar, camera bag, and so on. There's even a speaker, so you can ping it to find your wallet, keys, or TV remote in the house.

Bluetooth trackers, with a range of 30 feet, had been around. But AirTags have a range of *the developed world*. That's because they quietly broadcast an ultra-wideband wireless signal that any iPhone, iPad, or Mac can pick up—a surveillance network of a billion nodes. If one happens to pass within 300 feet of the thing you've lost, *you* get notified, in your Find My app. The location and the signal are encrypted and invisible to Apple or anyone else, and the person whose iPhone "found" your lost thing will never even know what a good deed they've done.

Most AirTag tales have happy endings. People recover their luggage, retrieve their phones, remember where they left their backpacks. Relieved caregivers worry less about their loved ones with dementia.

But inevitably, unsavory characters began planting AirTags to stalk their exes, and for Apple, the PR fallout was not good.

AirTags already had some safeguards: If your iPhone began noticing a nearby AirTag that wasn't yours, the AirTag chirped to get your attention, and a notification appeared on your iPhone screen: "AirTag Found Moving With You. The location of this AirTag can be seen by the owner."

Apple tuned the chirp to make it louder and chirp sooner. In 2022, it released an Android app that could detect an AirTag moving with *those* phones' owners; in 2024, it collaborated with Google on a technology standard that would allow *any* company's unwanted tracker to be detected by *any* company's phone.

Technology vs. human nature: It's never easy.

Fitness

From the beginning, Apple intended the watch to assist with your health and fitness. For example, it would "tap" you on the wrist once an hour to suggest that you get up and move around for a minute. (Sitting all day is really bad for you.)

But how is a device on your wrist supposed to know what's going on in your brain (stages of sleep), your lungs (breathing rate), or your blood (oxygen saturation)?

Answer: indirectly.

For example, to measure your heart rate, Apple adapted a technique called PPG (photoplethysmography): During a workout, bright green LED lights on the back of the watch flash through your skin hundreds of times a second. When your heart beats, the surge of blood absorbs more green light; sensors on the back of the watch measure how much light is reflected, thereby learning your pulse rate.

To calculate your stages of sleep (REM cycle, deep sleep, light sleep), the watch studies data from your skin temperature, heart rate, and the tiniest movements of your arm, tracked by an accelerometer (motion sensor) and gyroscope (rotation sensor)—together, only a tenth of an inch square.

And to determine how these indirect data streams correlate to real biological signals, Apple turned itself into a medical research company.

Conference rooms became sleep labs. Volunteers wore mesh headgear and electrodes to measure their breathing, pulse, and movement. (The sleep-tracking feature arrived in 2020.) To learn how many calories you were burning—a critical metric for assessing fitness—Apple hired doctors, set up a lab, and collected tens of thousands of hours' worth of human motion data—the largest such study ever conducted. Employee volunteers went about their days wearing calorimetry gear: face masks connected to backpacks that measured their CO_2 output.

"We also rented an apartment to measure real daily activities," says Lynch. "We had people going in there, and they just did all-day activities, like preparing dinner and doing household tasks, while wearing this backpack and this mask." Apple became the world's largest purchaser of metabolic carts that measure calorimetry.

As the ship date approached, hundreds of Apple employees wore the watch, reporting bugs and design glitches on Radar, Apple's internal bug-tracking system.

The Shower Algorithm

When you build something new, you get new problems, too. That was COO Jeff Williams's experience when, as part of the core Watch team, he wore a prototype into the shower.

Even this early model was designed for water resistance. What it could not do, however, was tell the difference between a finger touching the screen and high-speed water drops. "As soon as I got in the shower, it started doing all kinds of stuff on the watch," he says.

Random taps might be annoying, but what damage could they really do? "And then all of a sudden it dawned on me: Holy cow, we have Smart Replies!"

When someone texts or emails you, the watch offers one-tap, context-sensitive replies to save you typing, like "OK," "Let me call you later," or "Sounds good."

"I mean, what if your boyfriend or girlfriend just texted you, 'Should we break up?'" Williams says. "And the shower answers . . . 'Sounds good!'?"

Until that moment, the billions of touchscreens Apple had shipped were concerned only with detecting *whether* they'd been tapped—not *by what*. But this new device was designed to be worn in the shower. And failure to distinguish a finger from a droplet could mean disaster.

Alarmed, Williams called Steve Hotelling, head of the touchscreen engineering group. "Houston, we got a problem," he said.

Hotelling's Apple job description had not, so far, involved plumbing. But he threw himself into the construction of a shower simulator, in order to study the characteristics of water droplets as they struck the watch's screen: their size, positions, frequency, and duration of contact. Eventually, his team wrote new algorithms that did a far better job at distinguishing fingers from water drops.

To date, no known Apple Watch has broken up with anyone in the shower.

The Rings

A device on your body can do one useful thing that a phone can't: take biometric readings. Apple hired Jay Blahnik, a trainer and health educator who'd collaborated with Apple while at Nike, to oversee the fitness initiatives.

First, he had to contend with the "10,000 steps" trope that many Americans thought should be their daily goal. It was a made-up stat from a Japanese marketing campaign in 1965, never based on any study. "It was also really overshooting for a lot of people. The average American at the time was doing like 2,500 or 3,000 steps a day," says Blahnik. "That is literally the worst possible thing you could do for people: setting them up for failure."

Furthermore, the *kind* of steps matter—brisk ones are better than languorous ones—and the timing does, too. "You could get thousands of steps in by noon and then sit around all day at your desk, and that's not healthy, either." He wanted the watch to measure quantity, intensity, *and* frequency of your movement.

Your goal: Close the rings every day.

The team tried out animated bar graphs, checkboxes, number odometers. In the end, though, the goal wasn't to get you hung up on numbers; it was to provide a simple motivational dashboard.

They settled on a circular graph. Three concentric colored rings track your movement through the day, your exercise, and the amount of time you stand. Your goal is to close the rings each day—or, for extra credit, to send any of them beyond a full revolution into a second lap around the ring.

A triathlete might have a higher goal than a first-time jogger—but everyone liked closing the rings. "People like the feeling of completion," Blahnik says.

Fashion

Ive felt deeply that the watch should be a fashion item. He treasured the idea that, with different metal cases, bands, and digital faces, everyone's watch could be different. "Generally, if you wear the same thing, it's because you're in prison or in the army," he says.

The watch, he decided, would come in a choice of aluminum, stainless steel, or gold (for $10,000 to $17,000). In 2012, Apple had acquired a dozen engineers from a Chicago high-performance alloy company called QuestTek to assist Ive's pursuit of the perfect metals in iPhones and other products. Now their task was to turn gold—ordinarily a soft, dentable metal—into something rugged enough for a watch. By melting computer-calculated amounts of copper, palladium, and silver into the gold, the QuestTek team produced an alloy twice as tough as regular gold. The high-end models' screens would be made of synthetic sapphire, far tougher than glass.

For the watchband, Ive proposed materials like silicone, fine metal mesh, and leather, for which a dedicated soft-goods team contacted tanneries all over the world.

To develop the marketing and sales of the watch, Ive hired former executives of TAG Heuer, Burberry, and Yves Saint Laurent. Ive flew to New York to give *Vogue* editor Anna Wintour a personal demo. He collaborated with French luxury fashion house Hermès to design a band. He arranged a pop-up display at the famous Paris fashion store Colette. And in 2016, Apple sponsored the Met Gala, the exclusive fundraising event at New York's Metropolitan Museum of Art.

Apple, Ive wanted to make clear, was serious about fashion.

The Birth of Taptics

On a phone, notifications get your attention with little vibrations. Traditionally, that buzzing sensation—called haptic feedback—was produced by a tiny, off-center weight spinning furiously inside the phone.

It seemed all wrong for the watch: too loud, too aggressive. "Having a phone in your pocket or your bag make that buzzing sound, that's okay," Duncan Kerr says. "But for this thing that's touching your skin, it just felt awful."

The designers worked for weeks to tame the rotary buzzer: shorter buzzes, lower voltage. Meanwhile, at every meeting, Product Design—whose job it is to turn the design studio's ideas into mass-producible products—displayed the watch's latest internal layout. And week after week, its mockups showed a hole exactly the size of the iPhone's spinning-motor system. "They were just waiting for the day when we'd throw up our hands and go, 'Okay, we can't do it any other way—put it in!" Kerr says.

But in the end, his team settled on a linear actuator: a tiny metal plate that slid rapidly back and forth along a track, precisely controlled by electromagnets. This more sophisticated design meant the watch could vary both the strength and the patterns of its vibration. It could even "tap you on the wrist"—thus its name, the Taptic Engine.

It could be so subtle that at one point, the team designed it to produce an infinitesimal click with each jump of the analog watch face's sweep-second hand, just as on a traditional watch held up to your ear. (That feature wound up consuming too much power.)

The Taptic Engine was so obviously superior to the spinning-weight system that it made its way into the iPhone, too. Today, it's responsible for the little click you feel when you tap a letter key, adjust a slider, touch the camera's shutter button, and so on.

Launch

In 2014, an Ebola outbreak ravaged West Africa, Microsoft bought the maker of Minecraft, and George Clooney married Amal Alamuddin. But in Cupertino, Tim Cook took the stage at Apple's favorite spot for important unveilings: the Flint auditorium. What he was about to reveal, he promised the 2,000 people in the auditorium—and the 90 million online—would be "the next chapter in Apple's story."

Into the watch's tiny, jewellike case, 1.8 x 1.5 inches, Apple had somehow managed to pack a battery, processor, speaker, microphone, memory, Wi-Fi and Bluetooth antennas, accelerometer, gyroscope, barometer, heart-monitor system (green and red LEDs, infrared lens, photo sensor), and the Taptic Engine.

Tim Cook shows off the Apple Watch's different looks, models, and configurations.

After the unveiling, Apple invited the audience of tech and fashion journalists to view the watches up close in a gorgeous, white, temporary two-story-high fabric building that Apple had constructed outside the Flint Center. Setting up this facility had been a $25 million undertaking that involved removing all the trees beforehand, and replanting them after the event.

The reporters could see immediately that the watch was light-years better than any of the clunky efforts that had come before it. The screen was nicer, the software more refined, the case more like jewelry. The most common question, though, was: "What is the Apple Watch *for*? Why do I need it?"

It could *do* all kinds of things: track your fitness, pay for things, take calls, play music, display notifications without requiring your phone, answer texts, take phone calls, show photos, operate Siri, trigger your phone's camera remotely, show maps, ping your lost phone under the couch cushions. And, yes, it could tell the time, to within 50 milliseconds.

But all of that was also on your phone. Was it worth $350 (or $17,000) to move some of your phone's functions to your wrist?

Reviewers noted that the battery barely lasted a day. The rival Fitbit didn't have a color screen and didn't do anywhere near as much, but it lasted five days on a charge.

U2Gate

Steve Jobs was a fan of the Irish rock band U2. In 2004, he orchestrated a mutual promotion. U2 would appear in a "Silhouettes" iPod ad for free, playing their song "Vertigo"; in exchange, Apple would promote their album, offer an exclusive iTunes Store "box set" of U2's music, and sell a U2 edition of the iPod. In 2006, Apple created a special red iPod whose sales proceeds would benefit lead singer Bono's campaign to fight AIDS in Africa.

U2 even performed live at the Apple Watch launch event. And then, standing side by side with Tim Cook, Bono made an astonishing announcement. The band had just recorded its first new album in five years—*Songs of Innocence*—and Apple would be giving a free copy of the whole album to every one of its 500 million iTunes users.

To make the deal, Apple had paid the band an unspecified fee and promised to spend $100 million promoting the album. Apple assumed that everyone would be delighted.

The mistake wasn't *offering* the album; it was force-downloading it. Within hours, millions of people discovered a new album on their Macs and phones that they hadn't asked for. Many had never even heard of the aging band. They didn't appreciate being deprived of *choice* in their musical selections. To make matters worse, they found that they could not delete the U2 album. They started calling it the U2 virus.

"Woke up this morning to find Bono in my kitchen, drinking my coffee, wearing my dressing gown, reading my paper," went one tweet.

"The free U2 album is overpriced," went another.

Within days, Apple developed a web page and a little app that would delete the U2 album for good. Bono apologized. "I had this beautiful idea and we got carried away with ourselves," he said on Facebook.

Above all, navigating this thing was nonintuitive; it was not a prime example of saying no to ideas. Between the three case materials, two sizes, and six bands, there were 96 different options on day one—plus a choice of 11 customizable watch faces.

First-year sales were disappointing. "Skeptics who were saying, 'Gosh, anything the watch can do, a phone can do better, 'cause it's got a bigger screen,' and stuff like that," says Jeff Williams. After years of celebrating the sales figures for iMacs, iPods, iPads, and iPhones, Apple declined to say how many watches it had sold. Within the executive team, there was even talk of canceling it.

But Ive reminded the executives that Apple had had slow-selling product launches before—remember the Mac? The iPod? The iPhone? With patience and steady improvements year by year, they had turned into monster sellers.

In the years following the watch's introduction, the screen got bigger and brighter. The case got thinner and more watertight. The speed got better. The battery began to last longer and charge faster. New features came along: GPS. An altimeter that could track stairs you climbed. A compass. A thermometer. A cellular version, so you didn't need your phone to get calls and texts. A truly massive Ultra model for harder-core athletes and workers with twice the battery life (36 hours), more precise GPS, divers' features, louder audio, and a ruggedized case.

As Ive had predicted, the watch gradually caught on. It wasn't a world-changing phenomenon like the iPod, iPad, or iPhone. But it became the bestselling smartwatch, then the bestselling *watch*, in the world, an easy choice as a graduation or Christmas gift. (That would be the $350 model, not the gold one, which Apple stopped selling after 16 months.)

Health

The original Apple Watch had fitness-tracking features. But in successive versions, Apple began taking the watch deeper into *medical* uses. "Traditionally at Apple, it's always been design and engineering—and now there's a third leg of the stool, which is our clinical team and our health team," says medical director Sumbul Desai. "Now we have physicians or other clinical experts sitting at the table, with design and engineering."

It's not an easy realm. The FDA strictly regulates the marketing of medical devices; the last thing it wants is emergency rooms flooded by consumers alarmed unnecessarily by their gadgets. Securing FDA authorization for a gadget entails years of studies.

In 2018, the fourth-generation watch gained the FDA's okay for its ability to show an ECG (electrocardiogram) graph, a critical tool for spotting heart problems.

In 2020, the sixth-generation watch added a pulse oximeter, which measures the percentage of oxygen in your blood, a key indicator of respiratory and circulatory health sometimes called the fifth vital sign, along with temperature, blood pressure, pulse, and breathing rate. Suddenly, you could monitor your oxygen saturation without needing a hospital-style SpO2 fingertip clip. (Thanks to a complicated patent lawsuit, Apple was forced to shut off that feature in late 2023 on the latest watch models in the U.S.; but in August 2025, Apple engineered its way out of the stalemate. It moved the data processing and results display to the iPhone, in the Health app, instead of right on the watch. That workaround satisfied regulators and returned the "fifth vital sign" to Apple Watch fans.)

2018 also introduced the watch's ability to detect episodes of atrial fibrillation, the world's most common heart-rhythm abnormality. It's when the top chambers of the heart (the atria) *quiver* instead of pumping. The blood just sits, eventually clotting. In 35 percent of A-fib patients, pieces of that clot break off, find their way to the brain, and cause a stroke.

If you know you have A-fib, you can take medicine for it—but *how* would you know? A-fib is often intermittent, so tests in the doctor's office might miss it.

The watch doesn't want to alarm anyone unnecessarily; its A-fib notification doesn't pop up until it senses multiple episodes in 24 hours. And it took a heart study of 400,000 watch wearers, conducted with Stanford University, to convince the FDA and other countries' regulators that the technology was solid.

But now, "we hear almost every day from individuals who tell us that their physicians tell them, 'I'm so glad you showed up when you did, because this really could have ended much differently,' " says Desai.

In 2025, the Series 9 and later watches gained the ability to detect hypertension—chronic high blood pressure. Hypertension is a big deal; 1.3 billion people have it, but only 23 percent of them know it; the rest are at far greater risk—as much as ten times the risk—of heart attacks and strokes. Since even a doctor's office test can miss it, hypertension has become known as the "silent killer."

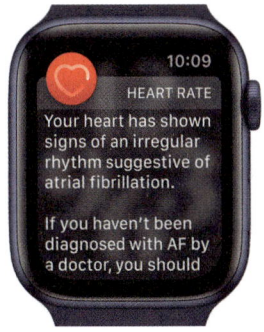

The Apple Watch notifies its wearers when there may be trouble.

Apple's system uses AI to study the shape of the pressure waves generated by your heartbeats over 30 days. Bigger, sharper wave shapes generally mean stiffer arteries—and higher blood pressure. Apple expects that, thanks to this feature, a million people a year will learn that they have hypertension, and will have the opportunity to do something about it.

Today, the watch can alert you to a long list of additional medical problems: high or low heart rate, sudden changes in sleep patterns, and sleep apnea (in which your breathing periodically stops while you're asleep—a possible indicator of serious diseases). It tracks and forecasts menstrual cycles. It can also alert you about changes to your walking steadiness, which could be a sign of conditions like declining vision, arthritis, Parkinson's disease, or depression.

In 2024, a new app called Vitals began taking a more holistic look at your health. If, overnight, it discovers that two or more of your metrics are out of their usual range (heart rate, breathing, wrist temperature, blood oxygen, sleep duration), it warns you in the morning. Chances are pretty good that either you're getting sick—or you had too much to drink.

45. Vision Pro

By 2015, it was clear that the Tim Cook era wasn't the Steve Jobs era. For one thing, the Jobs pattern—another earth-changing device every three years—seemed to be over. Maybe *nobody* could replicate Jobs's freakish ability to see the future, or maybe all the the obvious ideas had already been executed.

"I knew what I needed to do was not to mimic him," Cook says of Jobs. "I would fail miserably at that, and I think this is largely the case for many people who take a baton from someone larger than life. You have to chart your own course."

But shareholders certainly weren't crying. In the holiday quarter of 2015, Apple made $18 billion—more than it had ever earned in a quarter, more than *any company ever* had made in a quarter.

Apple's profit for that year ($53 billion) was more than that of Facebook, Alphabet, and Microsoft *combined*. AAPL stock hit an all-time high: $120 a share. In July 2016, Cook mentioned that Apple had just sold its one billionth iPhone. (The two billionth iPhone sold in 2021; the three-billionth in 2025.)

Cook had begun steering Apple to greater social responsibility, too. There is no record of Steve Jobs ever giving anything to charity; in fact, when he returned to Apple in 1997, he shut down the company's philanthropic programs and never reinstated them.

But immediately upon becoming CEO, Cook introduced an Apple grant program that contributed to nonprofits in education, healthcare, racial equity, criminal justice, technology, and the environment. He also introduced charitable matching: If any employee made a gift to charity, up to $10,000, Apple would kick in an equal amount, and for every hour of employee volunteer work, Apple makes a $25 donation to the same agency. By 2022, these programs had raised $880 million for 44,000 nonprofits, and 76,000 employees had volunteered 2.1 million hours.

In 2021, Apple also instituted a $100 million racial-justice initiative, with elements like grants and scholarships for historically Black colleges, a tech campus in Atlanta serving those colleges, a coding school for Detroit kids, and venture capital funding for Black and Brown entrepreneurs.

China Sales

Apple had three factors to thank for its historic profits in 2015: the growth of subscription services like Apple Music, a very popular iPhone 6 family—and China.

In the early 2000s, Apple didn't have much presence in China. Chinese politics made it difficult for American companies to open stores, and Apple had assumed that the low-income population would not have had much taste for luxury phones.

But when Apple did manage to open one in July 2008, in Beijing, it was mobbed by people who were prepared to spend a quarter of their entire annual salary on an iPhone. When the iPad came out in 2010, the Beijing store sold its entire two-day supply of iPads—$3.7 million worth—in five hours. The store had to close the Genius Bar to make room for more cash registers.

Shoving and fistfights in Apple Stores weren't uncommon. One 17-year-old boy sold his kidney on the black market to buy an iPhone and an iPad. Many of the "customers" were scalpers, who snapped up iPhones to resell elsewhere in China.

When the iPhone 4S went on sale in 2012, the line at the Beijing store began forming five days in advance. When a fearful staff didn't open the doors, people shook the handles so hard that the glass shattered. John Ford, head of Apple's Chinese retail efforts, hired hundreds of security guards just to maintain order. Even with a buying limit of two phones per person, armored trucks had to come by the store several times a day to haul away the cash.

In 2012, as China's rising middle class continued to seek Apple goods as status symbols, sales in China hit $22.5 billion. In the next three years, it rose by 2,830 percent. "Absolutely staggering," Tim Cook said.

In took ten years for Apple to solve the Chinese distribution problem: by building more stores, by teaming up with cell carriers, by launching an online store, until there were so many distribution points that the scalpers' profit margin evaporated.

Today, Apple sells $70 billion of electronics in China each year. After the U.S., it's Apple's biggest market in the world.

TDG

Still, it was clear that the public, the media, and even shareholders *missed* new, category-defining blockbuster products. New platforms were *exciting*.

In 2015, Mike Rockwell, senior VP of R&D at audio/video tech mecca Dolby Labs, got a tempting offer from Apple hardware-engineering head Dan Riccio: Come to Apple and help us figure out Apple's next big thing.

Mike Rockwell
Born: Glendale, CA
Before Apple: Dolby
Apple: 2015–present

Rockwell's hobbies include astrophotography, metal- and woodworking, and music writing/playing.

Rockwell set up shop in the old Mariani 1 building, on the same floor where the iPhone had been developed. "You could feel the history," he says.

He had no idea what the next big thing would *be*, so he named his new group generically: the Technology Development Group (TDG), staffed with engineers and designers from both inside Apple and hired from outside.

The group's first focus was virtual-reality (VR) headsets: goggles equipped with small screens in front of your eyes. Sensors detected when you turned your head so that your view changed realistically.

Headsets were hot in the Valley. Facebook had just bought a VR headset company called Oculus, which had produced only a prototype, for $2 billion. The hope was that these headsets would be a natural for games and simulations.

In *virtual* reality, the goggles are opaque, blocking your view of the real world; all you see are the little screens. Apple considered that approach too isolating. When you strap on a VR headset, you can't see other people, and they can't see your eyes. If somebody taps you on the shoulder, you jump out of your skin.

Apple preferred *augmented* reality (AR), where you still see your surroundings. The graphics of your game, movie, or app seem to float right there in the room with you.

The 2016 Oculus Rift headset.

But in 2015, both VR and AR goggles were in their infancy. First, they weren't self-contained; they had to be wired to a PC. Second, the pixels were so coarse, it was like looking through a screen door; the screens weren't sharp enough to read text or do anything besides playing games. Third, the lag between moving your head and seeing a change in your view made a lot of people sick.

"It was all kind of bad," Rockwell says.

It seemed to him that if Apple could solve those problems, a headset could be an amazing tool for entertainment and work.

Resolution

The low-resolution screens seemed like a good first problem to fix. But no available screen had pixels small enough to serve as high-resolution lenses for headsets; the closest thing Apple could find was a high-definition microLED screen used in military optics.

Unfortunately, these screens were much too small to fill your field of view; in a headset, each would appear the size of a tea bag five feet away. Rockwell's team tried tiling together six of those screens so that they filled more of your view—but the seams were distracting.

Then one of the engineers stumbled onto a technology called fiber-optic lenses: a bundle of fiber-optic cable strands, each incredibly fine at one end, thicker at the other. The bundle could conduct light from that tiny military microLED screen to your eyes, enlarging the image in the process. And if six of *those* were tiled, there were no visible seams.

The clarity was so fine, the team members found that they could read text or browse the web. When they watched videos, it felt like being in a theater with a hundred-foot screen. They had just built a prototype of the world's first 1080p high-definition headset.

It weighed 50 pounds and had to be supported by a crane arm; the whole contraption looked like a periscope. Even so, it seemed transformative.

Latency

The fraction-of-a-second lag between the time your head moves and the time the screen updates itself is called *latency*. That's the disconnect that often made people feel seasick in these headsets.

Rockwell's team commissioned studies. How short did the lag have to be to avoid nausea? The answer: below 12 milliseconds (thousandths of a second). It took "a lot of clever hackery," but the TDG managed to get the lag down to 11.

Personas

From the earliest days of the Vision Pro project, its engineers loved the idea of Spatial Facetime: video calls where your conversation partners seemed to appear in the room with you.

But how could they see *you*, on their Macs, iPhones, or iPads? "There's no videoconferencing camera looking at you," Rockwell points out. "And even if there were, you're wearing something over your eyes."

The solution was the most difficult engineering challenge of the entire project. You would appear to your conversation partners as a Persona: a computer-generated, *3D hologram of your head*.

Computer-generated people have always been hard to get right. We've evolved to be finely tuned observers of fellow human faces. So when a CGI version gets it only *almost* right, the effect is uncomfortable and cringey. Seeing them sends us to, as roboticists and CGI artists call it, the "uncanny valley." Anyone who's ever seen the 3D people in *The Polar Express*, or the computer-generated Princess Leia in *Star Wars: The Rise of Skywalker*, knows the problem. And *those* filmmakers had 300,000 times the computing horsepower of the Vision Pro.

At first, Rockwell's group considered adapting Memoji, the personalized, animated cartoon avatars that had been available in Messages since 2018. But in practice, seeing them at life size was more alarming than delightful.

With three years to go before Apple hoped to ship its headset, now called Vision Pro, the Persona technology still wasn't working. The general feeling was that they'd have to kill the feature.

But Rockwell, following in the footsteps of the iPhone keyboard team 15 years earlier, pulled the Stop handle. Every engineer and designer was to set aside whatever they were doing. For the

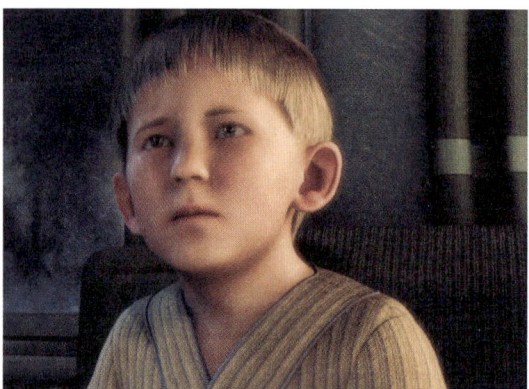 

Two approaches that Apple wanted to avoid: uncanny-valley characters as in
The Polar Express *(left) and life-size versions of Memoji.*

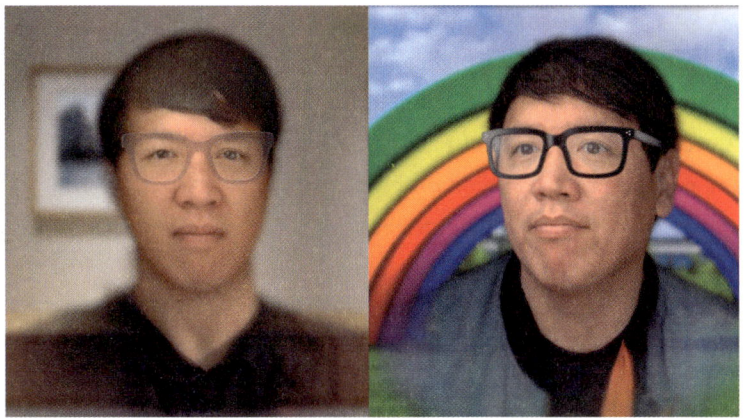

Norman Chan, from Mythbusters, *demonstrates the original Persona (left) and the far more realistic version that Apple released in 2025.*

next six weeks, everyone had a single assignment: Solve the 3D people problem, using only the computing power in the headset.

That challenge produced four fresh ideas. One of them, a machine-learning AI idea that didn't use any of the traditional Hollywood movie approaches, seemed especially promising. Unfortunately, machine-learning algorithms require training with thousands of data examples—in this case, scanned heads of real people.

To collect those examples, Apple built photogrammetry rigs—rings of cameras pointing inward—and photographed thousands of people inside them, much as the Face ID team had done five years earlier. Once again, the goal was to represent the full variation of humanity: every face shape, skin color, and hairstyle. In the end, they collected several *petabytes'* (millions of gigabytes') worth of head data.

By 2024, Personas were ready to go. During setup, you were directed to face the front of your Vision Pro headset so that its cameras could see you. Then, after some computation, it could produce a slightly blurry, computer-generated version of you. It wasn't perfect, but it wasn't uncanny—and after another year of improvements, it became downright realistic.

EyeSight

The TDG engineers, meanwhile, were fighting the isolation problem on another front. What would other people in the room see when they looked at you?

They made the entire front of the goggles a screen. It would display the top half of your face—rather, the Persona version of it—so that other people could see *you*.

But it couldn't be a standard screen, showing a video feed of your face inside the goggles; it looks freakish to have your eyes apparently pasted on the outside of the glass. Instead, they developed a screen that appeared to be deeper inside, where your actual face would be. It was a *lenticular* screen, meaning that it looked different, in the correct perspective, from the positions of everyone in the room.

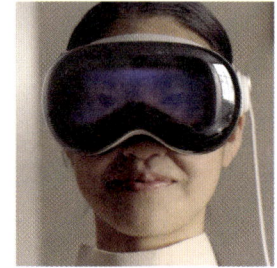

EyeSight: a dim, CGI version of your eyes.

Vision Pro

Vision Pro was by far the most ambitious, difficult, advanced product Apple had ever released. In fact, it was arguably the most technologically advanced consumer-electronics product ever sold.

During its eight years of development, the project racked up nearly 5,000 patents. Vision Pro had twelve cameras, six microphones, two infrared beams and receivers, a Lidar scanner for room mapping, two 4K micro-OLED screens, and dual-driver speakers on the headband next to each ear. It incorporated a custom chip, the Apple R1, to process all that sensor data, *and* two tiny fans to cool it, *and* a noise-cancellation system so you wouldn't hear the fans. It had eye tracking, hand tracking, head tracking, environmental mapping, and motorized iris scanners to authenticate your identity.

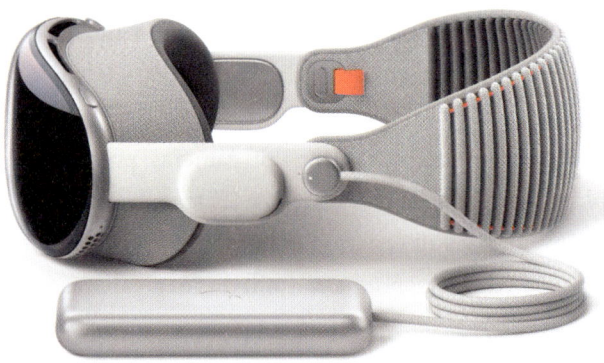

The finished Vision Pro included the computer itself (left), a light shield, a headband, and a battery on a cord.

When you saw the room (or office, or plane) around you, you weren't looking through glass; you were actually seeing a live video feed from tiny exterior cameras. But it *felt* like you were looking through glass, because each pixel of each screen was 7.5 microns across—about the size of a red blood cell.

By the time Apple announced the Vision Pro on June 5, 2023, the hype engine was in overdrive. It was the company's first new platform since the Apple Watch, eight years earlier. Despite delays, cost overruns, and internal bickering about its prospects, Cook had been a strong believer in the project throughout.

Even the haters had to admit that the experience was breathtaking. Your apps, photos, and

The graphics of your game, movie, or app seem to float within your physical space.

movies floated convincingly on huge, clean, lovely, vertical planes in front of you. There was no plastic controller; you just *looked* at something to point to it, and then tapped your thumb and finger together to "click" something. You could scroll a window by pinching and pulling down in the air, as on a window-shade cord.

The Vision Pro could run any of 1,000 iPad apps, or its own apps. Each app's window floated in its own six-foot-tall "panel" in front of you. You could watch movies on a 100-foot screen even while cramped in a coach airplane seat, or work on your Mac while viewing a virtual gigantic monitor.

Movies and photos can appear at gigantic size.

Reviewers found the Vision Pro to be visually stunning, thoroughly immersive, and technologically gobsmacking—but also heavy and uncomfortable. Even without the battery, which Apple off-loaded to a belt pack, the headset put 1.4 pounds on your nose and forehead.

For most people, though, the greatest discomfort was the price: $3,500 and up. Meta's Quest headset wasn't anywhere near as advanced, but it cost only $500.

You couldn't wear glasses with the Vision Pro; instead, you had to buy a pair of ZEISS magnetic inserts for $100 to $150, meaning that you couldn't hand your goggles to a friend to try. The carrying case, a $200 option constructed with over 150 parts, was the size of a small backpack.

Once early adopters snapped up Vision Pros, sales slowed dramatically. Hundreds tweeted that they were returning their headsets to the Apple stores.

As with the Apple Watch, Apple would not say how many Vision Pros it sold, but analysts estimated that it was 400,000 in the first year. (First-year sales for the iPhone were 1.4 million, the Apple Watch 12 million, the iPad 15 million.)

But Apple didn't need reviews to tell them that Vision Pro was heavy and expensive. None of this was a surprise. "At $3,500, it's not a mass-market product," Cook said. "Right now, it's an early-adopter product."

From his first onstage mention of the Vision Pro, he had called it "the beginning of a journey." The initial plan was to develop another Vision Pro with lower price, weight, and bulk.

For Tim Cook, though, the holy grail was not a headset at all. "Few people are going to view that it's acceptable to be enclosed in something, because we're all social people at heart," he said in 2016. Apple's ultimate ambition was to bring the magic and productivity of augmented reality to ordinary-looking *glasses*.

For the Vision Pro group as 2026 dawned, smart glasses came into focus as the next big small thing.

Project Titan

In 2014, the automotive industry was undergoing rapid change. Cars were going electric; they were becoming more computerized; they were on a trajectory to become fully self-driving; and, thanks to apps like Uber and Lyft, car transportation was morphing into a service.

It occurred to Dan Riccio, senior VP of hardware engineering, that Apple had world-class expertise in all four areas: battery tech, software, AI, and services. So in February 2014, he pitched Tim Cook and the board on an ambitious but potentially world-changing project: an Apple car.

The logic was sound. Like computers, cars are a massive industry—a $2 trillion market—ripe for innovation. Like music, they're personal and beloved. Like phones, we use them constantly.

And Apple was already a world leader in hardware and software design, cameras, sensors, customer service, environmental responsibility, supply-chain management, and retail. Best of all, Apple already had more than a billion ardent fans. If anyone could build a gorgeous, safe, electric car, it was Apple.

Cook loved the idea. "I don't know what kind of car you have today, but the interface probably isn't in the top-ten list of what you love about your car," he said at a conference. "The industry is at an inflection point for massive change."

No public photos exist of Apple's revised Titan design, but this discontinued Canoo Lifestyle van approximates its look and shape.

Riccio and Apple's corporate-development team focused on two possible approaches. First, they could jump-start the project by buying Tesla, whose electric, semiautonomous Model S was already a hot commodity. At about $20 billion, it would be a huge acquisition, fraught with questions about merging the companies' cultures and Elon Musk's role.

Cook preferred the alternative: developing the car in-house. He gave the green light to Apple's biggest undertaking yet, to be called Project Titan.

Riccio handpicked executives in product design, electrical design, software design, and program management, and invited former hardware director Bob Mansfield back from retirement to lend his wisdom.

In various unmarked Sunnyvale warehouses, a new, stealth car company took shape. Apple's top engineers leapt at the chance to join the project. Engineering VP Steve Zadesky, who'd built and led the first iPod and iPhone teams—and had been a Ford engineer before joining Apple—began assembling a dream team. Within 18 months, he'd hired a thousand engineers and designers from companies like BMW, Ford, Tesla, and Jaguar Land Rover.

The initial plan was to deliver a beautiful electric vehicle in five years, priced at around $50,000. At first, it would self-drive only on highways, but the team intended to upgrade its self-driving abilities year by year until it became fully autonomous. They often compared the mission to that of the iPhone: You can't get to the iPhone X, with its edge-to-edge screen, without first shipping the iPhone 1.

Jony Ive, a car aficionado and collector himself, offered two of his top designers to the project; Marc Newson joined the brainstorming, too. They came up with a white, rounded minivan with a glass roof and sliding doors, like a futuristic descendant of the Volkswagen Microbus from the seventies.

But in January 2016, Ive sat down to give the Titan work a hard look—and he wasn't satisfied. The microbus design seemed derivative. It wasn't *iconic* like the iMac, iPhone, and AirPods.

On that day, the Titan project took a dramatic pivot. Instead of working toward an enhanced version of the electric cars already on the road, Ive wanted to create something that had never existed: a revolutionary, fully autonomous vehicle, with no steering wheel or pedals at all, only a joystick or touchscreen as backup. It could be a luxury living room on wheels.

Some of the engineers considered this moonshot too ambitious for a 1.0 product. They'd gone from working on a well-defined project with a plausible ship date—to taking on one of the world's most challenging AI problems with no definable finish line. Some, including Zadesky, left Apple.

But the designs for the new car were thrilling. The exterior would be symmetrical, front and back, made of carbon fiber and new aerodynamic alloys, with a glass roof and whitewall wheels with black centers. Some designs featured gullwing doors, which open upward.

Ive's team built prototype interiors with two rows of seating, which could recline or even rotate to face each other, and high-end materials like stainless steel, white fabric, and wood. The car would feature the world's greatest car-cabin experience: Road-noise soundproofing with stunning immersive audio. You could adjust the darkness of the tinted windows; there were even experiments with making them augmented-reality screens that could identify streets or shops passing by. Whatever you were doing on your Mac, iPhone, or iPad would transfer effortlessly to the built-in touchscreens.

Apple outfitted a set of Lexus cars to test its self-driving software.

"It was light-years beyond anything that's out there today," says a Titan executive.

In April 2017, Apple equipped a fleet of 70 luxury SUVs with lidar sensors (like radar, but with lasers) and its self-driving software. With a backup human behind the wheel, it began test-drives around California and on a test track it had bought near Phoenix. A set of modified VW Transporter vans began carrying Titan team members among Apple's campuses—autonomously.

In 2014, Apple executives had visited British sports-car maker McLaren to discuss a potential partnership or acquisition. Maybe McLaren's core technology could accelerate the Titan project; maybe its headquarters could become an Apple innovation hub in the UK. But in the end, McLaren's CEO was more interested in Apple becoming an F1 racing sponsor than an innovation partner; no deal materialized.

Over the next three years, Apple discussed partnerships with BMW, Jaguar Land Rover, Mazda,

Ive After Apple

After Steve Jobs's death, Jony Ive's workload steadily grew. In 2012, he added all software design to his portfolio; in 2014, he took on Titan, Apple's electric-car project. All of this was on top of his work on Apple Park, the new corporate campus.

The Apple Watch project, though, had been particularly draining. Steve Jobs had been his friend, creative partner, and defender of ideas; in his absence, Ive had taken a lead role on marketing the watch, and found himself having to work to sell his ideas to the new management.

So in 2015, soon after the watch launched, Ive told Cook that he was exhausted. They came up with a new, part-time role for Ive: chief design officer. It would free him of day-to-day managerial responsibilities and let him focus on his design work.

In 2019, after four years in that role—and 27 years at Apple—Ive finally stepped away. He and Marc Newson founded their own San Francisco–based design firm, LoveFrom, with a handpicked team of designers, engineers, and creative people. Its work spans three categories: work they do for the love of it (unpaid), work for clients (including Ferrari, Airbnb, and OpenAI), and work for themselves, which includes buying and renovating buildings in LoveFrom's Jackson Square neighborhood in hopes of cultivating a creative haven.

In 2025, OpenAI bought io, another Ive startup dedicated to creating a family of screenless or wearable AI devices—glasses or pendants, perhaps—for $6.5 billion. For Ive, it would be a chance to address the distraction and digital addiction that accompanied the iPhone era. "I shoulder a lot of the responsibility for what these things have brought us," he says.

LoveFrom's first big client was Apple itself, which offered Ive a $100 million consulting contract. In 2022, the two companies decided not to renew it.

Ferrari, and Koenigsegg. The deal that came closest to completion, though, was the one with Mercedes. The two companies would jointly design and engineer the first car; Apple would do the software, Mercedes would handle the manufacturing and distribution. But once again, the discussions fell apart—this time over concerns about melding two very different cultures, who would have the final say on design, and how the co-branding would work.

In 2023, one final acquisition possibility arose: Rivian, the California maker of electric SUVs. It was a critical crossroads for the project.

The Titan executives presented Apple's executive team with three options. They could continue developing a fully autonomous vehicle alone, in-house; partner with or buy Rivian to get cars on the road, and deliver full self-driving later; or develop self-driving software that Apple could license to other car companies.

Cook dismissed the last option; Apple is a product company, not a licensing company. He left the decision about the first two options to the executive team.

Silicon Valley was already a graveyard for failed electric-car startups: Canoo, Faraday, Byton, Zap, Aptera, and Better Place. Nobody had ever created a fully autonomous car. You can train AI to stop at red lights or stay in the lane on highways, but there are also thousands of judgment calls. How aggressively should you merge? That pedestrian waiting at the crosswalk—is he going to cross, or is he just texting? Is it safe to drive over that black thing in the road?

In the decade since Titan was conceived, the project had gone through changes in leadership and direction. It had produced breakthroughs in battery technology, glass chemistry, chip design, audio technology, sensors, touchscreens, navigation software, adaptive ride dynamics, safety features, wireless technologies, and charging technologies. Yet the sheer immensity of the task meant that the finish line seemed to keep receding. Meanwhile, electric-car sales had been slowing, and ride-sharing services like Uber turned out to be barely profitable.

And so, in the end, the executive team voted not to buy Rivian—and, soon thereafter, to exit the car business altogether. On February 27, 2024, in an all-hands video call, COO Jeff Williams and then–Titan operations leader Kevin Lynch announced that Apple was canceling the project.

About a third of the Titan team members, mostly the pure vehicle engineers, were laid off. The rest moved into Apple's software, AI, and hardware engineering groups, where they brought fresh technologies—and hundreds of patents—to Apple Intelligence, Vision Pro, and future products. In that regard, says a member of the team, "in no way was Titan a failure."

With a total investment of about $10 billion, Titan was the biggest skunkworks in Apple's history, and its conclusion, both inside and outside Apple, was heartbreaking. "If we'd stuck to our original plan [getting a first-generation car on the road, then working toward full autonomy], we'd be on our third-generation car by now," says the staffer.

But many of those who worked on it are proud of the work they did—and of Apple for greenlighting the project and making the attempt. "At some point, when a company gets big, it usually stops taking risks," says a Titan executive. "Folks worry that they have more to lose than to gain by taking a risk. And that's the death knell of companies."

46. Silicon

Over the years, the Steve Jobs "make the whole widget" philosophy—hardware and software, end to end—paid off handsomely. It enabled far greater privacy and security. It made possible Apple-only features like Face ID, AirDrop, and Handoff. Above all, it differentiated Apple's unified, coherently designed products from the fragmented marketplaces of Windows and Android.

But Apple had never made a key component of the widget: the processor. From its very founding, Apple's machines had incorporated brains from MOS Technology, Motorola, IBM, and Intel. In time, Apple paid dearly for that reliance. When those partners didn't innovate, fell behind, or missed a deadline, Apple suffered.

So in 2008, when it was clear that the iPhone was the next big thing, Jobs asked hardware head Bob Mansfield to look into an ambitious idea that no consumer-electronics company had ever attempted: taking over the design of its own processors.

Mansfield's favorite candidate for the job was Johny Srouji, a rising star at IBM in Austin, where he was managing the latest PowerPC project. Srouji agreed to take the job based on only a vague description of what it would entail.

Johny Srouji
Born: Haifa, Israel
Schooling: Technion Israel Institute of Technology (BS, MS)
Before Apple: IBM, Intel
Apple: 2008–present

Srouji speaks fluent Arabic, Hebrew, French, and English.

Staffing Up

The chips in a modern computer are among science's greatest achievements. Each is a tiny slab of silicon, densely etched with tens of billions of tiny transistors. A high-end CPU (central processing unit) chip can handle trillions of calculations per second.

The eternal goal of chip design is to make the individual transistors smaller, faster, and less power hungry. It's fair to say that chipmakers have made some progress in this quest: Modern transistors are only 3 nanometers wide, the width of a DNA molecule. A human hair is 30,000 times thicker.

But designing silicon, as Jobs was proposing, is depressingly difficult.

"Silicon design is unforgiving. Super, super, super hard. I can't emphasize it enough," Srouji says. If you're writing an app and you discover a bug, you just fix it and distribute a new version. But if you discover a bug in your chip design, you're looking at a massive product recall.

It takes three or four years, furthermore, to deliver each new generation of a processor; the new iPhones that fans buy in 2026 contain processors that were designed in 2022. For Srouji's team, that meant placing long-term bets they couldn't afford to get wrong. "I told them from the get-go: There is no backup plan," he says. "Burn the bridges."

Getting to SoC

If you opened up a 2008 computer, you'd find a traditional circuit board studded with soldered chips: CPU, graphics chip, Wi-Fi chip, memory and storage controllers, and so on.

But in smartphones, all of those functions are integrated into one, much bigger silicon slice called an SoC—a system on a chip. SoCs are complex and hard to design flawlessly, but the savings in space and battery power are worth the trade-off.

An Apple A4 system-on-a-chip, featured here in an Apple TV.

Samsung had been making the iPhone's SoCs since the original model in 2007. Srouji's assignment was to create something better—from scratch—on a tight timeline. He professed to love the challenge. "Hard is good. Easy is a waste of time," he likes to say.

He couldn't go from 0 to 60 in a single attempt. So for the first version of the SoC, the only homegrown piece would be the CPU (processor). Srouji had inherited a team of engineers who had been working on Mac chipsets; now, to gain CPU expertise in a hurry, Apple acquired Palo Alto Semiconductor, a 150-person boutique chipmaker whose specialty was fast, low-power chips. Apple sealed the deal in April 2008 for $278 million, and soon had chip-testing facilities running in unmarked buildings in nearby towns.

By 2010, the Apple silicon team had completed, tested, and debugged their first custom system on a chip, which they called the Apple A4. It was thinner, smaller, faster, and less power hungry than the Samsung design that it replaced, and, ultimately, less expensive. It was based on the same high-speed, low-power ARM architecture that Apple itself had helped to launch in 1990.

The A4 debuted in the first iPad. Apple used the same chip in the iPhone 4, the fourth iPod Touch, and the second Apple TV.

Taking charge of its own chip destiny brought Apple a long list of benefits. First, the obvious: Apple would no longer be beholden to anyone else's schedule. Second, the chips that Apple had been buying always included elements that had been requested by *other* Samsung customers. Now Apple could ditch the features it didn't need, saving time, money, space, and power.

Finally, there was a security advantage. If Apple controlled the whole *stack* (the entire set of technologies, from low-level hardware to the highest-level software), it didn't have to worry about security weaknesses in third-party components. It could also build in new security features—like its walled-off Secure Enclave, which stores your fingerprint or Face ID face profile.

Year by year, Srouji's team designed more of the SoC's elements, buying other chip companies as needed, including Intrinsity in 2010 (CPU performance optimization), key parts of Dialog Semiconductor in 2018 (power management circuitry), and Intel's smartphone wireless division in 2019 (to develop its own cellular modem). "We went at it one by one, based on what we thought is the highest urgency," Srouji says. After ten years, "we were owning and building the best SoC in the world."

Each iPhone chip generation—A5, A6, and so on—was faster and more efficient than its predecessor, while still incorporating more circuitry for more features, like fingerprint sensor, video calling, Siri, and Face ID. Soon, Srouji's team was designing chips for almost everything in Apple's product line: iPad, iPhone, Apple Watch, HomePods, AirPods, AirTags, and so on.

There was only one shoe left to drop, and it was a doozy: a custom chip for the Mac.

The M1

In principle, an Apple system-on-a-chip could bring to a Mac all the same benefits it had brought to phones and tablets: better speed, lower power consumption, tighter security, smaller size, lower cost. But Apple had already dragged Mac fans through two previous processor transitions: to PowerPC in 1994, and to Intel in 2006.

Apple had been growing frustrated by Intel's work for years. It had intended to include Intel's Skylake processor in the next Macs, but was appalled at the bugginess of its prototype. "Our buddies at Apple became the number one filer of problems," says a former Intel engineer. "When your customer starts finding almost as much bugs as you found yourself, you're not leading into the right place."

(Intel's CEO Paul Otellini had also passed on making the processors for the iPhone; the price Apple was offering seemed too low. "I should have followed my gut," he says now. "The world would have been a lot different if we'd done it.")

Still, when Cook announced that Macs would incorporate Apple's first Mac chip, the M1, the announcement alarmed developers. *This* again?

Once again, all software programs would have to be reworked to run on the new chip architecture. But at this point, Apple's engineers were veterans of processor transitions. This time, they

Making Chips

Apple now *designs* its own chips—but doesn't actually *make* them.

Manufacturing semiconductors requires massive, multibillion-dollar clean-room facilities, where the air is 1,000 times cleaner than a surgical room. An eyelash, a speck of dust, or even the wrong color light can ruin the delicate silicon wafers.

Today, only about 12 percent of the world's chipmaking takes place in the U.S.; the vast majority is in East Asia, and mostly at TSMC (Taiwan Semiconductor Manufacturing Company), the world's biggest chipmaker.

American companies like Apple, AMD, NVIDIA, and Qualcomm routinely design their own chips, and then hire TSMC to make them. TSMC even makes some chips for Intel, one of whose founders co-*invented* the integrated circuit.

Samsung's plants manufactured Apple's A4, A5, and A6 chip designs. But in 2013, Apple, too, switched to TSMC.

The pandemic showed how fragile the Asian supply chain can be. If anything were to happen to Taiwan—earthquake, cyberattack, conflict with China—Apple and the rest of the industry would be in big trouble.

In 2022, Congress passed the CHIPS and Science Act, which earmarked $52 billion in funding to rekindle the chipmaking industry in the U.S.; by 2025, 22 new plants had broken ground, including three TSMC plants on American soil.

Apple likes to make the whole widget—but when it can't, it likes to have backups.

were ready with tools that made adapting developers' apps relatively easy. And to buy the software companies time, Apple once again created an emulator—a software layer that translated older apps' instructions for the new chips. This version was called Rosetta 2, a nod to the original Rosetta emulator for Intel chips.

The traditional problem with emulators is that they slow down your software. But the M1 chip was so fast that, *even in the emulator*, older apps ran faster on the new Apple processor than they had run on Intel chips in the first place.

The new M1 laptops were 50 to 70 percent faster than the old models—and the battery could now run 20 hours on a charge. Apple laptops became charge-twice-a-week machines.

The reviews of the first M1 Macs were glowing. "Chip transitions are devilishly hard and don't usually go smoothly," noted the Verge's critic. "This MacBook Air not only avoids almost all of those pitfalls, but it gleefully leaps over them."

Cult of Mac's reviewer downloaded the Microsoft suite, ran Photoshop, and opened 650 browser tabs—simultaneously. "The machine is as fleet as a greyhound," he wrote. "I've never seen anything like it."

"The new M1-powered MacBook Air is hilariously fast, and the battery lasts a long-ass time," wrote Ars Technica. "Just go ahead and sell your old MacBook Air immediately and get this thing instead."

The only real loss was that without Intel inside, Macs could no longer run Windows.

It took two and a half years for Apple to replace every Mac model with an Apple silicon version. In time, the M1 chip gave way to successors—M2, M3, M4, and so on—and to bigger, more powerful, pricier variants (M3 Pro, M3 Max, and M3 Ultra, for example).

Even after all of this, Apple still does not make the *whole* widget. It still buys screens, batteries, memory, camera lenses, and thousands of other parts from other suppliers. But the processor—the system on a chip—is the beating heart of every device.

47. Apple Intelligence

In late 2022, the biggest headlines in the technology world—if not the *whole* world—were about artificial intelligence. AI had been around for decades, but a company called OpenAI had introduced stunning breakthroughs in *generative* AI: AI that could create artwork or writing based on your requests.

You could ask ChatGPT to write bedtime stories, scripts, invoices, essays, lyrics, jokes, speeches, recipes, therapy sessions, software code. It was a phenomenal tutor, explainer, and summarizer. It could answer questions about photos. Its companion site, Dall-E, could create photos, paintings, logos, cartoons, diagrams—any image you described.

ChatGPT became the fastest adopted software in history. Within two months, 100 million people were using it. By 2025, ChatGPT was processing 2.5 billion requests a day.

Fears loomed of mass unemployment, disinformation, deepfakes, and unmanned military killbots. Within months, half of college students were using ChatGPT to cheat on their homework, leading to measurable declines in memory, critical thinking, creativity, and reading and writing skills.

But AI breakthroughs also offered hope. A Google AI project called AlphaFold decoded the submicroscopic shapes of all 200 million known proteins, the first step in curing Alzheimer's, Parkinson's, ALS, and cystic fibrosis. Another learned to predict floods days in advance. Country singer Randy Travis, silenced by a stroke in 2013, put out a new album, his voice flawlessly regenerated by AI.

Apple had been using artificial intelligence in its products for years—in Face ID, Siri, and other features—without ever using the term. Since 2017, every iPhone had contained what Apple calls a Neural Engine, a dedicated part of Apple's system-on-a-chip for AI functions. In 2018, Apple hired a superstar away from Google: John Giannandrea, whose teams had driven the AI in Google products like Gmail, Photos, and Translate.

But the *generative* AI revolution seemed to catch Apple by surprise. Once ChatGPT exploded onto the scene in 2022, its rivals were quick to respond. In November 2022, Microsoft released Copilot. In early 2023, Google offered Bard (later renamed Gemini). In late 2023, Meta unveiled Meta AI. Thousands of startups sprang up overnight to extend the idea to videos, songs, and software code.

But from Apple—nothing.

"We're perfectly fine with not being first," Cook would say. "As it turns out, it takes a while to get it really great."

Small Features

As the June 2024 Worldwide Developers Conference approached, a year and a half after the ChatGPT era began, expectations ran at fever pitch.

Craig Federighi
Born: 1969, San Leandro, CA
Schooling: University of California, Berkeley (BS, MS)
Before Apple: NeXT, Ariba
Apple: 1996–1999, 2009–present

Despite video evidence to the contrary, Federighi is not actually an expert in parkour.

Software head Craig Federighi—as always, Apple's most entertaining presenter—took the stage to reveal that Apple had decided not to create a chatbot like ChatGPT at all. Instead, its approach was to scatter AI features across the operating system, so that they popped up exactly where they'd be most useful. These features, he said, would work on the most recent iPhones, and would be called Apple Intelligence.

He demonstrated a few examples:

- **Writing Tools.** Highlight text, tap a button, and get your writing proofread, summarized, or rewritten to be more friendly, concise, or professional.

- **Call transcriptions.** Record a call, then view a transcript or summary of what was said.

- **Image Playground.** A new app dedicated to generating graphics that match your typed request.

- **Notification summaries.** Concise summaries of notifications to save you reading.

- **Visual Intelligence.** Aim the camera at something, and then tap either the Search button, to search Google for

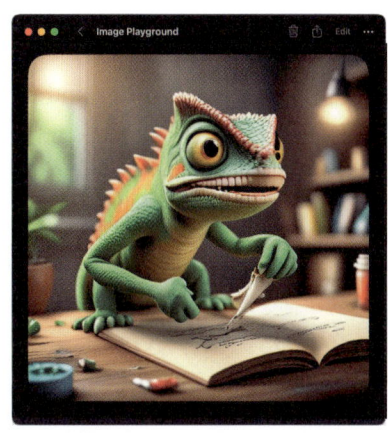

"An angry chameleon author rips up his manuscript."

details about it, or Ask, to submit it to ChatGPT to ask questions about it ("How many calories in this?" "What artist painted this?" "Interpret these stacked parking signs. Can I can park here right now?"). If you took a picture of a flyer announcing an event or performance, the phone would offer to add it to your calendar.

- **Photos.** Type a description of a slideshow you'd like made from your photo collection ("Date nights with Casey"), watch it instantly. Remove things (telephone pole, photobombing tourists) by scrubbing them out with your finger.

These features arrived in a series of software updates through the spring of 2025. Another round came in the fall of 2025, such as:

- **Screen queries** let you ask ChatGPT questions about whatever is on your screen, for example, or to search sites like Google and Etsy for similar images or products.

- **Junk-call screening.** When you get a call from a number that's not in your Contacts, the phone answers it silently, in the background. Only after the caller provides their name and the reason for their call does your phone ring—and you can gauge whether or not to pick up by reading the caller's name and purpose.

- **Hold assist.** If the phone detects hold music, you're offered a Hold button. Now you can go about your business. When an agent finally comes on the line, your phone rings to let you know.

- **Real-time translations** of conversations in Messages, FaceTime, and phone calls. If you're wearing recent AirPods, you can even hear *spoken* translations of your conversation partner's utterances. If both parties are wearing them, you each hear the entire conversation in your native language, *Star Trek*–style.

After witnessing the mind-altering, landscape-changing power of ChatGPT and its rivals, many critics found Apple's nips and tucks "kinda helpful, kinda funny, but mostly forgettable," as the Verge put it. "If you're expecting AI fireworks, prepare for AI . . . sparklers," wrote Joanna Stern in the *Wall Street Journal*. Giannandrea encouraged patience, calling the first round of Apple Intelligence features "the first of a many, many-year journey."

It was a disappointing reception for Apple's AI teams, especially because they had gone to such efforts to deliver AI features without AI's worst aspects:

- **Cheating.** Apple didn't want to encourage students (or anyone else) to use AI for cheating. "We literally had ethicists involved in our discussions on this," Federighi says. That's why Apple Intelligence can summarize or rewrite blocks of text, but won't write first drafts for you.

*The Clean Up tool proposes what to remove (left),
or you can scribble things out with your finger.*

- **Deepfakes.** The potential for deepfakes—computer-generated photos or videos of real people doing and saying things they never actually did or said—is one of the most terrifying aspects of the AI revolution. If bad guys can generate evidence of things that never happened, what happens to news reporting, trust, and democracy?

  To avoid making the problem worse, Apple's image-generation tools can produce only cartoon-styled images, and don't respond to requests for pictures that include weapons, blood, or full-body shots of people.

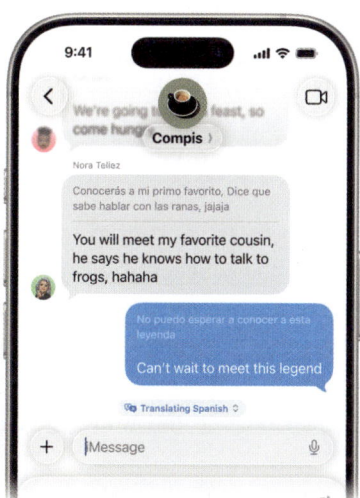 

Real-time translation can handle both text messages and spoken utterances.

- **Privacy.** When you ask OpenAI or Google Gemini for a piece of writing or art, your query is transmitted to its servers, where it can use your transaction to train further AI features. By contrast, Apple wanted to maintain its privacy track record. "You should not have to hand over all the details of your life to be warehoused and analyzed in someone's AI cloud," Federighi said.

  The holy grail would be doing *all* the AI processing right on your phone so that no data ever left your possession. But the amount of AI magic a phone can perform is limited; the iPhone's on-phone AI system has three billion parameters (a measure of learning ability and complexity), compared with 175 billion for ChatGPT. In practice, that means that an iPhone can handle small AI requests, like Image Playground images, on the spot. But more complex jobs, like creating summaries of audio recordings, require crunching of your actual, unencrypted data by powerful servers in data centers.

  The solution: Apple would design and build a new kind of data center, whose servers were

Private Cloud Compute

For complex generative AI requests, Apple's engineers didn't want to rule out the option of using external servers—but that would mean the end of "What happens on your iPhone, stays on your iPhone," as one Apple slogan had put it in 2019.

The solution was to invent a new kind of server that stores nothing. "Your data comes in, you get your response, and the data gets erased," says Federighi. "The system does not have the ability to send your data somewhere else. It doesn't have the ability to write that data down anywhere where anyone could pick it up."

No one, not even a rogue Apple engineer, can access your data, not even in an emergency. Apple calls the system Private Cloud Compute, and asserts that it's every bit as secure as data that never left your phone.

That's what Apple *says*. But Big Tech companies have lied before. Why should anyone believe Apple's story?

The answer is unheard-of in the AI industry: Apple makes its server system available for inspection by anyone who wants to pore through its code—mainly, security researchers. After a first round of hiring security firms to inspect the software themselves, the company opened its AI servers' secrets to everyone, friend and foe. And to make it worth the researchers' while, Apple instituted a "bug bounty" program: For any new security bug a researcher reports, Apple awards them up to $250,000. If the bug is in Private Cloud Compute, the awards go up to $1 million.

To date, nobody's found anything worth reporting.

incapable of saving any data. Apple called the initiative Private Cloud Compute.

For requests that even Apple's AI could not yet handle, Apple also built in an escape hatch: a button that would send your query off to ChatGPT. You'd tap to grant permission, and you'd get ChatGPT's generated reply instead of Apple's. Even then, Apple asserted, your data would remain your own; its deal with OpenAI precluded capturing and using your data.

Siri offers to ask ChatGPT for you.

- **Massive power use.** Generative AI has become famous for being a power hog—an enormous problem at a time when the planet's goal is to burn less fuel. Apple's data centers run on 100 percent renewable power.

Fixing Siri

Critics may have found the first round of Apple Intelligence features underwhelming. But the demo of a new Siri blew them away.

Apple's numbers showed that by 2024, people were asking questions of Siri 1.5 billion times a day, but it had no numbers on how many complained about it. Over the years, the equivalent feature on Android had become far smarter and more useful. And ChatGPT, of course, was *freakishly* good as an all-knowing, conversational assistant.

But at the end of Apple's WWDC June 2024 video keynote, Federighi introduced a feature that even ChatGPT couldn't touch: a Siri that knew *you personally*. It would be able to see, for example, what was on your screen. The examples he provided were tempting:

- **Someone sends you an address in Messages.** You could say, "Add this address to this person's Contacts card."

- **A friend texts you about a new album.** You: "Play that."

- **You want to share photos.** You: "Send Casey the photos from Saturday's barbecue."

- **You're filling in an application form.** You say: "Fill in my driver's license number." Siri finds the photo of your license, if you have it, auto-extracts the number, and fills it in.

Siri would even be able to see into your apps: your calendar, email, messages, and so on. "Imagine that I am planning to pick my mom up from the airport, and I'm trying to figure out my timing," said AI lead Kelsey Peterson. She spoke to her phone to demonstrate. "Siri, when is my mom's flight landing?"

In a blink, Siri found the flight details in an email her mom had sent, checked the flight's progress on a flight-tracking website, and displayed her updated landing time.

Peterson went on: "What's our lunch plan?" Siri mined the restaurant details from her chat history with her mom. And finally, "How long will it take us to get there from the airport?" The instantaneous answer: 21 minutes.

Never once did Peterson open the Mail, Messages, or Maps apps, go online, or search for anything. "This year marks a new era for Siri," she said.

This was useful. *This* would save real time. This was *beyond* ChatGPT; it was AI as your agent. And only your iPhone had access to your data. None of it went to Apple; none of it left your phone.

Fans lost their minds. Investors drove Apple's total stock valuation up $200 billion, the biggest gain in Apple's history.

Siri Struggles

But inside Apple, the project was in trouble. The software teams had identified two different roads for achieving the "more personalized Siri": a smaller rewrite, which Federighi would later refer to as version 1, and a more complete overhaul, version 2. As WWDC 2024 approached, V1 was working well enough that Apple expected to deliver the update in six or nine months.

"So we announced it as part of WWDC [2024]," says Federighi.

The more Apple worked on the V1 Siri, however, the more it looked like a dead end; it was giving wrong answers far too often. It became apparent that it wouldn't be releasable on the original schedule. Federighi's team had no choice but to resume work on the more time- consuming V2 overhaul.

The central problem: Apple does not knowingly ship features that *sort of* work. But when generative AI doesn't know the answer to your question, it often makes one up. Those fabricated facts are known as hallucinations.

They were already biting Apple's rivals. In May 2024, Google's search-summary feature, called AI Overviews, had advised one citizen that adding "about ⅛ cup of non-toxic glue to the sauce" to pizza sauce would prevent cheese from sliding off. It told another that, according to a 2018 study, "parachutes are no more effective than backpacks" when jumping from an aircraft. A sheepish Google had to turn off the feature for a few months to make fixes.

Shortly after Apple Intelligence debuted, hallucinations haunted Apple, too, when its notification-summary feature occasionally went awry. "Luigi Mangione shoots himself," went one

BBC News summary, a reference to the murderer of a healthcare CEO in December 2024. It wasn't true. Apple quickly turned off the summarizing feature for news and entertainment apps.

The notion that Siri might supply bad information put the engineers under excruciating pressure. Nobody—not Google, not OpenAI, not Microsoft—had solved the hallucination problem, and it was dogging Apple's V1 version of the new Siri, too. It was giving wrong answers far too often.

It became apparent that V1 wouldn't be releasable on the original schedule. Federighi's team had no choice but to resume work on the more time-consuming V2 overhaul.

And yet Apple had already begun advertising the smarter Siri on billboards, in print, and on TV. In one TV ad, actor Bella Ramsey, at a party, gets a wave from a guy whose name she can't recall. She asks Siri: "What's the name of the guy I had a meeting with a couple of months ago at Café Grennell?" Siri tells her that his name is Zack Wingate.

But there was still no sign of the smarter, personalized Siri, even a full year after Apple's original demo, even after Google rolled out similar features in its Pixel phones.

Over the years, Apple had acquired considerable institutional wisdom about the dangers of preannouncing products. That's how Sculley had gotten into trouble with the Newton. Then there was the business of AirPower, a wireless charging mat that could charge your iPhone, Watch, and AirPods case simultaneously. Apple unveiled it in September 2017 and continued to reference its upcoming release for over a year—but, when engineering challenges proved insurmountable, Apple canceled it altogether.

In this case, the Zack Wingate ad disappeared from YouTube. "It's going to take us longer than we thought to deliver on these features, and we anticipate rolling them out in the coming year," Apple said in a statement.

The fallout was brutal. Class-action lawsuits accused Apple of advertising features it didn't have. Critics wondered if they'd ever even existed. "What Apple showed regarding the upcoming 'personalized Siri' at WWDC was not a demo," wrote Apple blogger John Gruber. "It was a concept video. Concept videos are bullshit, and a sign of a company in disarray, if not crisis." (Apple maintained that the demo had not been staged.)

Apple, once the world's most valuable company, sank to number three as NVIDIA and Microsoft stock surged, driven by their AI breakthroughs.

There was some executive shuffling—Apple reassigned Siri to Vision Pro godfather Mike Rockwell—and quick acquisitions of at least seven AI companies. The new target for the new Siri was the second quarter of 2026.

"In the longness of time," Cook said of the brouhaha, "I don't think it will be even a footnote."

48. Headwinds

Every company wants happy customers, a reputation for excellence, and healthy profits. That's how a company grows.

But growth has downsides, too. It's harder to manage 165,000 employees than 100.

When you're that big, for example, you can't release a product until you have millions of units ready to sell on launch day. If some high-tech part isn't yet available in that quantity, you have to watch in frustration as your smaller competitors introduce something similar in smaller numbers.

Above all, when you're a massive corporation, you become a target. Everything you do falls under scrutiny from regulators, competitors, customers, and the press.

As Tim Cook guides the company into its second half century, he faces no shortage of mega-company challenges.

Slowing iOS Sales

For more than a decade, Apple sold more iPhones and iPads every year—but then the trajectory flattened. iPhone revenue for the holiday 2024 season, for example, was down 3.5 percent from 2022.

Apple still releases a new iPhone every year, but the improvements are generally minor. The modern iPhone fan upgrades their iPhone every four years instead of every two, as they did in the 2010s.

Apple also points out that its phones are built tougher, so people use them longer. One marketing executive notes indignantly that in the early years, people made fun of all the iPhones with cracked screens—a phenomenon, he maintains, that actually signaled quality. "If you dropped our competitors' phones hard enough to break the glass, that phone *stopped working*. You dropped an iPhone enough to break the glass, it kept working, which is why you would see iPhones out there with broken glass and not competitors."

App Store Charges

When Apple opened its iPhone App Store in 2008, it offered app authors this deal: Apple would keep 30 percent of all app sales. Apple, after all, was testing every app for privacy, security, content,

TaxGate

For decades, global brands like Apple took advantage of tax strategies offered by the Irish government, which hoped to attract foreign investment. One complex arrangement, nicknamed the Double Irish, involved setting up two Irish entities: one in Ireland, the other managed from a tax haven like Bermuda—a "stateless" company under Irish law. These shell companies could hold hundreds of billions of dollars of profits, untaxed by the U.S. until it was repatriated (brought back to the U.S.).

Thanks to this arrangement, Apple spent years paying only low single-digit taxes on its overseas earnings, at a time when the U.S. corporate tax rate was 35 percent.

These tax shelters were technically legal. But the European Union objected to the arrangement, and, in 2016, ordered Apple to pay about $15 billion in back taxes to Ireland.

Apple, the world's largest taxpayer, rejected the implication that it behaved shadily. "Apple believes comprehensive international tax reform is essential, and for many years has been advocating for simplification of the tax code," it said in a statement.

It soon got its wish. In 2018, the Tax Cuts and Jobs Act cut the U.S. corporate tax rate down to 21 percent—and allowed repatriation of offshore money to come in at 15.5 percent. Apple took the opportunity to bring back cash that it had been storing overseas—$274 billion—and paid $38 billion in U.S. taxes on it.

---

and quality before listing it so that the world's iPhones wouldn't fill up with buggy crud. Apple also saved app developers the costs of credit card transactions, hosting, and customer service.

Apple didn't want buggy crud to creep in from other sources, either. So the App Store would also be the *only* allowable source of iPhone apps.

Over the years, the number of apps grew to two million, and Apple paid out over $155 billion to developers. The financial models of most apps shifted: Now the app is often free, but you pay for *in-app purchases*—when you buy a new game level, for example, or a subscription to a service.

As the App Store's dominance grew, however, so did developer complaints. 30 percent struck them as an awfully big bite, especially for little companies. In 2020, Apple lowered the rate to 15 percent for software companies that earned less than $1 million in a year.

But developers were also annoyed by the 30 percent fee on in-app purchases, resulting in some bizarre side effects: For example, Netflix, Spotify, and the *New York Times* designed their apps so that you *couldn't* buy a subscription in their apps. You had to sign up on their websites—dodging the 30 percent Apple commission—and then return to their apps to enjoy your movies, music, or news. You couldn't buy an ebook in the Kindle app, either; you had to duck out to Amazon's website, buy the book, and then return to the app to read it.

As the 2020s wore on, the App Store model fell under attack on multiple fronts.

- **Epic Games,** maker of the white-hot battle game Fortnite, staged a rebellious stunt. It created a software patch that let fans buy Fortnite's in-game currency directly, bypassing the app stores—and their 30 percent fees.

    Apple and Google promptly took the game off their stores, and met Epic in court.

    Apple won on most counts. But the judge did rule that Apple had to permit, within apps, *links* to sites where fans could buy their ebooks, sign up for subscriptions, and so on.

    Apple's initial reaction to the court order was to permit those links—but announced that it would still charge a 27 percent commission on any resulting purchases within seven days. The U.S. district court judge was not pleased. "This is an injunction, not a negotiation," she said. Apple finally relented fully. Today, apps like Spotify, Kindle, and Patreon include links that take you instantly to their subscription and purchase websites—and Apple collects nothing.

- **U.S. v. Apple.** As Big Tech became Bigger Tech, the public, and the officials they elected, grew distrustful of their power. In 2024, the U.S. government brought antitrust suits against Google, Meta, Amazon—and Apple.

    The Department of Justice and 16 states accused Apple of giving its own products special treatment. For example, texts to fellow Apple users enjoyed messaging features like read receipts and full-resolution photo sharing. When texting Android phones, those features were missing. (Plus—green text bubbles! Ew.) Similarly, pairing Apple Watches and AirPods with an iPhone was far faster and more effortless than with rival brands. Apple also prevented third-party apps from accessing the iPhone's NFC chip, effectively making its Apple Wallet app the only Tap to Pay option at checkout.

    To Apple, these objections seemed absurd. Of *course* it had designed its products to work seamlessly together! That's precisely why Apple makes the whole widget: so that everything works smoothly together, with tight privacy and security. Throwing the door open to other parties could invite a real mess.

    The U.S. case is likely to take years to resolve. In the fall of 2025, however, Apple and Google both caught a break. A federal judge ruled that Google could continue making exclusive deals like the one it has with Apple: Google.com comes as the default search page on every iPhone and iPad—and in return, Google pays Apple about $20 billion a year.

- ***EU v. Apple.*** Starting in the 2010s, the European Union began staking its claim as the world's most aggressive tech-company regulator. It passed a series of laws that govern competition, data privacy, AI guardrails, and content moderation (cleaning up toxic user posts). Its courts investigated or fined Meta, Microsoft, Amazon, Google, and Apple.

    In 2022, for example, the EU passed the Digital Markets Act (DMA), a sweeping law designed to shut down anticompetitive practices. It took particular aim at Alphabet (Google), Amazon, ByteDance (TikTok), Meta, Microsoft, and Apple. It included a long list of requirements that these companies would have to fulfill by March 2024. For example, Apple would have to permit customers to install apps they got from any source.

    If Apple hoped to continue operating in the 27 EU countries, it had little choice but to comply. So in March 2024, Apple's EU customers gained the option to download apps from app stores other than Apple's. Banks can now use the iPhone's wireless NFC chip for their own payment systems. And because Apple wasn't allowed to favor Safari, its own web browser, you encounter a pop-up menu screen offering you a choice of browsers the first time you try to open a web page.

    In 2025, an Australian court, too, found that Apple and Google's app store exclusivity was illegal; China, the UK, and Spain began investigations of their own.

    Apple expressed its unhappiness with these findings by pointing out that "the new options for processing payments and downloading apps on iOS open new avenues for malware, fraud and scams, illicit and harmful content, and other privacy and security threats."

    It didn't have to mention its other incentive for fighting these battles: The new law would bite into the App Store commissions that had been earning Apple $27 billion a year.

China Censorship

It's safe to say that the Chinese government is not quite as progressive as Western governments. Its "Great Firewall"—a system of filters, surveillance, and regulations—controls what Chinese citizens can see and read. In mainland China, for example, you can't access Google, YouTube, Facebook, X, Instagram, WhatsApp, Wikipedia, Gmail, Google Maps. Nor can you read foreign news outlets like the *New York Times*, the *Wall Street Journal*, Bloomberg, BBC, Reuters, or CNN.

The only way to break through this firewall is to use a VPN (virtual private network) app, which masks your location as you connect to the internet.

In 2017, the Chinese government asked Apple and other tech companies to remove VPN apps from their app stores, as well as the *New York Times* app and Skype. Apple eventually complied.

China has since asked for the removal of hundreds more apps; Apple complies or even proactively removes them from the Chinese app store. They've included encrypted messaging apps, human rights apps, gay dating apps, tools for organizing protests, apps that let you edit your selfie, sex position guides, document-sharing apps, and apps about the Dalai Lama. Apple even removed the Taiwanese flag emoji from Apple iPhones in mainland China and Hong Kong.

For human rights organizations, senators, and journalists, it seemed as though Apple was enabling Chinese censorship and repression so that it could keep making money in China.

Tim Cook defends Apple's compliance. "From my American mindset, I believe strongly in freedoms," he said in 2017. "But I also know that each country in the world decides their laws and their regulations. And so your choice is: Do you participate? Or do you stand on the sideline and yell at how things should be? My own view very strongly is: You show up and you participate. You get in the arena, because nothing ever changes from the sideline."

China Reliance

For years, Cook's enormous investment in Chinese manufacturing looked like a brilliant move. In return for supplying billions in equipment investments and high-tech manufacturing expertise, Apple gets advanced manufacturing with low cost, incredible speed, and impressive quality.

But if anything goes wrong—such as an earthquake, pandemic, or political unrest—Apple has few options. And Apple has faced steadily stringent demands from Chinese president Xi Jinping as he tries to make China less reliant on foreign technologies and assert greater power over the West. Apple is now required to store Chinese customer data on state-owned servers, for example.

In 2025, Donald Trump threatened a 25 percent tariff on phones made outside the United States, which Cook estimated would cost Apple $900 million a year. If the iPhone were made in the U.S., analysts calculated that its price would be about $3,500.

The tariff threat was enough to send Apple stock tumbling; in a single day, it lost $773 billion in stock value, capping a 21 percent drop for the year. It was also a surprise, considering the efforts Cook had made to earn Trump's favor: He had personally contributed $1 million to Trump's election campaign and joined Trump for calls and dinners.

The common wisdom is that Chinese manufacturing is cheap because labor is cheap. And that's true. "If you put in health and benefits and all those things," says an American manufacturing expert in China, "you're talking six, seven times more expensive on an hourly basis for direct labor, let alone management and supervisory level."

But the bigger issue is China's manufacturing talent, accumulated know-how, and parts infrastructure, which are light-years ahead of any other country.

"You need a thousand rubber gaskets? That's the factory next door. You need a million screws? That factory is a block away," says a former Apple manufacturing executive. "You need that screw made a little bit different? It will take three hours."

The talent needed to bring manufacturing back to the U.S. just doesn't exist. The number of American tool and die makers—the tradespeople who design, build, and repair precision metal-parts equipment—began to decline in the late sixties. Today, in the U.S., "you could have a meeting of tooling engineers, and I'm not sure we could fill the room," Cook said in 2017. "In China, you could fill multiple football fields."

Still, Apple has been diversifying its manufacturing since 2017. Its new manufacturing lines in Vietnam will make most U.S.-bound Macs, iPads, AirPods, and Apple Watches, and a new Foxconn plant in India will make most U.S.-bound iPhones. One Mac model—the Mac Pro—has been built in a Texas plant run by Flextronics since 2013.

In 2025, Apple committed to spending $600 billion in the U.S. over four years: to build a new server factory in Houston, to fund a manufacturing academy in Michigan, to invest in silicon engineering, and to expand its data centers in North Carolina, Iowa, Oregon, Arizona, and Nevada. In truth, Apple had already committed much of that money, but the announcement was seen as a way to please Trump and avoid the tariffs. So was the Apple-designed, gold-based trophy that Cook presented to Trump in the Oval Office—a gesture you might see as either sickeningly subservient or shrewdly strategic, depending on your politics. In any case, it all seemed to work; Trump, for the time, backed off of the tariff threat.

The 2013 "trash can" Mac Pro (left) and 2019 tower Mac Pro are manufactured in Texas.

Even so, 80 percent of Apple's manufacturing still takes place in China, where the supply-chain ecosystem for advanced electronics manufacturing is decades ahead of the U.S. Could Apple bring its manufacturing back to the U.S.?

Depending on which analyst you ask, the answer is either "a fairy tale," "not feasible," or "a pipe dream."

49. Apple: The Next 50 Years

During Steve Jobs's second tenure at Apple, he drove new products into existence with unprecedented frequency and impact. iMac, iPod, iTunes, iPhone, iPad—all within 12 years.

From the moment of his passing, the trillion-dollar question was: Can Apple keep up that pattern without him?

We now know the answer: No. Apple's post-Jobs hardware introductions haven't approached either the rate or the societal impact of the i-Products in those 12 years.

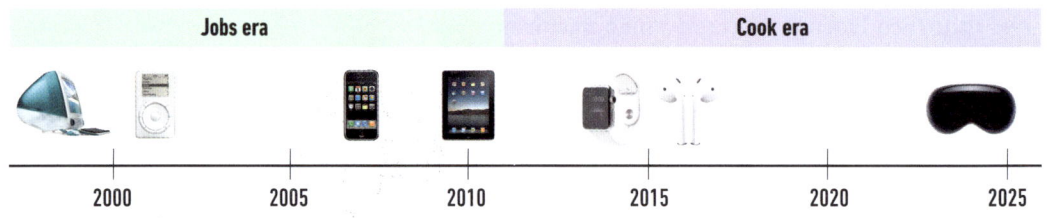

To many observers, that's an alarming development. *Fast Company*: "Is Apple Falling Behind on Hardware?" Bloomberg: "Apple's AI and AR Struggles Show It Has Lost Some of Its Product Edge." *Forbes*: "At Its Core, Apple Is No Longer Innovative."

To be sure, the company did get big, which can make nimble innovation harder to pull off. "Company insiders worry that Apple, despite its years of gravity-defying profits, is hamstrung by the political infighting, penny pinching and talent drain that often bedevil large companies," noted the *New York Times* in 2025.

But there's another possibility, too: that Jobs's hardware blockbusters owed some of their success to beautiful timing. Their arrival coincided with early-2000s breakthroughs in miniaturization, battery, processor, and wireless technologies—and a generational shift toward acceptance of technology. Maybe Jobs picked all the low-hanging fruit. Maybe even he wouldn't have been able to keep up that success streak. No other company has introduced an iPhone-level gadget since his death, either.

It could be that the era of frequent, successive new-category hardware breakthroughs is over. Maybe the world has moved on to software, services, and AI—exactly where Apple has been focusing.

Of course, as Phil Schiller notes, "Apple is doomed" editorials have appeared in every era of Apple's first 50 years. A phone without a keyboard? Nobody wants that! Detached white earbuds?

You'll lose them! The iPad? Who needs one if we already have a phone? The watch? Who needs one if we already have a phone?

"If we launch a product and there aren't people saying, 'Oh, what a stupid idea—that'll fail,' then I'm worried that we didn't push the envelope enough," he says.

Apple, as it enters its second half century, is equipped with certain advantages. It's among the most profitable companies on earth, with annual revenues close to $400 billion. It employs some of the world's best hardware engineers. It runs an unbelievably effective chip division. Its products work seamlessly together, meaning that few of its customers are inclined to leave the ecosystem.

So where, exactly, will Apple go in the coming decades?

Predicting the future of technology is a fool's game, of course. (Just ask 20th Century Fox executive Darryl Zanuck, who said about television in 1946, "People will soon get tired of staring at a plywood box every night"—or Ken Olsen, founder of DEC, who said in 1980, "There is no reason an individual would ever want a computer in their home.")

And the further out anyone tries to predict the future of technology, the more often they're wrong.

But who can resist taking a stab at Apple's future anyway? You can make educated guesses based on published reports, the technologies that are getting the most scientific interest and investment dollars today, and common sense.

Home AI

According to tips leaked to business reporters, Apple is toying with new lines of smart home products, voice controlled and powered by AI. There's talk, for example, of smart home cameras that use Face ID to unlock your front door, or detect who's in a room and adjust the lights or the music accordingly.

Apple teams are also experimenting with simple robotics, including a smart speaker with a motorized arm that keeps its screen facing you as you move around the room. AI-driven Siri would be its personality, now with an on-screen, animated character to represent it.

(Of course, Apple has always fostered internal experiments and skunkworks; many of them don't see the light of day.)

Phones

iPhones still represent half of Apple's revenue, and they sit in the pockets of 1.4 billion people. Apple therefore makes changes very carefully.

As Apple's 50th birthday approaches, the buzz is all about folding phones, which can expand to tablet size when you need more area.

The first-generation folding phones from other companies didn't sell well, representing only 1.5 percent of the market. They were very expensive—$1,500 to $3,800. Their screens showed a distracting hinge line down the middle. And nobody knows how well their flexible screens will hold up over time. Reports are that Apple's folding iPhone prototypes have stronger hinges and less visible fold lines.

In the short term, Apple continues to work on making iPhones thinner—a useful exercise that paves the way for the folding iPhone—as evidenced by the 2025 iPhone Air, about a quarter of an inch thick. At this point, the limiting factor is the USB-C connector; in time, Apple may get rid of jacks altogether. You can already charge an iPhone by clicking a magnetic ring onto the back panel; will anyone miss having to use both hands to plug in a cord every night?

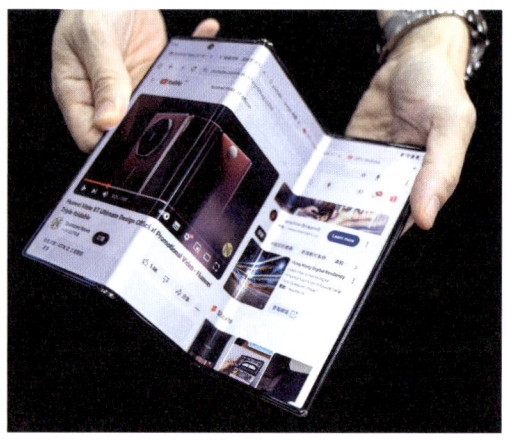

Huawei's trifold phone starts at $3,800.

Upcoming iPhones will feature cameras more sensitive in low light and better battery life (as new chemistries come along). Apple's designers would love to eliminate the capsule-shaped cutout at the top, too, by moving the Face ID sensors and the front camera underneath the screen.

But what about the longer time horizon? According to Eddy Cue, "You may not need an iPhone ten years from now, as crazy as it sounds."

In the post-phone era, augmented-reality glasses or earbuds could save us the constant retrieval of phones from pockets and purses. Our AI assistant will always be there, hands-free and ready to help us.

We know that companies like Neuralink, Synchron, and Blackrock Neurotech are already eyeing the far end of the next 50 years, when we enter the sci-fi fantasy zone: AR contact lenses and brain-implanted computers. At that point, you're not typing or dictating questions to your phone—you're *thinking* them to your implant. You'd get memory enhancement, real-time language translation, and direct access to the universe of online data, right in your brain. Or even telepathy with other people with Apple iBrains.

The notch, as it was once known, has become a capsule-shaped hole; eventually, it may disappear altogether.

Of course, even if the technology is achievable, we'd first have to get past some obvious show-stoppers. How many people would actually *want* an implant, considering that every new version requires another brain surgery? And how would you charge the battery? And what happens if your implant got hacked—or taken over? (Hey—it happened in *Ghost in the Shell*!)

Glasses

Apple's most urgent hardware dream is to pack all of the Vision Pro's magic into a lightweight pair of glasses.

By 2025, two categories of tech glasses had emerged:

- **Smart glasses** are normal-looking glasses with tiny microphones, speakers, and a camera, but no screen. In 2024, companies like Meta+Ray-Ban and Amazon offered them. Without needing your hands, you can take or place phone calls, take pictures or shoot videos, hear notifications read aloud, listen to music, and speak to your voice assistant. A light appears when you're using the camera, to avoid the privacy freakout that killed the Google Glass project in 2012.

 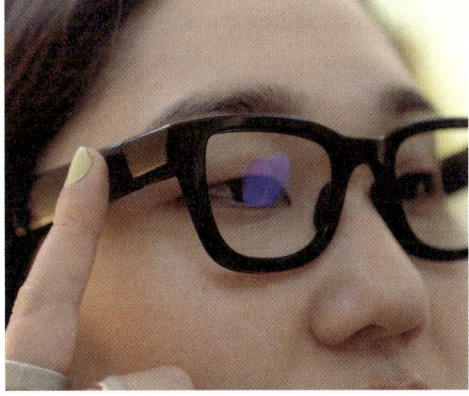

Smart glasses like these from Meta and Ray-Ban (left) have cameras and speakers. AR glasses like these from Google (right) add an optional translucent screen in front of one eye.

- **AR glasses** add screens inside the lenses. In 2025, Meta, Snapchat, and Google each displayed prototype AR glasses.

  In an onstage demo of Google's Android XR glasses, the built-in AI answered spoken questions about what the camera had seen earlier. ("Gemini, what was the name of the coffee shop

Prototype Google XR glasses can give you real-time instructions (left), translate Korean, and guide you to a destination.

on the cup I had earlier?" "That was Bloomsgiving, a vibrant coffee shop on Castro Street.") A screen built into one lens can display turn-by-turn navigation instructions or real-time English subtitles for a foreign-language conversation partner.

Meta began selling its own single-screen AR glasses in 2025. Like the Google prototypes, they represent very early technology: dim screens, narrow field of view, faint speakers, short battery life. But they hint at tantalizing possibilities. If you have trouble hearing, you can read live subtitles for the conversation. Someday, you might look at a historical site and see a "living" overlay of what it looked like centuries ago. On a city street, you might look at a restaurant's facade and see its reviews, ratings, and latest health department report.

Apple doesn't need any encouragement; Tim Cook loves the idea, and the Vision Products Group has launched an AR glasses initiative of its own, with a goal to release the screenless glasses first.

The technical challenges are immense: How do you squeeze high-quality cameras, microphones, speakers, processors, dual screens, and batteries into a lightweight pair of glasses? The Google and Meta prototypes suggested that the answer is "with great difficulty." But if anyone knows how to build a lot of technology into tight spaces, it's Apple.

Macs

At Apple, the Mac's market-share anxiety is long gone. Today, the Mac's share is about 15 percent; among young people, it's over 50. Apple's home-designed silicon initiative has worked miracles on speed, battery life, and silence; even the Macs that have fans rarely have to spin them up.

Today, close to 90 percent of all Macs sold are laptops, and that number is going up. A laptop's portability is a bonus, as is its ability to disappear almost completely from the decor when not in use.

Apple has been steadily redesigning the iPad software to look and work more like the Mac's. (Does that mean that they're destined to merge? "Still a big fat no," says global marketing head Greg Joswiak.)

When futurists describe the evolution of the clamshell computer, they mention unfolding screens that could give us vastly bigger canvases for our data. The keyboard/mouse user interface will become far less important as AI-driven agents take over the operations. Telling Siri to "find the photos from the barbecue and send them to Dad" is far simpler and quicker than bumbling around yourself on the trackpad.

The hot topics in processor science are quantum computing, neuromorphic chips, and optical processors. All of them are concepts for vastly more powerful "brains" than the transistors-on-silicon model we've been using for 50 years.

All of this presumes, of course, that anyone will want stand-alone computers in the first place. If those AR glasses are good enough, the entire concept of a computer you *sit down* to use will seem like a relic.

That's already occurred to Apple. In one 2017 ad, a mom asks her middle school daughter, busy on her iPad, "Whatcha doing on your computer?" The kid's reply: "What's a computer?"

AirPods

Until 2020, getting hearing aids required testing and consultation with a doctor or audiologist—and only six companies on earth sold them. As a result, the average price of hearing aids was $4,700, and only 20 percent of people who needed them ever got them. Studies showed higher rates of hospitalization, depression, and dementia among people with hearing loss who *don't* get hearing aids.

In 2022, a new FDA ruling changed the game. For the first time, hearing aids would be available without an exam or prescription. The theory was that many more makers would enter the field, which would drive prices down.

Apple was among them. The AirPods Pro 2 and later can double as clinical-grade hearing aids, complete with a setup test that tailors them to your hearing.

Compared with "real" hearing aids, they offer shorter battery life and, for some people, discomfort after a few hours. And, of course, there's the social problem: The telltale white AirPod stems seem to announce, "Don't talk to me—I'm listening to music."

But reviewers praised its effectiveness for mild-to-moderate hearing loss, especially with properly tweaked settings. And at $250, they're now in reach for far more of the people who need them.

Computational Photography

If the metric is "number of photos taken," the iPhone is the world's most popular camera by a huge margin. In many cases, it's hard to tell an iPhone photo from one taken by a big, heavy, long-lensed SLR camera.

But since the iPhone camera is not big, heavy, and long-lensed, how is that quality possible?

Hardware tricks help, like packing more lenses onto the back: one lens apiece for wide-angle, standard, and zoom shots.

But the biggest answer has to do with math. Using algorithms to improve a photo is known as computational photography. When you press the shutter, the phone silently takes *multiple* photos and then compares or stacks them.

The iPhone 4 (2010) introduced high-dynamic range (HDR) photography, where the camera secretly snaps *three* photos at different exposures and then combines the best parts of each, resulting in brighter brights and darker darks.

Then came Portrait mode, a mathematical re-creation of the blurry-background look associated with professional portraits. And Night mode, where the camera quietly takes 100 photos and, by combining them, gathers enough light to create an image that's even brighter than it looks to your eye.

At this point, *every* iPhone shot is computational. In a fraction of a second, the software distinguishes skin, teeth, hair, glasses, clothing, and background as separate elements, and exposes, sharpens, and color-balances each individually. Backlighting no longer turns faces into silhouettes; the camera selectively merges multiple photos at different exposures into one.

The front camera on the 2025 iPhones presented another twist on computation: a big, *square* sensor instead of the usual rectangle. This arrangement lets you snap a wide, horizontal selfie even while holding the phone vertically—a more comfortable grip. Furthermore, with the vertical grip, your gaze at the screen is centered with the camera lens, so you don't get the shifty, avoiding-eye-contact look that results from looking at the screen while the lens is off to one side (which is what happens when you hold the phone horizontally).

"Even though we are building these absolute state-of-the-art masterworks of technology and engineering," says camera software VP Jon McCormack, "our goal is for nobody to realize that. Our goal is that it's basically 'Shutter button—go.'"

But hearing is only the beginning. At some point, Apple realized that it had introduced one of the most popular wearable computers ever made. Hundreds of millions of people were now wearing tiny machines next to their faces containing battery, processor, microphone, and speaker. What else could it do?

For starters, AirPods can provide the same real-time language translation that the Mac and the iPhone do—but instead of reading the translations, you *hear* them spoken directly into your ear after a momentary delay.

If there were cameras on the AirPods' stems, they could become smart glasses without the glasses. The cameras would see the world around you; you'd speak your questions about what you're seeing; and Siri would speak the answers, courtesy of the visual intelligence technology that's already in the latest iPhones. "What does this sign mean in English?" you could ask. "Who owns this building?" "What is this weird bug?"

The AirPods Pro have FDA approval to be sold as hearing aids.

In principle, the cameras might also see your hands, watching for gestures that would control your music or your phone.

Then there are the medical possibilities. With enough advances in shrinking electronics, AirPods could detect sudden falls and crashes, just as the Apple Watch does. Better yet, your ear canal is an excellent spot for biometric sensors; it's loaded with blood vessels, and its temperature isn't affected by movement or breezes. AirPods can already monitor your heart rate; future models could replace or enhance the Apple Watch's sensors by measuring your body temperature and blood-oxygen level. They'd be excellent at detecting changes in your walking gait over time, giving you early warning about neurological diseases. In theory, they'd also have everything they'd need to alert you when you're getting overwhelmed by stress or anxiety—data about your pulse, breathing, and voice stress—and could propose a guided calming exercise.

Watch

In the beginning, the Apple Watch was essentially a Fitbit. It tracked your workouts and measured your pulse.

But the back of the watch is a relatively big surface pressed against your skin, and Apple packed it with sensors: a skin thermometer, red, green, and infrared LEDs to shine light through your skin, and light sensors to measure the reflection. There are sensors inside the watch—microphone, accelerometer (motion sensor), and gyroscope (rotation sensor)—and an electrode in the crown, permitting ECG (electrocardiogram) heart tests.

After thousands of hours of testing, Apple's medical and sensor teams learned to correlate the readings from all of these sensors with a long list of biological processes, including indirect measurements of body parts you wouldn't think had any connection to your wrist: your breathing, blood-oxygen level, stages of sleep (isn't that your *brain*?), and so on.

But more alerts are coming. The Apple Watch's blood-oxygen (SpO_2) sensor could, in theory, play a part in flagging incipient conditions like COPD, asthma, pneumonia, pulmonary embolism, COVID, heart failure, altitude sickness, and anemia.

Continuous Glucose Tracking

Apple has been working on a blood sugar (glucose) monitoring technology since 2010. The company has bought startups, patented technologies (optical and radio-frequency sensors), won patents, and worked on prototypes. If such a thing comes to market, it would be an enormous breakthrough; your blood sugar level is a useful window into your metabolic health.

Diabetes is a leading cause of blindness, kidney failure, amputations, and heart attacks, and affects over 500 million people worldwide. The six million people with type 1 diabetes, for example, have to monitor their glucose levels either with multiple finger-prick blood tests every day or by wearing an arm patch. (If they don't, and their glucose levels rise, they could experience coma or death in 24 hours.)

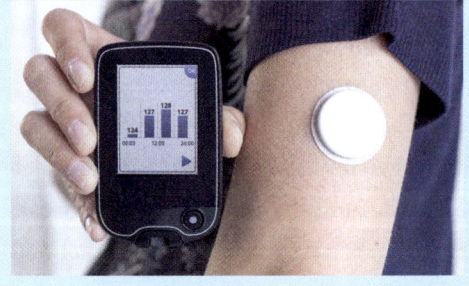

But continuous glucose monitoring (CGM) on a watch could eliminate a lot of expense and hassle, and save thousands of lives.

Actually, continuous monitoring would benefit more than diabetics. Spikes and drops in glucose affect *everyone's* energy, mood, and mental clarity. Millions of people are getting into personalized nutrition, where they learn how certain foods affect their own blood sugar in real time. And a watch-based system could also alert you to prediabetes, where blood sugar levels are elevated, but not yet in the diabetic range.

Google and Samsung have also been working on detecting glucose at your wrist, but it's been a tough slog; differences in skin tone, temperature, and hydration can throw the readings off. No technologies have yet been accurate enough to win FDA approval.

But if they ever get there, they'll sell a lot of watches.

In university studies, researchers have established that smartwatches are technically capable of detecting infectious diseases, anemia, and type 2 diabetes. During the pandemic, a study of 100,000 Fitbit wearers revealed that heart, temperature, and breathing data could alert wearers to a COVID infection three days *before* any symptoms emerged. There's even preliminary work on detecting *cancer* from smartwatch data.

Knowing your blood pressure in real time would be another useful stat, especially if it warned you before yours reached dangerous levels. (The Apple Watch's hypertension alerts require 30 days of data.) And for people who already have hypertension, heart failure, diabetes, or kidney disease, a continuous readout would be enormously helpful in adjusting lifestyle or medications.

Your blood pressure can also change just before a stroke or heart attack. In those cases, knowing sooner than later is better.

Apple holds at least eight patents for blood pressure–tracking technologies, most involving special watchbands. Perfecting them and winning FDA approval may come in 10 years, 20, or maybe never.

Meanwhile, there's plenty of low-hanging fruit in the medical-detection realm; there's no reason the Apple Watch couldn't offer to listen for your snoring, as the Fitbit does, and gently vibrate to nudge you into a shallower sleep stage, for the sake of your romantic partner's rest.

One thing is for sure: More health assistance will come from Apple's watches, AirPods, and headsets/glasses. The logic is unassailable: If you're going to wear a device on your body, it may as well tell you useful things *about* your body.

50. Throughlines

Apple's greatest achievements rarely come from inventing new technologies; its gift, especially Steve Jobs's, has always been recognizing the potential of fledgling technologies and simplifying, refining, and democratizing them. That's how Apple brought us the first color computer, mouse, menus, windows, proportional fonts, Wi-Fi, CD-ROM, laser printing, home networking, video editing, pay-by-phone, and multitouch into the mainstream, as well as the first laptop trackpad, backlit laptop keyboard, voice assistant, fingerprint sensor, 3D face recognition, eSIMs, magnetic power connector, LiDAR, and app store.

Along the way, Apple transformed the music biz, of course, but also retail, cellphones, banking, publishing, health, and chipmaking. Apple's stuff tore down the gatekeepers to creativity: No longer did you need a recording studio to produce a great song, or a movie studio to put out a movie. Millions of people now make music, art, film, and apps using tools Apple makes.

For years, Apple fans felt like a tribe of underdogs. They alone appreciated beauty, simplicity, and pleasure in computing, concealing their disgust every time they heard someone say, "But I can get a Dell for $400!" But finally, in time, Apple's products became cool, aspirational, influential. Those iPods and AirPods, those iMacs and MacBook Airs gave Apple the aura of lifestyle brand.

And then, above all: the iPhone, the towering mother of the modern tech era. It put us into a continuous state of connectivity. It marked the end of boredom.

But it also launched social media, and social media wrought its own changes upon the land. It unlocked new landscapes of news and entertainment, empowered grassroots movements, and amplified marginalized voices. But social media apps also polarize us politically, rewire our brains, and contribute to digital addiction.

Apple's very bigness gives it a colossal influence. Competitors now think about the beauty and pleasure their products deliver instead of just speeds and feeds. Apple's environmental efforts have set a bar for its rivals and taught hundreds of component suppliers to clean up their own acts. And deep down in almost every Silicon Valley tech executive's brain, in every investor's office, at every all-hands meeting, and on every keynote stage, they're thinking: "How would Steve do this?"

There are many long-lived companies, but few are recognizable today in anything more than name. Samsung started out selling dried fish and noodles. Wrigley used to make soap and baking powder. The network-equipment company Nokia started out as a paper mill; for a while there, its primary business was making cellphones.

And then there's Apple. It was born, as Jobs used to say, at the intersection of technology and the liberal arts, and it hasn't moved a foot. Its first product was an effort to design a better machine; so is its latest product.

By way of explanation, Apple points to its corporate values. A surprising number of them have guided the company's spending, decision-making, and focus for decades—some for all 50 years. Talk to enough executives and employees and 15 themes recur.

Focus

Apple's entire product line no longer fits into a 2 x 2 grid. But for a nearly $4 trillion company, Apple sells remarkably few things. In a given year, only about two dozen products make up its catalog: five Mac models, four iPads, three Watches, three AirPods, two Apple TVs, two HomePods, three iPhones, and Vision Pro. Contrast with, for example, Sony or Samsung, who make thousands of products.

"We try to pick the one or two things that we think we can put the whole company behind and get it a better chance of success, because we're all invested in that thing," says Phil Schiller. "I think that's really hard for most companies to do."

Secrecy

Jobs knew that *surprise* adds impact to every product or feature revelation, resulting in more news coverage and customer buzz. Leaks, on the other hand, let other people frame the story that he wanted to tell, and gave competitors more time to copy what Apple had done.

He engineered a deeply rooted culture of secrecy. To this day, Apple engineers work on pieces of the whole, often without even knowing what the product is going to be. Prototypes have fake cases and ever-changing code names to prevent employees from adding it all up. Design and engineering labs are locked off, with entry only by security badge on a need-to-enter basis. Your level of disclosure on a project determines which meetings, or even parts of meetings, you're allowed to attend. Apple employees don't discuss their work at home or even in hallways. Leak, and you're fired.

There are some downsides to all of this: Resentments sometimes result, and information silos can sometimes hinder collaboration.

The secrecy job is harder now; Apple relies on more external contractors, and most employees work out of the office two days a week. Counterfeiters and leakers have offered Chinese factory workers as much as three months' salary to smuggle the enclosure of a new Apple product out of the plant. Today, those workers pass through TSA-style checkpoints as they leave work each day.

None of this is unusual, of course; Google, Microsoft, Facebook, and Amazon are also NDA-based, leak-averse cultures. Their policies aren't as extreme as Apple's—but then again, their products rarely inspire as much frenzy as Apple's.

Beauty

Steve Jobs fell in love with Japanese art in high school, and with calligraphy at Reed College. By the time he cofounded Apple, he was obsessed with design.

"I think people would've been shocked to have seen the amount of time, the level of discussion and arguments on just an *icon*," says Eddy Cue, who spent weeks with Jobs designing the iTunes music store. "If I wrote a biography, the title would be *Off by One Pixel*."

Jobs and Jony Ive were even notoriously invested in the backs and even *interiors* of the products. "You *sense* care even if you can't *see* care," Ive says. "To work hard on what's internal, to make it beautiful even if it's covered—why would you do that? I do believe people can look at an object and know the biography of that object. '*That* was developed entirely driven by money, *that* was developed entirely driven to be there quickly—and *this* was driven because somebody gave a damn about me.'"

Nor was it just about aesthetics. "Most people make the mistake of thinking design is what it looks like," Jobs said. "People think it's this veneer—that the designers are handed this box and told, 'Make it look good!' That's not what we think design is. It's not just what it looks like and feels like. Design is how it *works*."

The tidy guts of an Apple Watch (left), MacBook Air, and Power Mac. (Not to scale.)

RoundRects

The round-cornered shape of the very first Mac's screen, developed at Jobs's insistence, became the hallmark of almost everything Apple has produced since: every screen, keyboard, key, trackpad, laptop, iPod, iPhone, iPad, Apple Watch, AirPods case, Mac mini, HomePod, Apple TV (and its remote), Mac OS X window, and iOS panel. Even Apple's $19 polishing cloth has those rounded corners.

The corner size varies on an Apple Watch, an iPhone, and an Apple polishing cloth.

Round corners convey safety and comfort. They also distribute impact better than sharp corners—a plus for phones, tablets, and laptops.

In Apple's book, not every corner curves the same way. In general, things you use closer to your body (Watch, AirPods case) have broader curves than things far from you (iMac screen).

The hallowed RoundRects, as programmers called them, were so important to Jony Ive that he used Bezier curves—mathematically defined arcs—to precisely specify the rate of curvature in those corners, to control the transition between straight line and curve. By contrast, the corners of the iPhone's app icons were simple quarter circles. The software didn't match the hardware. "They drove me crazy," he says.

In 2012, Ive was put in charge of iOS. As he redesigned the phone's software for iOS 7, fixing the dissimilar corners was near the top of his To Do list.

iOS 6 apps had quarter-circle corners (left). Ive redesigned them to match the corners of the phone itself.

Throughlines • 533

Best, Not First

Apple was not the first company to introduce a personal computer. Or a music player. Or tablets, touchscreen phones, computer retail stores, fingerprint recognition, face recognition, pay-by-phone technology, or smartwatches. Yet by waiting out the mistakes of its hasty competitors, Apple came to dominate in all of those categories.

> **The Greatest Apple Time-Savers of History**
>
> Apple's new products and OS overhauls get the headlines. But often, it's the tiniest features that save the most time and spare you the biggest annoyances:
>
> **Two-factor autofill.** When you log into a bank, medical portal, or streaming service, the website texts or emails you a six-digit code, which you're supposed to memorize or copy and then enter to prove that you're you. But Apple machines can paste that code into the box automatically.
>
>
>
> **Activation Lock.** By 2013, 14 percent of all reported crimes in New York City were stolen iPhones. Even if your phone was password-protected, a thief could just erase it and resell it.
>
> So in iOS 7, Apple invented Activation Lock. The simple concept: Your phone *can't* be erased without your Apple ID password. The bad guy, now unable to erase the phone, can't resell it; it's now a worthless brick. Overnight, iPhone thefts worldwide dropped by more than half.
>
> **MagSafe.** Tripping on a laptop's cord used to send the computer crashing to the floor. But on the first MacBook Pro, Apple added magnets. Now if anybody tripped on the cord, the cable fell harmlessly to the floor. It was a simple, beloved idea that disappeared from the 2019 laptop models—but returned in 2021 by popular demand.
>
> **Optimized Storage.** Photos take up a lot of disk space. Take enough of them, and you fill up your Mac or phone. But if you turn on Optimize Storage, your original, full-size photos and videos get stored online—and you're left with screen-sized copies on your Mac or phone, occupying a tiny fraction of the original space. Now you can carry ten times as many photos on your Mac or phone, ready for showing people. And yet when you share, edit, or print a photo, its original high-res version seamlessly downloads so you can work with it.

"We would *love* to be first," says Phil Schiller. "However, our bar for when it's good enough to ship is a different bar than others. Sometimes, internally, there's a lot of angst over that, like: 'Oh, we have this idea, and it's going to be three years away. And here's another company, that they're going to beat us to market with, and dammit, I wish they didn't, because we want to be first.' But we're not going to ship it until we think it's good enough to ship."

Mail Drop, introduced in 2014, put an end to the hell of emailing large files—and having them bounce as undeliverable. Now you can attach files up to 5 GB to an outgoing message. If it's too big for your recipient's email system, Apple courteously offers them a Download button to receive it.

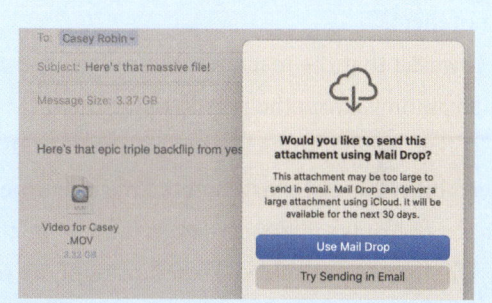

Head gestures. AirPods have no trackpad or screen. So how are you supposed to take calls, dismiss notifications, and answer texts? Very cleverly: Nod your head "yes" to take the call or prepare for a reply, shake your head "no" to decline.

PDF signatures. You can teach the Mac's Preview app what your signature looks like—by holding it up to the camera. Thereafter, when someone emails you a PDF document to sign, you can add your autograph with a single click. Better yet: When you hit Save, Preview asks if you'd like the signed PDF auto-attached to a *reply* to the original email. You've just saved yourself 786 fussy steps.

iPhone Mirroring. This 2024 feature lets you see your iPhone on your *Mac's* screen, in its own window. Now you can use your keyboard to type into phone apps, click things with your trackpad, and copy and paste between Mac and phone—even if your physical iPhone is lying somewhere else in the house.

Throughlines • 535

Simplicity

"Simplicity is the ultimate sophistication," went the headline of the first Apple II brochure. Apple's machines, especially in the Steve Jobs era, have often been case studies in the *removal* of parts: cables, jacks, seams.

That ethos came, of course, from the top. "You know: only silver and gray cars, only black turtlenecks. He had no furniture in his house," remembers iMovie and iPhone author Glenn Reid. "He just didn't like complexity and therefore you shouldn't, either. He may be the only person I ever worked with who advocated taking things *out* of products. Like, 'What does that do? Get it out of there.'"

It wasn't that the machines themselves were simple. The art was *concealing* complex technical underpinnings from the person using them.

Today, Apple products are nowhere near as simple as they used to be. "The principle of discoverability has been lost," wrote interface expert and Apple Fellow Don Norman—and the 1993 *founder* of Apple's interface-design office—in 2015. "The screens offer no assistance in remembering whether one should swipe left or right, up or down, one finger or two."

Then again, two generations have now grown up with computers. Technology doesn't *have* to be as simple as it was decades ago. Apple's competition is no longer a typewriter; it's the long feature lists of *other* companies.

Simplicity at today's Apple comes in other forms: the sleek looks of its hardware, the shared design conventions across different devices, and the thousands of small, step-saving tweaks.

Sustainability

Apple, it turns out, is a hardware company. Every stage of its business affects the earth: mining raw materials, consuming power to build and ship them, contending with the detritus when they're discarded.

Apple published its first corporate environmental policy, focused mostly on reducing toxic materials, in 1990, the Sculley era. It would take a 125-page book to document all of the progress Apple has made since. In fact, that book—the annual *Environmental Progress Report*—is a PDF online.

The first goal is to dig up less raw material, which means using recycled materials—and designing products so that those materials can be recycled again. In 2024, by weight, one-quarter of everything Apple shipped was made of recycled material: 99 percent of the tungsten, 71 percent of the aluminum, 53 percent of the lithium, and nearly 100 percent for cobalt, tin, gold, and rare-earth elements.

The next goal is to reduce the amount of power Apple generates by burning coal and oil. Apple has established two goals for 2030:

- **100 percent clean energy for the value chain**, meaning everything involved in making products: acquiring raw materials, manufacturing, transportation, selling, recycling, or disposing of them. Apple even includes, in this category, the power its parts suppliers use *and* the electricity customers use to run its products.

Apple's biggest source of carbon pollution is the manufacturing in China, where coal still powers 53 percent of the grid. To fight that problem, Apple has funded 500 MW worth of solar and wind farms, works with advocacy groups to remove coal subsidies, and contributed

> ### A Greener Apple
>
> In 2006, the eco-advocacy group Greenpeace launched a campaign to shame tech companies who were using PVC plastic, which contains carcinogens and isn't recyclable, and brominated flame retardants (BFRs), which are linked to cancer and brain-development delays in kids. When Greenpeace ranked the 14 biggest tech companies, Apple ranked near the bottom. (In first place: Dell. *Dell!*) Greenpeace launched a petition and held protests at Apple Stores.
>
> In April 2007, Jobs responded with an open letter called "A Greener Apple," in which he announced that Apple would phase out PVC and BFRs within a year. Irked that Apple wasn't getting credit for the steps it had already taken, he said that Apple would release annual sustainability reports from that moment on.
>
> The engineering team hadn't seen Jobs's announcement coming. They were using PVC and BFRs to make hard drives, optical drives, processors, and other components. They convened a war room with all the key department heads.
>
> Of course, the toxic materials didn't come from Apple executives; they came from the company's thousands of suppliers. Apple would have to pressure *them* to clean up their acts. "We had to start going out to these suppliers and saying, 'Okay, what's your plan?'" says John Ternus, Apple's head of hardware engineering. "Every single one of them, we were the first company that ever came to them!"
>
> The toughest component to fix was power cords. Apple was shipping millions of them, each made of PVC plastic. "PVC was God's gift to power cords," says Ternus. "It's this perfect mixture of flame-resistant and flexible."
>
> But Apple and its suppliers did ultimately concoct formulations that phased out PVC and BFRs—by the end of 2008, exactly as Jobs had promised.

$400 million to Chinese renewable-energy initiatives. Two-thirds of Apple's power in China is now renewably generated.

- **Carbon neutrality for its entire existence on earth**, meaning everything listed above *plus* all its buildings, trucks, and other facilities.

   Apple's stores, offices, and data centers have been running on 100 percent renewable power (solar, wind) since 2018. "Any place you see an Apple logo is running on clean energy," says Lisa Jackson, Apple's head of environment, policy, and social initiatives. But going carbon neutral for *everything*—"upstream and downstream," as climate scientists say—is a massive, decades-long task.

Lisa Jackson
Born: 1962, Philadelphia, PA
Schooling: Tulane (BS), Princeton (MS)
Before Apple: New Jersey EPA commissioner, U.S. EPA administrator
Apple: 2013–present

Jackson was the shortest-serving chief of staff to the governor of NJ (two weeks!) before leaving to become EPA administrator in 2009.

Being carbon neutral isn't as good as using 100 percent clean energy. You can be a polluter and still be carbon neutral; you just have to offset your emissions by buying *carbon offsets*. Those are investments in projects that remove or prevent atmospheric carbon, like planting trees or building wind or solar farms.

Apple aims to reduce its global greenhouse-gas pollution by 75 percent (compared with 2015 levels) by 2030, and *90* percent by 2050. To cover whatever pollution is left, Apple will use carbon offsets. For example, the Apple Restore Fund, to which Apple has contributed $280 million, pays for the restoration of hundreds of thousands of acres of forest in South America.

Finally, there's the disposal problem. Apple's primary approach is encouraging customers to return their used electronics, cables, cases, and accessories to Apple. If the thing has any value left, you get a credit toward your next purchase. Apple refurbishes and resells about 16 million products a year; the rest it turns over to its high-speed industrial disassembly robots, which pluck apart and sort the reusable components.

Apple's environmental practices still have critics. They often point out that carbon offsets let rich companies buy their way out of responsibility for pollution, and that Apple sometimes takes steps toward sustainability only in response to outside pressure.

For example, the more repairable a device is, the longer it remains usable and out of the

The Packaging Problem

Plastic never biodegrades; it just crumbles, over time, into microplastic particles, which wind up in our food, water, livers, kidneys, and brains. It's an environmental disaster, and Apple's packaging—billions of boxes and protective material a year—weren't helping.

But by 2024, Apple had eliminated foam and plastic from all packages for all new products. Now all of it is made of paper and cardboard. "There's the little protective films on the screen—you gotta come up with substitutes for that. Even the inks have polymers in them," says Lisa Jackson. "You just gotta grind through it one thing at a time."

Apple is also shrinking the *size* of the boxes, so that more of them fit into a given shipping pallet, and switching to shipping by sea, which pollutes 95 percent less than air transport.

Apple is teaching its suppliers and vendors how to make the same switch, and other tech companies have also gone plastic-free. As Tim Cook often tells his sustainability executives: "We want to be the ripple in the pond."

landfill. Apple's products used to be designed with oddball screws and fasteners so that only authorized Apple dealers could make repairs. After years of haranguing by right-to-repair groups like iFixit, Apple opened up its manuals, parts, and tools to independent shops in 2019, and to do-it-yourselfers in 2022. And beginning with the iPhone 16, many components are designed for easy swapping out.

Since 2015, Apple's carbon pollution has dropped by 60 percent—and during that same period, its revenues are up 65 percent. "We have to break in people's minds the idea that this can't

Daisy, Apple's iPhone disassembly robot in the Netherlands, disassembles 200 iPhones an hour.

happen in a capitalist world where profit is a motive," says Jackson. "You can break the link between consumption of natural resources and revenue. It can be done; it's just a ton of work."

Closed Systems

A closed system, Jobs always argued, offers a better experience, greater security and reliability, and tighter integration of features.

But the evidence was mostly against him. The Apple II, with its slots, was a mega-bestseller; the closed Mac was not. The iPhone became a culture-changing hit only after Apple opened it up to app developers. Microsoft, whose Windows was available to all hardware makers, won the PC war. Google, who used the same formula with Android, beat the iPhone in worldwide phone sales.

But Jobs held firm to his beliefs to his dying day. "What is best for the customer—integrated versus fragmented? We think this is a huge strength of our system versus Google's," he said in 2010.

Over the decades, the meanings of "open" and "closed" have become more nuanced. Google wound up embracing some aspects of Apple's "closed" systems—with its app store, for example, and its Chrome and car operating systems.

In the end, though, the debate remains unwon. Apple still leans toward closed systems, Google still leans toward open ones; Google sells more phones, Apple makes more money.

Small Acquisitions

Big tech companies routinely acquire other tech companies for staggering amounts. Microsoft, for example, spent $75 billion to buy Activision, $26 billion on LinkedIn. Facebook bought WhatsApp for $19 billion, Oculus for $2 billion, Instagram for $2 billion. Google bought Motorola Mobility ($12.5 billion) and Nest ($3.2 billion).

Apple makes frequent acquisitions, too. But it buys small companies you haven't heard of—whose technologies or teams it can absorb quickly, quietly, and for far smaller amounts. In fact, in its entire history, Apple has spent more than $1 billion on acquisitions only three times. The 2014 purchase of Beats Music for $3 billion was the rare exception to its "pay very little" and "never buy a known brand" rules.

Such surgical-strike acquisitions minimize the likelihood that the acquired team will encounter culture clashes or duplication of functions. Small deals also help maintain secrecy, avoid the kind of regulatory scrutiny that ties up big ones, and don't require shareholder approval.

Above all, though, this strategy is about developing, owning, and controlling its own technologies. Apple, as you may have heard, likes to make the whole widget.

Accessibility

"It sure is nice to get out of that bag," said the first Macintosh on the day Steve Jobs unveiled it in 1984. Jobs may not even have been aware that he had just introduced Apple's first accessibility feature. But it took only a year for Apple's institutional commitment to accessibility—design for people with disabilities—to arrive.

In 1984, Alan Brightman came to Apple's education department from the nonprofit world, where he'd spent years working at camps and schools for disabled kids.

The Apple II was already a hit in the disabled community. Its command-line interface meant that you could operate it even if you were blind or had limited mobility. Add-on cards and programs made it friendly to all kinds of other people.

But the Macintosh was another story.

Soon after his arrival, Brightman sent Jobs a memo. The Macintosh, he wrote, was supposed to be "the computer for the rest of us"—but it was more like "a computer for roughly 80 percent of the rest of us."

Jobs responded by creating the Office of Special Education—and Brightman was the department.

One day in July 1985, he invited a dozen of the original Mac engineers to sit in front of their Macs. As John Sculley observed, Brightman asked them to type a memo using only a pencil held in their teeth.

They didn't get far. They couldn't even turn the computer on; the power switch was a rocker switch on the back panel. There was no way to use the mouse or to insert a floppy. Blind people couldn't use the graphic interface at all.

Within minutes, the engineers had encountered 35 Mac features that were obstacles to people with disabilities. "That was probably the most significant day of my entire career in the computer industry," Brightman says. "It set a fire."

Soon after that meeting, the Mac software team began implementing one accessibility feature after another:

- **The silent beep.** If you set the speaker volume to zero, error beeps turned into menu bar flashes for the benefit of deaf people.

- **Sticky keys.** If you had only one hand available, you could press keyboard combinations (involving the Shift, Option, or Command keys) one key at a time instead of simultaneously.

- **Mouse keys** let you move the cursor across the screen using the number keypad, as though the number keys were directional arrows.

- **Slow Keys.** When one engineer tried to operate the Mac wearing work gloves, he kept hitting keys by accident. His solution was Slow Keys, which registered a key press only if you left your finger on the key for, for example, a full second.

The Mac gained CloseView, a screen magnifier, in 1988, and an on-screen keyboard in 1990. In 1992, Sculley demonstrated Casper on *Good Morning America*: voice control of the Mac, an ancestor of Siri.

But the feature that made the biggest dent in the universe was VoiceOver. It was Mac software that read aloud whatever's on the screen; when it landed on the iPhone in 2009, it was a sensation. Suddenly a phone that seemed worthless to blind people—a black, featureless slab—was, as the *Atlantic* put it, "one of the most revolutionary developments since the invention of Braille."

On the phone, VoiceOver speaks anything your finger touches: icons, words, even status icons at the top. As you go, the voice tells you what you're tapping. "Mail—14 new items," or "45 percent battery power." You don't even have to lift your finger; you can just slide it around, getting the lay of the land. In time, blind people get so good at VoiceOver that they crank up the speaking rate; people who haven't gotten used to it can't even make out what it's saying.

Over the years, a long list of features joined VoiceOver on the greatest hits list:

Apple's advertising began featuring accessibility campaigns in the 1980s.

- **Eye tracking** lets you control iPhone or iPad just by looking at it. A pointer follows your gaze; when you pause, you "tap" what you're gazing at. It's a feature that used to require $5,000 in stand-alone equipment.

- **Personal voice** creates an AI-generated clone of your voice. It's intended for people with conditions like ALS that risk the eventual loss of speech.

- **Sound recognition.** The phone flashes its LED to notify a deaf person when it hears sirens, a baby crying, a doorbell ringing, or their name being spoken.

- **Live Listen** turns the iPhone into a remote microphone, transmitting voices from across the restaurant table (or the lecture hall) to your AirPods. Today, real-time captions can appear on your phone or watch: subtitles for the real world.

- **Door Detection.** If you're out and about, the iPhone's voice can announce when you're approaching a door, what its sign says, and how to open it. ("Door 12 feet away. Text: Ray's Pizza. Pull handle, swing.")

- **Scene description.** The phone can also describe what its camera is seeing: people, furniture, and scenes. "A vase full of flowers on a white table in front of a fireplace."

Today, members of Apple's accessibility team are involved early in the design process for every new product.

The Curb-Cuts Effect

The 1990 Americans with Disabilities Act (ADA) mandated that cities must build ramps into the curbs at street intersections. The original idea was to accommodate people in wheelchairs, but the ramps became unexpectedly popular among people with bikes, strollers, rolling luggage, and so on. That phenomenon—when the larger population embraces a feature originally designed for accessibility—became known as the curb-cuts effect.

Soon after the Apple Watch came out, accessibility director Sarah Herrlinger began getting emails from people who had amputations or dexterity issues. "I'm using my nose to answer the call or to start a workout," they'd say. "I wish there were a better way for me to use this thing."

Apple's sensor teams began exploring whether the Watch's accelerometer and gyroscope data could pick up muscle and tendon movements. They came up with a feature called AssistiveTouch, where the watch detects finger pinches or fist clenches, and translates those gestures into virtual taps on the screen.

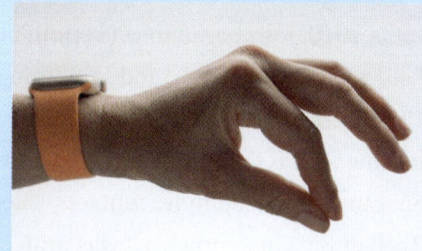

To Herrlinger's surprise, influencers on TikTok and Instagram began making videos about this cool new "secret" feature. Now, they said, you could operate your watch when your arms were full of groceries, skiing with thick gloves on, or holding your toddler's hand.

Many Apple accessibility features have followed the same trajectory: The Magnifier app helps with tiny print. The iPhone's Back Tap feature can turn on the flashlight—quicker than fumbling with buttons. VoiceOver turns out to be great for listening to articles or emails hands-free while you're getting ready for the day.

"I always tell people, go into the accessibility features as a life hack," says Herrlinger. "There are things in there that can help you just be more productive."

16 to 20 percent of the world's population identifies as having impairments to vision, hearing, speech, movement, and so on; over 65, the number is 50 percent. And almost everybody *will* experience some kind of disability.

"It's not charity," says accessibility head Sarah Herrlinger of her team's efforts. "This is the human experience. There is nothing more human than having a disability."

Limited Time, Tiny Teams

As American author Elbert Hubbard wrote in 1911: "Two necessities in doing great and important work: a definite plan, and limited time."

Nobody believed it more than Steve Jobs. He routinely gave his team impossible deadlines—and they routinely made them. He also believed that no team should exceed 100 people; anything more is too bloated to be nimble.

Physical isolation can give that little team the feel of an independent startup. The groups that created the Mac, Newton, QuickTime, iPhone, Vision Pro, and Star Trek (Mac OS on Intel chips) all worked in separate buildings. "It was a little bit like the old Mac pirate-flag thing," says QuickTime engineer Peter Hoddie. "Put the people somewhere else where they're comfortable but a little bit isolated, and let them go do their thing."

Privacy

Data privacy was not always central to Apple's mission. The Newton's operating system, for example, was designed as a "data soup," where every app had access to every other app's data.

But in 2010, Jobs began to get serious about data privacy. "We worry that some, you know, 14-year-old is gonna get stalked and something terrible's gonna happen because of our phone," he told the *Wall Street Journal*'s Walt Mossberg.

Starting with iOS 4 (2010), Apple provided on/off switches for every app's access to your location, camera, microphone, contacts, calendar, and so on. The concern then wasn't snooping by Big Tech or Big Government; it was protection against viruses and spyware.

2010 was also the year of Siri. "This was a foundational shift," says senior director of user privacy Erik Neuenschwander. "This was the first time Apple was going to be processing user data on our servers." Until that moment, Apple had had only small amounts of customer data to protect, like iTunes Store transactions and iCloud syncing—relatively straightforward data-management tasks. Siri would involve *recordings of people's voices.*

He assigned two senior engineers to conduct a security audit of the Siri system; soon after, Apple formed its first formal privacy engineering team.

In the 2010s, the Big Tech companies' handling of data became a front-burner issue for the public. In 2018, the world learned that British consulting firm Cambridge Analytica had been secretly harvesting personal data from up to 87 million Facebook users to target them with political ads. The same year, an Associated Press investigation found that Google was still tracking your phone's location even if you'd turned off Location History in your settings.

Facebook and Google, of course, *need* to collect data. Their business is targeted advertising. But Apple, primarily a hardware company, never had any incentive to collect data—a point Tim Cook began to make publicly as CEO.

The safest place for your data to be is on your phone. If your data never gets sent anywhere, it can't be intercepted or inspected as it passes through the internet or data centers.

San Bernardino

On December 2, 2015, an Islamic terrorist and his wife opened fire on his coworkers at the Department of Public Health in San Bernardino, California, killing 14 of them. The police killed the couple in a shootout later that day.

The FBI hoped that the shooter's work phone, an iPhone 5C, could give them clues about his motives. But the previous year, Apple had introduced a small feature with a huge impact: full phone encryption. Only the phone owner's fingerprint or passcode could unlock it. Apple's engineers had no access to it.

To make matters even more secure, Apple designed the phone to erase itself after ten wrong password guesses.

The FBI asked Apple to create a special version of iOS that would bypass the lockout feature. Apple refused. "We feel very strongly," Cook said, "that you can't have a back door that's only for the good guys."

When a judge ordered Apple to obey, the standoff became national headlines. Americans were split between those who believed that Apple should help law enforcement and those who feared the opening of Pandora's box. It was a Goliath-vs.-Goliath battle: the world's biggest company against the U.S. government.

Just as Apple's deadline to comply approached, the issue became moot. The FBI managed to get into the phone on its own. It had found an Australian hacking company with expertise in "exploit chains": code that strings together minor known security bugs to forge a larger path into an encrypted device.

The FBI paid the hackers $900,000. The software vulnerability they'd used was promptly fixed. The court order was dropped.

And the FBI found nothing useful on the phone.

In the beginning, the iPhone wasn't powerful enough to interpret Siri voice recordings; they *had* to be sent to an Apple server. But by 2020, the iPhone could process Siri requests itself, requiring no internet connection and no data transmission at all.

Apple began designing more and more data tasks to remain on the phone: face recognition in Photos, speech dictation, autocomplete suggestions, Maps searches and routing, and so on.

Text messages and email were still transmitted, of course. But even then, Apple offered alternatives. Messages, for example, encrypts your texts with other iPhone owners, as indicated by blue text bubbles in the app. "If you send me a message, then nothing in the middle can read it," Neuenschwander says. "It's encrypted between your device and my device."

In time, Apple extended end-to-end encryption to almost everything stored outside the phone: online backups, notes, photos, reminders, web bookmarks, voice memos, and so on.

The Whole Widget

When Jobs described the advantages Apple enjoyed by making "the whole widget," he meant the hardware and the software. He was making a contrast to PCs, where one company made the machine and another the operating system—a recipe for headaches.

But in the last 25 years, Apple has steadily increased the *amount* of the widget it makes, especially in areas where the company got burned by relying on other companies.

When Adobe refused to create a Mac version of Premiere, Apple wrote iMovie. When Intel couldn't deliver improved chips on schedule, Apple started its own silicon division. To reduce its reliance on Qualcomm's wireless modems, Apple bought Intel's entire wireless division. Each time, Apple assumed more control over its products and its future.

Education

Apple's first computers dominated in schools. Part of the reason, of course, is that it had no choice. IBM won the corporate marketplace; where else could Apple go?

But Jobs often fretted about the state of public education. He'd speak at length about how teacher unions were turning schools into bureaucracies, how parents were tuning out of school involvement, and how a school-voucher system would improve public education.

In the eighties, his "Kids Can't Wait" program put Apple IIe's in about 9,000 California schools; the University Consortium offered Macs to college kids at half price.

But over the years, cheap PCs ate away much of Apple's dominance in K–12 schools. So in

Steve Jobs Goes to Congress

In 1982, Steve Jobs had a very Steve Jobs idea: Apple should give a free computer to every school in the United States.

He had read that donations of computers to *universities* qualified as corporate tax deductions, which would take much of the sting out of Apple's expenditure on those 100,000 computers.

When he ran into U.S. congressman Pete Stark on a flight, he proposed a bill that would expand the tax break to include donations to *K–12* schools. With Jobs's input, Stark drafted HR 5573, the Computer Equipment Contribution Act. "It was one of the most incredible things I've ever done," Jobs said.

He flew to Washington to lobby for the bill. "I actually walked the halls of Congress for about two weeks," he said. "I met probably two-thirds of the House and over half of the Senate myself, and sat down and talked with them."

The bill passed in the House, but never came to a Senate vote. Finance Committee chair Bob Dole "killed it," Jobs remembered. California, however, loved the idea—and passed Jobs's law. A company could now get a 25 percent tax credit for computer donations to schools.

In January 1983, Apple offered a fully loaded Apple IIe to every California elementary and high school that had at least 100 students. Total retail value: $21 million. With the tax credit (and an actual cost of $5.2 million), Apple's total outlay would be only $1 million.

In the end, 90 percent of California schools took Apple up on its "Kids Can't Wait" offer. In many schools, that Apple IIe was the first computer to enter the building. Soon, schools were buying more Apple II's, and so were the kids' parents.

In 1997, Congress passed the 21st Century Classroom Initiative Act; today, companies can deduct the value of computers they donate to K–12 schools. It had taken only 15 years.

2002, Jobs asked John Couch, former Lisa project manager, to return as Apple's first VP of education. Couch gave a free copy of Mac OS X to every teacher in North America; initiated annual back-to-school discounts for college kids; started iTunes U, a central hub for free educational lectures provided by universities; and offered free weekly classes for kids—on topics like creating music, editing videos, and coding—in Apple Stores.

Today, the company offers grant programs, free classes for teachers, coding workshops, and other resources. Jobs would likely have approved. The last industry he hoped to revolutionize, even as he was dying, was textbooks.

Excellence and Execution

Every company, of course, claims to strive for excellence. But speak to enough current and recent Apple employees, and you keep hearing the same thing. "I've worked at multiple other companies, and Apple really does things in a different way than other technology companies," says Vision Pro and Siri leader Mike Rockwell. "The way that we approach development, the level of focus on getting it right, the willingness to go the extra mile to get it right—I've never seen that with another company."

"Sometimes it almost felt like we spent as long on the packaging as we did on the product," adds designer Duncan Kerr. "How it's presented, how you touch things. The hierarchy of what you see first, and then how you pick up the thing and take it out. If that's really cared for—you feel that, you know?"

The public may not be aware that the Apple Watch volunteers lived in a rented apartment wearing oxygen backpacks, or that the Face ID team attended twins conferences, or that sound engineers fretted over the audio characteristics of the AirPods' case snapping shut. But it's all part of a deep-seated thread of Apple's business plan.

"If you make products that people love, you don't have to worry about where the money comes from, because that takes care of itself," says 50-year Apple veteran Chris Espinosa. "And that's been there for 50 years."

For employees, the mandate to achieve excellence translates into an expectation of long hours and high-level effort. "Apple's a hard place to work," says Allen Olivo, who spent two stints at Apple—once as marketing/comms director and once as retail marketing head. "People talk about it as a very creative company and innovative company, but it's also a very strong execution company. You have to execute really, really well."

Some people burn out; turnover among engineers in the Valley is notoriously high. But those who fall under the spell of Apple's mission, especially in the top executive ranks, tend to stay for decades.

Olivo says that his health and relationships suffered during his Apple years. But like many Apple veterans, he describes them as the most fulfilling of his life.

"I remember when we launched the iMac at Flint Center. The night before, we set up," he says. "And I remember sitting in my little balcony that I had, and the lights were down low, and it was eight o'clock at night, and I was just sitting and staring, just thinking: 'Wow, what a moment this is! Look at what I get to do, and we get to do, and look at what's going to happen tomorrow!'

"Maybe once or twice in your life you get that chance, where you can move the temperature just a little bit and say: 'Yeah. I was there.'"

Acknowledgments

In 2024, the Computer History Museum in Mountain View, California, invited me to emcee a celebration of the Mac's 40th anniversary. The original creators of the Mac joined me onstage: Bill Atkinson, Andy Hertzfeld, Susan Kare, Chris Espinosa, Steve Capps, Bruce Horn, Mike Murray, Dan'l Lewin, Guy Kawasaki, and many more. They told the most incredible stories about their tiny team, working on an impossible deadline, with a mission to change the world. It was a magical night of emotion, humor, and PTSD.

A few weeks later, my wife, Nicki, shook me awake in the middle of the night. An idea had come to her! In a dream! I should write a book about Apple's first 50 years. Imagine how many amazing stories *that* could tell!

Sometimes, you really do get good ideas when you're asleep. Or at least Nicki does.

A long list of wonderful people did other favors for this book, just because they thought it was a cool project: Penny Ahlstrand, Garth Beagle, Leslie Berlin, Norman Chan, Dmitrii Eliuseev, Dan Farber, Tom Frikker, Robin Diane Goldstein, Kris Gunnars, Elise Houren, Yukari Iwatani Kane, Jeff Kenoff, Mary Ellen Manock, Tim McLaughlin, Doug Menuez, Tripp Mickle, Bert Monroy, Brad Myers, Madisun Nuismer, Sarah O'Brien, David Schwartz, Diane Sculley, Ken Shirriff, Len Shustek, Phil Simpson, Gina Smith, Rick Smolan, Phillip Torrone and Limor Fried, Kyle Wiens, Janet Wozniak, and the great people at the Stanford University archives.

I'm deeply grateful to the teams at Apple who helped arrange interviews, unearth archival material, and answer technical and historical questions—and especially to the interview subjects themselves, the current executives, designers, and engineers whose work probably got delayed by these interviews.

Some participants went *miles* beyond the call of duty. Robert Brunner shipped me, across the country, his entire collection of 1990s-era PhotoCDs of the industrial-design group's prototypes. John Greenleigh, whom Apple hired to photograph its products for 24 years, opened up his photo vault to me, which involved scanning stuff for hours. Satjiv Chahil took down the *Mission: Impossible* PowerBook poster from his downstairs bathroom wall and had it scanned at a service bureau.

Bill Atkinson hosted me for an entire day at his house, let me *fly home* with his precious album of 1979–1981 Lisa screen Polaroids, and called a few times when he thought of more great stories.

At the time, neither of us knew that he had pancreatic cancer. He died only a few months later; my heart broke.

Chris Espinosa, the only person who's been at Apple for all 50 years, was a gold mine. His memory is a steel trap, *and* he's a great storyteller, *and* he made himself available to answer questions, shoot down myths, and help with technical understandings. And Woz himself responded to every historical question within a matter of hours. The dude's an absolute legend.

Collector Jeremy O'Connor now owns the 23 notebooks that Del Yocam filled with neat handwritten notes during his ten years at Apple. He scanned anything I needed, shared the audio of his own Yocam interviews, and introduced me to Del.

My researcher, history PhD Steven Rodriguez, is incredible. He was able to dig up just about any article or stat I asked for, no matter how old, dusty, or obscure.

I also want to toast Hansen Hsu: former Apple engineer, Cupertino native, computer historian, tech sociologist, and a curator at the Computer History Museum. Who could possibly be better qualified to be this book's tech editor? He saved me from humiliating gaffes, tracked down interview subjects, sent me incredible transcripts from CHM's vault of oral histories, and spent hours sleuthing out answers to decades-old mysteries.

Crary Pullen, this book's photo editor, worked tirelessly with Tara Farrell Conlon to chase down the sources and rights to the 360 photos in this book. Lori Paximadis did a stellar job creating the endnotes that follow, and Olivia Noble beautifully debugged the transcripts.

At Simon & Schuster, my editor, non–tech nerd Priscilla Painton, cheerfully immersed herself in this manuscript, Johanna Li saw it over the finish line, Felice Javit kept an eye on legalities, copy editor Rob Sternitzky polished the prose, copyediting director Jonathan Evans turned a flood of feedback into a coherent final draft, and Paul Dippolito accommodated my much-more-complicated-than-usual ideas for the book's layout. As always, my agent, Jim Levine, performed nobly as my champion and defender.

Above all, this book owes its existence to my friends and family; the *CBS Sunday Morning* gang; my kids, Kell, Tia, and Jeffrey; my stepdudes, Max and Farley; and above all, to my brilliant, beautiful Nicki. They expressed nothing but support and patience during my long exile to book-writing Siberia, and warmly welcomed me when I returned.

Sources

To write this book, I interviewed 150 main characters from Apple's life story. Virtually everyone was generous with their time and happy to help straighten the record. (Many of them mentioned how crazy it drives them when some bogus story appears in a book or an article—and then, in a game of author telephone, it gets repeated in every *subsequent* book and article until it becomes part of the standard lore. I've tried to fix a lot of those.)

I know this looks like just a big list of people, but each name here represents an actual conversation—sometimes several, some many hours long, often profoundly moving:

Early years: Bill Atkinson, Allen Baum, Alan Brightman, Steve Capps, John Couch, Andy Cunningham, Chris Espinosa, Bill Fernandez, Bill Hambrecht, Andy Hertzfeld, Susan Kare, Guy Kawasaki, Larry Kenyon, Daniel Kottke, Dan'l Lewin, Jerry Manock, Jeff Moffatt, Mike Murray, Charles Pfister, Paul Terrell, Steve Wozniak, Del Yocam.

Interregnum: Gil Amelio, Fred Anderson, Robert Brunner, Satjiv Chalil, Gary Davidian, Hugh Dubberly, Hartmut Esslinger, Jean-Louis Gassée, Peter Hoddie, Jon Krakower, Carlos Montalvo, Paul Mercer, Ike Nassi, Allen Olivo, Glenn Reid, John Sculley.

Modern era: Molly Anderson, Kate Bergeron, Jay Blahnik, Tim Bucher, Deidre Caldbeck, Eddy Cue, Sumbul Desai, Kaiann Drance, Alan Dye, Tony Fadell, Craig Federighi, Scott Forstall, Myra Haggerty, Sarah Herrlinger, Jony Ive, Lisa Jackson, Greg Joswiak, Duncan Kerr, Bruce Leak, Kevin Lynch, Jon McCormack, Erik Neuenschwander, Deirdre O'Brien, Bas Ording, Garrett Rice, Mike Rockwell, Jon Rubinstein, Phil Schiller, Joe Schorr, Ken Segall, Xander Soren, Johny Srouji, Michael Tchao, John Ternus, Avie Tevanian, Jeff Williams.

For decades, the Computer History Museum (CHM) has been conducting an oral history project: long, in-depth interviews with important figures in the computing world. CHM's Hansen Hsu, David Murphy, and Massimo Petrozzi made fantastic materials available to me, including the oral histories of Robert Belleville, Bruce Daniels, Gary Davidian, Adele Goldberg, Joanna Hoffman, Rod Holt, Bruce Horn, Dan'l Lewin, Jerry Manock, Mike Markkula, Regis McKenna, Ike Nassi, Rich Page, Chuck Peddle, Caroline Rose, Bertrand Serlet, Larry Tesler, Avie Tevanian, and Steve Wozniak.

Books

Amelio, Gil, and William L. Simon. *On the Firing Line: My 500 Days at Apple.* HarperBusiness, 1998.

Carlton, Jim. *Apple: The Inside Story of Intrigue, Egomania, and Business Blunders.* Times Business, 1997.

Chafkin, Max. *Design Crazy: Good Looks, Hot Tempers, and True Genius at Apple.* Fast Company Press, 2013.

Couch, John, with Jason Towne. *My Life at Apple: And the Steve I Knew.* SelectBooks, 2021.

Fadell, Tony. *Build: An Unorthodox Guide to Making Things Worth Making.* HarperBusiness, 2022.

Fisher, Adam. *Valley of Genius: The Uncensored History of Silicon Valley (As Told by the Hackers, Founders, and Freaks Who Made It Boom).* Twelve, 2018.

Gassée, Jean-Louis. *Grateful Geek: 50 Years of Apple and Other Tech Adventures.* Tavo Reno, 2023.

Hamm, Steve. *The Race for Perfect: Inside the Quest to Design the Ultimate Portable Computer.* McGraw Hill, 2008.

Hertzfeld, Andy. *Revolution in the Valley: The Insanely Great Story of How the Mac Was Made.* O'Reilly, 2011.

Hiltzik, Michael. *Dealers of Lightning: Xerox PARC and the Dawn of the Computer Age.* HarperBusiness, 2000.

Isaacson, Walter. *Steve Jobs.* Simon & Schuster, 2011.

Kane, Yukari Iwatani. *Haunted Empire: Apple After Steve Jobs—Insights into Tim Cook's Leadership, Product Development, and the Future of Apple.* HarperBusiness, 2014.

Kocienda, Ken. *Creative Selection: Inside Apple's Design Process During the Golden Age of Steve Jobs.* St. Martin's Press, 2018.

Kunkel, Paul. *AppleDesign: The Work of the Apple Industrial Design Group.* Graphis Press, 1997.

Lapsley, Phil. *Exploding the Phone.* Grove Press, 2013.

Lashinsky, Adam. *Inside Apple: How America's Most Admired—and Secretive—Company Really Works.* Business Plus, 2012.

Levy, Steven. *Insanely Great: The Life and Times of Macintosh, the Computer That Changed Everything.* Penguin Books, 2011.

———. *The Perfect Thing: How the iPod Shuffles Commerce, Culture, and Coolness.* Simon & Schuster, 2006.

Linzmayer, Owen. *Apple Confidential 2.0: The Definitive History of the World's Most Colorful Company.* No Starch Press, 2004.

McGee, Patrick. *Apple in China: The Capture of the World's Greatest Company*. Scribner, 2025.

Mickle, Tripp. *After Steve: How Apple Became a Trillion-Dollar Company and Lost Its Soul*. HarperCollins, 2022.

Moritz, Michael. *Return to the Little Kingdom: Steve Jobs, the Creation of Apple, and How It Changed the World*. Overlook Press, 2009.

Pfiffner, Pamela. *Inside the Publishing Revolution: The Adobe Story*. Adobe Press, 2002.

Rose, Frank. *West of Eden: The End of Innocence at Apple Computer*. Viking, 1989.

Schlender, Brent, and Rick Tetzeli. *Becoming Steve Jobs: The Evolution of a Reckless Upstart into a Visionary Leader*. Crown Business, 2016.

Sculley, John, with John A. Byrne. *Odyssey: Pepsi to Apple . . . A Journey of Adventure, Ideas, and the Future*. Harper & Row, 1987.

Segall, Ken. *Insanely Simple: The Obsession That Drives Apple's Success*. Portfolio, 2013.

Wayne, Ronald G. *Adventures of an Apple Founder*. 512k Entertainment, 2010.

Wozniak, Steve, with Gina Smith. *iWoz: Computer Geek to Cult Icon*. W. W. Norton, 2007.

Young, Jeffrey. *Steve Jobs: The Journey Is the Reward*. Scott Foresman, 1988.

Notes

ABBREVIATIONS:

CHM — Computer History Museum
CHMOH — Computer History Museum Oral History, searchable at https://computerhistory.org/oral-histories
Triangulation (podcast)

CHAPTER 1: TWO STEVES

David Sheff, "Playboy Interview: Steven Jobs," *Playboy*, Feb. 1985; ITday Japan, "201004 Global Digicon Salon 004 Bill Fernandez Interview," YouTube, Oct. 4, 2020. **The Woz Prank Hall of Fame:** Wozniak, ch. 8. **The Blue Box:** Lapsley, foreword; Fisher, p. 87. **Educations:** Steve Jobs, Stanford University commencement speech, June 12, 2005. **The Breakout Collaboration:** Wozniak, ch. 9; undated correspondence between Bill Hunter and Wozniak, https://web.archive.org/web/20110612071502/http://www.woz.org/letters/general/91.html.

CHAPTER 2: APPLE I

Wozniak, ch. 10. **WESCON:** Charles Ingerham "Chuck" Peddle, CHMOH, June 12, 2014. **Conversor 4000:** Young, ch. 6; "Steve Wozniak: The Homebrew Computer Club and the Apple I," ComputerHistory.org; Macintosh Anniversary, "Mac@30 Rod Holt, Daniel Kottke and Woz Discuss Early Apple," YouTube, Jan. 27, 2014. **Entrepreneurs:** Wozniak, ch. 11; Young, ch. 6.

CHAPTER 3: APPLE COMPUTER COMPANY

Isaacson, ch. 5; @cx404v1, "Revealed: The Truth About Apple's Name & Logo," YouTube, May 23, 2013; Wayne, pp. 102–4. **Byte Shop:** Wozniak, ch. 12; Young, ch. 6; *Triangulation* 235: Bill Fernandez. **Building Apple I:** *Triangulation* 235: Bill Fernandez; *Triangulation* 31: Daniel Kottke; interview with Paul Terrell; Wozniak, ch. 12; ITday Japan, "201004 Global Digicon Salon 004 Bill Fernandez Interview," YouTube, Oct. 4, 2020; Tim Harwick, "Rare Apple-1 Computer in Koa Wood Case Fetches $500,000 at Auction," *MacRumors*, Nov. 10, 2021. **Ron Wayne Bows Out:** Wayne, pp. 105–7.

CHAPTER 4: APPLE II

Wozniak, ch. 13; Apple Archive, "Apple II Forever—Apple," YouTube, May 24, 2023; *Steve Jobs: The Lost Interview* (video), 2012. **The Power Supply:** Ray Holt, CHMOH, May 26, 2023. **Slots:** Wozniak, ch. 13. **Making the Case:** Kunkel, p. 13. **Regis McKenna:** Kunkel, p. 14; Regis McKenna, "My Biggest Mistake: Regis McKenna," *Independent*, Oct. 10, 1992. **Birth of a Logo:** Will Burns, "Rob Janoff and the Fascinating True Story Behind His Original Apple Logo Design," *Forbes*, Mar. 26, 2018. **Finding Financing:** *Something Ventured* (documentary), 2011.

CHAPTER 5: IN BUSINESS

Mike Markkula, CHMOH, May 1, 2012; *Something Ventured* (documentary), 2011; Wozniak, ch. 13; Isaacson, ch. 6. **Leaving HP:** Mike Markkula, CHMOH, May 1, 2012; Wozniak, ch. 13. **Apple Inc.:** Michael Swaine & Paul Freiberger, *Fire in the Valley: The Birth and Death of the Personal Computer* (Pragmatic Bookshelf, 2014), 240; "Interview with Chris Espinosa," June 13, 2000, Making the Macintosh, Stanford University; Mike Markkula, CHMOH, May 1, 2012; Isaacson, ch. 6. **The West Coast Computer Faire:** Steven Levy, *Hackers: Heroes of the Computer Revolution* (Anchor, 1984); Mike Markkula, CHMOH, May 1, 2012; Swaine & Paul, *Fire in the Valley*. **The Faire Prank:** Wozniak, ch. 13. **On Sale:** Carl Helmers, "An Apple to Byte," *Byte*, March 1978, 18. **Staffing Up:** Jay Yarow, "Exclusive: Interview with Apple's First CEO Michael Scott," *Business Insider*, May 24, 2011. **Apple II Evolves:** Wozniak, ch. 14. **Badge Numbers:** Yarow, "Exclusive: Interview with Apple's First CEO Michael Scott"; Isaacson, ch. 6. **Disk II:** Wozniak, ch 14. **VisiCalc:** "Interview: 1990," All About Steve Jobs.

CHAPTER 6: APPLE III

Interview with Del Yocam. **The Fan and the FCC:** "An Evening with Former Apple Industrial Designers Robert Brunner and Jerry Manock," CHM, June 4, 2007; Wendell Sander, CHMOH, May 30, 2024. **IPO:** Marilyn Chase, "Apple Computer Registers Its First Offer, of 4.5 Million Shares at $14 to $17 Apiece," *Wall Street Journal*, Nov. 7, 1980, 7; Isaacson, ch. 9; *Triumph of the Nerds* (documentary), 1996; Sculley, p. 188. **The Woz Plan:** Triangulation 31: Daniel Kottke; Wozniak, ch. 15. **Heat:** Wendell Sander, CHMOH, May 30, 2024; Marilyn Chase, "Technical Flaws Plague Apple's New Computer," *Wall Street Journal*, Apr. 15, 1981, 31. Levy, *Insanely Great*, ch. 5; Sheff, "Playboy Interview: Steven Jobs."

CHAPTER 7: LISA

Bitmapped Graphics: Interview with Bill Atkinson. **Xerox PARC:** Larry Tesler, CHMOH, Dec. 12, 2016; Mac History, "PARC Scientist Larry Tesler Recalls Jobs' Famous Xerox Visits," YouTube, Nov. 11, 2011; "Steve Jobs," Oral and Video Histories, Smithsonian Institution, Apr. 20, 1995. **The Button Wars:** Ray Holt, CHMOH, May 26, 2023; Levy, *Insanely Great*, ch. 4; Chafkin, ch. 1. **Keyboard Shortcuts:** "MacPaint Interview and Demonstration, with Bill Atkinson and Andy Herzfeld," CHM, Sept. 27, 2013. **The Apple Mouse:** Fisher, p. 133. **Hello World:** Larry Tesler, "The Legacy of the Lisa," *Macworld*, May 1985.

CHAPTER 8: CRASHES

Isaacson, ch. 8; Couch, ch. 3. **Woz's Plane:** Wozniak, ch. 16. **Black Wednesday:** Hertzfeld, p. 53; Mike Markkula, CHMOH, May 1, 2012; Swaine & Paul, *Fire in the Valley*, p. 261. **Welcome, IBM:** Couch, ch. 5; Sheff, "Playboy Interview: Steven Jobs." **Lisa Lands:** Philip Faflick, "The Year of the Mouse," *Time*, Jan. 31, 1983, 50. Sheff, "Playboy Interview: Steven Jobs"; Bruce Daniels, CHMOH, Aug. 1, 2023.

CHAPTER 9: MACINTOSH

"A Conversation with Jef Raskin," *Ubiquity*, July 2003; Jef Raskin, "The Mac and Me: 15 Years of Life with the Macintosh," white paper, 1994; Luke Dormehl, *The Apple Revolution: Steve Jobs, the Counterculture, and How the Crazy Ones Took Over the World* (Virgin, 2012), ch. 7; Sculley, p. 286; Hertzfeld, p. 36; Daniel Terdiman, "Recollections of the Mac's Creators," CNET, Jan. 22, 2009; Joanna Hoffman, CHMOH, Jan. 22, 2018; Andy Hertzfeld, "Good Earth," Folklore.org, Oct. 1980; Couch, ch. 5. **Burrell's 'Stache:** Hertzfeld, p. 43. **Pirates:**

Interview with Chris Espinosa; "Jef Raskin on 'The Book of Macintosh,'" Making the Macintosh, Stanford University, Apr. 13, 2000; Couch, ch. 5; David Sheff, "Steve Jobs," DavidSheff.com; "A Conversation with Jef Raskin"; Raskin, "The Mac and Me"; Isaacson, ch. 10; Sheff, "Playboy Interview: Steven Jobs." **Raskin's Amanuensis:** Raskin, "The Mac and Me." **Staffing Up:** Joanna Hoffman, CHMOH, Jan. 22, 2018; Fisher, p. 153. **The Apple Bicycle:** Couch, ch. 1; Isaacson, ch. 10; Hertzfeld, pp. 75, 76. **Reality Distortion:** Interview with Guy Kawasaki; Hertzfeld, p. 58; Levy, *Insanely Great*, appendix; CHM, "CHM Live: Insanely Great: The Apple Mac at 40," YouTube, Jan. 24, 2024; Hertzfeld, p. 125; "An Interview: The Macintosh Design Team," *Byte*, Feb. 1984, 74; Couch, ch. 5. **Defining the Mac:** Private correspondence with Mike Murray; Couch, ch. 5. **The Case:** "An Interview: The Macintosh Design Team," *Byte*. **Twiggy and Sony:** Hertzfeld, pp. 247–48; Rose, ch. 5.

CHAPTER 10: SCULLEY

Interview with John Sculley; Sculley, pp. 80, 129, 135, 136, 217, 233; Mike Markkula, CHMOH, May 1, 2012. **The Plant in Fremont:** Isaacson, ch. 17. **Volatility:** Interviews with Fred Anderson, Guy Kawasaki, Andy Hertzfeld, Steve Wozniak; Sculley, p. 283; Chafkin, ch. 5; Kocienda, ch. 1; Glenn Reid, CHMOH, Nov. 9, 2022; Ray Holt, CHMOH, May 26, 2023; *Triangulation* 235: Bill Fernandez; Levy, *Insanely Great*, appendix; Joanna Hoffman, CHMOH, Jan. 22, 2018.

CHAPTER 11: SOFTWARE

Interview with Steve Capps; heczTV, "What an Interview with Steve Jobs Feels Like (Intimidating Personality)," YouTube, Jan. 23, 2022; CHM, "CHM Live: Insanely Great: The Apple Mac at 40." **City Fonts:** Joanna Hoffman, CHMOH, Jan. 22, 2018. **128K:** Interviews with Andy Hertzfeld, Steve Capps. **MonkeyLives:** "MacPaint Interview and Demonstration, with Bill Atkinson and Andy Herzfeld," CHM, Sept. 27, 2013. **"Business" Cards:** Hertzfeld, p. 138. **Consistency:** Joanna Hoffman, CHMOH, Mar. 9, 2018. **Saving Lives:** Hertzfeld, p. 249; Larry Kenyon, "Mac Boot Time," *Life Stories*, May 30, 2023. **Manuals:** "Interview with Chris Espinosa," June 13, 2000.

CHAPTER 12: MARKETING THE MAC

Sculley, p. 235. **Software Evangelism:** Interview with Guy Kawasaki; CHM, "CHM Live: Insanely Great: The Apple Mac at 40." **The IBM Challenge:** Dan Farber, "Macintosh's 100-Day Marketing Blitz: After the Applause, Confusion," CNET, Jan. 23, 2014. **Influencers:** Interview with Mike Murray; Hertzfeld, p. 325. **"1984":** Sculley, pp. 241, 243; Bloomberg Originals, "The Real Story Behind Apple's Famous '1984' Super Bowl Ad," YouTube, Dec. 3, 2014; CHM, "CHM Live: Insanely Great: The Apple Mac at 40"; Fisher, pp. 161, 163; Gail E. Schares, "How Apple Spent $2.5 Million for Ad Space," *San Francisco Chronicle*, Nov. 8, 1984; Hertzfeld, p. 204. **Computer of the Year:** Isaacson, ch. 13. **$2,500:** Hertzfeld, p. 297.

CHAPTER 13: INSANELY GREAT

CHM, "CHM Live: Insanely Great: The Apple Mac at 40"; Fisher, p. 158. **Flint:** Interviews with Andy Hertzfeld, Mike Murray; Sculley, p. 252; Dan Farber, "The Macintosh Turns 30: Going the Distance," CBSNews.com, January 22, 2014; Chafkin, ch. 2. **Aftermath:** Erik Sandberg-Diment, "Software for the Macintosh: Plenty on the Way," *New York Times*, Jan. 13, 1984; "Apple Introducing Macintosh Office, Planned '2nd Standard' for Business," *Los Angeles Times*, June 18, 1985; Isaacson, ch. 12; Sculley, p. 278. **Postpartum:** Levy, *Insanely Great*, ch. 8, appendix; Hertzfeld, p. 399; Isaacson, ch. 17; Jim Walsh, "What's an 'A' Player?," GlobalLogic, Nov. 9, 2017. **Slump:** Interviews with Andy Hertzfeld, Bill Atkinson; Levy, *Insanely Great*, ch. 8; Sculley, p. 277; Jimmy Maher,

"A Computer for Every Home?," *Digital Antiquarian*, July 28, 2013. **Apple Fellows:** Larry Tesler, CHMOH, Feb. 17, 2017. **Snow White, Red Book:** Interview with Hartmut Esslinger; Hartmut Esslinger, *A Fine Line: How Design Strategies Are Shaping the Future of Business* (Jossey-Bass, 2009). **Apple IIc:** Eware, "Apple II Forever," YouTube, Oct. 14, 2007; Thomas C. Hayes, "Apple Is Banking on New Portable: The IIc Computer," *New York Times*, Apr. 24, 1984; Thomas Neudecker, "Apple IIC: A Transportable with Limited Software, Aimed at Mass Market," *InfoWorld*, July 9, 1984.

CHAPTER 14: MACINTOSH OFFICE

Don Kennedy, "PCs Rated Number One," *PC Magazine*, Apr. 16, 1985, 42. **PostScript:** Pfiffner, pp. 19, 33. **Fonts:** Pfiffner, p. 50. **AppleTalk:** "Apple Introducing Macintosh Office, Planned '2nd Standard' for Business," *Los Angeles Times*. **Office Unveiled:** "Losing Its Shine?," *MacNeil-Lehrer NewsHour*, PBS, Apr. 5, 1985; Erik Sandberg-Diment, "Macintosh Marketing Overcomes Its Drawbacks," *New York Times*, March 26, 1985, C4. **Lemmings:** Sculley, p. 295; Daniella Hernandez, "Tech Time Warp of the Week: The Horrifying Apple Super Bowl Ad That Time Forgot, 1985," *Wired*, Jan. 31, 2014. **What Happened to the Lisa:** R. Anne Thayne, "Apple's Final Lisa Burial," *Herald Journal* (Logan, UT), Sept. 24, 1989. **Celebration:** Wozniak, ch. 13; Isaacson, ch. 17. **Ghosts of the Missing Macs:** Esslinger, *A Fine Line*.

CHAPTER 15: RIFT

Wozniak, ch. 18; Patricia Bellew, "Apple Computer Co-Founder Wozniak Will Leave Firm, Citing Disagreements," *Wall Street Journal*, Feb. 7, 1985; Sculley, pp. 338, 371. **Woz After Apple:** Jeanne DuPrau & Molly Tyson, "The Making of the Apple IIGS," *A+ Magazine*, Nov. 1986, 59; Bellew, "Apple Computer Co-Founder Wozniak Will Leave Firm"; Wozniak, ch. 20. **The French Connection:** Sculley, p. 318. **Power Shift:** Sculley, pp. 272, 275, 328, 330; Isaacson, ch. 17; Andrew Pollack, "The Restructuring of Apple Computer," *New York Times*, June 1, 1985. **Board Meeting:** Interviews with Del Yocam, John Sculley; Sculley, p. 315; Robert Belleville, CHMOH, Aug. 22, 2016. **E-Staff Meeting:** Isaacson, ch. 17; Sculley, p. 343. **Reorg:** Sculley, pp. 346–47; Isaacson, ch. 17; Hertzfeld, p. 407. **Summer of '85:** Interview with Mike Murray; Steve Jobs, Stanford University commencement speech, June 12, 2005; Gerald C. Lubenow & Michael Rogers, "Jobs Talks About His Rise and Fall," *Newsweek*, Sept. 30, 1985; Robert Belleville, CHMOH, Aug. 22, 2016; Isaacson, ch. 18; Patricia Bellew Gray, "Jobs Asserts Apple Undermined Efforts to Settle Dispute over His New Venture," *Wall Street Journal*, Sept. 25, 1985; Andrew Pollack, "Jobs Calls Apple Suit 'a Shock,'" *New York Times*, Sept. 25, 1985.

CHAPTER 16: NEW IDEAS

Michael Schrage, "Steven Jobs' Departure Fits 'Valley' Pattern," *Washington Post*, Sept. 19, 1985; Rose, ch. 22; Sculley, pp. 404, 405. **PageMaker:** Erik Sandberg-Diment, "Macintosh Marketing Overcomes Its Drawbacks," *New York Times*, Mar. 26, 1985; Pfiffner, p. 51; "Printer a Key Part of Apple's Office-Market Appeal," *Christian Science Monitor*, June 3, 1985; Robert Belleville, CHMOH, Aug. 22, 2016; Christine McGeever, "Development for Mac 'Won't Stop,'" *InfoWorld*, Sept. 16, 1985. **ATG:** Larry Tesler, CHMOH, Feb.17, 2017. **Apple IIgs:** DuPrau & Tyson, "The Making of the Apple IIGS." **The Apple Jonathan:** Kunkel, pp. 50, 51.

CHAPTER 17: HERESIES

Interview with Jean-Louis Gassée; Sculley, pp. 367, 380; Donna K. H. Walters, "Apple to Unveil Two More Powerful Macintosh PCs," *Los Angeles Times*, Mar. 2, 1987; Gassée, p. 106. **Mac II:** Steven Levy, "The Making of the Macintosh II," *Macworld*, May 1987, 55; DuPrau & Tyson, "The Making of the Apple IIGS"; Brenton R. Schlender, "Apple to Unveil Two McIntosh Models to Step Up Role in Business Computers," *Wall Street Journal*, Mar. 2, 1987. **Knowledge Navigator:** Folklorg, "John Sculley, the Knowledge Navigator," YouTube, Feb. 20, 2022; John Sculley's keynote address for EDUCOM 1987; Hugh Dubberly, "The Making of Knowledge Navigator," Dubberly Design Office, Mar. 20, 2007. **Apple vs. Microsoft:** Couch, ch. 6; Hertzfeld, p. 294. **The Loma Prieta Earthquake:** Interview with Paul Mercer.

CHAPTER 18: HIGH/RIGHT, LOW/LEFT

Mac Portable: ReDiscovered Future, "The Macintosh Portable Introduction—September 20, 1989," YouTube, Feb. 23, 2016; Hamm, p. 56. **Golden Age:** Interview with Jean-Louis Gassée. **Gassée Trouble:** Interviews with Jean-Louis Gassée, John Sculley; Gassée, pp. 118, 122; Carlton, p. 119. **Low and Left:** Cheryl England, "That Vision Thing," *Macworld*, Sept. 1991, 175.

CHAPTER 19: SYSTEM 7

Interview with Paul Mercer. **QuickTime:** "Press Play: The Origins of QuickTime," CHM, Feb. 28, 2018. **Pencil Test:** Alfred DiBlasi, "Pencil Test—Behind the Scenes," YouTube, May 6, 2007. **KanjiTalk:** Edward W. Desmond, "Byting Japan," *Time*, Oct. 5, 1992. **The Font Wars:** Interview with Carlos Montalvo.

CHAPTER 20: NEWTON

Scribe: This Does Not Compute, "Apple PenLite: The iPad Before the iPad," YouTube, June 21, 2024. **Sakoman:** Doug Menuez & Markos Kounalakis, *Defying Gravity: The Making of Newton* (Beyond Words, 1995); Folklorg, "Love Notes to Newton—Full Movie—Official," YouTube, Aug. 3, 2023. **Tesler Era:** Folklorg, "Love Notes to Newton." **General Magic:** *General Magic* (documentary), 2018; Fisher, p. 238; Fadell, part 2 intro. **The Speech:** Folklorg, "Love Notes to Newton." **The Newton-Sized Pocket:** Interview with Robert Brunner; Kunkel, p. 259. **The Birth of ARM:** Interview with Michael Tchao; Carlton, p. 232. **Handwriting:** Folklorg, "Love Notes to Newton"; John Markoff, "'Doonesbury' and Apple Hatch a Comic Surprise," *New York Times*, Dec. 18, 1995. **eMate 300:** Kunkel, p. 270.

CHAPTER 21: MOONSHOTS

Project Aquarius: Babbage, "The First 'Apple Silicon': The Aquarius Processor Project," *Chip Letter*, May 28, 2023. **Emulation:** Interview with Gary Davidian; Jonathan Weber, "Apple Computer's Sculley to Give Up CEO Position," *Los Angeles Times*, June 19, 1993.

CHAPTER 22: POWERBOOKS

Brunner: Interview with Robert Brunner. **PowerBook:** Interviews with Robert Bruner and Jon Krakower. **PowerBook Duo:** Jim Heid, "Apple's Dynamic Duo," *Macworld*, Dec. 1992, 194. **Mac-Like Things:** Interview with Satjiv Chahil.

CHAPTER 23: SPINDLER

Calvin Sims, "Silicon Valley Takes Partisan Leap of Faith," *New York Times*, Oct. 29, 1992; Mike Markkula, CHMOH, May 1, 2012; Brent Schlender & Michael H. Martin, "Paradise Lost: Apple's Quest for Life After Death," *Fortune*, Feb. 19, 1996; Carlton, pp. 206, 210; Weber, "Apple Computer's Sculley to Give Up CEO Position";

General Magic (documentary), 2018. **Michael Spindler:** Schlender & Martin, "Paradise Lost." **OpenDoc:** Uncle Tuna, "Apple OpenDoc Technology Intro," YouTube, Dec. 6, 2006; Carlton, p. 276.

CHAPTER 24: CLONES

Bart Ziegler, "First Agreement to Clone Mac Is Announced," *Wall Street Journal*, Dec. 29, 1994; Charles Piller, "First Clones," *Macworld*, Apr. 1995, 92. **The IBM + Apple Plan:** Mike Markkula, CHMOH, May 1, 2012. **Snapple:** Amelio & Simon, p. 14.

CHAPTER 25: AMELIO

Julie Pitta, "Apple of No One's Eye: CEO Amelio Takes On a Very Tough Assignment Today," *Los Angeles Times*, Feb. 5, 1996. **Dr. Amelio:** Amelio & Simon, pp. 6, 50; Stewart Alsop, "Apple of Sun's Eye," *Time*, Feb. 5, 1996; Schlender & Martin, "Paradise Lost"; Peter H. Lewis, "What Is an Apple Worth? Ailing Innovator's Luster as an Acquisition Dims," *New York Times*, Jan. 25, 1996. **The Anderson Plan:** Interview with John Sculley. **The Maglite Incident:** John Markoff, "Apple Chief Is Seen Retreating from Price Cuts," *New York Times*, Apr. 2, 1996. **The Shortest Honeymoon:** Amelio & Simon, pp. 30, 42; Carlton, p. 418. **Copland:** Amelio & Simon, pp. 95, 151. **Pippin:** Amelio & Simon, p. 91.

CHAPTER 26: NEXT

Gassée, pp. 145, 146–47; Schlender & Tetzeli, ch. 8. **NeXT Calls:** Interviews with Garrett Rice, Satjiv Chahil; Hansen Hsu, "Next: Steve Jobs' Dot Com IPO That Never Happened," CHM, Feb. 7, 2017; Carlton, p. 409; Amelio & Simon, p. 193. **The Deal:** Interview with Fred Anderson; Schlender & Tetzeli, ch. 8; Steve Lohr, "Creating Jobs: Apple's Founder Goes Home Again," *New York Times*, Jan. 12, 1997; Amelio & Simon, p. 202.

CHAPTER 27: TRANSITION

Keynote: Amelio & Simon, p. 211. **Restructuring:** Interview with Jon Rubinstein; Isaacson, ch. 24; Amelio & Simon, p. 239; Brent Schlender, "Something's Rotten in Cupertino," *Fortune*, Mar. 3, 1997; Linzmayer, 267. **Trimming:** Interviews with Jon Rubinstein, Avie Tevanian; Jon Rubinstein, CHMOH, Aug. 15, 2019. **Woolard:** Isaacson, ch. 24; Amelio & Simon, p. 265; Carlton, p. 431. **Era's End:** Interview with Michael Tchao; Schlender & Tetzeli, ch. 9.

CHAPTER 28: TURNAROUND

James Daly, "101 Ways to Save Apple," *Wired*, June 1997, 115; Louise Kehoe, "Sour Taste at Apple," *Financial Times*, July 11, 1997; Schlender & Tetzeli, ch. 9. **The Board:** Mike Markkula, CHMOH, May 1, 2012. **The Structure:** Interviews with Phil Schiller, Jon Rubinstein; Isaacson, ch. 25; Schlender & Tetzeli, ch. 9; TVArchive, "Steve Jobs - The Lost Interview," YouTube, Mar. 30, 2024. **The Quadrant:** Interview with Myra Haggerty; "Apple Internal—Introducing the Think Different Campaign," Archive.org, Sept. 23, 1997; JC, "Macworld 1998: Steve Jobs Talks About Apple's Return," YouTube, Oct. 20, 2011. **Killing the Newton:** Steven Levy, "An Oral History of Apple's Infinite Loop," *Wired*, Sept. 16, 2018; Isaacson, ch. 25. **Think Different:** Rob Siltanen, "The Real Story Behind Apple's 'Think Different' Campaign," *Forbes*, Dec. 14, 2011; Isaacson, ch. 25; Schlender & Tetzeli, ch. 9. **Macworld Boston:** Isaacson, ch. 24. **The Best People:** Interview with Fred Anderson; Schlender & Tetzeli, ch. 9. **Clone Death:** Isaacson, ch. 25. **iCEO:** Isaacson, ch. 25.

CHAPTER 29: IVE

Interview with Robert Brunner. **Jony:** Interviews with Robert Brunner, Jony Ive; Kunkel, p. 255. **The Collaboration:** Schlender & Tetzeli, ch. 9; Elisa Lipsky-Karasz, "Jony Ive on Life After Apple," *Wall Street Journal*, Nov. 2, 2022; Isaacson, ch. 26. **Project C1:** Isaacson, ch. 27. **The Bet:** Interview with Allen Olivo; Ken Werner, "Brian Berkeley Reflects on His Career at Apple, Samsung, and SID," Society for Information Display, April 6, 2021.

CHAPTER 30: IMAC

The Slot-Load Crisis: Isaacson, ch. 27. **The iMac Era:** Levy, *Perfect Thing*, pp. 131–32. **iMovie:** Isaacson, ch. 30; Jim Carlton, "Jobs Makes Headway at Apple, but Not Without Much Turmoil," *Wall Street Journal*, Apr. 14, 1998.

CHAPTER 31: KEYNOTES

The Making of Wi-Fi: McGee, ch. 10; John C. Dvorak, "The iBook Disaster," *PC Magazine*, July 26, 1999.

CHAPTER 32: MILLENNIUM

Hello, Dave: Ken Segall, "Behind HAL: Apple's Last Super Bowl Ad," KenSegall.com. **Mac OS X:** Interview with Avie Tevanian. **The First Demo:** Chafkin, ch. 6. **The Cube:** Interview with Jon Rubinstein; Steven Levy, "20 Years Ago, Steve Jobs Built the 'Coolest Computer Ever.' It Bombed," *Wired*, July 24, 2020.

CHAPTER 33: RETAIL

Interview with Ron Johnson; Isaacson, ch. 29; Rita Warkov, "Steve Jobs and Mickey Drexler: A Tale of Two Retailers," CNBC, May 22, 2012; Cliff Edwards, "Commentary: Sorry, Steve: Here's Why Apple Stores Won't Work," Bloomberg, May 21, 2001.

CHAPTER 34: IPOD

iTunes: Devin Leonard, "This Is War," *Fortune*, May 27, 2002. **Fadell:** Fadell, ch. 2.4. **The Race:** Levy, *Perfect Thing*, p. 42. **The Reveal:** Levy, *Perfect Thing*, pp. 51–52. **iPod for Windows:** Interview with Jon Rubinstein. **iTunes Store:** Apple Novinky, "Steve Jobs Introduces iTunes Music Store—Apple Special Event 2003," YouTube, Apr. 3, 2018; Chris Taylor, "Entertainment: Burn, Baby, Burn," *Time*, May 20, 2002. **Silhouettes:** Segall, ch. 5. **Generations:** Interviews with Tony Fadell, Jon Rubinstein, Jony Ive; Levy, *Perfect Thing*, pp. 101, 105, 137–38; Damon Darlin, "The iPod Ecosystem," *New York Times*, Feb. 3, 2006; Schlender & Tetzeli, ch. 15. **The Color-Changing Mac:** Interview with Duncan Kerr.

CHAPTER 35: BACK TO THE MACS

Interview with Jon Rubinstein. **iMac, Evolved:** Isaacson, ch. 34. **iMac G5:** "Apple's Intel Transition Bears Fruit," *InfoWorld*, Apr. 24, 2006, 56. **Intel:** Dan Clark & Nick Wingfield, "Apple Explores Use of Chips from Intel for Macintosh Line," *Wall Street Journal*, May 23, 2005. **Diagnosis:** Isaacson, ch. 35.

CHAPTER 36: IPHONE

The Tablet: Isaacson, ch. 36. **The First Phone:** Betsy Morris, "Steve Jobs Speaks Out," *Fortune*, Mar. 7, 2008. **P1 and P2:** Interview with Tony Fadell; Isaacson, ch. 36. **OS:** Morris, "Steve Jobs Speaks Out." **Secrecy:** Interview with Scott Forstall. **Two Weeks:** Interview with Scott Forstall; *Wall Street Journal*, "Apple's Secret iPhone Launch Team:

The Event That Began It All," YouTube, June 29, 2017. **The Keyboard Solution:** Kocienda, chs. 6, 8. **Kim's Lock-In:** Fred Vogelstein, "And Then Steve Said, 'Let There Be an iPhone,'" *New York Times*, Oct. 4, 2013. **The Network:** Interview with Eddy Cue. **How the iPhone Got Glass:** Thomas Ricker, "What Happened to the Original iPhone's Plastic Screen?" Verge, July 9, 2019; McGee, ch. 18. **The Show:** CHM, "CHM Live: Original iPhone Engineers Nitin Ganatra, Scott Herz, and Hugo Fiennes (Part One)," YouTube, June 20, 2017; Austin Carr, "Because of Steve Jobs's First Public iPhone Call, Starbucks Still Gets Orders for 4,000 Lattes," *Fast Company*, Mar. 4, 2013. **Aftermath:** Jenna Wortham, "Customers Angered as iPhones Overload AT&T," *New York Times*, Sept. 2, 2009.

CHAPTER 37: APPS

"Apple Event, September 2009." **Android:** Isaacson, ch. 39. **The iPhone Era:** "Global App Store Helps Developers Reach New Heights, Supporting $1.3 Trillion in Billings and Sales in 2024," Apple.com, June 5, 2025.

CHAPTER 38: SEQUELS

iPhone 4: Apple Archives Pro, "iPhone 4 Introduction with Steve Jobs—Apple WWDC 2010," YouTube, Jan. 31, 2021. **GizmodoGate and AntennaGate:** Charles Arthur, "Steve Jobs Solves iPhone 4 Reception Problems: '"Don't Hold It That Way,'" *Guardian*, June 25, 2010. **AirPods:** Interviews with Molly Anderson, Kate Bergeron. **iPhone X:** Interview with Alan Dye.

CHAPTER 39: iPAD

Jim Goldman, "Apple's Jobs and His Health: Take Accurate over Being First," CNBC, June 11, 2008; "Apple Media Advisory," Jan. 14, 2009. **Prototypes:** Interview with Scott Forstall; Victoria Woollaston, "Happy Birthday iPad! Apple Unveiled Its 'Revolutionary' Tablet on This Day 5 Years Ago," *Daily Mail*, Jan. 27, 2015; Chafkin, ch. 12. **Keynote:** Martin Peers, "Apple's Hard-to-Swallow Tablet," *Wall Street Journal*, Dec. 30, 2009; "Women Mock the iPad, Calling It iTampon," *Los Angeles Times*, Jan. 27, 2010; John D. Sutter, "iPad Name Draws Feminine Hygiene Jokes," CNN Money, January 28, 2010; John Paczkowski, "Who's Your Daddy? iPad Rewriting Adoption Records," All Things D, Oct. 5, 2010; Apple Archives Pro, "iPad 2 Introduction with Steve Jobs—Apple Special Event 2011," YouTube, Feb. 4, 2021; "Report: iPad Lowering Number of Pilot Injuries," Today's iPhone, May 1, 2022; Stephen Shankland, "Steve Jobs' Letter Explaining Apple's Flash Distaste," CNET, Apr. 29, 2010.

CHAPTER 40: CHINA

"Runner Up: Tim Cook, the Technologist," *Time*, Dec. 19, 2012; Kit Eaton, "Tim Cook, Apple CEO, Auburn University Commencement Speech 2010," *Fast Company*, Aug. 26, 2011; Tim Cook, "Tim Cook Speaks Up," Bloomberg, Oct. 30, 2014; "Runner Up: Tim Cook, the Technologist," *Time*; McGee, ch. 10; Josh J. Tyrangiel, "Tim Cook's Freshman Year: The Apple CEO Speaks," Bloomberg, Dec. 6, 2012; interview with Jeff Williams; Adam Lashinsky, "The Genius Behind Steve," CNN Money, Nov. 10, 2008. **Foxconn:** Interview with a U.S. supplier in Shenzhen; Austin Carr & Mark Gurman, "Apple Is the $2.3 Trillion Fortress That Tim Cook Built," Bloomberg, Feb. 9, 2021. **Crisis:** Philip Elmer-Dewitt, "Steve Jobs on Foxconn Suicides," *Fortune*, June 1, 2012.

CHAPTER 41: APPLE PARK

Origins: Steven Levy, "Apple Park's Tree Whisperer," *Wired*, June 1, 2017; Witold Rybczynski, "The Untold Story of Apple Park," *Architect*, Nov. 9, 2018; "The Shape of Things to Come," *New Yorker*, Feb. 23, 2015; Lewis Wallace, "How (and Why) Jony Ive Built the Mysterious Rainbow Apple Stage," Cult of Mac, May 9, 2019.

CHAPTER 42: LOSS

Jay Yarow, "What It Was Like When Steve Jobs Told Tim Cook He Was Going to Be the CEO of Apple," *Business Insider*, Dec. 6, 2012; Jony Ive, "Jony Ive on What He Misses Most About Steve Jobs," *WSJ Magazine*, Oct. 4, 2021; Owen Thomas, " 'When Steve Jobs Died, It Was the Saddest Day of My Life,' " *Business Insider*, May 29, 2012.

CHAPTER 43: SERVICES

Miguel Helft, "The Understudy Takes the Stage at Apple," *New York Times*, Jan. 23, 2011. **Skeuomorphism:** Marco della Cava, "Jony Ive: The Man Behind Apple's Magic Curtain," *USA Today*, Sept. 19, 2013. **Apple Pay:** Nathaniel Popper, "Banks Did It Apple's Way in Payments by Mobile," DealBook, Sept. 11, 2014. **Apple University:** Interview with Deirdre O'Brien. **Apple Card:** Evan Zimmer, "The Apple Card Was Rated No. 1 in Customer Satisfaction. Why?," CNET, Oct. 11, 2024. **Apple Music:** Apple Novinky, "Steve Jobs Introduces iTunes Music Store—Apple Special Event 2003," YouTube, Apr. 3, 2018. **Apple News:** Isaacson, ch. 38. **Apple Arcade:** Anita Balakrishnan, "Tim Cook: Goal Is to Double Apple's Services Revenue by 2020," CNBC, Jan. 31, 2017. **Apple TV+:** Interview with Eddy Cue; Tim Goodman, "Critic's Notebook: Apple and All of Its Money Just Got Real About the TV Business," *Hollywood Reporter*, June 21, 2017. **Apple One:** Kif Leswing, "Apple's Services Unit Is Now a $100 Billion a Year Juggernaut After 'Phenomenal' Growth," CNBC, Oct. 31, 2024.

CHAPTER 44: WATCH

Interview with Jony Ive; Levy, "An Oral History of Apple's Infinite Loop." **Prototypes:** Interview with Kevin Lynch. **Fashion:** Interview with Jony Ive. **Health:** Interview with Sumbul Desai.

CHAPTER 45: VISION PRO

Tripp Mickle, "How Tim Cook Has Made Apple His Own," *Wall Street Journal*, Aug. 7, 2020; Steven Russolillo, "Replacing Steve Jobs: How Apple CEO Tim Cook Has Fared Five Years Later," *Wall Street Journal*, Aug. 21, 2016; "Careers at Apple," Apple.com. **China Sales:** Commonwealth Club World Affairs of California, "Patrick McGee: Apple in China," YouTube, May 29, 2025. **Vision Pro:** Ben Cohen, "Tim Cook on Why Apple's Huge Bets Will Pay Off," *Wall Street Journal Magazine*, Oct. 20, 2024; Mark Gurman, "Apple's New Headset Meets Reality," Bloomberg, May 18, 2023. **Project Titan:** Wlievano, "Tim Cook Talks at WSJDLive About the Future of Cars," *Wall Street Journal* video, Oct. 20, 2015. **Ive After Apple:** Tripp Mickle, "Apple Ends Consulting Agreement with Jony Ive, Its Former Design Leader," *New York Times*, July 12, 2022.

CHAPTER 46: SILICON

The M1: Dave James, "Intel Insider Claims It Finally Lost Apple Because Skylake QA 'Was Abnormally Bad,' " *PC Gamer*, June 24, 2020; Alexis C. Madrigal, "Paul Otellini's Intel: Can the Company That Built the Future Survive It?," *Atlantic*, May 16, 2013; Dieter Bohn, "Apple MacBook Air with M1 Review: New Chip, No Problem," Verge, Nov. 17, 2020; Leander Kahney, "M1 MacBook Air Is an Instant Classic," Cult of Mac, Nov. 24, 2020; Lee Hutchinson, "Apple's M1 MacBook Air Has That Apple Silicon Magic," Ars Technica, Nov. 28, 2020.

CHAPTER 47: APPLE INTELLIGENCE

Cohen, "Tim Cook on Why Apple's Huge Bets Will Pay Off." **Small Features:** Verge, "Apple Intelligence So Far: Were Promises Kept?," YouTube, Jan. 30, 2025; Daring Fireball, "The Talk Show Live from WWDC 2024,"

YouTube, June 14, 2024; Apple, "WWDC 2024—June 10," YouTube, June 10, 2024. **Private Cloud Compute:** Interview with Craig Federighi; Marques Brownlee, "Talking Tech and AI with Tim Cook," YouTube, June 12, 2024. **Siri Struggles:** John Gruber, "Something Is Rotten in the State of Cupertino," Daring Fireball, Mar. 12, 2025.

CHAPTER 48: HEADWINDS

China Censorship: *Fortune*, "Tim Cook Discusses Apple's Future in China," YouTube, Dec. 6, 2017. **China Reliance:** Charles Duhigg & Keith Bradsher, "Why Apple Says It Can't Build an iPhone in the US," NBC News, Jan. 22, 2012; Jack Nicas, "A Tiny Screw Shows Why iPhones Won't Be 'Assembled in U.S.A.," *New York Times*, Jan. 28, 2019; Julia Shapiro & Alex Gangitano, "Trump, Apple Relationship Devolves with Threats," The Hill, May 24, 2025.

CHAPTER 49: APPLE: THE NEXT 50 YEARS

Interview with Phil Schiller; Tripp Mickle, "What's Wrong with Apple?," *New York Times*, Apr. 11, 2025. **Phones:** Mark Gurman, Lean Hylen, and Stephanie Lai, "Apple Eyes Move to AI Search, Ending Era Defined by Google," Bloomberg, May 7, 2025.

CHAPTER 50: THROUGHLINES

Beauty: Interviews with Eddy Cue, Jony Ive; Rob Walker, "The Guts of a New Machine," *New York Times Magazine*, Nov. 30, 2003. **RoundRects:** Ian Parker, "The Shape of Things to Come," *New Yorker*, Feb. 16, 2015. **Best, Not First:** Interview with Phil Schiller. **Simplicity:** Don Norman, "Why Are Apple's Products So Confusing? They Ignore Design Principles," LinkedIn, Aug. 12, 2015. **A Greener Apple:** Interview with John Ternus. **Closed Systems:** Erick Schonfeld, "Steve Jobs: 'Open Systems Don't Always Win,'" TechCrunch, Oct. 18, 2010. **Accessibility:** Liat Kornowski, "How the Blind Are Reinventing the iPhone," *Atlantic*, May 2, 2012. **San Bernardino:** *Wall Street Journal*, "Tim Cook Defends Apple's Encryption Policy," YouTube, Feb. 18, 2016. **Steve Jobs Goes to Congress:** "Steve Jobs," Oral and Video Histories, Smithsonian Institution, Apr. 20, 1995. **Excellence and Execution:** CHM, "CHM Live: Insanely Great: The Apple Mac at 40," YouTube, Jan. 24, 2024; Interviews with Duncan Kerr, Allen Olivo.

Photo Credits

Except as noted, product photos are courtesy of Apple. Vintage OS screenshots generated at InfiniteMac.org. Headshots featured in "bio boxes" supplied by their subjects, except as noted.

About the cover image: At first glance, you might recognize the famous iPod clickwheel, the iconic feature on the product that heralded Apple's rebirth as a global consumer electronics giant. At second glance, you might notice the ⏮ and ⏭ buttons, which might mean "review the past" and "scan ahead to the future"—exactly what this book attempts to do.

Chapter 1: Two Steves *Eighth-grade Woz*: Steve Wozniak. *Bill Fernandez's Cream Soda*: Bill Fernandez. *Jobs's boyhood home*: Turtix / Shutterstock. *Jobs in high school*: Archivio GBB / Alamy. *Woz with blue box*: Steve Wozniak. *Breakout game*: Atari.

Chapter 2: Apple I *MOS flyer and chip*: Public Domain. *Apple I*: Apple. *Steves at Homebrew*: Joe Melana / Apple / Computer History Museum (CHM).

Chapter 3: Apple Computer Company *Steves holding board*: Db Photo / DPA / Zuma Press. *Wayne*: Ron Wayne. *Signatures*: Abaca Press / Alamy. *Original Apple logo*: Apple. *Steves and Apple I*: Joe Melana / Apple. *Delivering Apple*: Open Road Films, courtesy Everett Collection. *Full Apple I setup*: Jimmy Grewal.

Chapter 4: Apple II *Karateka game*: Jordan Mechner. *Manock's Apple II design*: Apple. *Manock 1978*: Mary Ellen Manock. *McKenna and Jobs*: Roger Ressmeyer / Getty. *Apple II brochure*: Apple. *Specs for Apple logo*: Jerry Manock Collection.

Chapter 5: In Business *Apple's first office*: Google Maps. *Mike Scott*: Apple. *West Coast Computer Faire*: CHM. *Apple II (left)*: Phillip Torrone and Limor Fried. *Apple II (right)*: FozzTexx / Wikimedia Commons. *Apple II keyboard*: © Mark Richards. Courtesy of the CHM. *Drive II stacked*: Interfoto / Alamy. *VisiCalc*: apple2history.org / Wikimedia Commons.

Chapter 6: Apple III *Apple III interior jpg*: Invaluable LLC. *Apple IPO prospectus*: Wikimedia Commons / Public Domain. *Final Apple III*: CHM.

Chapter 7: Lisa *Couch*: Roger Ressmeyer / Corbis / VCG via Getty. *Atkinson*: Michel Baret / Gamma-Rapho via Getty. *Xerox PARC*: Getty. *Doug Engelbart*: SRI International. *Alto*: © Mark Richards. Courtesy of the CHM. *Tesler*: Ann E. Yow-Dyson / Getty. *"NO MODES"*: www.nomodes.com. *Lisa*: Courtesy of the CHM. *So advanced ad*: Apple. *Lisa team photo*: Lee Youngblood. *Kevin Costner*: YouTube.

Chapter 8: Crashes *Beech Bonanza*: Bill Larkin / Wikimedia Commons.

Chapter 9: Macintosh *Raskin*: Digibarn. *Tribble*: Bud Tribble. *Hoffman*: Hoffman. *Canon Cat*: Wikimedia Commons.

Chapter 10: Sculley *Sculley 1983*: Eric Sander / Liaison / Getty. *Fremont factory*: Apple. *Job and Sculley*: Ed Kashi / VII / Redux.

Chapter 11: Software *Kare*: Elena Dorfman / Redux. *Mr. Macintosh board*: Dan Farber / CNET. *Brushing woman*: incamerastock / Alamy. *Jobs in spare home*: Diana Walker / Briscoe Center for American History, UT-Austin. *Hertzfeld card*: CHM.

Chapter 12: Marketing the Mac *Macworld*: Wrights Media. *Apple Dating Game*: YouTube.

Chapter 13: Insanely Great *Macintosh people*: Apple Unknown. *Mac unveiling 1984*: YouTube. *First Mac*: John Greenleigh / Apple. *Mac boxes*: John Greenleigh / Apple. *Apple Fellows*: Michael Collopy / Apple. *Esslinger models*: Victor Goico, frogdesign. *Hartmut Esslinger*: Jürgen Schwarz / Imago. *ImageWriter II*: John Greenleigh / Apple. *Apple IIc*: Bilby / Wikimedia Commons. *First LCD screen*: Bilby / Wikimedia Commons. *Execs with IIc*: Sal Veder / AP.

Chapter 14: Macintosh Office *ImageWriter*: John Greenleigh / Apple. *Warnock & Geschke*: Adobe / Wikimedia Commons. *LaserWriter*: John Greenleigh / Apple. *Lisa in landfill*: Herald Journal / courtesy of USU SpecialCollections. *Big Mac*: Hartmut Esslinger.

Chapter 15: Rift *US Festival*: Ed Kashi / Liaison / Getty. *Apple e-staff*: Ed Kashi / VII / Redux.

Chapter 16: New Ideas *City Center*: Joseph Sohm / Visions of America, LLC / Alamy. *Apple IIgs*: Tony Diaz. *Jonathan*: Hartmut Esslinger.

Chapter 17: Heresies *Gassée car*: Ed Kashi / Liaison / Getty. *Mac Plus*: Felix Winkelnkemper / Wikimedia Commons. *Macintosh SE*: Apple. *Lawsuit*: Public Domain.

Chapter 18: High/Right, Low/Left *Mac IIx*: Apple. *Gassée builds IIcx*: YouTube. *Mac IIcx*: Pawel Pieczul / oldcrap.org. *Mac SE/30*: Wikimedia Commons. *Mac Portable*: Rama / Wikimedia Commons. *Portable on space station*: NASA. *Spindler*: Daniel Giry / Sygma via Getty. *Three cheap Macs*: Wikipedia.

Chapter 19: System 7 *Chahil*: Satjiv Chahil.

Chapter 20: Newton *Scribe tablet*: YouTube. *Swatches*: Paul Mercer. *Newton Cadillac*: Grant Hutchinson / Flickr. *Tchao and Capps*: Doug Menuez. *Newton*: John Greenleigh / Apple. *eMate 300*: Felix Winkelnkemper / Wikimedia Commons.

Chapter 21: Moonshots *Jaguar*: Robert Brunner. *Aquarius Cray*: John Greenleigh.

Chapter 22: PowerBooks *PowerBook 140*: © Mark Richards. Courtesy of CHM. *PowerBook 100*: Danamania / Wikimedia Commons. *PowerBook Duo (left)*: Dmitrii Eliuseev. *PowerBook Duo (right)*: Stephen Hackett. *Infinite Loop*: nvadingInvader / Wikimedia Commons. *PenLite*: YouTube. *Paladin*: Robert Brunner. *Mac-Like Things*: Phillip Torrone & Limor Fried.

Chapter 23: Spindler *Sculley and Clinton*: Consolidated News Pictures / Getty. *Exec staff*: Doug Menuez. *Power Macs*: Apple. *QuickTime 100*: Robert Brunner. *QuickTake 200*: Phillip Torrone and Limor Fried.

Chapter 24: Clones *Four clones*: Garth Beagle. *PowerBook 5300*: Serged / Wikimedia Commons.

Chapter 25: Amelio *Amelio*: Raymond Reuter / Sygma via Getty. *Fred Anderson*: Gary Parker. *Mission: Impossible poster*: Apple, courtesy Satjiv Chahil. *Bandai Pippin*: Evan-Amos / Wikimedia Commons.

Chapter 26: NeXT *BeBox*: Be, Inc. *NEXTcube*: All About Apple Museum / Wikimedia Commons.

Chapter 27: Transition *Spartacus prototypes*: Robert Brunner. *Twentieth Anniversary Mac*: Stephen Hackett. *Keynote 97*: Richard Wisdom / San Jose Mercury News / MCT / ZUMAPRESS.com / Alamy. *Phil Schiller*: Apple. *Greg Joswiak*: Apple.

Chapter 28: Turnaround *Gates*: John Mottern / AFP / Getty.

Chapter 29: Ive *Ive with 20th Anniv. Mac*: May Tse / SCMP / Newscom. *Ive and Jobs*: Art Streiber / August.

Chapter 30: iMac *PowerBook G3 "Wallstreet"*: Danamania / Wikimedia Commons. *Ive, Ruby with iMacs*: Susan Ragan / AP.

Chapter 31: Keynotes *Prepping for sales meeting*: Photographer unknown. *iBook*: Apple.

Chapter 33: Retail *Apple Store*: Leon Neal / Getty.

Chapter 34: iPod *iPod prototype*: Tony Fadell. *B&O phone*: Bang & Olufsen. *Big iPod Prototype*: Cabel Sasser. *iPod 1G*: Chris Wilson / Alamy. *iPod Nano line*: D. Hurst / Alamy. *Zune*: BulbousSum / Wikimedia Commons.

Chapter 35: Back to the Macs *Parallels*: Parallels Inc. *Commencement speech*: Linda A. Cicero. Courtesy Stanford University / Apple.

Chapter 36: iPhone *iPhone team*: Jonathan Sprague / Redux. *FingerWorks tablet*: FingerWorks. *Projector on FingerWorks*: Apple. *ROKR phone*: Motorola. *Wallabies, Kocienda keyboards*: Ken Kocienda. *Jobs with iPhone*: Getty. *iPhone lines*: Michael Nagle / Getty.

Chapter 37: Apps *MacBook Air*: Kimberly White / Corbis via Getty. *Jobs with MacBook Air*: Tony Avelar / AFP via Getty.

Chapter 38: Sequels *Jobs and AntennaGate*: San Francisco Chronicle / Hearst Newspapers / Getty. *Siri*: Kevork Djansezian / Getty. *Portrait mode*: David Pogue. *AirPods (right)*: Jeremy Moeller / Getty. *Face ID testing*: Apple.

Chapter 39: iPad *iPad Jobs*: Justin Sullivan / Getty. *Cockpit*: Scott Eells / Bloomberg via Getty.

Chapter 40: China *Cook and Jobs*: David Paul Morris / Getty. *Foxconn*: Visual China Group via Getty.

Chapter 41: Apple Park *Apple Park aerial*: Brooks Kraft / Apple Inc. *Steve Jobs Theater*: Nigel Young / Foster + Partners.

Chapter 42: Loss *iPad Vigil*: Ano Shlamov / AFP via Getty.

Chapter 43: Services *Tim Cook*: Kevork Djansezian / Getty. *Apple Pay*: Shutterstock.

Chapter 44: Watch *Design Studio*: Jason Schmidt. *Watch unveiling*: Stephen Lam / Getty.

Chapter 45: Vision Pro *Chinese crowds*: AFP via Getty. *Oculus*: Oculus. *Persona demo*: Norman Chan. *Titan car test*: Andrej Sokolow / dpa (Photo by Andrej Sokolow / picture alliance via Getty). *Jony Ive today*: Carolyn Fong / *The New York Times* / Redux.

Chapter 46: Silicon *Chip fab*: Adam Glanzman / Bloomberg via Getty.

Chapter 47: Apple Intelligence *Clean Up*: David Pogue.

Chapter 49: Apple: The Next 50 Years *Huawei trifold phone*: Lam Vik / Bloomberg via Getty. *Smart glasses (left)*: Josh Edelson / AFP / Getty. *Smart glasses (right)*: Vjeran Pavic / The Verge. *Glucose monitor*: Click and Photo / Shutterstock.

Chapter 50: Throughlines *Three teardowns*: iFixit. *Disassembly robots*: Apple.

The Beta Readers

Even after a typical book has been read and reread by the author, editor, tech reviewer, copy editor, proofreader, and indexer, you'd be surprised—or maybe not—at how many typos still find their way into print. For this book, I invited an army of beta readers to swarm over the chapters. They found and fixed a trove of glitches, which makes me both sheepish and grateful. Here are those noble volunteers:

Wati Aurora, Tim Bajarin, Tom Barclay, Nicky Bleiel, Maxine Bleiweis, Mary Bogart, Rachael K. Bosley, Patrick "Jethro" Bowman, Darren Branagh, Brandon Carson, Ed Casey, Bruce Colwin, Elbert Cuenca, Hamilton Davies, Denise Deverelle, Karen Donoghue, Michael Dunthorn, Emily L. Ferguson, Pamela Ficarra, Aaron Fields, Julie Taylor Fitzgerald, Debra Joy Frank, Biff Franks, Matt J. Fuller, Daniel Gesmer, Marc S. Goldberg, Edward S. Goss, Federico Guerra, Steve Hayman, Jeff Holck, Elena-Beth Kaye, Daniel D. Kelly, Tom Kerber, Steve & Diana Kohler, Jonathan Leblang, Chris Mansfield, Jim McDonough, Lisa Mertz, Mike Meyer, David Milazzo, Jeff Mines, Allen Mogol, Geoffrey Morris, Joe Neel, Brian O'Donovan, Vince Patton, Stefanie Pont, Brad Price, David Ramsey, Jim Rossman, Chris Ruter, Paul Sadler, Noah Salzman, Beth Slick, Kal Sostarecz, Laura Staff, MD, Becky Steenburg, James Swirczynski, Shay Telfer, Laura Tilsley Garcia, Steven D. Trigili, Susan M. Troccolo, Julianne Turé, Tony Vinayak, Beverly Voth, Mark Weir, Dora West, and Jon Zilber.

And then there's former Apple engineer and technical writer Scott Knaster, who started as a two-chapter beta reader but found so many sneaky lingering errors—and added so many bits of fantastic humor and historical color—that I invited him to eagle-eye the entire book. Fortunately for me—and for you—he accepted.

Index

A4, A5, A6 chips, 503, 504
accelerometer, 480, 528, 543
accessibility, 541–44, *542*
    Americans with Disabilities Act (ADA), 543
    curb-cuts effects, 543
    origins at Apple, 541
    VoiceOver, 542
ACM, *see* Markkula, Armas Clifford, Jr.
Acorn Computers, 226, 227, 323
Activation Lock, 534
Activision, 540
ADA (Americans with Disabilities Act), 543
Adams, Douglas, 145, 147, 293
Address Book app, 382
    see also Contacts app
Adler, Darin, 122, 222
Adobe, 210, 256, 332, 388, 440, 445, 459, 477
    Display PostScript, 217, 282
    Flash, 440
    founding, 155–56
    Photoshop, 119, 341, 388, 402, 505
    PostScript, 155–57, 179
    Premiere, 333, 546
    Type Manager, 216–17
ADP company, 269
Advanced RISC Machines Limited
    see ARM
advertising
    "1984" ad, 129–34
    accessibility campaigns, *542*
    bunny-suit ad, 326
    by IBM, 305, 385
    for iPod, 371
    for iTunes, 359–60
    for Lisa, 78, 79, 106
    for Macintosh computers, 106, *129*, 129–31, *133*, 133–34, 136, 140, 143, 344, 385
    for Macintosh Office, 159–60, 179
    by Pepsi, under Sculley, 105
    for personalized Siri, 513
    for PowerBooks, 273
    for Power Mac G3, 326
    during Super Bowl, 129–31, 133–34, 160, 344
    "I'm a Mac/I'm a PC" ads, 385
    "Lemmings," 159–60
    "Silhouettes" (iPod) ads, 371
    snail ad, 326
    "Switch" ads, 385
    "Think Different" ad, 2, 293, 304–9, 353, 460
    "Welcome, IBM" ad, 82–83
    "Y2K bug" ad, 344
Advertising Age, 160
agency model, 438
AI, *see* Apple intelligence; artificial intelligence
AIM (Apple, IBM, and Motorola) alliance, 233–36, 262
Airbnb, 420, 499
AirDrop, 501
AirPods, 1, 427–31, *429*, *430*, 516, 530
    accessibility features, 542
AirPods Pro, 431, *527*
AirPods Pro 2, 525–26, *527*
    future of, 525–27
    head gestures, 535
    manufacturing, 519
    real-time translations, 508
    silicon for, 503
AirPort, 338, *340*, 342, 348, 387, 428
    Air Port Express, 394, 464
AirPower, 513
AirTags, 479, 503
Akamai, 343
Alarm Clock desk accessory, 118
Alcorn, Al, 146, 234
Aldus, 179, 256
Alexio, Peggy, 95
Ali, Muhammad, 287, 293, 307
aliases, 210
Alias Wavefront program, 321
Alice (video game), 115, 141
Allied Linotype, 157
All One Farm, 18
Alphabet, 488, 517
    see also Google
AlphaFold, 506
Alps company, 100
Alsop, Stewart, 236
Alta Mesa Memorial Park, 458
Alto, 64–67, *65*, 71
Amazon.com, 417, 438, 516, 517, 523, 532
ambient-light sensor, 412
AMD, 233, 504

Amelio, Gil, 254, *265*, 266–74, 276, 278, 281, 283–96, *290*, 300, 302, 310, 311, 313–15, 318
America Online (AOL), 270, 367
Ames, Roger, 368–69
Anderson, Fred, 109, 269–72, 284, 285, 287, 291, 294, 295, 301, 309, 312–14, 345, 361, 394
Anderson, Molly, 429
Android (operating system), 418, 464, 501
   AI agent for, 511
   AirTag app, 479
   COVID-tracking collaboration, 455
   origins of, 418
   phones, 196, 214, 223, 423, 440, 455, 469, 516, 540
   XR glasses, 523–24, *524*
Annie project, 86
antennas, iPhone, 410, 422–23
AOL Time Warner, 367–69
Aperture app, 390, 443, 445
Apple (Apple Computer Company, Apple Computer, Inc., Apple Inc.)
   1985 French deal, 165
   administrative structure, 301–2
   CEO transitions, 38–39, 457
   changes over history, 1
   Chinese manufacturing, 448–51, 518–19
   consumer electronics projects, 246, 248, 276, 302–3, 357, 360–61, 472, 474, 494–96, 521
   corporate values (*see* corporate values)
   culture, 105, 106, 296, 301–2, 407
   current challenges, 514–19
   design language and hallmarks, 102–3, 148–50, 186, 238–40, 257, 533

desktop publishing, 178–79, 202, 216–17
digital hub strategy, 357, 389
earnings, 53–54, 330
financing, 18–19, 22, 25, 33–36, 270–71
founding, 2, 15–20, 38, 531
future hardware/products, 520–29
headquarters *see* office buildings
IBM/Motorola alliance with, 233–36, 262
incorporation, 38
influence of, 530
initial public offering (IPO), 53–57, *55*
Japanese market, 214–16
killing old technologies, 303–4, 322, 425, 445
licensing decisions, 202–3, 259–61, 312–13
logo, *20*, 20–21, 32–33, *33*, 42, 149, 381
market capitalization, 1, 325
market share, 1, 154, 202, 258, 329, 360, 418, 525
merger and acquisition plans, 165, 233, 250, 252, 262–65, 267, 281, 285–87, 290–92, 358, 390, 397, 428, 497, 499–500, 502, 540
moonshot era, 232–37
myths about, 2, 64–66, 130–31
number of products, 256, 258, 292–94, 302–3
plan to split into two companies, 251
power structure, 166
preannouncing products, 513
product placement in TV and film, 273
product quadrant, *302*, 302–3, 325, 327–29, 350, 381
product strategies, 197–201, 204–6, 271

profit/loss, 24, 106, 163, 171, 177, 179, 189, 206, 264, 269, 287, 314, 315, 325, 337, 345, 466, 540
profit margin, 47, 197
restructurings, 80, 251, 278, 290–94
revenue, 43, 144, 189, 206, 371, 420, 455, 463, 471, 474, 521, 539
sales, 49, *54*, 143, 145, 147, 153, 166, 167, 188, 201, 202, 205–6, 250, 256, 266, 329, 342, 343, 350, 371, 391, 414–15, 420, 422, 438–39, 466, 486, 488, 489, 496
sales conferences, 131–32
silicon chips made by, 503–5
stock options, 299–300, 345
stock valuation, 177, 188, 249, 264, 265, 300, 314, 345, 378, 413, 463, 466, 488, 512, 518
tax strategies, 161, 247, 515
throughlines, 530–548
turnaround, 299–315
see also Apple employees; corporate values; specific products or services
Apple I, 12–26, *15*, *16*, *19*, *22*, *24*
   Apple II vs., 27
   assembly, 23–26
   manual, 85
   marketing, 15–17, 23
   processor chips, 39
   sales, 1, 21–26
   schematics for, 12–13
   wooden case for, 26, *26*, 269
   Woz's first build, 13–15, 36
Apple II, 27–31, *43*
   accessibility features, 541
   Apple IIc, 150–53, 178, 199
   Apple II division, 92, 144, 163, 164, 166, 171, 183, 184

Apple IIe, 546, 547
"Apple II Forever" event, 150–53
Apple IIgs, 150, 180–81, 183, 188, 270
Applesoft BASIC, 81, 194
assembly, 49–50
case, 29–31, *30*, 98
color display, 27, 28
compatibility with, 51–53, 57, 183
competitors of, 139
design for, 26–29, 36, 42, 50, 53
disagreements over, 39
expansion slots and add-ons, 29, 43, 49
floppy disk drive, 44–47
Lisa and, 62, 63
Macintosh and, 87, 96
manual, 85, 123
marketing, 39–42, 168
for Markkula, 36, 104
as open system, 540
power supply, 27, 29
reviews, 42–43
sales, 48, 147
simplicity of, 536
VisiCalc for, 47–48
watermark in, 187
Apple III, 2, 50–60, *52, 58, 59,* 79, 86, 98
Apple Annual Report (1984), 128
Apple Arcade, 471, 473–74
Apple Card, 468
AppleCare warranties, 474
AppleCenter, 216
Apple Computer Company (Apple Computer, Inc.), *see* Apple
Apple Corps, 18
AppleDesign Powered Speakers, 248, *248*
Apple employees
 in Advanced Technology Group, 180, 210, 211, 221, 227, 231, 247, 294, 302, 340

badge numbers, 45
benefits for, 106, 153, 202
business cards, 122
crediting Apple turnaround to, 315
in Education Research Group, 180
in Industrial Design Group, 205, 220, 230, 238–40, 288, 316, 321
labor conditions in China, 450-51
number of, 30, 43, 54, 174, 245, 455, 463, 514
reorganizations, 82, 92, 144–45, 171, 206, 252–53, 269, 272, 273, 291, 312
staffing increases, 43, 244, 345, 514
team size as corporate value, 544
turnover, 109, 163, 172, 203, 299–300, 303, 450, 548
volunteering program, 488
see also specific personnel by name
Apple Fellows program, 146, 147, 158, 180
Apple Fitness+, 473
Apple Flat Panel Display, 151
Apple Inc., *see* Apple
Apple Intelligence, 356, 500, 506–13
 call transcriptions, 507
 Clean Up tool, *509*
 hold assist and spam-call screening, 508
 Image Playground, 507, 510
 personalized Siri, 511–513
 real-time translation, 508
 visual intelligence, 507
 writing tools, 507
Apple Interactive Television Box, 472
Apple Labs, 167, 168
AppleLink Personal Edition, 270
Apple Maps, 464–66, *466*
AppleMasters program, 293
Apple Music, 469–71, *470*, 473–74, 489
Apple Network Server, 292

Apple New Product Process, 319
Apple News and News+, 470–71
Apple One, 473–74
Apple One-on-One lessons, 354
Apple Park, 452–56, *453, 454,* 499
Apple Pay, 466–68, 471
Apple Pencil, 440
Apple PowerCD, 248, *248*
Apple Restore Fund, 538
Apple Scanner, 150
AppleScript, 180
Appleseed, Johnny, *see* Markkula, Armas Clifford, Jr.
Apple silicon, 501–505
Applesoft BASIC, 81, 194
Apple SOS, 51, 59
Apple Stage, 456
Apple Stores, 1, 345, 351–56, *354, 413,* 414, 416–17, 458, 460, 489, 537, 538, 547
AppleTalk, 158, 159, 220
Apple TV, 464, 472–74, *502, 503,* 531
 Apple TV+, 246, 248, 443, 471, 473–74
 RoundRects in, 533
Apple University Consortium, 128–29, 173, 467
Apple v. Microsoft, 195–96, 259
Apple Wallet, 516
Apple Watch, 1, 475–87, 494, 531, *532*
 accessibility features, 543
 AirPower and, 513
 Apple Fitness+, 473
 Apple Pay on, 466
 Apple Watch Series 9, 487
 Apple Watch Ultra, 486
 chip for, 503
 crown, 478
 excellence of execution for, 548
 fashion consciousness of, 482
 fitness tracking, 480–82, 527

Apple Watch (cont.)
   future innovations, 527–29
   Ive's design for, 475–77, 499
   launch and reviews of, 483–86
   Mach kernel in, 283
   manufacturing, 519
   medical alerts, 486–87, *487*, 528
   motivational rings, 481–82
   OS 26 for, 443
   pairing, 516
   project team, 223
   prototypes, 477–79
   RAM in, 97
   RoundRects in, 533, *533*
   sales, 496
   shower challenge for, 481
   Taptics Engine, 483
AppleWorks, 346
AppleWorld conference, 186, 187
apps
   iPad, 436, 495
   iPhone, *414*, 415–17, *417*, 436, 455, 514–15
   killer, 48
   Macintosh, 118, 337, 346
   skeuomorphism in, 465
   web, 416
   see also specific apps by name
App Store, 1, 420, 439, 443, 471, 514–18
Aptera, 500
Aqua interface, 347, 382, 400
Aquarius project, 234
AR (augmented reality), 522–25
   see also Vision Pro headset
ARM (Advanced RISC Machines Limited), 227, 271, 315, 503
Arnold Communications, 304
ARPANET, 14
artificial intelligence (AI), 180, 192
   Android XR glasses, 523–24
   cross-OS AI features, 507–8

   generative, 506–13
   hypertension detection, 487
   smart home products, 521
   time to market for Apple features, 506–8
   Titan project, 498, 500
   see also Apple Intelligence; Siri
Asimov, Isaac, 218
Askeland, Colette, *137*
Astarte, 358
Atari, 10, 11, 17–19, 22, 27, 28, 34, 86, 106
Atari ST, 183
Atkinson, Bill, 61–64, 67, 69–73, 75, 76, 86, 87, 90, 95, 96, 102, 103, 116, 118–19, 121, 146, 147, 189, 191, 207, 212, 222, 277, 411, 459
atrial fibrillation, 486–87
AT&T, 193, 222, 223, 342, 408, 411–18, 422
   Hobbit processor, 220, 223, 226, 227
Auburn University, 446
audiocassettes, data storage on, *24*, 24–26, 44–46
augmented reality (AR), 522–25
   see also Vision Pro headset
Auburn University, 446
Australia, 517
autocomplete, iPhone, 406
autocorrect, *404*, 406
autonomous vehicles, 497–98
A/UX operating system, 188
AV Foundation, 214

BabyMac prototype, 162
Back button, 404
Back Tap feature, 543
Baez, Joan, 459
Ballmer, Steve, 414
balloon help feature, 210
Bandai @World, 276

Bandley Drive buildings, 43–44, *44*, 49–50, 56, 57, 96, 97, 119, 121, 127, 138, 142, 172, 244, 245
Bang & Olufsen, 363
Bannister & Crun, 85–86
Barnes, Susan, 168, 169, *169*, 171, 173
Barram, Dave, 170
Battat, Randy, 220
batteries, 200, 264, 428, 429, 505, 522, 524
   AirPods, 428, 429
   iPad, 397
   iPhone, 414, 421, 427
   iPhone Air, 434
   iPod, 362, 364, 366
   Newton, 221, 229, 304
   PowerBooks, 240, 242, 263, 327
   Vision Pro, 494, 496
battery life
   Adobe Flash, 440
   AirPods, 526
   Apple silicon and, 502
   Apple Watch, 484, 486
   eMate, 230
   future iPhones, 522
   iPod, 366
   M1 laptops, 505, 525
   Mac Portable, 199–201
   PowerBook 5300, 263
   PowerBook G4, 380
   smart glasses, 523, 524
   solid-state drives, 319
Baum, Allen, 11, 13, 22, 37
BBC, 517
BBDO, 304
Be, Inc., 280–83, 285
Beatles, The, 18, 56, 209, 368, 369
Beats Music, 427, 469, 540
beauty, 31, 38, 95, 102, 530, 532
BeBox, *280*, 280–81
Belleville, Bob, 100–101, 167–69, *169*, 171, 172, 179, 417

Bell Labs, 265
Bennett, Susan, 423
BeOS, 280–81, 285
Berg, Jeff, 273
Bergeron, Kate, 429
Berners-Lee, Tim, 191, 343
Best Buy, 351
Better Place, 500
Bewkes, Jeff, 471
Bezier curves, 533
BFRs (brominated flame retardants), 537
BhutanTalk, 216
Big Mac prototype, 162, 173
biometric sensors, 527
bitmapped graphics, 62–65, 155
Bitstream, 217
BlackBerry, 400, 413, 417, 423, 440
Blackrock Neurotech, 522
Black Wednesday firings, 92
Blahnik, Jay, 481, 482
blood-oxygen (SpO2) sensor, 480, 486–7, 527–8
blood pressure tracking, 529
blood sugar monitoring, 528
Bloomberg, 517, 520
blue boxes, 9–10, 17
Bluetooth device pairing, 428, 429
BMG, 367–69
BMW, 497, 498
board of Apple
   board members, 46
   CEO hunt, 104
   changes under Spindler, 254
   compensation to Jobs, 345
   concern about Jobs, 166
   firing Amelio, 294–6
   firing Mike Scott, 82
   firing Sculley, 249–52
   firing Spindler, 265–7
   hating "1984" ad, 131–3
   Jobs demotion, 80

   Jobs joins, 309, 314
   Jobs replaces members, 300–301, 355–6
   Jobs vs. Sculley, 145, 169–70, 172–174
   Jobs's last meeting, 457
   NeXT deal, 285–92
   Original members, 54
   Sun offer, 264
Bohlin Cywinski Jackson firm, 354
Boich, Mike, 122, 125–26, 141, 261
Bomb app, 347
Bondi blue, *320*
Bonetti, Amy, 205
Bono, 369, 459, 485
Book of Macintosh (Raskin), 86, 89, 90
Boot Camp, 384, 391
Boston Computer Society, 78
Boston Globe, 329
Bowers, Ann, 467
Brady, Sheila, 208
Brainerd, Paul, 178–79
Brandt, Steven, 177
Branson, Richard, 307
Breakout (video game), 10, 17, 28
Bricklin, Dan, 48
Briggs, Jon, 424
Brightman, Alan, 541
Broadview, 250
Broedner, Walt, 222
brominated flame retardants (BFRs), 537
Brunner, Robert, 205, 225, *233*, 238–43, *239*, *246*, 257, 287, 288, 296, 316–18, 320
Buckley, Jim, 261
bug bounty program, 510
Bull, Bill, 95, *137*
Bunnell, David, 128
Burberry, 482
Bushnell, Nolan, 10, 34

Businessland, 143
BusinessWeek, 139, 235, 244, 268, 308, 355, 382, 446
butt-head astronomer, 253
ByteDance, 517
Byte magazine, 43, 48, 326
Byte Shop, 21–26, 31
Byton startup, 500

C. Itoh Electronics, 154
Calenda, Gifford, 208
Calendar desk accessory, 118, 465
Callas, Maria, 307, *307*
Call Computer, 14
call transcription, 507
Cambridge Analytica, 545
Campbell, Bill, 169–73, 177–79, 301, 460
Canon, 157, 215, 292, 360
   Cat, 90
   LBP-CX10 (print engine), 154–55
Canoo Lifestyle van, *497,* 500
Capps, Steve, 97, 110, 113–15, 118, 119, 138, 141, 145, 146, *146,* 207, 219, 221–22, 224, 226, *226,* 227, 229
car (Apple car) *see* Titan car project
Carbon apps, 346
carbon neutrality, 538
carbon offsets, 538
Carlton, Jim, 295
Carter, Matt, 95, 261
Case, Steve, 270
Casey, Don, 212
Casper voice control, 542
cassette tapes (data storage) *see* audio-cassettes, data storage on
Cavett, Dick, 136
CD-burning drives, 328, 357–58
CDMA standard, 409
CD-ROMs, 214, 258, 296, 328

cellphones, 398, 522, *522*
   see also iPhones
central processing unit (CPU) chip, 501–2
   see also silicon chip design
Centris family Macs, 256
CERN (European Organization for Nuclear Research), 343
CES, see Consumer Electronics Show
Chahil, Satjiv, 214, 215, 217, 246, 272, 273, 284, 287, 288, 290, 291
Chan, Norman, *493*
Chang, Gareth, 300
Channel Marketing, 355
Chat AV, 382
ChatGPT, 192, 506–8, 510–12, *511*
Chaudhri, Imran, 396, 399, 402
Chemcor glass, 408–9
Chen, Jason, 422
Chiat, Jay, 160
Chiat/Day, 130, 131, 133, 136, 159, 160, 304–9, 326, 329, 344, 359–60, 364, 371, 385
China, 1, 169, 489, 517–19, 537–38
Chinon, 257
CHIPS and Science Act, 504
Chooser desk accessory, 158
Christie, Greg, 404, 467
CHRP, see Common Hardware Reference Platform
Cinema Display, 349
Cingular, 408, 410, 411
Circuit City, 351
Cisco Systems, 412
City Center building, 180, *180*, 247, 264, 281, 309, 467
CL9 company, 164
Claris Corp., 190, 291
   ClarisWorks, 190, 255
Clark, Rocky Raccoon, see Wozniak, Steve
Classic apps, 346

clean energy, 537–38
Clintons, Bill and Hillary, 249, *249–50*
Clipboard desk accessory, 72, 76
clock and clock chip
   Apple III, 51, 57, 58
   Mac desk accessory, 118
clones (Macintosh) see licensing
closed systems, 89, 99, 185, 186, 416, 540
CloseView, 542
Clow, Lee, 130, 131, 134, 160, 304, 305, 306
CNN, 463, 517
Coach, 374
Cocks, Jay, 135
Cocoa apps, 346
CODA (film), 473
codecs (compression-decompression systems), 213
CodeWarrior, 256
Cognac project, 236–37
Coldplay, 460
Coleman, Debi, 95, 107, *137*, 168, 169, *169*, 171, 178
Colette fashion store, 482
Colligan, Bud, 192
Color Math program, 104
Comcast, 472
command key, 70, 121
Commodore, 33–34, 106
   Amiga, 183
   Commodore 64 model, 62, 153
   PET, 28, 34, 41, 42, 48
Common Hardware Reference Platform (CHRP), 262, 275, 278
Compaq, 165, 197, 256, 259, 260, 263, 304, 321, 345, 446, 448
   LTE laptop, 201
component software, 255
CompUSA, 256, 302, 311–12, 351
computational photography, 526

Computer City, 351
Computer Conversor 4000, 14, 17
Computer Equipment Contribution Act, 547
computer industry, collapse of, 106–7, 147, 165, 202
ComputerLand, 143, 302
Conflict Catcher, 274
Connectix QuickCam, 192
Connect service, 469
consumer electronics projects, 246, 248, 276, 302–3, 357, 360–61, 472, 474, 494–96, 521
   see also AirPods; iPads; iPhones; iPods
Consumer Electronics Show (CES), 46, 47, 224–28
Contacts app, iPhone, 399, 465
contact tracing (COVID) app, 455
continuous glucose tracking, 528
controller card, 46
Control Panel, 118
Cook, Bob, 161
Cook, Tim, 1, 393, *463*
   Apple Maps crisis, 466
   on AR glasses, 524
   as CEO, 394, 446–48, 457–58, 463–64, 488
   challenges at Apple, 514
   Chinese manufacturing and sales, 449–51, 489, 518, 519
   on data privacy, 545
   digital services under, 463–74, 489
   on G4 Cube, 350
   on generative AI, 507
   as interim CEO, 394, 435
   iPads under, *439*
   iPhone 6 Plus, 426
   Ive's role, 499
   on Jobs's death, 458, 460, *460*
   Lucent deal, 342

at Macworld Tokyo, 361
as operations head, 303–4, 351, 446–47
personality of Jobs vs., 446–47, *447*
on personalized Siri, 513
product demos/keynotes, 483–85, *484*, 504
public doubts, 463–64
return to office under, 455
on sustainability, 539
Titan car project, 496–97, 500
Vision Pro, 494, 496
cooperative multitasking, 274
Coopers & Lybrand, 269
Copland, Aaron, 274
Copland operating system, 274–79, *275*, 313, 346
CorelDRAW, 268
Corning, 408–9
corporate values, 530–48
  accessibility, 541–44, *542*
  beauty, 31, 38, 95, 102, 530, 532
  closed systems, 89, 99, 185, 186, 416, 540
  education, 44, 128–29, 169, 172, 191–93, 230, 340, 354, 380, 467, 546–47
  excellence in execution, 548
  focus, 38, 531
isolation of project teams, 544
  "make the whole widget" philosophy, 348, 440, 501, 504, 505, 540, 546
  mergers and acquisitions, 165, 233, 250, 252, 262–65, 267, 281, 285–87, 290–92, 358, 390, 397, 428, 497, 499–500, 502, 540
  privacy, 383, 510–11, 544–46
  product quality and time to market, 79, 506–8, 534–35
  RoundRects, 102–3, 533, *533*

secrecy, 65, 100–101, 135, 253, 308, 316, 401–2, 410, 475–76, 531–32, 540
simplicity, 32, 60, 121, 384, 530, 536
small team size, 544
sustainability, 149–50, 453, 511, 536–40
time constraints, 23, 46, 137–38, 225–28, 363–65, 368, 432, 544
Costner, Kevin, 79
Cotton, Katie, 308
Couch, John, 60–61, 78, *78*, 95–97, 193, 547
COVID pandemic, 455, 467, 473, 504, 529
CPU (central processing unit) chip, 501–2
  see also silicon chip design
Craigslist, 470
Cramer Electronics, 22, 23, 25
Cray X-MP/48 supercomputer, 234
Cream Soda computer, 6, *6*, 9, 12
Crichton, Michael, 293
Crisp, Peter, 54, 168, 250, 252, 254, 265, 266
Crow, George, 95, 100–102, *137*, 173
Crow, Sheryl, 369
CSR company, 401
Cue, Eddy, 303, 367–69, 394, *395*, 408, 409, 441, 463, 466, 469, 472, 473, 522, 532
Cuisinart, 29
Culbert, Mike, 224
Cunningham, Andy, 106
Cupertino, CA
  city council, 452
  science fair, 5
  school district, 6
  Stevens Creek Blvd, 37
  factory in, 59
  post office, 143

  see also Bandley Drive office buildings; Infinite Loop; Apple Park
curb-cuts effect, 543
cursor acceleration, 72
CU-SeeMe, 192

Daisy robot, *539*
daisy wheel printers, 154–55
Dalai Lama, 518
Daniels, Bruce, *78*, 84
Dark Mode, 276
DARPA, 155, 423
DAs, *see* desk accessories
Dashboard widgets, 382–83
Data General, 184
Dating Game parody, 132
daughterboard, 53, 57–58
Davidian, Gary, 213, *236*, 236–37, 256
Dawson, Bill, 122
DayStar, 261
De Anza office buildings, 168, 194, 201–2, 212
deepfakes, 509
Dell, 197, 259, 260, 345, 451, 530, 537
Dell, Michael, 314, 459
Deloitte & Touche, 267
De Luca, Guerrino, 291, 301
Denman, Donn, 118
Department of Defense, 14
Department of Justice, 438, 516
Department of Public Health, 545
Desai, Sumbul, 486, 487
design
  beauty, 31, 38, 95, 102, 530, 532
  Espresso design language, 240, 257
  graphical user interface, 60–62, 64–68, 72–74, *74*
  language, 102–3, 148–50, 186, 238–40, 257, 533, *533*

design (*cont.*)
    simplicity, 32, 60, 121, 384, 530, 536
    software consistency, 122
    Snow White design language, 148–50
    see also Brunner, Robert; Ive, Jony; Esslinger, Hartmut; individual product names
desk accessories (DAs), 118, 183, *383*
Desktop, 75–76
desktop publishing, 178–79, 202, 216–17
DESTs (Distinguished Engineers, Scientists, and Technologists), 146
Details screen, iPhone, 404
Detroit Free Press, 382
device independence, 155
diabetes, 528
Dialog Semiconductor, 503
Diamond Rio MP3 player, 359
Diery, Ian, 215
digital cameras, 257, 389, 413, 420, 426–27, 522, 526
Digital Markets Act (DMA), 517
digital services, 463–74, 489
disabilities, people with, 541–44, *542*
Disk II drive, 46–47, *47*, 50, 99, 100
Disney, 458, 472, 474
    Disneyland, 53
Disney, Walt, 457
Distinguished Engineers, Scientists, and Technologists (DESTs), 146
DMA (Digital Markets Act), 517
Dock, Mac OS X, 347
Dolby Labs, 490
Dole, Bob, 547
Doonesbury (comic strip), 229
door detection, 543
DOS, 92, 127, 182, 186, 188, 193 219
dot-com crash, 343, 345, 362, 388
Double Irish arrangement, 515

Drance, Kaiann, 431
Dre, Dr., 469
Dresselhaus, Bill, 77
Drexel University, 129
Drexler, Mickey, 352, 355
Dreyfuss, Richard, 293, 306–7, 460
drivers, device, 322
Dubberly, Hugh, 193
Duke University, 93
Duo Dock, 245–46, *246*
DVDs, 349, 358, 380–82, 386, 389
Dye, Alan, 433, *476*
Dylan, Bob, 139, 224
Dylan language, 224
Dynabook, 435

EarPods, 429, *429*
Easter eggs, 117, 187, 210, 229, *229*, 276, 278
eBay, 223
ECG (electrocardiogram), 486, 528
EDGE network, 417–18
Edison, Thomas, 305, 307
education, 44, 128–29, 169, 172, 191–93, 230, 340, 354, 380, 467, 546–47
EDUCOM, 191–93, *192*
Eight Inc., 312
Eilish, Billie, 390
Einstein, Albert, 120, 305, 307
Eisenstat, Al, 139, 169–71, *251*
electric cars, 496–500
electrocardiogram (ECG), 486, 528
Electronic Data Systems, 267
Elias, John, 396, 397
Ellison, Larry, 292, 300–301, 320, 459
eMac, 379
email
    iCloud, 442
    iPad, 437, 439
    iPhone, 412
    iTools, 343

Macintosh Office server, 159
Mail app, 346, 465
MailDrop (attachments), 535
MobileMe, 441
Scribe, 218–9
Emagic, 390
eMate 300, 230, 320
EMI, 367, 368, 369
emulators, 52–53, 236–37, 392, 505
encryption, 479, 510, 518, 545–6,
Engelbart, Doug, 65, *65*, 66
ENIAC, 12
Ensoniq sound chip, 183
environmental policy, 536–40
    see also sustainability
Epic Games, 516
Epson, 293
Erickson, Tom, 218, 219
Erlicht, Jamie, 473
eSIM, 410
Espinosa, Chris, 16, 37–38, 41, 42, 45, 56, 91, 95, 123, *137*, 467, 548
Espresso design language, 240, 257
Esslinger, Hartmut, 148–50, *149*, 162, 182, 238, 240, 281, 318
Estridge, Don, 104
ET (executive team), 170, 301, 394, 397, 428, 432, 463
European Union, 426, 515, 517
eWorld online service, 253, 270
excellence, 548
expansion slots, 29, 43, 49, 99, 185, 198, 220, 540
Explorers Club, 7
ExxonMobil, 463
EyeSight, 493–94, *494*
eye tracking feature, 542

FAA (Federal Aviation Administration), 439
Facebook, 470, 485, 488, 490, 517, 532, 545

Face ID, 431–34, *433,* 467, 493, 501, 503, 506, 521, 522, 548
FaceTime, 492–93, 508
Fadell, Tony, 223, 361–64, *362,* 372, 375, *395,* 399–402, 476
Fairchild Semiconductor, 35, 39, 43, 168, 266
Fair Labor Association (FLA), 451
FairPlay copy protection, 368–69, 377
fans (air), 29, 51–52, 59, 186, 331, 349, 350
fans (Apple), 136, 150, 250, 252, 265, 287, 311, 315, 356, 458, 497, 530
　Apple II fans, 42, 150
　Apple Music fans, 470
　iPhone fans, 414, 425, 427, , 434, 512
　Mac fans, 143, 145, 198, 202, 344, 348, 351, 382, 385, 392, 503
　Newton fans, 228, 303–4
Fast Company, 520
Fat Mac (Macintosh 512K), 95, 147
FBI (Federal Bureau of Investigation), 545
FCC (Federal Communications Commission), 52, 57, 248, 410
FDA, 486, 487, 525, *527,* 528, 529
Federighi, Craig, 463, 466, 507, 508, 510–13
Fernandez, Bill, 6, *6,* 8–9, 22–24, 45, 49, 56, 95, 110, 112
fiber optics, 491
FileMaker, 190, 346
Filer dialog box, 75
file servers, 158, 159, 163, 166
Final Cut Pro, 390, 440
Finder, 75–76, 114, *114,* 138, 183, 188, 207, 220, 235, 346–48
Find My app, 425, *441,* 442, 479
fingerprint sensor, 421, 431
FingerWorks, 396–97, *397*

Fiorina, Carly, 374
FireWire, 331, 334, 348, 366, 367, 373
Fitbit, 484, 527, 529
Fitch, Johnathan, 182
fitness tracking, 480–82, 527
FLA (Fair Labor Association), 451
flash memory, 374, 377, 378
Flash video format, 440
flat screens, 192, 288, 289, 386–87
Flextronics, 519
Flint Center shareholder meetings
　Apple II vs. Macintosh teams, 163
　Apple Watch demo, 483–85, *484*
　cloning announcement, 259–60
　iMac demo, 325, 327–29, 548
　iMovie demo, 331–34
　Macintosh demo, 138–42, *140,* 325
floppy disk drive, 44–47, 50, 51, 59, 99–100, 125, 145, 186, 243, 322–24
Flyover, 464
focus, as corporate value, 38, 531
folders, 74, *74,* 277
folding phones, 424, 522, *522*
Folon, Jean-Michel, 117
Fontographer, 216
fonts, 11, 63, *63,* 115–16, 155–58, 183
Food and Drug Administration (FDA), 486, 487, 525, *527,* 528, 529
Forbes magazine, 474, 520
Ford, 49, 55, 497
Ford, Henry, 107
Ford, John, 489
Forrester Research, 268
Forstall, Scott, 112, 394, *395,* 400–402, 404, 405, 407, 413, 414, 416–17, 423, *424,* 436, 458, 459, 465, 466
Forsyth, Fred, 291
Fortune magazine, 244, 268, 292, 308
Foster, Norman, 354, 452

Foxconn, 448–51, *449,* 464, 519
France, 165, 184
Franklin Computer, 187
Frankston, Bob, 48
Fremont factory, 106, 107, 163, 167, 200, 206
frogdesign, 148, 150, 238, 316
front camera, 413, 522, 526
Fujifilm, 257
Fuller, Buckminster, 307

G4 Cube, *349,* 349–50, 379
Gable, Jim, 256
Galvin, Robert, 61
Game Center app, 465
Ganatra, Nitin, 411
Gandhi, Mohandas, 120, 305, 307
Gap, 352, 353
GarageBand app, 389–90, *390,* 440
Gassée, Jean-Louis, 162, 169, 170, 173, 177, 182, 184–88, *185,* 197–204, *198,* 206, 219, 223, 234, 238, 252, 271, 280–82, 285, 286
Gates, Bill, 127, 132, 143, 177, 183, 193–95, 255, 279, 283, *310,* 310–11, 385, 458, 459, 463
Gateway, 260, 261, 352, 356, 386
Gemmell, Rob, 148
General Electric (GE), 463
General Magic, 222–23, 362, 477
generative artificial intelligence, 506–13
Genesis MP, 261
Genius Bar, 353, 355, 356, 489
Georgia Institute of Technology, 265
Gerstner, Lou, 263
Geschke, Chuck, 155–56, 459
gesture interpretation, 396, 397, 403, 477, 495, 527, 535
"Get a Mac" ads, 385
Giannandrea, John, 506, 508

Index • 577

Gibbons, Fred, 132
Gilley, Tom, 247
Giugiaro, Giorgetto, 224, 239–40
Gizmodo (blog), 422
global search, 277
Gmail, 506, 517
Goldblum, Jeff, 273, 287
Goldman, Jack, 64
Goldman, Phil, 222, 223
Goldman Sachs, 251, 270–71, 468
Goldstein, Bernard, 250, 252, 254, 300
Goldstein, David, 355
Good Earth building, *37*, 37–39, 43, 68, 87, 89
Google, 427, 455, 459, 463, 470, 475, 479, 512, 516, 528, 532, 540
   Bard, 506
   Chrome, 540
   Docs, 192
   Gemini, 506, 510, 523–24
   Glass, *523*, 523–24
   Gmail, 506, 517
   Maps, 411, 412, 418, 464, 466, 517
   Pixel phone, 513
   Wallet, 468
   see also Android operating system
Gore, Al, 460
Gorilla Glass, 409, 421
Gou, Terry, 448–49
Goyo, Hashiguchi, 120
GPS, 418
graphical user interface, 60–62, 64–68, 72–74, *74*
Graziano, Joe, 235, *251*, 254, 264, 267
"Great Firewall" censorship, 517–18
"Greener Apple, A" (Jobs), 537
greenhouse-gas pollution, 537, 538
Greenpeace, 537
Greenspan, Alan, *249*
Grignon, Andy, 405
Gruber, John, 199, 513

Grubhub, 420
GSM standard, 408, 410
GS/OS, *181*
Guido, Tony, 182
Gypsy word processor, 65
gyroscope, 480, 528, 543

Haggerty, Myra, 1, 303, 395, 431–33
hallucinations, AI, 512–13
Hambrecht & Quist (H&Q), 54, 56
Hancock, Ellen, 278, 283–84, 286, 291
Hancock, Herbie, 153
handheld computers, 218–31
   see also Newton; Swatch handheld
Handoff feature, 443, 501
handwriting recognition software, 218, 228–29, 231
HangulTalk, 216
haptic feedback, 396, 483
hard drives, 51, 59, 77, 100, 321, 332, 357, 360–62, 365, 419
Hawkins, Trip, 67
Hayden, Steven, 130, 131, 159
HDR (high-dynamic range) photography, 526
head gestures, 535
headphone jack, iPhone 7, 426–27
Health app, 486 *see also* health features; medical alerts
health features
   AirPods as hearing aids, 525–27
   Apple Fitness+, 473
   Apple Watch features, 480–2, 486–7
   atrial fibrillation, 486–7
   biometric sensors, 527
   blood pressure tracking, 529
   blood-oxygen (SpO2) sensor, 480, 486–7, 527–8
   ECG (electrocardiogram), 486, 528
   fitness tracking (watch), 480–2
   glucose monitoring, 528

   rings (Apple Watch), 382
   sitting all day, 480
   sleep apnea, detecting, 487
   sleep tracking, 380, 528
   see also heart sensing; medical alerts; microphones
hearing aids, AirPods as, 525–27
heart sensing
   atrial fibrillation, 486–7
   ECG (electrocardiogram), 486, 528
   in AirPods, 431, 527–8
   pulse measurements (PPG), 480, 483, 486
   problem detection with Apple Watch, 487, 529
Heinen, Nancy, 301, 338, 345
Heinen, Roger, 235
Heller, Dave, 359
Henson, Jim, 128, 306, *307*
Hermès, 482
Herrlinger, Sarah, 543, 544
Hertzfeld, Andy, 88, 90, 92–96, 105, 109, 110, 112, 114–17, 119, 122, 136, *137*, 139, 144, 145, 147, 152, 171, 195, 207, 222, 261, 325, 411, 459
Hewlett, Bill, 7
Hewlett-Packard, *see* HP
Hierarchical File System (HFS), 207
Higa, James, 368–69
high-dynamic range (HDR) photography, 526
Hillman, Dan, 181
HinduTalk, 216
Hines, Gregory, 287, 293
Hintz, Jürgen, 254
Hitachi, 61, 317
Hoddie, Peter, 212, 213, 544
Hodgman, John, 385
Hoenig, David, 323
Hoffman, Joanna, 87–88, 91, 92, 95, 112, 122, *137*, 144, 145, 161, 222

Hold call button, 508
Holland, Sam, 234
Holt, Joe, 332
Holt, Rod, 27, 29, 43, 45, 46, 69, 92, 93, 95, 100, 101, 110, 112, 146
Homebrew Computer Club, 12, 15–17, *16,* 21, 24, 25, 30, 39, 85
Home button, *403,* 404, 407, 421, 431, 433, 437, 467
Home PNA, 340
HomePod, 474, 503, 531, 533
Honolulu, Hawaii, 132
Horn, Bruce, 95, 96, 114, 122, 141, 145
Hosiden company, 201
Hotelling, Steve, 481
Hovey, Dean, 52, 72
Howard, Brian, 85–88, *137,* 146
HP (Hewlett-Packard), 263, 293, 374, 451
    Apple employees who worked for, 7, 11, 13–15, 17–19, 24, 30, 36–37, 60, 61, 80, 121, 125, 184, 219
    branded stores, 352
    industry collapses for, 106, 345
    laptops, 199
    LaserJet printer, 157
    RISC chips from, 282
    softkeys, 61
    Sony floppy drive, 100
    video prototyping, 193
HTC, 418
Huawei phones, *522*
Hubbard, Elbert, 544
Hudson, Katherine, 254, 300
HyperCard software, 180, 189, 191
HyperTalk, 189
hypertension detection, 487, 529

IBM, 104, 281, 312, 451
    in AIM alliance, 232–36, 262

Apple computers vs. PCs, 84, 97, 101, 126, 140, 141, 184, 202, 330, 386, 546
Apple employees and board members from, 278, 301, 304, 446, 447
    as Apple rival, 82–84, 94–97, 107, 130, 139, 154, 165, 305
    branded stores, 352
    floppy drives, 46
    laptops from, *241*
    Mac features in software for, 193–96
    Macintosh Office for PCs, 158–60, 163, 165
    Microdrive, 360–61
    in OpenDoc partnership, 255
    PC clones, 259, 260
    PCjr, 153
    Pictureworld, 75
    Power PC G5 chips, 391, 501
    proposed mergers with Apple, 250, 252, 262–63
    software developers for PCs, 132, 177, 196 (*see also* Windows)
    in USB coalition, 321
    Xerox and, 139, 193–94
iBook laptop, 337–40, 342, 358, 381–82, *382,* 397, 399
iBooks app, 437, 438, 442, 465
iCards, 343
iChat AV, 346, 382, 383
iCloud, 442, 445, 457, 471, 474, 544
    iCloud Drive, 343
icons, 65, 74, 77, 91, *115,* 115–16, 532
iDisk, 343
iDVD, 346, 358, 388, 390
IEEE (Institute of Electrical and Electronics Engineers), 340, 342
iFixit, 539
iGesture NumPad, 396–97, *397*
Ihnatko, Andy, 438

Ikepod company, 475
iLife suite, 390, 391
iMac, 351, *351,* 358, 530
    C1 project, 1, 109, 319–24, *320, 327, 327,* 334, 350, 380, 412, 548
    consumer demand for, 329–30
    demo and introduction of, 323–24, 548
    design of, 319–21
    Foxconn as manufacturer of, 448, 449
    handle, 324
    hardware, 321–22
    iMac DV, 331–34
    iMac G4, 350, 384, *386,* 386–87, 391
    iMac G5, *387,* 387–88, 391
    with Intel, 392
    marketing, 329, 333
    naming, 322–23
    product quadrant for, 325, 327–29
    sales, 333, 485
    second and third-generation, 330–31
    slot-loading CD-ROM drive, 328
Image Playground, 507, 510
ImageWriter, *154,* 154–55
    ImageWriter II, 150, *150*
"I'm a Mac/I'm a PC" ads, 385
iMovie, 111, *331,* 331–34, 348, 357, 358, 388–90, 412, 422, 440, 546
    iMovie 2 and 3, 346
in-app purchases, 515
Independence Day (film), 273
India, 1, 11, 519
Industrial Design magazine, 239, *239*
Industrial Light & Magic, 197
inertial scrolling, 219, 399
Infinite Loop offices, 244–45, 301, 303, 338, 359, 392, 397, 401, 406, 453, 460, 475
InfoWorld magazine, 153, 388

infrared technology, 220, 224, 433, *433*
initial public offering (IPO), Apple, 53–57, *55*
injection molding, 40
Instacart, 420
Instagram, 420, 517, 540, 543
Instant Hotspot feature, 443
Institute of Electrical and Electronics Engineers (IEEE), 340, 342
Institutional Venture Partners, 34
integrated video, 198
Integrated Woz Machine (IWM), 100
Intel, 1, 31, 35, 168, 232, 254, 282, 321, 501, 503–5
   8080 microprocessor, 12, 13
   chip speed claims, 326
   Core 2 Duo processor, 419
   Mac chips made by, 391–92
   Pentium chip, 256, 326, 327
   Skylake processor, 503
   Star Trek project, 235, 391, 544
   wireless division, 546
intellectual property rights, 196, 259
Intelligent Electronics, 446
Interactive Television Box project, 246, 248
Interface Age, 31
International Typeface Corporation, 157
internet, *see* email; iCloud; iDisk; iTools; Safari; World Wide Web
Internet Explorer, 310, 311
Interpress, 155
Intrinsity, 503
Intuit, 301
Inventec, 364
io company, 499
iOS operating system, 401, 466
   development of, 401
   features of, in Macs, 442–43
   Flash for, 440

generative AI features, 507–8
iOS 4, 544
iOS 6, 464, 465, 533
iOS 7, 465, *533*, 534
   for iPhone 4, 422
   and OS 26, 443–44
   RoundRects, 533
   sales, 514
   update cycle, 421
Iovine, Jimmy, 469
iPad, 458, 463, 496, 503, 531
   accessibility features, 542
   Apple News, 470–71
   apps, 436, 470–71, 495
   Chinese sales, 489
   development, 2, 38, 192, 283, 304, 399, 426, 435–40
   Flash video, 440
   gesture interpretation, 477
   iOS 26, 443, 444
   iPad Air, Mini, and Pro *439*, *439*, 463
   keynote demo, 436–40
   manufacturing, 519
   photo syncing for, 445
   prototypes, 435–36
   RoundRects in, 533
   sales of, 485, 486, 514
   scrolling direction, 442
   software for Macs and, 525
iPhone, 429, 430, 440, 442, 458, 466, 503, 530
   accessibility features, 542–43
   AirPower, 513
   AirTag safeguards, 479
   Android competitors, 418
   Apple designed chips for, 503, 504
   Apple Intelligence, 507–10
   and Apple Watch, 477, 478
   apps, *414*, 415–17, *417*, 436, 455, 486, 514–15
   camera, 413, 522, 526

Chinese sales, 489
computational photography, 526
consumer demand and reviews, *413*, 413–15
cultural impact, 420
data privacy and, 546
"death grip," 422–23
design, 402–3, 405–7
development of, 2, 38, 192, 196, 283, 304, 356, 363, 379, 395, 398–415, *411*, *414*, 438
digital addiction, 499
disassembling, *539*
Foxconn manufacturing, *449*, 451
future of, 521–23
gesture interpretation, 477
glass screen, 408–9
Health app, 486
iOS 26 , 443
iPhone 3G, 417–18, 420, 435
iPhone 3GS, 421
iPhone 4, *421*, 421–23, 503, 526
iPhone 4S, 423–25, 458, 489
iPhone 5, 425, 463
iPhone 5C, 545
iPhone 5S, 421
iPhone 6, 425, 426, *434*, 489
iPhone 6 Plus, 425, 426
iPhone 7, 426–27
iPhone X, 431–34, *434*, 497
iPhone 16, 539
iPhone 16 Pro, *434*
iPhone 17 Pro, *434*
iPhone Air, 434, *434*, 522
jailbreaking, 416
Jobs keynote demo, 411–13
mapping software, 464–66
mirroring, 535
multitouch screen, 399–400
naming, 412
network for, 408–10
Neural Engine, 506

NFC chip, 468, 516, 517
notch, 522
on-screen keyboard, 403–6
operating system, 401
P1 project, 398, 400
personalized Siri, 511–13
photo syncing, 445
project team, 497, 544
RoundRects in, 533, *533*
sales, 1, 485, 486, 514
secrecy, 401–2, 410
silicon design for, 501, 502
skeuomorphism, 465
"Slide to unlock," 407
SoCs for, 502
Taptic Engine, 483
tariffs on, 518
update cycle, 421
USB-C charging, 426
Wallaby rig, 402, *403*
iPhoto, *388*, 388–90, 443, 445
IPO (initial public offering), Apple, 53–57, *55*
iPod, 372, 401, 417, 435, 530
advertising for, 371
competing products, 376
design of, 365
effect of, at Apple, 378–79
on Flash, 440
hard drive, 360–61, 365
and iPhone, 406, *406*, 412, 415
development of, 1, 2, 223, 360–79, *364*, 412, 438
iPod earbuds, 428–29
iPod mini, 373, 374, 375
iPod nano, 375, *375*, 377, 378, 449, 450, 463–64, 478
iPod photo, 374
iPod shuffle, 374–78, 376
iPod Touch, 356, 430, 436, 463, 503
music piracy and, 364–65

P1 project, 398, 400
project team, 361–63, 497
race to develop, 363–65
reveal of, 366
RoundRects, 533
sales, 485, 486
scroll wheel, 363
subsequent generations, 372–78
U2 Special Edition, 377
update cycle, 421
video iPod, 399
for Windows, 367, 369, 371, 373, 420, 467
Ireland, 515
iReview, 343
Irvine, Joseph, 161
Isaacson, Walter, 314, 394
Ishida, Bob, 220
iSight webcam, 382, 383
iTools, 343, 348
iTunes, 1, 412
2012 redesign of, 464
Autofill button in, 375
force-downloading U2 album, 485
in iLife suite, 390
iMac release, 328
iPod syncing, 365
iTunes 1.0, 359
iTunes 2, 389
iTunes U, 547
for Mac OS X, 346, 358–60
podcasts in, 370
project team, 274, 364, 388, 469
for Windows, 367, 369, 371, 373, 420, 467
iTunes Store, 18, 367–70, *370*, 375, 377, 378, 398, 469, 472, 485, 532, 544
iTV, 472
Ive, Jony, 316–19, *317*, *318*, 330
2000s Mac redesign, 380, 381
after Apple, 499

AirPort design, *340*
Apple Park design, 452, 456
Apple Watch design, 475–78, 482
chief design officer, 499
Cook and, 463
on executive team, 301, 394, *395*
iMac projects, 320–24, 328, 330, 386, 387
iPad design, 436
iPhone projects, *396*, 397, 401, 406, *406*, 407, 409, 410, 414, 421, 432
iPod projects, 364, 365, 374, 377
Jobs and, 112, 318–19, 328, 532
as lead designer for software, 465, 466
on loss of Jobs, 458, 460
MacBook Pro design, 450
Mac G4 Cube design, 349, 350
MessagePad 120 design, 231, 317
RoundRects for, 533, *533*
Titan project for, 497–98
Twentieth Anniversary Mac design, 287–89
Ivester, Gavin, 240, 242
IWM (Integrated Woz Machine), 100
iWoz (memoir), 164

Jackson, Janet, 216
Jackson, Lisa, 538–40
Jackson, Michael, 128
Jagger, Mick, 128
Jaguar Land Rover, 497, 498
Jaguar project, 233, *233*, 236–37
Janoff, Robert, 32–33
Japan, 120, 214–16, 218, 220, 270, 276, 361, 454, 532
JavaScript, 191, 319, 440
Joaquin, James, 226, 229
Jobs (film), 23, 142
Jobs, Clara, 6, 23
Jobs, Erin and Eve, 393

Jobs, Laurene, 285, 386, 393, 397, 414, 436, 458
Jobs, Patti, 6, 23
Jobs, Paul, 6, 23, 102
Jobs, Reed, 393, 436
Jobs, Steve, 7, 16, 19, 20, 22, 32, 63, 108, 152, 335, 395
  30th birthday party, 162
  1997 Macworld Expo, 289–90, *290*
  on advertising, 131, 133, 134, 304–8
  Amelio and, 272, 294
  on Android, 418
  Apple I assembly by, 23–24
  on Apple II design, 27, 29–31
  Apple after departure of, 177–78, 186, 245, 295–96, 426, 477
  on Apple board, 166, 171–73, 252, 300–301, 309, 457–58
  Apple CEO search by, 104, 105
  Apple corporate values, 501, 531–33, 540, 544, 546
  Apple IPO, 54–57, *55*
  Apple lawsuit against, 172–74
  Apple Park project, 452, 456
  Apple products after death of, 520–21
  on Apple products in schools, 44
  Apple Store concept, 352, 354–56
  badge number, 45
  butt-dialing problem, 407
  Byte Shop order, 21–22, 24
  cancer, 392–94, 435, 457, 467
  on CD-burning, 357
  as CEO, 2, 348–49, 391, 457–58
  childhood and adolescence, 6–8
  Congressional lobbying by, 547
  Cook as CEO vs., 394, 463, 488
  on creativity, 113–14
  on culture at Apple, 301–2
  death of and memorials, 458–60, *460*
  democratization of technology by, 530
  departures from Apple, 1, 2, 172–74, 203, 457
  design as passion, 238, 320–21, 354, 379, 406, 428, 436, 532, 533
  disobeying, 100–101, 416–17
  as division head, 144, 147, 148, 154, 156, 166–71, 184
  Easter-egg ban, 187–88
  end of licensing/clones, 312–13
  financing deals of, 25, 33–34, 36
  on Flash, 440
  founding of Apple, 15–20
  founding of NeXT, 281
  on Foxconn, 451
  Fremont factory, 107
  French deal, 165
  Gassée and, 184
  Gates vs., 127
  on Google Maps, 464
  "A Greener Apple," 537
  iMac project for, 322–24, 328
  on iMovie, 357
  as interim CEO, 313–14, 334, 344, 345
  on iOS, 401
  iPad project, 436, 470–71
  iPhone projects r, 398, 399, 401, 403–4, 408–10, 414, 416, 417, 421, 422
  iPod project for, 361–65, 367, 371, 372, 374, 375, 379
  on iTunes, 359, 360
  iTunes Store concept, 368–70
  Ive's collaborations, *318*, 318–21, 324, 475
  Japanese influence on, 120
  journalists and, 111, 135, 308
  keynote speeches and product demos, 78, 138–42, 152, 159, 309–11, 325, 327–29, 332, 335–43, 347–49, 357, 358, 361, 366, 371, 380, 392, 402, 411–13, 419, 420, 422–23, 435–40, 442, 457, 472, 541
  kindness of, 103, 459
  Lisa naming, 60
  on logo, 33
  in *Dating Game* parody, 132
  Macintosh Office for, 157–60
  as Macintosh project leader, 86, 89–98, 102, 115, 116, 122, 123, 125, 128, 136–38, 143, 190, 204
  Mac OS X, 382
  marketing stunts, 333
  Markkula and, 36, 38
  McKenna and, 31–32
  MobileMe project for, 440, 441
  multitouch demo for, 397
  NeXT era for, 281–86
  personality of Cook vs., 446–47, *447*
  at Pixar, 211
  planned coup at Apple, 169–71
  pre-Apple projects with Wozniak, 2, 9–10, 14, 28
  product downsizing, 230, 231, 255, 257, 302–4, 322, 325
  Raskin and, 85, 89, 91–92
  reality distortion field, 94, 123, 135, 284, 311, 369, 409, 446, 448, 463
  at Reed College, 11
  reorganization under, 312
  replacement of Apple's board by, 300–301
  Scotty and, 39
  Sculley and, 2, 106–8, 145, 155, 162, 163, 165–74

signature in Mac case, 95
silicon chip design project for, 501, 502
on skeuomorphism, 465
Stanford commencement speech, 393
at Steven's Creek office, 37–38
"sunflower" iMac, 386
supplier negotiations, 46–47, 61, 152, 193
on tablet computers, 397, 435–36, 437
on tech bubble and dot-com crash, 345, 388–90
"Think Different" ad narration, 306, 460
throughlines originating with, 1, 2
on U2, 485
as unpaid advisor at Apple, 1, 262, 283–87, 290–92, 294–96, 299, 392
on VisiCalc, 48
vision of, 177–78, 192, 295, 488, 520
volatility of, 108–12, 332
as VP of new product development, 50, 51, 53, 60–62, 69
watches opinion, 475
at West Coast Computer Faire, 42
Willy Wonka stunt, 333
Wozniak's first meeting with, 9
Woz's Zaltair prank, 41, 162
Xerox PARC visits, 64–67
Yocam and, 49
Johnson, Ron, 112, 351–56
Johnson & Johnson, 474
Jonathan prototype, 182
Jordan, Richard, 238
Joswiak, Greg, 296, 341, 367, 374, 377, 398, 419, 525
Juggernaut designs, 317
junk-call screening, 508

K48 project *see* iPad
Kahng, Stephen, 260
Kaleida Labs, 233, 236
Kamoto, Hidetoshi, 101
Kamradt, Alex, 14
KanjiTalk, 214–16, *216*
Kaplan, Jerry, 219
Kapor, Mitch, 132, 219
Kare, Susan, 97, 112, 114–16, 120, 121, 141, 145, 155, 187, 222
Kawasaki, Guy, 94, 110, 125–27, 142, 146, *146*, 271
Kay, Alan, 146, *146*, 147, 435
Keenan, Joe, 33
Kennedy, John F., 306
Kennedy, Kathleen, 293
Kenyon, Larry, 95, 101, 123
kernel, 207, 262, 275, 283, 346
Kerr, Duncan, 349, 379, 395–97, *396*, 399, 406, 477–78, 483, 548
keyboard, 60, 88, 186, 241, 400, 403–6, *404*
keyboard shortcuts, 70
Key Caps app, 118
Keynote program, 335
keynote speeches and product demos, 78, 138–42, 152, 159, 309–11, 325, 327–29, 332, 335–43, 347–49, 357, 358, 361, 366, 371, 380, 392, 402, 411–13, 419, 420, 422–23, 435–40, 442, 457, 472, 541
KidSafe tool, 343
"Kids Can't Wait" program, 546, 547
killer app, 48
Kincaid, Bill, 359
Kindle app, 516
King, Martin Luther, Jr., 307, *307*
Kitchen, Steve, 167
Kleiner Perkins, 34
Knaster, Scott, 122, 134, 568
"Knowledge Navigator" video, 191–93, *192*, 435

Kocienda, Ken, 109, 403, *404*, 405
Kodak, 257
Kossow, Al, 211
Kottke, Daniel, 11, 23, 56, 92, 95
Krakower, Jon, 241–42, *242*
Kubrick, Stanley, 344
Kuehler, Jack, 233
Kvamme, Floyd, 104

Lady Gaga, 456
Lam, Brian, 422
Lamiraux, Henri, 403
Landi, Marco, 291
landing areas, iPhone keyboard, *404*, 405
laptops, 151, 199–201, 230, 238–48, 327, 337–40, 380, 426, 442, 525
see also iBook; MacBooks; PowerBooks
LaserWriter, 154–59, *157*, 166, 173, 178, 179
LaserWriter II, 150
LaserWriter NTX, 216
Lasseter, John, 211, 459
latency, 429, 491
LCD screens, 200–1, 243, 263,
Leak, Bruce, 212, 213, *213*, 222, 223
"Lemmings" ad, *159*, 159–60, 163
Lennon, John, 56, 306, 307
Lennon, Sean, 128
Levinson, Art, 355
Levy, Steven, 187, 330
Lewin, Dan'l, 128–29, 173
Lewis, Delano, 254, 300
Lexus, *498*
LG, 448, 449, 475
licensing, 177, 195, 202–3, 214, 222, 227, 228, 232, 259–61, 276, 299, 312–13, 363, 500
Licht, Hildy, 49
lidar sensors, 498
Lightning connector, *425*, 425–27

Index • 583

Lightroom app, 445
LimeWire website, 357
Lin, Maya, 128
links, in-app, 516
Linux, 401
Liquid Glass interface, 444, *444*
Lisa, 2, 60–80, *77*
    advertising, *78*, 79, 106
    Apple III technology in, 59
    bitmapped graphics, 62–64
    Copland project, 274, 277
    debut, 76–79
    desktop, 75–76
    discontinuation and landfill, 161
    disk drive, 99–100
    failure, 84
    graphical user interface, 60–62, 68, 72–74, *74*
    keyboard shortcuts, 70
    lack of operational modes, 71
    Lisa 2/5, 161
    Lisa division, 68, 89, 94–98, 118
    Macintosh vs., 86, 91, 147
    manual, 123
    menus, 71–72
    mouse, 69, 72
    name, 60
    processor, 60–61
    project team, 80, 113–15, 118
    sales, 84
    user testing, 69–70
    writing Mac software on, 126
live listen feature, 542
Livingston, Sherry, 45
Lockheed, 5
Logic Pro software, 390, 440
Logitech, 247
logos
    Apple, *20*, 20–21, 32–33, 42, 339
    NeXT, 281
    QuickTime, 214
    PowerBook upside-down, 381

Loma Prieta, CA, 1989 earthquake, 194
Long, Justin, 385
loops, musical, 389
Los Altos, CA, 6–7, *7*, 11
Los Angeles Times, 142, 158, 437
Lotus, 132, 219
LoveFrom, 499
Lucasfilm, 112
Lucent, 342
Ludolph, Frank, 75, 76
Lunar Design, 238, 316
Lynch, Kevin, 222, 223, 477–79, 500

M-series processor chips, 503–5
Ma, Yo-Yo, 459
MacBook Air, 419, 464, 505, 530, *532*
MacBooks
    MacBook Air, 419, 464, 505, 530, *532*
    MacBook Pro, 392, 450, 464, 534
    see also PowerBooks
MacDraw, 121, 190
Mach kernel, 283, 299, 347
MacinTalk, 141
Macintosh (original), 85–103, *141, 142,* 256, 364
    accessibility features, 541–42
    advertising, 106, 129–31, *133,* 133–34, 136, 143
    Alice game for, 115, 141
    Apple III technology in, 59
    Bicycle name, 93
    case, 95, *98,* 98–99, *99*
    command key, 121
    demos for influencers, 105, 127–28
    desktop publishing system, 179
    disobedience of team, 99–101
    Flint Center unveiling, 137–42, 541
    floppy drive (Twiggy), 99–101
    Fremont factory, 107

    IBM PC vs., 126
    icons and fonts, 11, 115–16
    Jobs's takeover of project, 89–92
    keyboard shortcuts, 121
    Lisa vs., 147
    MacPaint program, 118–21
    manuals, 123–24, *125*
    marketing, 92, 93, 106, 125–36, 168
    memory constraints, 116–18
    Monkey DA debugging tool, 119
    original concept, 86
    original Jobs team, 92–96, 113–5
    original Raskin team, 87–9
    PC board aesthetics, 102
    price constraints, 89, 91, 96–98, 116–18, 136
    project team, 85–89, 92–97, 113–16, 137–38, 144–45, 411, 544
    QuickDraw algorithms for, 102–3
    Raskin's vision, 90
    release, 137–42, 541
    reviews, 142–44
    sales, 128–29, 143, 145, 147
    software for, 113–27, 132, 137–38, 184
    spelling of name, 86
    startup time, 123
Macintosh II, 150, 182, 187–89, 197, 205, 211, 219, 256
    Mac IIci, 198, 237
    Mac IIcx, 198, *198*
    Mac IIfx, 150, 197, 204, 226
    Mac IIgs, 219
    Mac IIsi, 205, *205,* 209, 256
    Mac IIx, 150, 197
Macintosh 512K (Fat Mac), 95, 147
Macintosh Classic, 95, 204–6, *205*
Macintosh computers, 531
    in Apple Stores, 353
    apps, 118, 337, 346

assembly, 448
CD-burning drives, 357–58
Centris series, 256
clones of, 259–61, *260*, 313
as digital hubs, 357, 389
emulator program, 236–37
eWorld, 270
future innovations, 525
with Intel chips, 391–92
iOS features, 442–43
iPods and, 365–67
iTunes Store for, 369
Japanese characters, 215
Lisa features, 77
M1 chip, 503–5
Macintosh division, 92, 144–45, 163, 166–71, 183, 184
manufacturing, 519
number of models, 274, 299
Performa series, 256, 258, 268, 292, 302, 321
photo syncing for, 445
processors in, 61, 249, 252
product quadrant for, 302–3
Quadra series, 256, 258
recording features on, 18
running Windows on, 391
signed cases, 95
software, 267–68
store-within-a-store, 311–12
"Switch" ad campaign, 385
syncing mobile devices and, 442
Windows on, 279, 384, 391, 505
see also specific models
Macintosh LC, 204–5, *205*, 209, 238, 240, 256, 258, 316
Macintosh Office, 154–60, 163, 179
Macintosh operating systems (Mac OS), 97, 98, 262, 267
  Copland project, 274–79, *275*, 313, 346
  Easter eggs, 277

for Intel chips, 235
KanjiTalk, 215
licensing (clones), 202–3, 259–61
Mac OS 8, 278, 309, 313, 347
Mac OS 8.5, 278
Mac OS 9, 278, 336–37
Mac OS X, 162, 336, 336–37, 345–49, 382–84, 391, 397, 400, 401, 466, 533, 547
Mac OS 10.1 "Puma," 382
Mac OS 10.2 "Jaguar," 382
Mac OS 10.3 "Panther," 382
Mac OS 10.4 "Tiger," 382–83, *383*, 392
Mac OS 10.5 "Leopard," 384
Mac OS 10.6 "Snow Leopard," 384
Mac OS 10.7 "Lion," 214, 442, 443
Mac OS 10.8 "Mountain Lion," 443, 464, 465
Mac OS 10.9 "Mavericks," 443
Mac OS 10.10 "Yosemite," 443, *443*
macOS 26, 443–44
Star Trek project, 235, 391, 544
System 1.0, 140
System 2, 140, 207
System 3, 140, 207
System 5, 207
System 7, 140, *208*, 208–10, 217, 220, 277–8, 313
Windows vs., 193–96, 208, 383
Macintosh Plus, 90, 95, 185–86, 188, 219, 256
Macintosh Portable, 150, 199–201, 204, 223, 241, 242
Macintosh SE, 95, 150, 186–87, 189, 201, 219, 256
Mac SE/30, 95, 150, 199, 204
Macintosh User Interface Guidelines, 122
Macintosh XL, 161
Mac-Like Things, 246, 248
"Macmaker" (song), 286

Mac mini, 350, *350*, 533
Mac OS, *see* Macintosh operating systems
MacPaint program, 118–21, 138, 141, 189, 190, 211, 220
Mac Pro, 464, 519, *519*
Mac Pro Server, 464
Macromedia Director program, 399, 402, 477
MacWorks, 190
Macworld Expo presentations
  1995, Spindler's absence, 264
  1997, Amelio and Jobs, 286–90, 309–11, *310*
  2009, Jobs's absence, 435
  Apple's withdrawal from, 339
  Be, Inc. attendance, 281
  G4 Cube, 349
  HyperCard, 189
  iBook, 337–40, 342
  iLife suite, 390
  iMacs, 330, 351, 380
  iPhone, 402, 408, 410–13, 420
  iTools, 343
  KanjiTalk, in Tokyo, 216, 361
  Mac as digital hub, 357
  MacBook Air, 419
  Mac IIcx, 198
  Mac OS X, 347–49
  Newton, 227
  SuperDrive, 358, 380
  "one more thing," 152, 314, 331, 339, 348
  see also keynote speeches and product demos
Macworld magazine, 128, *128*, 206, 245, 261, 349, 383
MacWrite, 119, 138, 141, 145, 190
Magic Cap, 222
Magic Keyboard, 428
Magic Mouse, 428
Magnifier app, 543

Index • 585

MagSafe power cord, 419, 425, 534
Mail app, 346, 465
Mail Drop feature, 535
Major, Anya, 130
"make the whole widget" philosophy, 348, 440, 501, 504, 505, 540, 546
Mandich, Mitch, 312
Manock, Jerry, *30*, 30–31, 40, 51–52, 69, 77, 92, 94, 95, *98*, 98–99, 112, *137*, 142, 148, 329
Mansfield, Bob, 394, 463, 477–79, 501
Maps app, 464–66, *466*
MapQuest, 466
Mariani buildings, 185, 194, 212, 244, 245, 490
marketing, 31–32, 38, 296, 341
    Apple I, 15–17, 23
    Apple II, 39–42, 168
    Apple IIc, 153
    Apple Watch, 499
    iMac, 329, 333
    Macintosh (original), 92, 93, 106, 125–36, 168
    Macintosh computers, 125–36, 206
    Macintosh Office, 179
    Newton, 223
    at NeXT Computer, 299
    at PepsiCo, 105
    see also advertising
Markkula, Armas Clifford, Jr. "Mike" (ACM), *35*, 35–36, 325, 341
    Amelio and, 266, 272
    Apple II and, 42–46, 104
    Apple III project, 58
    on Apple–IBM merger, 262, 263
    Apple incorporation by, 34, 37–39, 53
    badge number, 45
    on CEO search committee, 312
    as chairman of board, 54, 168, 252, 254, 264, 267, 300
    as interim CEO, 82
    IPO for, 53–54, *55*, 57
    Macintosh project, 86, 88–89, 91, 92, 97
    on managing succession, 295
    on "1984" ad, 133
    restructuring by, 80
    Sculley and, 104–6, 171–73, 250, 252
    Spindler and, 204, 265
    on West Coast Computer Faire, 42
    Woz's decision to leave HP, 36–37
Markoff, John, 271
Martindale, Jim, 45
Marvell, 401
Massaro, Don, 46–47
Matsushita, 222
Mayer, John, 390
Mazda, 498
McCammon, Mary Ellen, *137*
McCormack, Jon, 526
McIntosh Laboratory, 86
McKenna, Regis, 31–32, *32*, 38, 106, 127, 136, 170
McKinsey & Company, 291, 311
McLaren, 498
McNealy, Scott, 264–65
Media Arts Lab, 385
Medica, John, 122, 201
medical alerts, 486–87, *487*, 527, 529
Mega II chip, 181, 183
Memoji, 492, *492*
memory
    flash, 374, 377, 378
    in original Macintosh, 116–18
    protected, 77, 207, 275, 276, 348
    RAM, 21, 97, 116–18
    virtual, 210
menus, Lisa, 71–72, 76
Mercedes, 499
Mercer, Paul, 194, 207, 220, 364
mergers and acquisitions, 165, 233, 250, 252, 262–65, 267, 281, 285–87, 290–92, 358, 390, 397, 428, 497, 499–500, 502, 540
MessagePads, 229, 231, 250
    MessagePad 120, 231, 317
    MessagePad 2000, 303
Messages app, 492, 508, 511, 546
Meta, 1, 496, 506, 516, 517, 523, 524
    Meta AI, 506
    Meta+Ray-Ban smart glasses, 523, *523*
Met Gala (2016), 482
Metropolitan Museum of Art, 482
Metrowerks, 256
Meyerhöffer, Thomas, 230
microphones
    AirPods, 428
    Apple I, 26
    Apple Watch, 483, 527
    Beatles lawsuit, 18
    GarageBand, 389
    iPods, 373, 377
    Macintosh LC, 209
    PowerBook, 244
    smart glasses, 523
    Vision Pro, 494
Microsoft, 290, 414, 451, 458, 488
    ActiveSync, 418
    after Gates's departure, 463
    Amelio's deal with, 310–11, 313
    Apple rivalry, 94, 126–27, 193–96, 202, 210, 383, 513
    Applesoft BASIC, 81, 194, 253
    Apple software developed by, 127, 132, 141, 183
    Apple v. Microsoft, 195–96, 259
    Chart, 127
    Copilot and AI, 506, 513, 517
    culture, 532
    Excel, 48, 127, 186, 194
    File, 127

"I'm a PC" ads, 385
licensing model, 202–3, 259
mergers and acquisitions, 483, 540
Multiplan, 141
netcast/podcast debate, 370
Office suite, 6, 255, 268, 310, 505
OLE system, 255
and Pentium chip, 256
PlaysForSure, 376
PowerPoint, 194
and Switcher, 152
tablet computer with stylus, 397, 435
TrueImage, 217
USB for, 322
Word, 194
Xbox Kinect, 431
Zune, 376
see also Windows operating system
Milledge, Vicki, *137*
Miller, Andy, 114
minicomputers, 12
MIPS, 233
Mission: Impossible (film), 273, 274, 348
Mr. Macintosh (character), 117
MITS, 12, 41
  Altair 8800, 12, 15, 21
Mitterrand, François, 165
MobileMe service, 2, 440–42
Mobile PC, 244
modes, 71
Moffatt, Jeff, 14
Money magazine, 142
Monkey desk accessory, 119
MonkeyLives switch, 119
monospaced type, *63*
Monroy, Bert, 341
Montalvo, Carlos, 215–17
Montgomery Ward, 351
Morgan Stanley, 56
Moritz, Michael, 135

Morning Hair (print), 120
Mossberg, Walt, 329, 544
MOS Technology, 13, *13*, 39, 501
  6502 chip, 14, 21
Motorola, 222, 228, 232, 236, 244, 260–62, 313, 392, 398, *398*, 413, 501
  6800 microprocessor, 13
  6809 chip, 86
  68000 chip, 60–61, 76, 86, 91, 96, 136, 156, 220
  68020 chip, 188
  68030 chip, 182, 197, 199
  68040 chip, 256
  in AIM alliance, 232–36, 262
  Mobility, 540
  PowerPC G4 chip, 391
  RAZR phone, 398
  ROKR E1 phone, 398–99
Mountain View, 40, 41
mouse, 65, 69, 72, 76, 91, 99, 105, 121, 124, 183, 200, 329, 334, 396
Mouse Keys feature, 541
Mousing Around app, 125, *125*
movement trackers, 481–82
MP3 format and MP3 players, 357, 360–61, 372, 376
MPEG-4 format, 214
Muffly, Dave, 453
MultiFinder, 207, *207*
multitasking, 274, 278
multithreading, 275
multitouch technology, 395–97, *396*, 399–400, 436
Munsell system, 40
Murray, Mike, 95, 96, 125, 126, 128, 131–34, 160, 168, *169*, 171, 172, 203, 252, 459
MusicMatch Jukebox, 358, 367
MusicNet, 368, 369
music piracy, 364–65

Musk, Elon, 497
MVNOs (mobile virtual network operators), 408

Nagel, Dave, 210, *251*, 255, 278
Napster, 357
Nasdaq, 343, 345
Nassi, Ike, 258, 283, 296
National Computer Conference (NCC), 53
National Semiconductor, 31, 38, 57, 254, 266, 278
NCR (National Cash Register Co.), 340
NCs, *see* network computers
near-field communication (NFC) chip, 468, 516, 517
NEC company, 215
Nest, 540
Netflix, 1, 515
NetWare, 235
network computers (NCs), 319–21, 328, 332
Neuenschwander, Erik, 544, 546
Neural Engine, 506
Neuralink, 522
Newson, Marc, 475–77, 497, 499
Newsweek magazine, 134, 144, 308
"New to Macs" course, 354
Newton, 219–31, *221*, 228
  ARM's chip for, 227, 271
  CES demo, 224–28
  decision to kill, 294, 303–4
  design, 316, 317
  eMate 300, 230
  handwriting recognition, 180, 228–29
  infrared communication, 224
  launch, 147
  manufacturing, 364
  MessagePads, 229, 231, 250, 303, 317

Newton (cont.)
    myths, 2
    Newton, Inc. spinoff, 294, 303
    project teams, 219–24, 280, 440, 544
    in reorganization, 253
    reviews of, 229, 231
    Sculley and, 250, 272, 513
    secrecy protocol after, 308
    size (coat pocket), 225
    Swatch project and, 220
Newton, Isaac, 20
"New World" architecture, 262
New York Times, 79, 142, 178, 229, 308, 329, 349, 417, 463, 470, 515, 517
NeXT Computer, 112, 174, 238, 281–87, 295, 299, 312, 314, 315, 321, 335, 343, 346, 382, 400
    NeXTcube, 282, *282*
    NeXTstation, 282
    NeXTSTEP operating system, 282–85, 290, 346, 347, 392
NFC chip, *see* near-field communication chip
Ng, Stan, 364, 367
Night mode, 526
Nikon, 360
"1984" ad, *129,* 129–31, 133–34, 140, 160, 213, 304, *310,* 311
Nokia, 417, 437, 531
Norman, Don, 146, *146,* 536
Notepad desk accessory, 118
Notes app, 465
notification summaries, 507
Novell, 235
NuBus expansion card, 205
NuKernel, 275, 346
NVIDIA, 432, 504, 513

Obama, Barack, 458
Object Linking and Embedding (OLE), 255

object-oriented operating systems, 234–36, 290
O'Brien, Conan, 430
O'Brien, Deirdre, 303, 447, 455, 467
Oculus, 540
    Oculus Rift, 490, *490,* 540
Odyssey (Sculley), 192, 249
office buildings
    Apple Park, 452–56, *453, 454,* 499
    Bandley Drive buildings, 43–44, *44,* 49–50, 56, 57, 96, 97, 119, 121, 127, 138, 142, 172, 244, 245
    City Center, 180, *180,* 247, 264, 281, 309, 467
    De Anza buildings, 168, 194, 201–2, 212
    Good Earth building, *37,* 37–39, 43, 68, 87, 89
    Infinite Loop offices, 244–45, 301, 303, 338, 359, 392, 397, 401, 406, 453, 460, 475
    Mariani buildings, 185, 194, 212, 244, 245, 490
    Taco Towers, 68
    Texaco Towers, 89, 83, 94, 96, 117
    Valley Green, 239, 318, 349, 475
OfficeMax, 351
Office of Special Education, 541
OK button, 70
OLE (Object Linking and Embedding), 255
Olin, Laurie, 454
Olivo, Allen, 299, 324, 548
Olsen, Ken, 521
Olympic Games (1984), 143
Omidyar, Pierre, 223
Onion, The, 463
Ono, Yoko, 306, 307
OpenAI, 499, 506, 510, 511, 513
OpenDoc software, 255, 302
Open Firmware, 262

operating systems
    Apple SOS, 51, 59
    A/UX, 188
    backward compatibility, 51–53, 57, 183, 285
    DOS, 92, 127, 182, 186, 188, 193, 219
    GS/OS, *181*
    for Johnathan, 182
    licensing of Mac OS, 202–3, 259–60, 312–13
    NeXTSTEP, 282–85, 290, 346, 347, 392
    object-oriented, 234–36, 290
    OS 26, 443–44
    Taligent project, 234–35
    see also iOS operating system; Macintosh operating systems (Mac OS); Windows operating system
Oppenheimer, Peter, 394
Optimize Storage feature, 534
Oracle, 255, 300–301
Ording, Bas, 396, 399, 402
Orwell, George, 131
Ostrovsky, Alex, 370
O'Sullivan, Joe, 447
Otellini, Paul, 391, 504
Ovitz, Michael, 273
Oyama, Terry, 98, 148

P1 project (iPod + cellphone), 398, 400
P2 project *see* iPhone
P68 Dulcimer *see* iPod
Pachikov, Stepan, 228
packaging, product, 143, 539, 548
Page, Larry, 459
Page, Rich, 61, 67, 146, 173
PageMaker software, 178–79
Paladin fax/phone/Mac, 246, *248*
Palm, 413, 417, 440

PalmPilot, 407
Treo, 407
Palo Alto Semiconductor, 234, 502
Pandora, 474
Pantone system, 40
Paradigm, 222
ParaGraph, 228
Parallels, 391
Paris Apple Expo, 387
Passif Semiconductor, 428
Patreon, 516
PayPal, 468
PC cards, 192
PC Magazine, 342
PC World, 128, 244
PDAs (personal digital assistants), 225
PDF signatures, 535
Peddle, Chuck, 33–34
"Pencil Test" video, 211, 212
PenLite tablet, 247
Pennsylvania State University, 85
PepsiCo, 105–7, 163
    Pepsi Challenge, 105, 134
    "Pepsi Generation" ad, 105
Performa series Macs, 256, 258, 268, 292, 302, 321
Perlman, Steve, 223
personal digital assistants (PDAs), 225
Personal Interactive Electronics, 253
personal voice feature, 542
Personas, for Vision Pro, 492–93
Peterson, Kelsey, 512
Pfister, Chas, 26, *26*, 269
Philips, 222, 263, 362, 363, 364
phone phreaks, 9
Photo Booth app, 384, *384*
photoplethysmography (PPG), 480
Photos app, 445, 506, 508
Picasso, Pablo, 307, *307*
Ping service, 469

Pioneer company, 358
Pippin game console, 276
Pirates of the Silicon Valley (film), 35, 142, 337
Pixar, 112, 211, 291, 292, 295, 301, 306, 314, 459
pixels, 63
Pixo software, 364, 401
PlainTalk, 180
plastics, 29–31, 40–41, 149–50
platform independence, 275
Playboy magazine, 32
Podolny, Joel, 467
Poltrona Frau, 456
Pong (video game), 10, 11, 34
Porat, Marc, 222, 223
PortalPlayer, 364, 365
Portrait mode, 427, *427*, 526
Potel, Mike, 184
Powell, Gray, 422
PowerBooks, 252, 256, 258, 268, 273, 293, 296, 358, 392, 394
    PowerBook 100, 243, 448
    PowerBook 140, 243
    PowerBook 170, 243
    PowerBook 520/540, 244
    PowerBook 5300, 263–64
    PowerBook Duo, 245–47, *246*
    PowerBook G3, 327, 342
    PowerBook G4, 308, 380, 381
    TIM project, 240–44, *242*, 316, 317, 381
    see also MacBooks
Power Computing, 260–61, 313, 314
power cords, 419, 425, 534, 537
PowerLine, 340
Power Macs, 254, 256, 268, 328, 350, 532
    and iPhone development, 399
    Power Mac 6100, 237, 254, *254*
    Power Mac 7100, 253, 254, *254*, 260–61

Power Mac 8100, 254, *254*, 260–61, 263
Power Mac 9700, 292
Power Mac G3, 325, 326, 341, 380, *380*, 403
Power Mac G4, 358
Power Mac tower, 450
PowerPC chip, 233, 236–37, 244, 252, 261, 262, 275, 276, 281, 296, 341, 391, 392, 403, 501–3
power supply, Apple II, 27, 29
power use, by AI, 511
PPG (photoplethysmography), 480
preemptive multitasking, 274
PressPlay music store, 368, 369
Preview app, 535
Price Club, 256
price fixing, 438
PrimeSense, 431
printers
    daisy wheel, 154–55
    HP LaserJet, 157
    ImageWriter, 150, *150*, 154, 154–55
    LaserWriter, 150, 154–59, *157*, 166, 173, 178, 179, 216
    StyleWriter II, 316
privacy, 383, 510–11, 544–46
Private Cloud Compute server system, 510, 511
product placement, 273
product quadrant, *302*, 302–3, 325, 327–29, 350, 381, 531
product quality, time to market, 79, 506–8, 534–35
ProFile hard drive, 51, 59, 77
proportional type, 63, *63*
protected memory, 77, 207, 275, 276, 348
proximity sensor, iPhone, 412
pulse oximeter, 480, 486–7, 527–8
Purple Dorm, 401

Index • 589

Purple project *see* iPhone
Putnam, Dan, 179
Puzzle desk accessory, 209
PVC plastic, phasing out, 537

Quadra series Macs, 256, 258
quadrant (Jobs's four products), *302*, 302–3, 325, 327–29, 350, 381, 531
Qualcomm, 504, 546
Quantum Computer Services, 270
Quest headset, 496
QuestTek, 482
QuickBooks, 268
QuickDraw software, 73, 102–3, 189, 199, 212, 235
Quicken, 268
Quick Look feature, 384
QuickTake digital cameras, 257
QuickTime, 192, 210, 212–14, 235, 284–85, 332, 334, 358, 361, 544

R1 chip, 494
racial-justice initiative, 488
Radar bug-tracking system, 480
radio frequency interference (RFI), 30, 52, 149
Radio Shack, 41, 417
  TRS-80, 28, 41, 42, 48
Radius, 92, 261
RAM (random access memory), 21, 97, 116–18
  see also memory
Ramsey, Bella, 513
Rand, Paul, 281
Rashid, Richard, 283
Raskin, Jef, 63, 64, 67, 85–93, 95, 299
Ratzlaff, Cordell, 347
Ray-Ban, 523, *523*
reaction injection molding (RIM), 40
Reagan, Ronald, 89
RealJukebox program, 358

RealNetworks company, 362
real-time translations, 508, *509*, 527
Recording Industry Association of America (RIAA), 360
recycled material use, 536
Redse, Tina, 172
reduced instruction set computer chips, *see* RISC chips
Reed College, 11, 129, 393
Reekes, Jim, 209
Regis McKenna, Inc., 31–32
  see also McKenna, Regis
Rehorst, Victor, 304
Reid, Glenn, 109–10, 332, 333, 389, 536
Reiling, Bob, 39
repairable products, 538–39
Retina screen, iPhone, 422
Reuters, 517
Revolution in the Valley (Hertzfeld), 88
RFI *see* radio frequency interference
Rhapsody language, 347
RIAA (Recording Industry Association of America), 360
Riccio, Dan, 432, 435, 490, 496–97
Rice, Garrett, 282–84
Rickard, Jay, 181
Ride, Sally, 254
Rihanna, 390
RIM (reaction injection molding), 40
Ringewald, Erich, 234–35
RISC (reduced instruction set computer) chips, 233, 234, 236–37, 249, 252, 256, 282, 296
  see also Power Macs
Rivian, 499–500
Robbin, Jeff, 274, 359, 364, 369, 371, 469
Robertson, Alice, 14, 23, 56
Robertson, Brian, 137
robotics, 521
ROKR phone, 398–99

Roche, Gerry, 104–5
Rock, Arthur, 54, 168, 172, 250, 252, 254
Rockwell, Mike, 490–92, 513, 548
Rockwell company, 266
Roizen, Heidi, 274, 291
Rolling Stones, 368, 369
Rosetta emulators
  for Intel chips, 392
  Rosetta 2, for M1 chips, 505
Rosetta handwriting recognition software, 231, 247
Rosing, Wayne, 76, *78*, 172, 180
Rossmann, Alain, 143, 261
Rothmuller, Ken, 67
rounded rectangle shape (RoundRects), 102–3, 533, *533*
Rubin, Andy, 223, 418
Rubinstein, Jon, 291–93, 301, 302, 319, 321, 322, 328, *330*, 335, 342, 350, 360–63, 367, 372, 375, 380, 381, 386, 394, 401, 449
Rutan, Burt, 287

Safari web browser, 346, 382, 400, 412, 517
Sagan, Carl, 253
Sahota, Shaan, 307, *307*
Sakoman, Steve, 219–20, 223, 228, 280
Samsung, 377, 401, 407, 418, 425, 427, 437, 451, 475, 502, 503, 528, 530, 531
San Bernardino, CA, shooting, 545
Sander, Wendell, 43, 50, 52, 53, 57–58, 100, 146, 222
San Francisco Symphony, 162
San Jose Mercury News, 6
Sara project (Apple III), 2, 50–60, *52*, *58*, *59*, 79, 86, 98
Saturday Night Live (TV series), 377
Satzger, Doug, 109

Scalise, George, 301
Scavenger app, 77
scene description feature, 543
Schiller, Phil, 146, 293, 301, 302, 304, 319, 336, 338, 341, 361, 363–67, 380, 387, 394, *395*, 426–27, 433, 439, 455, 457–58, 477, 520–21, 531, 535
Schlein, Philip, 54, 168
Schmidt, Eric, 418
Schorr, Joe, 445
Schultz, Howard, 463
Scott, Mike "Scotty," 38–39, 42–45, 49, 54–57, 67, 88, 91–92, 325
Scott, Ridley, 130
Scott, Tony, 159
Scotts Valley, CA, 56
Scrapbook desk accessory, 118
screen queries, iPhone, 508
Screen Sharing feature, 384
Scribe, *218*, 218–19
scrolling
   direction of, 442
   inertial, 219, 399
   scroll bars, 72–74
   scroll wheel, *366*, 373, 399
SCSI port, 185, 321–2
Sculley, John, *104, 108,* 109, 189, 244, 258, *335*
   Adobe's relationship with, 217
   after Jobs's departure, 177–80, 182
   AIM alliance, 232–36
   Amelio vs., 272
   Apple IIc project, 151–53, *152*
   Apple profit/loss for, 189
   at AppleWorld conference, 186
   ARM investment by, 271, 272, 315
   business card, 122
   Claris Corp. and, 190
   Cue and, 367
   environmental policy under, 536
   firing of, *249*, 249–52, *251*
   French deal for, 165
   Gassée and, 184, 201–3
   hiring, 104–7
   Jobs and, 2, 106–8, 145, 155, 162–74
   "Knowledge Navigator" video, 192–93, 435
   Macintosh demo, 139, 142
   Macintosh marketing by, 128, 130, 131, 133, 134, 136
   Macintosh Office concept, 156–60
   Macintosh reviews and sales, 142–45, 147
   Microsoft deal for, 194, 195
   Newton project, 218–25, 227–29, 246, 304, 513
   number of Mac models, 256
   overseas operations, 215
   PowerBook project, 240, 242
   product placement strategy, 273
   product strategy, 204–6, 271
   QuickTime pitch to, 210
   Spindler vs., 254
Sears, 143, 256, 302, 351
SEC, *see* Securities and Exchange Commission
secrecy, 65, 100–101, 135, 253, 308, 316, 401–2, 410, 475–76, 531–32, 540
"Secrets of the Little Blue Box" (*Esquire*), 9
Secure Enclave feature, 503
Securities and Exchange Commission (SEC), 54, 345, 435
security, 503
Segall, Ken, 305–7, 322–23, 344, 371
self-driving cars, 497–500
selfies, 384, 422, 526
sensors
   accelerometer, 480, 528, 543
   Apple Watch, 527–28
   for autonomous vehicles, 498
   gyroscope, 480, 528, 543
   iPhone, 412–13, 431
   tilt sensor, iPhone, 412–13
   Touch ID, 421, 431
   see also heart sensing
Sequoia Capital, 34
Serlet, Bertrand, 384, 407
Sharp, 220, 228
Sherlock, 278
Shugart Associates, 46, 47
Sidhu, Gursharan, 146, *146,* 158
Sigman, Stan, 410
silencer switch, iPhone, 407
silent beep, 541
"Silhouettes" ad campaign, 371
silicon chip design, 234, 501–5
Silicon Valley Ballet, 164
Siltanen, Rob, 305
SIM (subscriber identity module) card, 409, 410, 422
simplicity, 32, 60, 121, 384, 530, 536
Simpson, Mona, 436
Sinbad, 287
Singleton, Henry, 54, 168
Siri, 423–25, *424*, 431, 433, 458, 466, 469, 503, 506, *511*, 525, 544–46
   origins, 423–5
   personalized Siri, 511–513
Sketchpad desk accessory, 118
skeuomorphism, 465
sleep apnea, detecting, 487
Sleep mode, 77, 200, 365
sleep stages, calculating, 480
"Slide to unlock" 407
slot-loading CD drive, 328, 331
Slow Keys feature, 542
Smalltalk language, 67, 71
smart glasses, 496, 523, *523*, 527
smart home products, 474, 521
Smart Replies, 481

Smith, Burrell, 86–88, 91, 95, 101, 102, 122, *137*, 139, 144, 145, 261, 417
Smith, Dan, 75, 76
Smith, Megan, 223
Snapchat, 523
snoring detection, 529
Snow White design language, 148–50, 186, 240
social media, 530
social responsibility, 488
SoC (system on a chip) technology, *502*, 502–3
soft keys, 61–62
software consistency, 122
software evangelism, 125–27, 179
Software Publishing, 132
solid-state drives (SSD), 419
Solomon, Doug, 284
Sondheim, Stephen, 128
Songs of Innocence (U2), 485
Sony, 120, 263, 360, 451
    comparing Apple to, 419, 437, 531
    floppy drive, 100–101, 136, 145, 417
    General Magic and, 222, 223
    MP3 player, 361, 364, 372
    as PowerBook manufacturer, 243, 264, 448
    PTC-300 handheld, 220
    smartwatches, 475
    Sony Music, 331, 367–69
    Sony Pictures Television, 473
    Trinitron, 42, 148, 150
    Walkman, 148, 322–23, 360
Soren, Xander, 388, 389, 455
Sosumi sound, 209
SoundJam program, 359
sound recognition feature, 542
Spartacus (Twentieth Anniversary Macintosh), 287–90, 316, 386
Spatial FaceTime, 492–93

Spindler, Michael, 204, 215, 235, 248, *251*, 251–55, 258–66, 296, 472
SpO2 (blood-oxygen) sensor, 480, 486–7, 527–8
Spotify, 469, 474, 515, 516
Spotlight search, 382
Sprint, 408
SRI International, 423
Srouji, Johny, 5015
SSD (solid-state drive), 419
Stanford University, 129, 160, 171, 285, 354, 400, 428, 452, 453, 487
    Jobs commencement speech at, 393
    Linear Accelerator Center, 15–16
    Stanford Hospital, 459
Stanton, Andrew, 211
Starbucks, 412, 463
Stark, Pete, 547
Starkweather, Gary, 146, *146*
StarMax clones, 261
Star Trek project, 235, 391, 544
Stern, Joanna, 508
Steve Jobs (film), 142
Steve Jobs Theater, *453*, 454, 456
Stevens Creek Boulevard office (Good Earth building), *37*, 37–39, 43, 68, 87, 89
Stickies app, 346
Sticky Keys feature, 541
stock options, 56, 269, 299–300, 345
Stony Brook University, 85
store-within-a-store concept, 311–12
Strickland, Don, 260
StyleWriter II, 316
subscription services, 463–74, 489
Sullivan, Kevin, 203, *251*
"sunflower" iMac G4, 350, 384, *386*, 386–87
Sun Microsystems, Inc., 193, 232, 233, 264–65, 267, 278, 282, 319

Sunnyvale, CA, 5
Sun Remarketing, 161
Super Bowl ads, 129–31, 133–34, 160, 344
SuperDrive, 197, 199, 358
SuperMacs, 261
Supreme Court, U.S., *195*, 196
Susman, Galyn, 211
sustainability, 149–50, 453, 511, 536–40
Sutter Hill firm, 34
Swatch company, 324
Swatch handheld, 220
Swift, Taylor, 469
"Switch" ad campaign, 385
Switcher program, 152
Synchron, 522
system on a chip (SoC) technology, *502*, 502–3

tablet computers, 395, 397, 399, 435–36
    see also iPad
Taco Towers building, 68
Taggart, Adam, 329–30
TAG Heuer, 482
Tague, Nancy, 211
Taiwan Semiconductor Manufacturing Company (TSMC), 504
Taligent, 234–36
Tamaddon, Sina, 359
Tandy, 41
    TRS-80, 28, 41, 42, 48
Tangerine firm, 317
Tanimoto, Craig, 305
Taptics Engine, 483
Tap to Pay technology, 468, 516
Target, 351, 468, 474
Tax Cuts and Jobs Act (2018), 515
tax strategies, 161, 247, 515
TBWA/Chiat/Day, *see* Chiat/Day

Tchao, Michael, 223, 224, *226*, 227, 228, 296, 439, 440
tech bubble, 343, 345, 362, 388
Tech Interactive museum, 164
Tecot, Ed, 122
Teledyne, 54, 168
Ternus, John, 537
Terrell, Paul, 21–24, *23*, 26, 31
Tesla, 497
Tesler, Larry, 62, 67–71, 78, *78*, 146, 180, 221–24, 226, 227, 247
"Test Drive a Macintosh" promotion, 134
Tevanian, Avie, 283–85, 291, 292, 294, 301, 302, 346, 347, 380, 382, 384, 394, 401
Texaco Towers office, 89, 93, 94, 96, 117
Texas Instruments, 106, 364
TextEdit app, 346
text messages, 409–10
ThaiTalk, 216
"Think Different" ad, 2, 293, 304–9, 353, 460
30-pin connector, 373, 425, *425*, 478
32-bit addressing, 210
Thomas, Brent, 130
Thomas, Johann, 329–30
Tidal music service, 469
Tiffany, Louis Comfort, 94
TikTok, 420, 517, 543
tilt sensor, iPhone, 412–13
time constraints, 23, 46, 137–38, 225–28, 363–65, 368, 432, 544
Time Machine, 384
Time magazine, 94, 135, 153, 216, 244, 268, 308, 370
time-saving features, 534–35
time to market, product quality and, 79, 506–8, 534–35
Time Warner, Inc., 367–69, 470–72
TIM project, *see* PowerBooks

titanium cases, 380
Titan car project, 496–500
T-Mobile, 408
Top 100 retreats, 323–24
Toshiba, 260, 263, 362, 451
 T1100 PC clone, 199
Touch ID fingerprint sensor, 421, 431
touchscreens
 butt-dialing problem, 407
 eMate 300, 230
 iPod Touch, 356, 430, 436, 463, 503
 multitouch technology, 395–97, *396*, 399–400, 436
 PenLite tablet, 247
 shower problem, 481
 Swatch handheld, 220
 see also Apple Watch; iPads; iPhones; multitouch technology
trackball, 243, 244
trackpads, 244, 442
Translate app, 506
translations, real-time, 508, *509*, 527
Trash folder, 75–76, *76*
Travis, Randy, 506
Tribble, Bud, 87, 88, 94, 95, 117, 173, 335, 455
Trudeau, Garry, 229
TrueType, 210, 217
Trump, Donald, 518
TSMC (Taiwan Semiconductor Manufacturing Company), 504
Turbo Mac, 144
turnaround by Jobs, 299–315
Turner, Ted, 128, 307
Twentieth Anniversary Macintosh, 287–90, 316, 386
21st Century Classroom Initiative Act, 547
Twiggy disk drive, 99–100, 125
two-factor autofill feature, 534
2001: A Space Odyssey (film), 344, 435

U2 (band), 377, 485
Uber, 420, 496, 500
Ubillos, Randy, 146
Umax, 261, 313
"Umbrella" (song), 390
uncanny valley, 492, *492*
Undo command, 119, 121
Unicode, 215
United Kingdom, App Store investigation, 517
United States, manufacturing in, 519, *519*
United States v. Apple, Inc., 516
universal serial bus, *see* USB
University Consortium, 128–9, 143, 173, 546
University of California, Berkeley, 9, 92, 164
University of California, Los Angeles, 269
University of California, San Diego, 85, 86, 87
University of Chicago, 129
University of Minnesota, 178
University of Southern California, 35
University of Washington, 87
Unix, 162, 173, 188, 282, 283, 292
unlimited-data subscriptions, 410, 415
unlocking iPhone, 407
UPS, 281, 436
USB (universal serial bus), 321–22, 334, 366, 373, 426

USB-C, 426

user profiles, 277, 278
user testing, 69–70, 76
US Festival, 164
US West, 267

Valentine, Don, 34, 36
Valley Green building, 239, 318, 349, 475

Van Amburg, Zack, 473
Venmo, 420
Venrock, 54, 168
Verge, 508
Verizon, 408, 410, 422
Vertelney, Laurie, 218
"Vertigo" (song), 485
Viacom, 472
video iPod, 399
video prototyping, 193
Vietnam, 519
virtualizer programs, 391
virtual memory, 210
virtual private networks (VPNs), 517
virtual-reality (VR) headsets, 490–1
Virtual User app, 119
Visa company, 468
VisiCalc program, 47–48, *48*, 50
Vision Pro headset, *193*, 356, 490–96, *494*, *495*, 500, 513, 531
    EyeSight for, 493–94
    latency challenges, 491
    launch and reviews, 494–96
    OS 26, 443
    project teams, 490–91, 544
    resolution challenges, 491
    Spatial FaceTime with, 492–93
Visual Intelligence, 507–8
Visual Voicemail, 410, 418
Vitals app, 487
Vivendi/Universal, 367, 368, 369
VMware, 391
Voice Memos app, 465
VoiceOver, 383, 542, 543
Volkswagen Transporter, 498
Vonnegut, Kurt, 128
Vorrath, Kim, 405
VPNs (virtual private networks), 517
VR (virtual-reality) headsets, 490–1

Wadsworth, Jack, 54–55
Walker, Vaughn, 196

Wallaby rig, 402, *403*
Wallstreet (PowerBook G3), 327, 342
Wall Street Journal, 54, 173, 308, 439, 470, 517
Walmart, 468
Warhol, Andy, 128
Warnock, John, 155–56, 217, 459
Warren, Jim, 39, 41
Wayne, Ron, 19–20, *20*, 25, 32, 38
Waze app, 466
Weakest Link, The (TV series), 424
wearable computers, 475–77, 499, 527
    see also Apple Watch; AirPods
web apps, 416
webcams, 382, 383
WebObjects suite, 282, 285, 352
WebTV, 223
Weeks, Wendell, 408–9
Welch, Jack, 463
WESCON (Western Electronics Show and Convention), 13, 39
West Coast Computer Faire, 39–42, *40*
Westerman, Wayne, 396, 397
Westinghouse, 281
Wharton School of Business, 105
"What Makes Apple Apple" (course), 467
WhatsApp, 517, 540
White, Chris, 418
Whitney, Tom, 67, 88
Whitney Museum of American Art, 153
Whole Earth Catalog, 393
Wi-Fi, 192, 338–40, 342
Wigginton, Randy, 16, 37–38, 41, 42, 45, 47, 56, 59, 92, 95, 138, 145
Wikipedia, 517
Williams, Jeff, 361, 408, 447–48, 481, 485, 500
Williams, Robin, 305, 306
windows, overlapping (regions), 73, 103, 189, 444, *444*

Windows operating system, 315, 501
    ActiveSync, 418
    announcement, 139
    Apple Music for, 469
    File Explorer, 75
    Hollywood program, 190
    on Intel chips, 202, 256, 505
    iPods and iTunes Store for, 367, 369, 371, 373, 420, 467
    laptops running, 240, 241, 380, 419
    licensing of, 202–3, 222, 259
    Mac OS vs., 193–96, 208, 383
    market share for PCs, 196, 202, 258, 540, 546–47
    Office for Mac OS vs., 268, 310
    OLE, 255
    QuickTime for, 212–3
    running on Macs, 279, 384, 391, 505
    tablet computer prototype, 435
    USB for PCs with, 322
    virus attacks, 382
    Windows 1.0, 193
    Windows 2.0, 195
    Windows 3.0, 232, *232*
    Windows 95, 259, 274, 347
    Windows Media Player, 358
    Windows Mobile, 423
    Windows NT, 262, 279
    Windows Phone 7, 440
    Windows XP, 382
Winer, Dave, 299
Wingate, Zack, 513
Wintour, Anna, 482
Wired magazine, 349, 389
Woman Combing Her Hair (Goyo), 120
Woodside Design, 167
Woolard, Ed, 294–95, 300, 312, 356
WordPerfect program, 255
Worldwide Developers Conference (WWDC) presentations
    Amelio's opening address, 274, 278

on Apple Intelligence, 507, 511, 512
on iPhone 4, 422
on Mach-based OS, 283
on Mac OS X, 336–37, 383, 384, 392
on QuickTime, 210, 212, 213
on Scribe, 218
World Wide Web, 192, 319, 343, 345
Wozniak, Francis "Jerry," 5, 17
Wozniak, Janet, 164
Wozniak, Margaret, 5, 17
Wozniak, Steve, *5, 16, 19, 20, 22,* 152, 164, 325, 411, 459
  1981 plane crash, 80–81, 103
  1997 Macworld Expo, 287, 290, *290*
  on advertising, 31, 133
  Annie project, 86
  Apple I creation, 13–15, 23–6, *16*, 36
  Apple II add-on design, 43, 46–47, 49
  Apple II creation, 26–29, 36, 42, 50, 53
  on Apple II and III, 59
  on Apple II programs, 44
  as Apple Fellow, 146
  badge number, 45
  childhood and adolescence, 5–6
  departure from Apple, 163, 164
  departure from HP, 36–37
  Disk II controller card, 99, 100
  "Easter eggs" buried by, 187
  founding of Apple, 15–20, 25
  HP job, 11, 13–15, 17–19, 24, 36–37
  Jobs's initial meeting, 9
  Jobs's relationship, 112
  pranks, 8, 41, 162
  pre-Apple projects with Jobs, 2, 9–10, 14, 28
  signature in Mac case, 95
  16-bit Apple II project, 181
  at Steven's Creek office, 37–38
  in *Time* "Man of the Year" issue, 135
  at West Coast Computer Faire, 39, 41
  Woz Stock Plan, 56
Wright, Frank Lloyd, 307
Writing Tools, iPhone, 507
WWDC presentations, *see* Worldwide Developers Conference presentations
Wyle, Noah, 337

X, Chinese access to, 517
Xbox Kinect, 431
Xerox Corporation, 68, 71, 109, 139, 194
  Palo Alto Research Center, 63–69, *64,* 72, 77, 87, 96, 113, 155, 180, 195, 215
  Star, 66
Xi Jinping, 518

"Y2K bug" ad, 344
Yaeger, Larry, 146, 231
Yocam, Del, 49–50, 144, 168, 170, 171, 177, 178, 254
York, Jerry, 301
YouTube, 111, 210, 211, 358, 393, 464, 513, 517
Yves Saint Laurent, 482

Zadesky, Steve, 409, 497, 498
Zanuck, Darryl, 521
Zarem, Abraham, 66
ZEISS inserts, Vision Pro, 496
Zenith, 260
zero-draft molding, 150
Zhang, Hannah, 412
Zilog, Z80 processor for, 41

About the Author

David Pogue is a seven-time Emmy Award winner for his stories on *CBS Sunday Morning*, a five-time TED speaker, a host of 20 *NOVA* specials on PBS, and a *New York Times* bestselling author. He's written about Apple for his entire career, including 13 years as a *Macworld* columnist, 13 more as the *New York Times* tech columnist, and 20 years as the #1 bestselling author of books about Macs and iPhones. He lives with his family in New York.